T0327608

The Bayesian Way

The Bayesian Way

Introductory Statistics for Economists and Engineers

Svein Olav Nyberg
University of Agder
Grimstad, Norway

Registered Office(s)
John Wiley & Sons, Inc., 111 River Street, Hoboken, NJ 07030, USA

Editorial Office
111 River Street, Hoboken, NJ 07030, USA

For details of our global editorial offices, customer services, and more information about Wiley products visit us at www.wiley.com.

Wiley also publishes its books in a variety of electronic formats and by print-on-demand. Some content that appears in standard print versions of this book may not be available in other formats.

Library of Congress Cataloging-in-Publication Data

Names: Nyberg, Svein Olav, author.
Title: The Bayesian way : introductory statistics for economists and engineers / Svein Olav Nyberg.
Description: 1st edition. | Hoboken, NJ : John Wiley & Sons, 2018. | Includes index. |
Identifiers: LCCN 2017060721 (print) | LCCN 2018007080 (ebook) | ISBN 9781119246886 (pdf) |
    ISBN 9781119246893 (epub) | ISBN 9781119246879 (cloth)
Subjects: LCSH: Bayesian statistical decision theory. | Economics–Statistical methods. |
    Engineering–Statistical methods.
Classification: LCC QA279.5 (ebook) | LCC QA279.5 .N93 2018 (print) | DDC 519.5/42–dc23
LC record available at https://lccn.loc.gov/2017060721

Cover image: © Javi Tejedor Calleja /
Cover design by Wiley

Set in 10/12pt WarnockPro by Aptara Inc., New Delhi, India

10 9 8 7 6 5 4 3 2 1

This book is dedicated to my two beautiful daughters Kira Amalie and Amaris Beate.

Thank you for being patient with Pappa.

Contents

Preface

What could possess a man to write a statistics textbook? They are, after all, bountiful. Well, they are, but when I started lecturing the subject, I found certain elements missing in all of the books I had available. And as I sat there contemplating what was lacking, I started writing down what I thought should have been there, and – well, I really never decided to write this book; *it* decided to make me its author.

There are two schools of statistics: bayesianism and frequentism. The key difference is how they interpret the concept of *probability*. Even though there is a lot of common ground, this difference ultimately gives rise to different methods and interpretations of results. We will explore both schools in this book, but with an emphasis on the Bayesian way.

So why should we choose the Bayesian way, when it is the one less travelled by? In the spring of 2016, The American Statistical Association published a joint proclamation against misuse of p values,[1] a popular frequentist measure of the validity of a scientific result. The reason for the problem is simple: users simply don't understand the concept! In addition, it is open to abuse (so called "p hacking") if you *have* understood it. The result of all this is scientific reports that don't hold water.

The ASA members were not in agreement as to the solution of the problem, but many suggested teaching applied users of statistics in the Bayesian school rather than in the predominant frequentist school. The two schools have each their advantage: The frequentist school has formulas that are quick to calculate manually, with the use of tables. The Bayesian theory is more unified, and thereby easier to comprehend. With our current access to tools for calculation, ease of manual calculation is no longer important, so the time is ripe for the Bayesian way.

---

[1] See *Nature* March 7th, 2016: http://www.nature.com/news/statisticians-issue-warning-over-misuse-of-p-values-1.19503

We must, however, remember that frequentism still is the dominant school. With that in mind, this is not a "purist" Bayesian text, but is rather "Bayesian in a frequentist way", making it easier for students to translate results to and from the frequentist framework and jargon.

Which background do you need to read this book? What do you need to know? Not much beyond high school math, and you will understand a lot with even less of a background. When you get to the chapters with the probability distributions, it helps to understand what a function is: that the function is not identical to a single formula, but that it is better to think of it as its graph: given an x as input on the horizontal axis, you get an f(x) as output on the vertical axis. To be able to do more, a little college math helps. That is, differentiation and integration, and basic linear algebra: multiplying matrices and vectors, and the inverse of a matrix. For understanding, the fundamental theorem of calculus is the key: if $F(t) = \int f(t)dt$, then $f(t) = F'(t)$.

A textbook has three dimensions: theory, understanding, and application. Since this book is aimed at applied rather than theoretical studies, proofs have been strongly toned down, and have been included only where a proof would build understanding for a practical user. But mostly, we aim to build understanding through illustrations, explanations, stories and examples. The basics of statistics are the same across disciplines, so the same material spans economics and engineering, as well as science, and even plain fantasy! But we have also emphasized plain drilling, since many practical users need simple assignments without too much text.

The ∗ mark on some sections means that this section is more theoretical or advanced, and may require a bit more of your mathematical background or interest.

You can't write a book without making mistakes. So we appreciate feedback, and will publish errata and other useful information on the book's web site, http://bayesians.net

A book like this is not written in isolation, and many have given useful comments and suggestions. I would particularly like to thank Billy Case, Steffen Monrad and Torbjørn Bratten for helpful comments and insights throughout the writing of this book, and Kira Nyberg, Abbot Sōzen and Nicholas Caplin for their illustrations. I would also like to thank Bjørn Olav Hogstad, Nils Johannesen, Aksa Imran, Øystein Rott, Tore Nordseth, Asle Olufsen, Trygve Pedersen, Arvid Siqveland, Jannicke Bærheim, Jostein Trondal, my wife Elin and my father Arne Olav for reading and commenting. I would also like to thank Erik Yggeseth, Sondre Glimsdal, Hans Grelland, Alireza Borhani, Mathias Pätzold, Odd Aalen, and Tom Lassen for examples, inspiration and other direct help. Thanks are also due to my students for having put up with being taught from preliminary versions of the book, all the while giving patient feedback for further development, and thanks to Yvonne Goldsworthy for insisting that I turn my notes into a book. This book was first published in

Norwegian, and then later translated into English with a few modifications. I would like to thank those who have helped me with the English-language version: Branislav Vidakovic, Jim Farino, Beatrice Kondo, Michael LaTorra, Gunvor Myklebust, Mark Tabladillo, Nathan Bar-Fields, Hugh Middleton, John Conway, Tom Chantler, and Hans Jakob Rivertz. I would also like to thank Michael Brace for his careful copyediting, and Brani Vidakovic for making sure the English version came into existence in the first place.

This book would not have been what it is without you.

Svein Olav Nyberg
Grimstad, August 2017

1

Introduction

CONTENTS

Modern statistics has two roots: the probability theory Pascal and Fermat invented in Holland in the late 1600s to solve gaming related problems, and Prussian state book keeping in the eighteenth century. The name comes from the latter; and *statistics* does indeed originally mean "pertaining to matters of the state" – state bureaucracy and gambling – an odd match indeed. Yet, they share a need for meticulous reckoning. Prussian bureaucrats kept track of every oar in the Prussian navy. Not 4218 oars, but 4217. Accuracy was a virtue! But the same applies in gambling: calculate your odds wrong, and your fantastic winning gambling strategy is the hole where you flushed last month's salary.

Statistics has matured and won new territories since its inception.

- Florence Nightingale originated the techniques of descriptive statistics we all know so well: techniques to visualize data that would otherwise be dry and lifeless. With a few simple charts, she showed how unsanitary conditions in the lazarets caused higher mortality rates in wounded soldiers.
- In physics, we use statistics to describe large systems when it is infeasible to keep track of all the elements. This is *statistical mechanics.*
- Physicists have also discovered that small particles act like microscopic roulette wheels: their actions are unpredictable, and we can only state probabilities for what they *might* do. This is *quantum mechanics.*
- On the internet, you often get ads tailored to your preferences. This tailoring is made based on usage statistics about what you look for and which links you click.

The Bayesian Way: Introductory Statistics for Economists and Engineers, First Edition.
Svein Olav Nyberg.
© 2019 John Wiley & Sons, Inc. Published 2019 by John Wiley & Sons, Inc.

- At the other end of the internet, your email client most probably filters unwanted promotion emails. These filters are often *Bayesian* spam filters, based on statistics of what you mark as junk.
- Medical research uses advanced statistics to tell whether and how well drugs work, how epidemics spread, and the connections between environment, genetics, other factors, and health.
- Meteorology is statistical. It rarely makes sure predictions, but does instead seek to find the probabilities that the temperature, wind speed, and rain fall will be within a certain interval.
- Advanced computer games often employ (Bayesian) statistical learning strategies so that, while the player learns the game, the game also learns the player's strategies in order to counter them better.
- Financial mathematics employs advanced probability theory to calculate the best pricing for buying and selling.

1.1 Parallel Worlds

According to the statistician P. A. Fisher, statistics has a threefold purpose:

(1) collecting and ordering data;
(2) systematizing, and summarizing in a few key numbers what the data are saying;
(3) using the collected data to estimate data we have yet to collect.

This sets up two parallel worlds, interacting and reflecting one another. The first world, the world of data, is the world we live in. The second world is the idealized world of our probability calculations, where the properties of probabilities mirror those of data, like idealized versions. These two worlds interact in that we set up the probabilities in World 2 based on the data from World 1, and then project probabilities of possible data back from World 2 to World 1.

These worlds mirror one another both in structure and in concepts. In World 1, we have tables of proportions, and in World 2 we have tables of probabilities. Indeed, the probabilities themselves behave exactly like proportions. In World 1, we draw graphs of how the height of military recruits is distributed in a given year, whereas in World 2 we draw graphs of what we consider to be the probability distribution of the recruits' heights.

The simplest and also oldest model of probability comes from Pascal and Fermat's model for games. The model calculates the probability P of an event A by finding a symmetrical set of *possible* states; the probability of event A is then the quotient of the *positive* states (those associated with A) to the *total* number of possible states:

$$P(A) = \frac{positive}{total}.$$

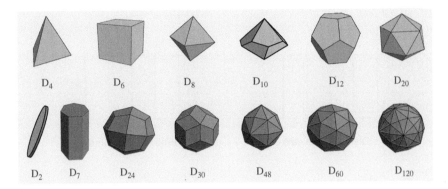

Figure 1.1 Dice D_k are named by the number of their surfaces.

This formula shows us probability as a kind of proportion. Proportions are numbers between zero and one, or percentages between 0 and 100%. A percentage is $\frac{1}{100}$, or if the probability is $p = 0.237$, you find the percentage-wise number by multiplying by 100%, that is: $p = 23.7\%$. In Chapter 5 we will look at the different definitions of probability in more detail. What they all have in common is that probability is a number between zero and one, and that it behaves like a proportion.

Consider a symmetrical n-sided die D_n to get an instructive example of how to calculate probabilities the way Pascal and Fermat did: these dice may be found in most games shops, and come in varieties from the four-sided D_4, via the common cube D_6, and further on to D_8, D_{10}, D_{12}, and D_{20}. See Figure 1.1.

If you paint five of the sides of a D_{20} red, the probability of getting *red* when you toss that die is $P = \frac{positive}{total} = \frac{5}{20} = 0.25 = 25\%$.

Trials in the real world rarely give us the precise proportion of our probabilities. The data are what they are. But the visual descriptions are the same. We will see more on this in Chapters 2 and 5. Let us look at the tosses with a four-sided die: outcomes of real-world trials to the left, and probabilities of outcomes in the ideal world to the right.

Data (Chapter 2)	Probabilities (Chapter 5)
33 tosses gave "1", 39 tosses gave "2", 42 tosses gave "3", 36 tosses gave "4".	$p(1) = 0.25, p(2) = 0.25,$ $p(3) = 0.25, p(4) = 0.25.$

Frequency table (+ proportions).

Outcome	Positives	Proportion
1	33	$33/150 = 0.22$
2	39	$39/150 = 0.26$
3	42	$42/150 = 0.28$
4	36	$36/150 = 0.24$
SUM	150	1

Probability table.

Outcome	Probability
1	0.25
2	0.25
3	0.25
4	0.25
SUM	1

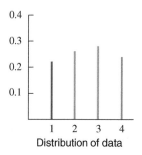
Distribution of data

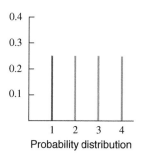
Probability distribution

1.2 Counting Positives

It is important to count "positives" in the right way. First of all, the possibilities have to be equal, or *symmetrical*. Consider the case where I write the integers "1" through "5" on some equally large cards, and put them in a hat: is the drawing of each number equally probable? What if I told you that there were 906 cards, and that I wrote "1" on 900 of them, and that "2", "3", and "5" were represented with two cards each – while I completely omitted "4"? Are the alternatives still equal? Of course not. The probability of drawing each *card* is the same, but the probability of each *number* is not. We say alternatives are symmetrical when you can swap two elements without altering the dynamics of the system. That is: no element is preferentially treated by the system compared to any other alternative.

Try now a simple coin, and flip it twice. What is the probability of no heads? It is tempting to enumerate the possibilities: no heads, one heads, two heads, and conclude that since there are three possibilities, $P(0 \text{ heads}) = \frac{positive}{total} = \frac{1}{3}$. But are the alternatives symmetrical? No, for if we inspect a bit more closely, we see that no heads means you flipped the sequence TT, whereas one heads could mean either HT or TH – that is: two different sequences. Two heads again means HH, which is realized only through a the single sequence of two heads. So, breaking this down to the single flips that we take as symmetrical, we see that the symmetrical options for two flips are: TT, TH, HT, and HH. This gives us

$$P(0) = P(TT) = \frac{|\{TT\}|}{|\{TT, TH, HT, HH\}|} = \frac{1}{4}.$$

We illustrate this through two sets of tables (two coin flips in Table 1.1, and two tosses of D_4 in Table 1.2).

Precise estimates and tests for such probabilities are the concerns of the later, more advanced chapters on statistical inference (basic inference in Section 13.2, estimates in Section 15.7, and testing in Section 14.2.2). The exploration of how to count the positives and the possibles is called *combinatorics*, and is explored in Section 4.

Table 1.1 Tables for two coin flips

Outcomes				Number of Heads				Probability table		
	H	**T**			**H**	**T**		Heads	Ways	Probability
H	*HH*	*TH*		**H**	2*H*	1*H*		0	1	1/4
T	*HT*	*TT*		**T**	1*H*	0*H*		1	2	2/4
								2	1	1/4

Table 1.2 Tables for two tosses of a four-sided die D_4

Outcome

		Die 1			
		1	**2**	**3**	**4**
	1	1, 1	2, 1	3, 1	4, 1
	2	1, 2	2, 2	3, 2	4, 2
	3	1, 3	2, 3	3, 3	4, 3
Die 2	**4**	1, 4	2, 4	3, 4	4, 4
	5	1, 5	2, 5	3, 5	4, 5
	6	1, 6	2, 6	3, 6	4, 6

Probability table

SUM	Ways	Prob.
2	1	1/24
3	2	2/24
4	3	3/24
5	4	4/24
6	4	4/24
7	4	4/24
8	3	3/24
9	2	2/24
10	1	1/24
TOT	24	1

Total

		Die 1			
		1	**2**	**3**	**4**
	1	2	3	4	5
	2	3	4	5	6
	3	4	5	6	7
Die 2	**4**	5	6	7	8
	5	6	7	8	9
	6	7	8	9	10

1.3   Calculators and Web Support

In our times, most of us have access to decent or even advanced statistical tools. The most common, and surprisingly advanced, tools are spreadsheets like Microsoft® Excel. We have chosen to focus on three common calculators from Casio®, Hewlett Packard, and Texas Instruments™, and on Mathematica®, which is freely and very easily accessible through the free front end Wolfram Alpha: http://www.wolframalpha.com.

We will also strive to link to other tools and resources at the book's web page, and if you would like to contribute a program or a manual, please contact us at http://bayesians.net.

1.4   Exercises

1   Review: Read the chapter.
  – What is the purpose of statistics?
  – What is a D_8?
  – Given a D_4 and a D_6: in how many ways can you get a total of five?

2   Find the symmetrical alternatives when you flip a coin three times. What is the probability of two heads and one tails?

3   Find the symmetrical alternatives when you toss two D_6. What is the probability that their sum is three? What is the probability their sum is seven?

4   If we could make pancakes forever, and it turns out that the proportion of burned pancakes stabilizes at 0.137, would that matter for the *probability* that pancakes are burnt?

Part I

Foundations

2

Data

CONTENTS	

By *data* we mean a collection of a given type of values. The values are most commonly numbers, but can be anything we have received in response to our queries or measurements. Non-numerical values are called *categorical* data, which simply means information about membership of a category. One example of this is if our query is about preferences for a political election; the data would then be the names of a political party like *Democrat, Republican, Libertarian,* or *Green* in the USA, but also the categories *Don't know, Others,* and *Blank.* With categorical data, numbers come into play only when we are counting the number of hits in the different categories. This is as opposed to *numerical data,* which is the most common form of data, where the data are themselves numbers. Examples of such data are the *times* for a 60 yards dash, where the data are the times, and not the runners themselves or their names. Or the waist *circumference* of diabetic teenagers, where again the data are simply the number of inches in each measurement.

The *population* is the total of all possible *values* – including the ones that are not measured. We have two models of this: the rather concrete *urn* model, and the more abstract *process* model. We start with the urn model.

The urn model of a population is a finite set, an "urn" that contains little notes with values written on them. In an election, the population is the political preferences of the voters, and not the voters themselves. In an urn model of the population of the 2016 US election, the population would be the 139 million individual votes, and have values "Green", "Libertarian", "Republican", or

The Bayesian Way: Introductory Statistics for Economists and Engineers, First Edition.
Svein Olav Nyberg.
© 2019 John Wiley & Sons, Inc. Published 2019 by John Wiley & Sons, Inc.

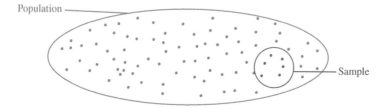

Figure 2.1 *Population and sample.*

"Democrat". So if, for instance, there were 4 042 291 votes for the libertarian party, the population contains 4 042 291 values "Libertarian". If you are looking at diabetic teenagers' waist circumferences, the population is the total set of waist measurements that could have been collected. So the population might contain a million "36.0 inches", and none of "20.0 inches". What matters to us are the values, and how many there are of each.

The *process* model of a population differs from the urn model in the same way that dice differ from a deck of cards. We abstract away the number of instances of each value, and look instead at the *proportions* for each value. For the US election mentioned above, four million then becomes 3.2%. For the value "17" on a D_{32} die, the proportion 3.125% is all we have, as there is no fixed number of dice tosses. When we later in this book will be talking about *sampling from a probability distribution*, we are referring to the process model.

The *sample* consists of the data we have actually collected. In an urn model, we justify sampling by appealing to cost: the sample will usually be a lot smaller than the population, as illustrated in Figure 2.1. So if we can draw sufficiently reliable conclusions by sampling a thousand values rather than doing an exhaustive measurement of several millions, then we should be sampling. In a presidential election, polls are often conducted by asking a few thousand voters for their preference. The pollsters then draw a fairly reliable conclusion about the political preferences of the entire population.

We will need to *index* our data. The most common way of indexing is enumeration, $x_1, x_2, \ldots, x_n$, but other indexes like *time* and *location* might at times be more expedient. If you are looking at stock prices, like the ones in Figure 2.2 from the Oslo Stock Exchange (OSE), then it is better to write the price of an Orkla stock, at 12:00 on the 16th of September 2008, as $x_{2008.09.16.12:00}$, than to enumerate it as x_{3127} if it was your 3127th observation. But if no special factors come into play, enumeration $x_1, x_2, \ldots, x_n$ is the default choice.

2.1 Tables and Diagrams

We often compress our observations into groups of equal value, noting only the number of observations for the value. We express this in a frequency table, counting how many there are of each kind, and a bar chart.

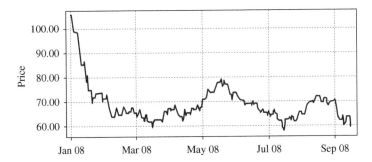

Figure 2.2 Orkla stock price. Source: OSE 2008.

Example 2.1.1 Nathan counts the different books on Suzie's science bookshelf. He makes the frequency table and bar chart in Figure 2.3 from his measurements. □

We have two basic types of data: numerical data and categorical data. When the data values are numbers, the horizontal axis becomes a value axis, whereas the vertical axis marks (relative) frequency.

Example 2.1.2 We asked 150 households how many TVs they owned. Our data are collected in Figure 2.4. □

We are frequently more interested in the relative frequencies (proportions) $p_k = a_k/\sum a_k$ than in the absolute frequencies a_k themselves. When polling agencies report probable voter distribution for the next election, most of us are more interested in hearing that Jill Stein got 1% of the polled votes than in knowing that exactly 18 of the 1800 respondents said they would vote for Stein.

Value v_k (subject)	Frequency a_k
A: Astronomy	3
P: Physics	5
EE: Electrical Engineering	16
M: Mathematics	6
S: Statistics	9
E: Economics	1
Sum, n	40

(a) Table.

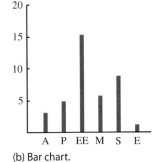

(b) Bar chart.

Figure 2.3 The books on Suzie's bookshelf.

Value v_k (TVs)	Frequency a_k
1	33
2	39
3	42
4	36
Sum	150

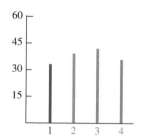

Figure 2.4 Number of TVs in a household.

Example 2.1.3 (Continuation of Example 2.1.2) We find the proportions of how many households own how many TVs by normalizing the frequency table. That is, by dividing the category frequency by the total frequency, as shown in Figure 2.5. □

Example 2.1.4 We have data from adherents.com about the adherents of the largest religions. We display the data in Figures 2.6 and 2.7.

In this chart, the relative size of each religion is very visible, since the size of the pie slices is proportional to the number of adherents of each religion. But if we are interested making the smaller religious groups visible, we must abandon proportional representation. We choose a method that maintains the ordering by size, but not proportionally, by taking the logarithm of the number of adherents. The largest religions remain largest, but a factor of 10 is now only 1 unit, and a factor of 100 is 2 units higher.

So a logarithmic vertical axis brings forth the smaller religions in this chart. □

2.1.1 Cumulative Data

In many applications, it is more useful to know how many of the values fall within a certain interval, or over or below a given value, than it is to know how

Value v_k (TVs)	Relative frequency p_k
1	33/150 = 0.22 = 22%
2	39/150 = 0.26 = 26%
3	42/150 = 0.28 = 28%
4	36/150 = 0.24 = 24%
Sum	1

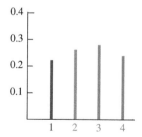

Figure 2.5 Relative frequency of number of TVs in a household.

■ Christianity	2 100 000 000
■ Islam	1 500 000 000
■ None	1 100 000 000
▓ Hindu	900 000 000
■ Traditional Chinese	394 000 000
░ Buddhism	376 000 000
▓ Tribal religion	300 000 000
■ Traditional African	100 000 000
■ Sikh	23 000 000
■ Juche	19 000 000
░ Spirits	15 000 000
▓ Judaism	14 000 000
■ Baha'i	7 000 000
░ Jainism	4 200 000
■ Shinto	4 000 000
■ Cao Dai	4 000 000
▓ Zoroaster	2 600 000
░ Tenrikyo	2 000 000
■ Traditional European	1 000 000

(a) Colour codes.   (b) The pie chart.

Figure 2.6 Pie charts are good for visualizing proportions.

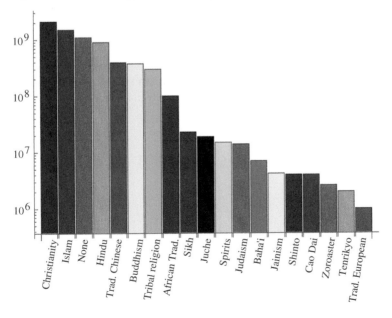

Figure 2.7 Bar chart with logarithmic vertical axis.

Table 2.1 Cumulative frequency table for number of TVs in household

Value	Frequency	Cumulative frequency		Value	Cumulative frequency
1	33	33		1	33
2	39	39 + 33 = 72		2	72
3	42	42 + 72 = 114		3	114
4	36	36 + 114 = 150		4	150

many have a certain exact value. It is, for instance, of more interest to know how large a proportion of drivers have a blood alcohol content above 0.5‰, than it is to know how many of them have precisely 0.73‰. The most common way of stating these numbers is the *cumulative* frequency: the number who are at or *below* a given threshold value.

Example 2.1.5 (Continuation of Example 2.1.2) We asked 150 households how many TVs they had. We want to know how many households had three TVs *or fewer*, and the same for the other possible values. In Table 2.1, we expand the frequency table with an extra column for the cumulative frequency, and then form the table with the cumulative frequency alone.

The cumulative frequency chart is related to its table in the same way that the bar chart is related to the regular frequency table. In Figure 2.14, we illustrate how we construct the cumulative frequency diagram 2.8c from the frequency chart 2.8a: ☐

2.2 Measure of Location: Mode

It is frequently expedient to say what is a "typical value" for a given set of data. This is often known as a *measure of location*. The main measures of location are the *mean* $\bar{x}$ and the *median* $\tilde{x}$. The median is literally in the middle, with half

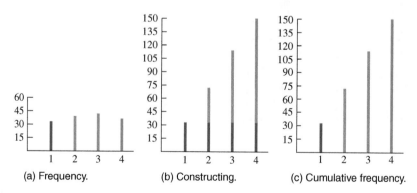

(a) Frequency. (b) Constructing. (c) Cumulative frequency.

Figure 2.8 Construction of the cumulative frequency diagram.

of the observations on either side, whereas the mean is weighted by distance, like the center of gravity of a group of equally heavy objects.

But there is one more measure of what is "typical", and it's particularly useful for categorical data. You can't perform arithmetic on categorical data, or even sort by size, so there is no mean or median. But the *mode* can be defined regardless of the nature of the data. It is simply the most frequently occurring value. In the religions example, the mode is "Christianity", since it has the highest number of adherents. In the TV survey, the mode is three, since the most common number of TVs in a household is three.

2.3 Proportion Based Measures: Median and Percentile

For proportion based measures, we order our observations by giving them new indexes by rising order of magnitude: $x_{(1)}, x_{(2)}, \ldots, x_{(n)}$.

Example 2.3.1 Let $x_1 = 9, x_2 = -1, x_3 = 7, x_4 = 5, x_5 = 2$. Switching to ordering indexes, we get $x_{(1)} = -1, x_{(2)} = 2, x_{(3)} = 5, x_{(4)} = 7, x_{(5)} = 9$. □

The median is the "observation in the middle": there are as many observations above it as below it. We calculate the median $\tilde{x}$ of the observations $x_{(1)}, \ldots, x_{(n)}$ like this: If n is odd, there is one observation in the middle, which is the median. If n is even, there are two observations in the middle; the median is then the middle value between these two.

Definition 2.3.2 *The median $\tilde{x}$ of the observations $x_{(1)}, x_{(2)}, \ldots, x_{(n)}$ is given by*

$$\tilde{x} = \begin{cases} x_{([n+1]/2)} & \text{if } n \text{ is odd} \\ \frac{1}{2}(x_{(n/2)} + x_{(n/2+1)}) & \text{if } n \text{ is even.} \end{cases}$$

Example 2.3.3 Your observations are 2, 200, 10, 78, 5. What is the median?

Answer: We order the five observations by value (see Figure 2.9): $x_{(1)} = 2, x_{(2)} = 5, x_{(3)} = 10, x_{(4)} = 78, x_{(5)} = 200$. Then

$$\tilde{x} = x_{([5+1]/2)} = x_{(3)} = \underline{\underline{10}}.$$ □

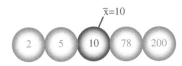

Figure 2.9 Odd numbers: the median is the middle observation.

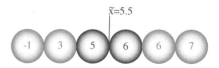

Figure 2.10 Even numbers: the median is the average of the two middle observations.

Example 2.3.4 Your observations are 3, 6, −1, 7, 6, 5. What is the median?

Answer: We order the six observations by value (see Figure 2.10): $x_{(1)} = -1$, $x_{(2)} = 3$, $x_{(3)} = 5$, $x_{(4)} = 6$, $x_{(5)} = 6$, $x_{(6)} = 7$. Then

$$\tilde{x} = \frac{1}{2}\left(x_{(6/2)} + x_{(6/2+1)}\right) = \frac{1}{2}(x_{(3)} + x_{(4)}) = \frac{1}{2}(5 + 6) = \underline{\underline{5.5}}.$$ □

The formula remains the same if we have grouped data having the same value. We just need to remember to count each value as many times as it was observed, and find the middle of the sorted observations.

Example 2.3.5 What is the median number of TVs in the households from Examples 2.1.2, 2.1.3, and 2.1.5?

To answer this calls for a revisit of the cumulative table, as shown in Table 2.2).

There are 150 measurements in total, an even number. The median is then the average of $x_{(75)}$ and $x_{(76)}$. We see from the ordered table that $x_{(75)} = 3$ and $x_{(76)} = 3$, so $\tilde{x} = (3 + 3)/2 = \underline{\underline{3}}$. □

2.3.1 Measure of Proportion: Percentile

The median $\tilde{x}$ divides the data so that at least 50% of them are smaller than or equal to $\tilde{x}$, and at least 50% are larger than or equal to $\tilde{x}$. The generalization of this is the *p*th *percentile*, a value that divides the data so that at least *p*% of them are smaller than or equal to it, and at least $(100 - p)\%$ of them are larger than

Table 2.2 Using the cumulative table to find the median

Value	Cumulative frequency	Observations, ordered
1	33	$x_{(1)}$ to $x_{(33)}$ have 1 TV.
2	72	$x_{(34)}$ to $x_{(72)}$ have 2 TVs.
3	114	$x_{(73)}$ to $x_{(114)}$ have 3 TVs.
4	150	$x_{(115)}$ to $x_{(150)}$ have 4 TVs.

or equal to it. There are, however, several different versions of the percentile, so we must choose one standard among the many possible.

We will follow the National Institute of Standards and Technology, NIST. The reasoning leading up to this standard goes like this: The median is the most important percentile, the 50th percentile. For an even number of observations $\tilde{x}$ is an average. But let us refine that perspective by allowing non-integer ordering indexes: If we have $n = 4$ observations, $\tilde{x} = x_{((50/100)\times(n+1))} = x_{((50/100)\times(4+1))} = x_{(2.5)}$. This puts $x_{(2.5)}$ right in the middle of $x_{(2)}$ and $x_{(3)}$, giving us $x_{(2.5)} = 0.5x_{(2)} + 0.5x_{(3)}$ by linear interpolation. We then define the pth percentile as $x_{((p/100)\times(n+1))}$, and calculate it by linear interpolation where necessary. We will mostly get $x_{(\kappa)}$ for a non-integer κ. For instance $x_{(3.7)}$. We calculate $x_{(3.7)}$ by linear interpolation, like this: $x_{(3.7)} = (1 - 0.7)x_3 + 0.7x_4$. The general formula and method are as follows.

Definition 2.3.6 *The pth percentile for an ordered list $\{x_{(1)}, \dots, x_{(n)}\}$ is*

$$P_p = x_{(\kappa)} = x_{(h)} + d \times (x_{(h+1)} - x_{(h)}),$$

where $\kappa = \frac{p}{100} \times (n + 1)$, a number with integer part h and decimal part d, meaning $\kappa = h.d$. If $\kappa < 1$ or $\kappa > n$, use respectively $\kappa = 1$ or $\kappa = n$ instead.

Mathematica: Quantile[$list$, $\frac{p}{100}$, {{0, 1}, {0, 1}}]
Excel: PERCENTILE.EXC(*marked cells*, $\frac{p}{100}$)

Method 2.3.7 How to calculate the pth percentile P_p, in detailed steps in the table below (with illustration in Figure 2.11):

Method	Example
0. Find the data and the desired percentile.	Data: $x_1 = 5.5$, $x_2 = 7.42$, $x_3 = -14.7$, $x_4 = 22.8$, $x_5 = 1.12$, $x_6 = 5.02$, $x_7 = 1$. Find the 70th percentile for these data.
1. Identify p and n.	$p = 70$ and $n = 7$.
2. Order the data by rising value.	$x_{(1)} = -14.7$, $x_{(2)} = 1$, $x_{(3)} = 1.12$, $x_{(4)} = 5.02$, $x_{(5)} = 5.5$, $x_{(6)} = 7.42$, $x_{(7)} = 22.8$.
3. $\kappa = \frac{p}{100} \times (n + 1)$, with integer part h and decimal part d.	$\kappa = \frac{70}{100} \times (7 + 1) = 5.6$, so $h = 5$ and $d = 0.6$.
4. $P_p = x_{(\kappa)}$ $= x_{(h)} + d \times (x_{(h+1)} - x_{(h)})$.	$P_{70} = x_{(5.6)}$ $= x_{(5)} + 0.6 \times (x_{(6)} - x_{(5)})$ $= 5.5 + 0.6 \times (7.42 - 5.5) = \underline{6.652}$.

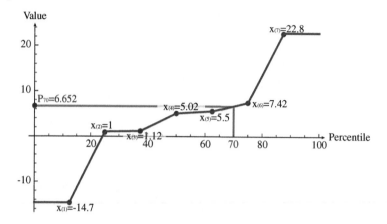

Figure 2.11 Percentile.

Definition 2.3.8 *The median $\tilde{x} = P_{50}$ is one of three important percentiles known as the* quartiles: $Q_1 = P_{25}, Q_2 = P_{50}, Q_3 = P_{75}.$ *The* interquartile range $Q_3 - Q_1$ *is often used as a measure of how spread out the data are.*

2.3.2 Measure of Location: Mean

The *mean* is a popular measure, and is the value we obtain if we obtain is we divide the total evenly between the objects we have measured. If, for instance, Peter, Paul, Ewan, and Tom earn respectively 1, 2, 4, and 5 ounces of gold per month, they would each have earned 3 ounces of gold per month if it had been divided evenly, since $\frac{1+2+4+5}{4} = 3.$ So 3 is the *mean* of the numbers 1, 2, 4, and 5.

Definition 2.3.9 *Given values $x_1, \ldots, x_n$, the* sum *is*

$$\Sigma_x = \sum_{k=1}^{n} x_k.$$

We use this for the following formal definition of the mean.

Definition 2.3.10 *The* **mean** $\bar{x}$ *of the observations $x_1, x_2, \ldots, x_n$ is defined as*

$$\bar{x} = \frac{\Sigma_x}{n}.$$

Example 2.3.11 "Ye olde milky bar" wants to know the mean number of milk shakes sold per day, for a given week. The number of milk shakes sold on the different weekdays are $x_1 = 163$, $x_2 = 178$, $x_3 = 167$, $x_4 = 191$, $x_5 = 175$. The mean number of units sold per day is then the total divided by the number of days:

$$\bar{x} = \frac{x_1 + x_2 + x_3 + x_4 + x_5}{5} = \frac{163 + 178 + 167 + 191 + 175}{5} = 174.8. \qquad \square$$

If we have many measurements of equal value, it may make sense to group the data by value. We define v_k to be the measured *values*, and a_k to be the number of observations with value v_k. Then

$$\bar{x} = \frac{\Sigma_x}{n} \quad \text{where} \quad \Sigma_x = \sum_k a_k v_k \quad \text{and} \quad n = \sum_k a_k. \qquad (2.1)$$

Example 2.3.12 We want to find the mean number of TVs in Example 2.1.2. We use formula 2.1 on the data from the table, and get

$$\bar{x} = \frac{33 \times 1 + 39 \times 2 + 42 \times 3 + 36 \times 4}{33 + 39 + 42 + 36} = 2.54. \qquad \square$$

We get a related formula if we use the *relative* frequency p_k rather than the frequency a_k:

$$\bar{x} = \sum_k p_k \times v_k, \quad \text{where} \quad p_k = \frac{a_k}{\Sigma a_k}. \qquad (2.2)$$

The relative frequency formula is particularly useful when there is no canonical base unit to break the measurements down into, as in the following example:

Example 2.3.13 In the hydro power plant *Vanna* they measure the water flow through the turbines at millions of cubic meter per hour. We have the following data from 14:00 to 15:00 on a given day: For the first 17 minutes the flow was 0.83 million m^3/h. From 14:17 to 14:45, it was 1.13 million m^3/h, after which it fell to 0.98 million m^3/h for the last quarter hour. What was the mean water flow at Vanna during that hour?

So how large a proportion of the hour did each period take? The first period lasted 17 minutes, so $p_1 = \frac{17}{60}$, and $v_1 = 0.83$. The second period lasted

28 minutes, so $p_2 = \frac{28}{60}$ and $v_2 = 1.13$. Finally, the third period lasted for 15 minutes, so $p_3 = \frac{15}{60}$ and $v_3 = 0.98$. The mean flow is then

$$\bar{x} = p_1 v_1 + p_2 v_2 + p_3 v_3 = \frac{17}{60} \times 0.83 + \frac{28}{60} \times 1.13 + \frac{15}{60} \times 0.98 = \underline{1.0075}.$$

$\square$

2.4 Measures of Spread: Variance and Standard Deviation

Variance and standard deviation are measures of how spread out the data are around the mean. The standard deviation is the square root of the variance, so we always calculate the variance first. There are two types of variance, the *sample* variance and the *population* variance. We write σ_x^2 for the population variance, and s_x^2 for the sample variance, meaning σ_x and s_x are the respective standard deviations. We recall that a *population* is all the possible values in our scope, whereas the *sample* is our actual measurements.

Definition 2.4.1 *For values* $x_1, \ldots, x_n$, *the sum of the squared deviations is*

$$SS_x = \sum_{k=1}^{n} (x_k - \bar{x})^2.$$

Given a finite population, the *population variance* is simply the average squared deviation for the population.

Definition 2.4.2 *The* population variance *of a population whose values are* $x_1, x_2, \ldots, x_n$, *is*

$$\sigma_x^2 = \frac{SS_x}{n}.$$

The quantity $\sigma_x = \sqrt{\sigma_x^2}$ *is the* population standard deviation.

However, we rarely measure entire populations, so the population variance is mostly a creature of theory. What we tend to do instead, is to measure a smaller *sample* from the population, and approximate the variance by calculations from that sample. The best approximation to the population variance of the total, is the *sample variance*. The sample variance is similar to the population variance, but to compensate for how the sample mean deviates from the population mean, we divide by $n - 1$ instead of n.

Look in Section 16.1 for a more detailed exposition of why we divide by $n - 1$ rather than by n.

Definition 2.4.3 *The* sample variance *calculated from the values* $x_1, x_2, \ldots, x_n$, *is*

$$s_x^2 = \frac{SS_x}{n - 1}.$$

The sample standard deviation *is* $s_x = \sqrt{s_x^2}$.

The most useful formula for SS_x is the following one, using the sum of squares.

Definition 2.4.4 *The* square sum *of the values* $x_1, x_2, \ldots, x_n$, *is*

$$\Sigma_{x^2} = \sum_{k=1}^{n} x_k^2.$$

As we add more data, we just need to add the new values x_i to Σ_x, and x_i^2 to our new friend Σ_{x^2} for book-keeping. This way, we don't need to recalculate all the $(x_k - \bar{x})^2$ against a $\bar{x}$ that will of course change value as new data arrive.

Rule 2.4.5

$$SS_x = \Sigma_{x^2} - n \cdot \bar{x}^2 = \Sigma_{x^2} - \frac{\Sigma_x^2}{n}.$$

Example 2.4.6 "Ye olde milky bar" wants to find the variance and standard deviation of the number of milk shakes sold per day. They use one week's measurements. They want to

(1) find the variance and standard deviation of the number of servings within the week itself, i.e. the week itself is the population;
(2) estimate the variance and standard deviation of the number of servings over a larger time period, i.e. the week itself is a sample from the greater time period.

We recall our data, $x_1 = 163$, $x_2 = 178$, $x_3 = 167$, $x_4 = 191$, $x_5 = 175$, and remember from example 2.3.11 that $\bar{x} = 174.8$. We then get

(1) $SS_x = (163^2 + 178^2 + 167^2 + 191^2 + 175^2) - 5 \cdot 174.8^2 = 472.8$;

(2) *population variance and standard deviation for the week itself:*
- $\sigma_x^2 = \frac{SS_x}{5} = \frac{472.8}{5} = 94.56$
- $\sigma_x = \sqrt{94.56} = 9.7;$

(3) *Sample variance and standard deviation for the week as an approximation to a larger time frame:*
- $s_x^2 = \frac{SS_x}{5-1} = \frac{472.8}{5-1} = 118.2$
- $s_x = \sqrt{118.2} = 10.9.$ $\square$

We have the following simplified formula when the data are given with relative frequencies.

Rule 2.4.7

$$\sigma_x^2 = \left(\sum_k p_k \times v_k^2 \right) - \bar{x}^2.$$

We revisit the continuous measurements of the water flow at the *Vanna* hydro power plant in the following example.

Example 2.4.8 We will now find the variance and standard deviation of the water flow at *Vanna*. From Example 2.3.13, we have $p_1 = \frac{17}{60}$ and $v_1 = 0.83$, and then $p_2 = \frac{28}{60}$ and $v_2 = 1.13$, and finally $p_3 = \frac{15}{60}$ and $v_3 = 0.98$. Mean flow was $\bar{x} = 1.0075$. So

- $\sigma_x^2 = \left(\frac{17}{60} \times 0.83^2 + \frac{28}{60} \times 1.13^2 + \frac{15}{60} \times 0.98^2 \right) - 1.0075^2 = 0.016\,1187;$
- $\sigma_x = \sqrt{0.016\,1187} = 0.126\,96.$ $\square$

2.5 Grouped Data

We often group data into larger groups of values, either for expediency, or because the data already is present in such groups. One example is clothes sizes, where we are typically presented with values of the kind "49–51 cm" rather than a precise, single value.

But the accuracy of measurements also naturally groups the data into such groups. If we measure the height of recruits for a military batallion, our accuracy will be in the order of 1 cm, so a measurement of "175 cm" really means all heights from 174.5 to 175.5 cm, and so on. Or it could be deliberately divided into even coarser groups, since a single cm matters little. We then display our data in a *histogram*. Notice that histograms differ from bar charts in that they

Height in cm	Number of recruits
150−170	88
170−180	1032
180−190	811
190−205	69
Sum	2000

(a) Grouped frequency table.

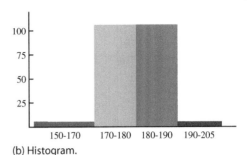

(b) Histogram.

Figure 2.12 Indre Istindfjord, 1959.

mark entire intervals packed back to front, rather than individual values. Do also note that with the histogram, we use area instead of height.

Example 2.5.1 As an example, here are the heights of recruits from Indre Istindfjord in 1959, grouped into groups of uneven width for the purpose of illustration in Figure 2.12. □

How do we treat grouped data? We have the following two options.

(1) *Representation*
We treat the data in the interval between x and y as discretely represented by the interval midpoint $\frac{x+y}{2}$, and calculate as if the midpoint was the actual value. With this option, we use the formulas we have already established for point data. When we ignore measure inaccuracy, and for instance just record the height of all men between 173.5 and 174.5 cm as "174 cm", this is what we actually do, and it works fine as long as the intervals are not too wide. See Figure 2.13.

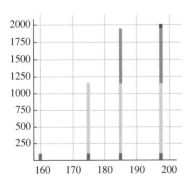

Figure 2.13 Cumulative bar chart when data are treated by interval midpoint.

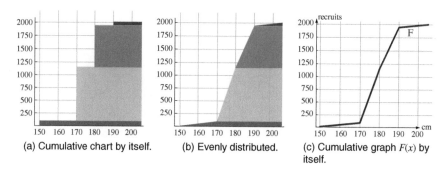

(a) Cumulative chart by itself.   (b) Evenly distributed.   (c) Cumulative graph $F(x)$ by itself.

Figure 2.14 Cumulative bar chart when data are treated as evenly distributed over interval.

(2) *Continuous*

We consider the data in the interval between *x* and *y* to be evenly distributed over the interval. At the beginning of the interval, we will therefore include none of the data, but will then progress along a straight line until the end of the interval, where all the data in the interval are included.

We will explore the continuous alternative. The cumulative graph in Figure 2.14 provides the information we need, and is also easy to set up directly from the values at the ends of the interval.

Example 2.5.2 We want to know the number of recruits at 183 cm or shorter, so we draw the cumulative graph for the data in Figure 2.15, and read (red line) that there is slightly in excess of 1400 recruits at 183 cm or less.   □

We can study *proportions* in the same way.

Example 2.5.3 We want to find the proportion of recruits between 172 and 178 cm. We may do this in two ways, both illustrated in Figure 2.16. We either

Height in cm	(Cumulative) frequency
Up to 150	0
Up to 170	88
Up to 180	1120
Up to 190	1931
Up to 205	2000

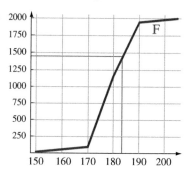

Figure 2.15 Finding percentiles when data are treated as evenly distributed over their respective intervals.

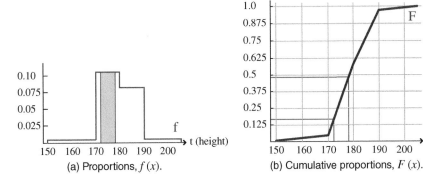

(a) Proportions, $f(x)$. (b) Cumulative proportions, $F(x)$.

Figure 2.16 Finding proportions when data are treated as evenly distributed over interval.

find the area under the graph of histogram 2.16a between 172 and 178, divided by the total area below the graph. Or: we simply read the difference between the values of the cumulative graph 2.16b at 172 and at 178. □

The cumulative approach is of course also just an approximation, giving us half and quarter recruits. But it enables us to deal with large data sets, and is therefore often preferable to exact calculations even where these are possible. They are also an early model of another type of proportion: *probabilities*. For even though observed recruits never will be evenly smeared over an interval, the *probabilites* for the different heights will be distributed with a density given by a function not too dissimilar from these continuous distributions of data.

2.5.1 Measures of Location and Spread for Grouped Data

Question: What were the mean, and the sample variance and standard deviation of the height of the recruits from Indre Istindfjord in 1959?

We will answer this question both by representation and by the continuous model.

Example 2.5.4 Answered through *representation*: the first group, from 150 to 170, has $a_1 = 88$ data points, and is represented by the midpoint $v_1 = 160$. For the second group, $v_2 = 175$ and $a_2 = 1032$. The third group has $v_3 = 185$ and $a_3 = 811$. Finally, for the fourth group, $v_4 = 197.5$ and $a_4 = 69$. The total number of recruits is $n = a_1 + a_2 + a_3 + a_4 = 2000$. This gives an average height of

- $\Sigma_x = 88 \times 160 + 1032 \times 175 + 811 \times 185 + 69 \times 197.5 = 358\,343$
- $\bar{x} = \frac{\Sigma_x}{n} = \frac{358\,343}{1999} = 179.171.$

We calculate (sample) variance and standard deviation as follows:

- $\Sigma_x^2 = 88 \times 160^2 + 1032 \times 175^2 + 811 \times 185^2 + 69 \times 197.5^2 = 64\,305\,706$
- $SS_x = \Sigma_x^2 - \Sigma_x^2/n = 64\,305\,706 - \frac{358\,343^2}{2000} = 101\,032$
- $s_x^2 = \frac{SS_x}{n-1} = \frac{101\,032}{1999} = 50.54$
- $s_x = \sqrt{50.5413} = 7.11.$

$\square$

For the *continuous* model, we need to modify the formulas used to calculate for variances. The modifications are needed in order to take into account that the data already are spread out by virtue of being evenly distributed over the interval. It turns out that the needed correction is a factor of $\frac{1}{12}$ times the width of the interval, like this:

Let interval k be $I_k = (l_k, u_k)$. The interval *midpoint* is $v_k = (l_k + u_k)/2$ and the interval *width* is $b_k = u_k - l_k$. With this notation, the modified formulas may be stated as the following rule.

Rule 2.5.5 *For numerical data grouped into intervals where interval k has midpoint v_k and width b_k, we get that*

$$\Sigma_x = \sum_k a_k \times v_k \tag{2.3}$$

$$\Sigma_{x^2} = \sum_k a_k \times \left(v_k^2 + \frac{1}{12} \times b_k^2\right). \tag{2.4}$$

Example 2.5.6 We return our attention to the example of the recruits: Σ_x and $\bar{x}$ are the same for both models, so $\Sigma_x = 358\,343$ and $\bar{x} = 179.171$

$$\Sigma_{x^2} = 88 \times \left(160^2 + \frac{1}{12} \times 20^2\right) + 1032 \times \left(175^2 + \frac{1}{12} \times 10^2\right)$$

$$+ 811 \times \left(185^2 + \frac{1}{12} \times 10^2\right) + 69 \times \left(197.5^2 + \frac{1}{12} \times 15^2\right)$$

$$= 64\,325\,291$$

$$SS_x = \Sigma_{x^2} - \frac{\Sigma_x^2}{n} = 64\,325\,291 - \frac{358\,343^2}{2000} = 120\,438$$

$$s_x^2 = \frac{SS_x}{n-1} = \frac{120\,438}{1999} = 60.25$$

$$s_x = \sqrt{60.25} = 7.76.$$

$\square$

Table 2.3 Cumulative frequency table and cumulative relative frequency table

k	Height in group k	Cumulative frequency, $\tilde{F}(u_k)$	Relative cumulative frequency, $F(u_k)$
0	Up to 150	0	0
1	150 to 170	88	0.044
2	170 to 180	1120	0.56
3	180 to 190	1931	0.9655
4	190 to 205	2000	1

2.5.2 Median and Percentile for Grouped Data

Definition 2.5.7 For grouped data we define the percentile P_p as the t value at which $F(t) = \frac{p}{100}$. The median is P_{50}.

Method 2.5.8

(1) Set up a table of cumulative relative frequencies. In the relative frequency column, for interval k, you will have $F(u_k)$ (which is equal to $F(l_{k+1})$).

(2) Locate the group k for which $F(l_k) \leq \frac{p}{100} \leq F(u_k)$.

(3) $P_p = l_k + \dfrac{\frac{p}{100} - F(l_k)}{F(u_k) - F(l_k)} \times (u_k - l_k)$.

Example 2.5.9 Find the 37.4th percentile for the Indre Istindfjord recruits in 1959.

(1) We do this through the cumulative table (Table 2.3), and the diagram shown in Figure 2.17.

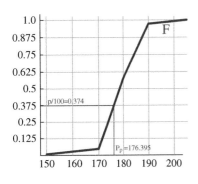

Figure 2.17 Finding the percentile through a cumulative graph.

(2) $\frac{p}{100} = 0.374$, which is between $F(l_2) = 0.044$ and $F(u_2) = 0.56$, so $k = 2$.

(3) Since $l_2 = 170$ and $u_2 = 180$, then

$$P_{37.4} = l_2 + \frac{\frac{p}{100} - F(l_2)}{F(u_2) - F(l_2)} \times (u_2 - l_2)$$

$$= 170 + \frac{0.374 - 0.044}{0.56 - 0.044} \times (180 - 170) = 176.395.$$

□

2.6 Exercises

1 **Review:** Read the chapter.
   (a) Explain in your own words what Σ_x, Σ_{x^2}, SS_x, $\bar{x}$, and $\tilde{x}$ are. Why do we have different formulas for these quantities?
   (b) What is the difference between *population* and *sample*, and how are they connected?

2 **Measures of location:** Find all the three measures of location (mode, median, mean), and decide which is the most suitable one for the situation.
   (a) The members of Femund Fishers' Union are located as follows:
      i. 24 are from Drevsjø in Engerdal, which has postal code 2443
      ii. 6 are from Ålesund, which has postal code 6020
      iii. 19 are from Røros, which has postal code 7374
      iv. 1 are from Bodø, which has postal code 8092
   (b) A class of 50 enginering students had the following income distribution 3 months after graduating:
      i. 5 were unemployed, so their income was 0.
      ii. 41 had yearly salaries (in NOK) of respectively 340 000, 341 000, 342 000 ... in increments of 1000 up to 380 000.
      iii. The last four earned respectively 613 000, 727 000, 958 000 and 70 000 000.
   (c) For a class of 100 economists, the wage distribution was as follows 3 months after graduation:
      i. 51 earned nothing.
      ii. Of the 49 remaining, 24 earned 312 000, while 25 earned 478 000.

3 The position number $\omega(letter)$ tells us where in the alphabet the given *letter* is located. For instance: $\omega(\text{"b"}) = 2$. Find the mean, median and population standard deviation for the vowel position numbers in the English alphabet.

4 For the data sets below:
   • Calculate the median and the interquartile range.
   • Find the mean and the sample standard deviation.
      (a) $\{-1, -3, 4\}$
      (b) $\{-0.2, 9.6, -0.1, 11.1, 1.3, -0.2, 11.1, -0.8, 0.4\}$

(c) $\{60, 66, 70, 103, 138, 34\}$

(d) $\{0.971\,49,\ 0.659\,64,\ 0.345\,81,\ 0.515\,90,\ 0.928\,81\}$

5 We have written the data sets below as pairs of lists: $v = \{v_1, \ldots, v_n\}$ and $a = \{a_1, \ldots, a_n\}$, meaning a_1 observations of value v_1 etc.
- Set up a frequency table and bar chart.
- Set up a cumulative frequency table and a cumulative bar chart.
- Calculate the median and the interquartile range.
- Calculate the mean and the sample standard deviation.
- Mark these measures on the horizontal axes of your charts.

(a) $v = \{5.6, 5.8, 5.1, 6.4, 5.2, 6.3, 5.0, 5.8, 6.0\}$ and $a = \{3, 2, 1, 2, 9, 8, 7, 4, 8\}$

(b) $v = \{2, 3, 5, 7, 11\}$ and $a = \{3, 4, 2, 6, 4\}$

(c) $v = \{0.620\,362,\ 0.230\,49,\ 0.375\,471,\ 0.035\,230\,2,\ 0.562\,372,\ 0.485\,507\}$ and $a = \{109, 130, 73, 61, 9, 74\}$

6 You are in charge of a joint purchase of retro sports jackets for local FC Bayern Munich supporter club. The sizes correspond to chest measurements, and are (in cm): S=87-94, M=94-102, L=102-110, XL=110-121, XXL=121-133, 3XL=133-145. A few of the members are interested in making orders, and the total is 23S, 161M, 93L, 211XL, 131XXL and 42 3XL. Use the formulas for grouped data for calculations on the chest measurements in your local FCB supporter club.

(a) Create a table and a cumulative table, and draw the histogram and the cumulative graph.

(b) For each interval, find the interval limits l_k and u_k, and calculate the widths b_k and the average value v_k.

(c) Find the median and the two quartiles.

(d) Find mean and the sample variance.

7 We have written the data sets below as pairs of lists, $a = \{a_1, \ldots, a_n\}$ and $I = \{(l_1, u_1), \ldots, (l_n, u_n)\}$, meaning a_1 observations in interval I_1 etc.
- Make the cumulative table and graph.
- Calculate median and interquartile range.
- Find the mean and the sample standard deviation.
- Mark these measures on the horizontal axes of your charts.

(a) $I = \{(12, 24)\}$ and $a = \{100\}$

(b) $I = \{(0, 30), (30, 60), (60, 90)\}$ and $a = \{48, 96, 48\}$

(c) $I = \{(0, 0.07), (0.07, 0.14), (0.14, 0.28), (0.28, 0.56), (0.56, 1.00)\}$ and $a = \{24, 38, 48, 66, 74\}$

8 A randomized survey among the supporters of the Scottish football team Heart of Midlothian FC yielded the following numbers in the different age groups: 0-12 years: 1, 13-18: 9, 19-34: 41, 35-50: 58, 51-64: 33, 65-80: 2. Use the formulas for grouped data in the following calculations:

(a) Create a table and a cumulative table, and draw the histogram and the cumulative graph.

(b) For each interval: Find the limits l_k and u_k, and the width b_k and the average value v_k. (Remember you are dealing with stated age here.)

(c) Find the median and the two quartiles.

(d) Find the mean and the sample standard deviation.

9 **The Big Gummy Worm Project:** This is a practical exercise, where you need
- a big bag of gummy worms or equivalent
- a measuring tool (e.g., a ruler or measuring tape)

For each gummy worm,
- stretch the gummy worm over the measuring tool until it snaps
- write down the colour, and the length at which it snapped

When you are done measuring and eating gummy worms, gather the data into tables: one joint table, and one per colour or colour group (depending on how many tables your teacher has told you to make). Remember that these are grouped data. For each group:
- Create the frequency table
- Create the cumulative frequency table
- Draw a diagram for each table. (bar chart or histogram; which one do you think is suitable?)
- Draw a cumulative diagram for each table. (cumulative bar chart or cumulative graph; which one do you think is suitable?)
- Calculate the median and the mean for each group, and mark on the horizontal axis of the diagrams.
- Calculate the population standard deviation for whole bag, and the sample standard deviation for each colour group. Explain the difference and the connection.

3

Multivariate Data

CONTENTS	

3.1 Introduction

We are frequently interested in finding statistical relations between different variables. We often have one or more controlled variables, and are looking to see what will happen to the uncontrolled variables. An example of this is if we study the impact of fertilizer on plant growth. The fertilizer is a controlled variable, whereas the biomass is an uncontrolled measurement variable, giving us a pair (fertilizer, biomass). We can also study the relation between a set of all uncontrolled variables. For instance, by looking at the triplet (age, height, weight) in random humans.

Example 3.1.1 For a little experiment, I asked my Facebook friends for their height h (cm) and weight w (kg), and got the data summarized in Table 3.1 and Figure 3.1. □

There is no theoretical limit to how many variables we can include in our joint measurements, and with current technology the practical limit is also as good as gone for most applications. We can handle data sets with millions of joint variables. But to build a fundamental intuition and understanding, two- and three-joint variables suffice.

The Bayesian Way: Introductory Statistics for Economists and Engineers, First Edition.
Svein Olav Nyberg.

Table 3.1 Height and weight data

h (cm)	188	189	170	163	172	182	168	182	190	181	170	183.5	187	181	162
w (kg)	102	160	100	67	58	94	56	57	83	83	57	79	93	94	79

Example 3.1.2 I asked for some personal data for a given week:

• total number of visits to or by friends (both ways);
• total number of dreams remembered;
• total number of nightmares.

The data are summarized in Table 3.2.

Displaying multivariate data can take some ingenuity, but is reasonably easy for two and three dimensions. If we have a computer, we can rotate the plot to see where the data points are located. On paper, we help out with effects like reference planes and indications of distance from that plane. We see this in Figures 3.2a and 3.2b. □

3.2 Covariance and Correlation

In addition to measures for the individual variables, such as the mean, median, and variance, we have measures of the relations between the variables. The two most important ones are *covariance* and *correlation*. *Covariance* is a bit like *variance*; we will see that the variance of x is the covariance of x with itself ($\sigma_{xx} = \sigma_x^2$). We define the following.

Definition 3.2.1 *For sets of pairs* $\{(x_k, y_k)\}_{k=1}^n$,

$$SS_{xy} = \sum_{k=1}^n (x_k - \bar{x})(y_k - \bar{y}).$$

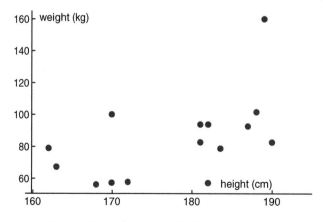

Figure 3.1 Plot of height versus weight data.

Table 3.2 Personal data for a week

Person	Visits	Dreams	Nightmares
A	4	2	2
B	12	2	0
C	5	1	1
D	11	0	0
E	1	3	2
F	0	5	3
G	6	0	0
H	8	3	1
I	0	7	7
J	0	7	7
K	0	1	0
L	7	0	0
M	6	36	24
N	1	2	0

Definition 3.2.2 *For n pairs of data, $\{(x_k, y_k)\}_{k=1}^{n}$ we have*

covariance

$$\sigma_{xy} = \frac{1}{n} SS_{xy} \quad \text{(population)}$$

$$s_{xy} = \frac{1}{n-1} SS_{xy} \quad \text{(sample)}$$

correlation

$$\rho_{xy} = \frac{\sigma_{xy}}{\sigma_x \sigma_y}$$

$$r_{xy} = \frac{s_{xy}}{s_x s_y}.$$

As with variance, the sample covariance is calculated from a subset of the population as an approximation and estimate of the population covariance from the entire population, so the population covariance is mostly a theoretical ideal, and what we calculate is the sample covariance. Interestingly, though, the sample and the population correlation are always identical.

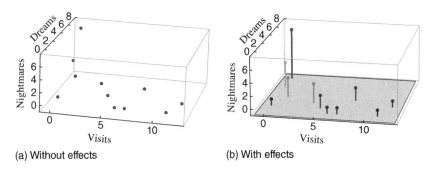

(a) Without effects   (b) With effects

Figure 3.2 Seeing data in 3D with and without helpful effects.

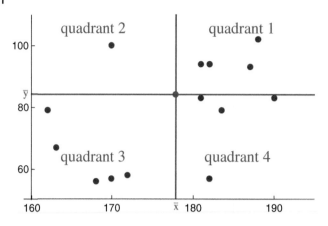

Figure 3.3 Dividing data into the four quadrants around the mean point $(\bar{x}, \bar{y})$.

To visualize what covariance and correlation are about (Figure 3.3), we divide the pairs into four quadrants, where the dividing lines are $\bar{x}$ and $\bar{y}$, respectively.

In quadrants 1 and 3, the product $(x_k - \bar{x})(y_k - \bar{y})$ is positive. For pairs in quadrants 2 and 4, it is negative. SS_{xy} is the sum of these products, so the covariances are positive if the products of the pairs in quadrants 1 and 3 outweigh the products in quadrants 2 and 4, and negative if it is the other way around.

Positive covariance is primarily characterized by a data cloud that stretches from quadrant 3 to quadrant 1, while being narrower in the direction of quadrant 2 to quadrant 4. With negative covariance we have the opposite, and if the covariance is zero, it is generally equally wide along both diagonals (Figure 3.4).

So what is the difference between covariance and correlation? It is this: correlation tells you how strongly the variation of the two variables are linked, whereas the covariation tells you a mix of the strength of the link and the size of the variations. The correlation varies from a minimum of -1 to a maximum

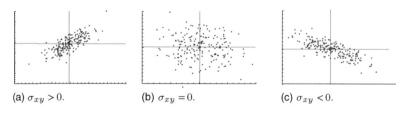

(a) $\sigma_{xy} > 0$. (b) $\sigma_{xy} = 0$. (c) $\sigma_{xy} < 0$.

Figure 3.4 Typical appearances for the three covariance values $\sigma_{xy} < 0$, $\sigma_{xy} = 0$, and $\sigma_{xy} > 0$.

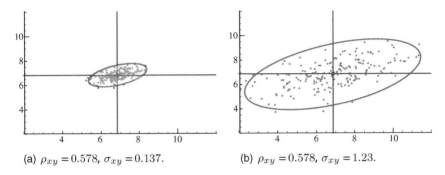

(a) $\rho_{xy} = 0.578$, $\sigma_{xy} = 0.137$. (b) $\rho_{xy} = 0.578$, $\sigma_{xy} = 1.23$.

Figure 3.5 Contrasting correlation and covariation.

at 1. The closer the correlation is to 1, the closer the cloud of data points will be to an increasing straight line, as we see in Figure 3.6. On the opposite end, near −1 the cloud is very nearly on a decreasing straight line. However, around 0, the x and y values are spread independently, and do not gather around any clear line.

The covariance includes the *spread* of the data cloud. The formulas make this precise: covariance σ_{xy} equals correlation ρ_{xy} times the measures of spread, $\sigma_x \sigma_y$. Thus the two always have the same sign. We notice the difference if we multiply the values by a constant n: the correlation remains unchanged, whereas the covariance is multipied by a factor of n^2.

In Figure 3.5b, we have expanded by a factor of $n = 3$, making the covariance $n^2 = 9$ times as big. The elliptical demarcations are just aids to mark the data clouds.

Which one of these two you will be employing, depends as always on context, and sometimes these quantities will simply be intermediate steps in a bigger calculation. As a rule of thumb, if we are interested in how *tight* the relation between the x and y values is, you are looking for the correlation, whereas if it's the *magnitude* that counts, you are interested in covariance.

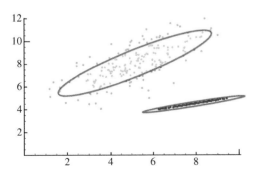

Figure 3.6 Lower plot: $\rho_{xy} = 1$, $\sigma_{xy} = 0.13$; upper plot: $\rho_{xy} = 0.71$, $\sigma_{xy} = 1.78$.

In the same way as we used Σ_{x^2}, the sum of squares, to calculate the variance, we employ the *product sum*, Σ_{xy}, to calculate the covariance, as follows.

Definition 3.2.3 *For n data pairs $\{(x_k, y_k)\}_{k=1}^{n}$, the product sum is*

$$\Sigma_{xy} = \sum_{k=1}^{n} x_k y_k.$$

The rule for calculating SS_{xy} is then simplified to the following rule.

Rule 3.2.4 *For n data pairs $\{(x_k, y_k)\}_{k=1}^{n}$,*

$$SS_{xy} = \Sigma_{xy} - n \times \bar{x} \times \bar{y} = \Sigma_{xy} - \frac{\Sigma_x \times \Sigma_y}{n}.$$

Example 3.2.5 In Table 3.3, we look at the height–weight data from Example 3.1.1.

Table 3.3 Height and weight data, together with some calculated values

h	w	h × w
188	102	19 176
189	160	30 240
170	100	17 000
163	67	10 921
172	58	9 976
182	94	17 108
168	56	9 408
182	57	10 374
190	83	15 770
181	83	15 023
170	57	9 690
183.5	79	14 496.5
187	93	17 391
181	94	17 014
162	79	12 798
2 668.5	1 262	226 386

$\Sigma_h = 2\,668.5$

$\Sigma_w = 1\,262$

$\Sigma_{hw} = 226\,386$

$SS_{hw} = \Sigma_{hw} - \dfrac{\Sigma_h \times \Sigma_w}{n}$

$\phantom{SS_{hw}} = 226\,386 - \dfrac{2\,668.5 \times 1\,262}{15}$

$\phantom{SS_{hw}} = 1\,877.09$

so

$\sigma_{hw} = \dfrac{SS_{hw}}{n} = \dfrac{1\,877.09}{15} = \underline{125.047}$

$s_{hw} = \dfrac{SS_{hw}}{n-1} = \dfrac{1\,877.09}{14} = \underline{133.979}.$

The standard deviations are $\sigma_h = 9.205\,43$, $\sigma_w = 25.684\,9$, $s_h = 9.528\,53$ and $s_w = 26.586\,4$. Then

$$\rho_{hw} = \frac{\sigma_{hw}}{\sigma_h \sigma_w} = \frac{125.047}{9.205\,43 \times 25.684\,9} = \underline{\underline{0.528\,9}}$$

$$r_{hw} = \frac{s_{hw}}{s_h s_w} = \frac{133.979}{9.528\,53 \times 26.586\,4} = \underline{\underline{0.528\,9}}.$$

We notice that $\rho_{hw} = r_{hw}$, as they should be, since they are always equal. □

3.3   Linear Regression

We study statistics with a twofold aim: to gain an overview of what is already the case, and to make predictions about the future in order to inform our choices. An old technique for both of these is *linear regression*. Linear regression is the idea that there is an underlying linear relation between two quantities: an *explanatory variable x*, and a *response variable y*, and that we can recover this linear relationship from the data, even if they are not on a straight line. The relation is in other words like this:

$$y = a + bx + \varepsilon,$$

where ε is noise or inaccuracy in how the straight line $a + bx$ predicts the *y* value.

Linear regression proceeds in three steps, as illustrated in Figure 3.7.

How *good* our guesses are, is the subject of Chapter 17.1. But for now, we can see from the illustrative example in Figure 3.7 that the guess will be a good but not perfect alignment with the original, underlying relation.

So why do we bother to use linear regression to find *y* values from *x* values, rather than just measuring the *y* values directly? Some common reasons are as follows.

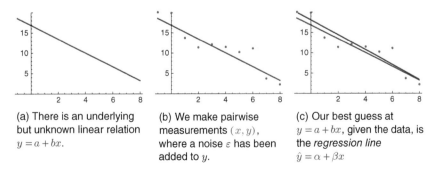

(a) There is an underlying but unknown linear relation $y = a + bx$.

(b) We make pairwise measurements (x, y), where a noise ε has been added to y.

(c) Our best guess at $y = a + bx$, given the data, is the *regression line* $\hat{y} = \alpha + \beta x$

Figure 3.7 The three steps of linear regression.

- Measurement of x may be cheap, nearby, now, or low-risk, whereas measurement of y may be expensive, far away, in the future (or distant past), or downright dangerous.
- Where the x measurement is easily available, the y measurement is possible only through destroying the object under study. An example of that is the measurement of body fat. Exact measurement of body fat y means literally picking the person apart, and is possible on dead people only, whereas measurement of the thickness of a skinfold, or electrical conductance in the body, is fast and readily done.

Example 3.3.1 "Well kept 20 room oil rig for sale. Panoramic view of the ocean, and own helicopter landing site." This was the actual text when the Norwegian oil company StatOil sold the oil rig Veslefrikk on the Norwegian equivalent of Craigslist, Finn.no, for 1 krone (roughly 12 US cents).

You are the project manager for a new Seeland, heading a team that has already restored a few other, similar oil rigs. Before restoring, you check the rigs for what you call "major faults", and you have noticed that the restoration price seems to have something reasonably close to a linear relation to the number of major faults. You have restored five platforms for a new district of Seeland so far:

x: Major faults	20	4	5	19	10
y: Restoration (millions of NOK)	121.56	104.23	108.08	119.01	110.16

Throughout the next pages, we will be answering the following questions regarding the oil platform restorations.

(a) What is the linear regression line between restoration price y and number of major faults x? (Answered in Example 3.3.4.)
(b) What are the residuals ε_i, the differences between predicted and actual price of restoration? (Answered in Example 3.3.10.)
(c) What is the standard error for the regression line? (Answered in Example 3.3.10.)
(d) You find a total of 14 major faults in Veslefrikk. What does the regression line predict the restoration cost to be? (Answered in Example 3.3.4.) □

We start out by looking at what it means to find a regression line for a set of data pairs. We know the regression line is a straight line; but which one should we choose? We will look at four candidates, four linear functions $\tilde{y}(x)$, and for each of them mark the *residuals* $\varepsilon_i = y_i - \tilde{y}(x_i)$, which are the differences between the values predicted by the regression lines and the actual measurements. The residuals are illustrated in Figure 3.8.

Our first definition of and formula for linear regression originated with Adrien-Marie Legendre, in his book on the orbits of comets. In his book, he

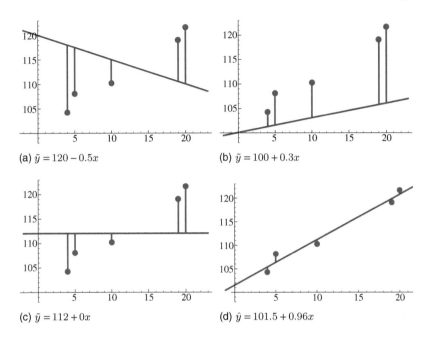

Figure 3.8 The regression line is the straight line closest to the data.

defined the straight line best adapted to the data as the one with the smallest total *squared* distance to the actual measurements. With him, we *define* the simple linear regression line in the following way.

Definition 3.3.2 *Given a set of data pairs* $\{(x_k, y_k)\}_{k \in I}$, the linear regression line *is the straight line* $\hat{y}(x) = \alpha + \beta x$ *minimizing*

$$\sum_{k \in I} \varepsilon_k^2 = \sum_{k \in I} (y_k - \hat{y}(x_k))^2,$$

where $\varepsilon_k = y_k - \hat{y}(x_k)$ *is the difference between the measured y value, and the one predicted by the line. We call x the* explanatory variable, *and y the* response variable.

There are several ways to calculate the coefficients α and β for the regression line $\hat{y} = \alpha + \beta x$. Our chosen method is *matrix regression*, which is tidy, simple, and easy to generalize. Notice that matrix regression is not a separate *type* of regression; it is simply an efficient *method*.

Method 3.3.3 We start out with a set of data pairs $\{(x_i, y_i)\}_{i=1}^n$. We then define the *design matrix* X and the *response vector* y as follows:

$$X = \begin{bmatrix} 1 & x_1 \\ \vdots & \vdots \\ 1 & x_n \end{bmatrix} \qquad y = \begin{bmatrix} y_1 \\ \vdots \\ y_n \end{bmatrix}.$$

Then

$$\beta = \begin{bmatrix} \alpha \\ \beta \end{bmatrix} = \left(X^T X\right)^{-1} X^T y,$$

where α and β are the coefficients of the regression line $\hat{y}(x) = \alpha + \beta x$.

Example 3.3.4 We want to calculate the regression line for the oil rigs in Example 3.3.1, and then use that to estimate the cost of restoring Veslefrikk, which had 14 major faults. We get

$$X^T X = \begin{bmatrix} 1 & 1 & 1 & 1 & 1 \\ 20 & 4 & 5 & 19 & 10 \end{bmatrix} \times \begin{bmatrix} 1 & 20 \\ 1 & 4 \\ 1 & 5 \\ 1 & 19 \\ 1 & 10 \end{bmatrix} = \begin{bmatrix} 5 & 58 \\ 58 & 902 \end{bmatrix}$$

$$\left(X^T X\right)^{-1} = \begin{bmatrix} 5 & 58 \\ 58 & 902 \end{bmatrix}^{-1} = \begin{bmatrix} 451/573 & -29/573 \\ -29/573 & 5/1146 \end{bmatrix}$$

$$X^T y = \begin{bmatrix} 1 & 1 & 1 & 1 & 1 \\ 20 & 4 & 5 & 19 & 10 \end{bmatrix} \times \begin{bmatrix} 121.56 \\ 104.23 \\ 108.08 \\ 119.01 \\ 110.16 \end{bmatrix} = \begin{bmatrix} 563.04 \\ 6\,751.31 \end{bmatrix}$$

giving us

$$\beta = \begin{bmatrix} 451/573 & -29/573 \\ -29/573 & 5/1146 \end{bmatrix} \times \begin{bmatrix} 563.04 \\ 6\,751.31 \end{bmatrix} \approx \begin{bmatrix} 101.471\,3 \\ 0.960\,061 \end{bmatrix}.$$

The linear regression line is then

$$\hat{y}(x) = \alpha + \beta x = \underline{101.471\,3 + 0.960\,061x}.$$

The expected cost of restoring Veslefrikk is then

$$\hat{y}(14) = 101.471 + 0.960\,061 \times 14 \approx \underline{114.9 \text{ million NOK}}. \qquad \square$$

We state the connection with variance, covariance, and correlation without further ado as follows.

Rule 3.3.5 For $\{(x_k, y_k)\}_{k \in I}$ we have $\beta = \dfrac{\sigma_{xy}}{\sigma_x^2} = \rho_{xy} \dfrac{\sigma_y}{\sigma_x}$.

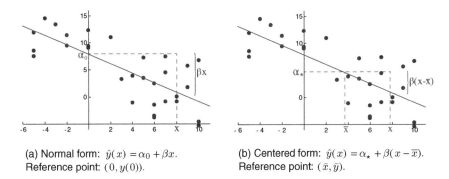

(a) Normal form: $\hat{y}(x) = \alpha_0 + \beta x$.
Reference point: $(0, y(0))$.

(b) Centered form: $\hat{y}(x) = \alpha_* + \beta(x - \bar{x})$.
Reference point: $(\bar{x}, \bar{y})$.

Figure 3.9 Two notations for the same line, with different reference points.

3.3.1 Centered Data

Statisticians have many tricks to modify data to make them more amendable to analysis. We will here employ the most useful one for the context of linear regression: centering of the x values. Centering simply means that we replace the x value by its deviation from the center, the mean $\bar{x}$, giving us $x^* = x - \bar{x}$. The resulting regression line will be the same, but we will write it in centered form. See Figure 3.9 for illustration.

When $x = 0$, a_0 is the y value; in other words a_0 is the y intercept. But α_*, on the other hand, is where the regression line intercepts the vertical line $x = \bar{x}$, and the value is $\alpha_* = \bar{y}$. We will use matrix regression, and show how we get the same line whether we use normal form or centered form.

Example 3.3.6 We continue Example 3.2.5, and find the linear regression line linking weight to height.

$$X^{\mathrm{T}}X = \begin{bmatrix} 15 & 2\,668.5 \\ 2\,668.5 & 475\,997 \end{bmatrix}$$

$$(X^{\mathrm{T}}X)^{-1} = \begin{bmatrix} 24.9651 & -0.139\,958 \\ -0.139\,958 & 0.000\,786\,72 \end{bmatrix}$$

$$X^{\mathrm{T}}w = \begin{bmatrix} 1\,262 \\ 226\,386 \end{bmatrix}$$

$$\beta = \begin{bmatrix} \alpha_0 \\ \beta \end{bmatrix} = (X^{\mathrm{T}}X)^{-1} \times X^{\mathrm{T}}v = \begin{bmatrix} -178.385 \\ 1.475\,65 \end{bmatrix},$$

which gives us the linear regression line $\hat{w}(h) = \alpha_0 + \beta h = \underline{-178.385 + 1.475\,65h}$.

□

We then calculate the regression line in centered form as follows.

Example 3.3.7 In centered form, we subtract $\bar{h} = 177.9$ from the height data. Our calculations then give us

$$X_*^T X_* = \begin{bmatrix} 15 & 0 \\ 0 & 1\,271.1 \end{bmatrix}$$

$$(X_*^T X_*)^{-1} = \begin{bmatrix} 0.066\,666\,7 & 0 \\ 0 & 0.000\,786\,72 \end{bmatrix}$$

$$X_*^T w = \begin{bmatrix} 1\,262 \\ 1\,875.7 \end{bmatrix}$$

$$\beta_* = \begin{bmatrix} \alpha_* \\ \beta \end{bmatrix} = (X_*^T X_*)^{-1} \times X_*^T w = \begin{bmatrix} 84.133\,3 \\ 1.475\,65 \end{bmatrix},$$

which yields $\alpha_* = 84.133\,3$ and $\beta = 1.475\,65$, and the linear regression line

$$\hat{w}(h) = \alpha_* + \beta \times h^* = \alpha_* + \beta(h - \bar{h}) = \underline{84.133\,3 + 1.475\,65(h - 177.9)}.$$

If we multiply the parenthesis, we get the same equation as in normal form:

$$\hat{w}(h) = 84.133\,3 + 1.475\,65(h - 177.9) = -178.385 + 1.475\,65h. \qquad \square$$

Notice that we will always have

$$X^T X = \begin{bmatrix} n & \Sigma_x \\ \Sigma_x & \Sigma_{x^2} \end{bmatrix} \quad \text{and} \quad X^T y = \begin{bmatrix} \Sigma_y \\ \Sigma_{xy} \end{bmatrix},$$

where we may recall from Rule 2.4.5 that $\Sigma_{x^2} - \Sigma_x^2/n$. If we are using centered form, this simplifies to

$$X_*^T X_* = \begin{bmatrix} n & 0 \\ 0 & \sum_{k=1}^{n}(x_k - \bar{x})^2 \end{bmatrix} = \begin{bmatrix} n & 0 \\ 0 & SS_x \end{bmatrix}. \tag{3.1}$$

3.3.2 Residuals and Standard Errors

We will look at the difference between the predicted values $\hat{y}(x_i)$ and the measurements y_i.

Definition 3.3.8 *Given data pairs* $\{(x_i, y_i)\}_{i=1}^{n}$, *we have*

- $\varepsilon_i = y_i - \hat{y}(x_i)$ (error/residual for measurement i);
- $SS_e = \sum_{i=1}^{n} \varepsilon_i^2$ (total squared error);
- $s_e^2 = \frac{1}{n-2} SS_e$ (standard error *for linear regression*, squared).

The best ways to calculate SS_e (and thereby also s_e), are the following.

Rule 3.3.9 *Given data pairs* $\{(x_i, y_i)\}_{i=1}^{n}$,

$$SS_e = \boldsymbol{y}^{\mathrm{T}}\boldsymbol{y} - \boldsymbol{\beta}^{\mathrm{T}}X^{\mathrm{T}}X\boldsymbol{\beta}$$
$$= \boldsymbol{y}^{\mathrm{T}}\boldsymbol{y} - \boldsymbol{\beta}^{\mathrm{T}}X^{\mathrm{T}}\boldsymbol{y}.$$

The errors remain the same, independently of whether we perform our calculations in normal or centered form, since the measurements and the regression line are identical, regardless of form.

Example 3.3.10 We revisit the oil rig example (Example 3.3.1).

Total squared error: we will calculate SS_e in two ways. First, we will calculate it by finding each individual error, squaring them, and then adding up the total:

i	x_i (error)	y_i (cost)	$\hat{y}(x_i)$	ε_i	ε_i^2
1	20	121.56	120.673	0.887 487	0.787 633
2	4	104.23	105.312	−1.081 54	1.169 72
3	5	108.08	106.272	1.808 4	3.270 32
4	19	119.01	119.712	−0.702 452	0.493 439
5	10	110.16	111.072	−0.911 902	0.831 566
				$\sum_{i=1}^{5} \varepsilon_i^2 = 6.5527$	

We then calculate SS_e by employing the matrix formulas. We see that $\boldsymbol{y}^{\mathrm{T}}\boldsymbol{y} = \sum y_k^2 = 63\,620.6$, so

$$SS_e = 63\,620.6 - \begin{bmatrix} 101.471\,3 \\ 0.960\,061 \end{bmatrix}^{\mathrm{T}} \times \begin{bmatrix} 5 & 58 \\ 58 & 902 \end{bmatrix} \times \begin{bmatrix} 101.471\,3 \\ 0.960\,061 \end{bmatrix} = \underline{6.5527}.$$

In other words, the two calculations yield the same, and $SS_e = \sum_i \varepsilon_i^2$.

Standard error: $s_e^2 = \frac{1}{5-2} \times 6.552\,7 = 2.184\,23$, so $s_e = \underline{1.477\,91}$.

We will pursue this example further later in the book, in Example 17.1.2. □

3.4 Multilinear Regression

We generalize the 2-variable linear regression for quantities (x, y) by letting x be a vector of k different quantities, where measurement i is the pair $(\boldsymbol{x}_i, y)$, and $\boldsymbol{x}_i = \langle x_{i,1}, x_{i,2}, \dots, x_{i,k} \rangle$. See Figure 3.10 for an illustration with $k = 2$.

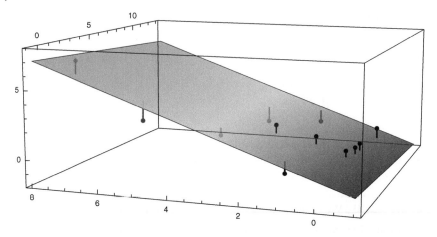

Figure 3.10 The difference between observed value y_i (point) and predicted value $\hat{y}(\boldsymbol{x}_i)$ (surface), with the distance marked as lines between point and surface.

Definition 3.4.1 *Given a data set* $\{(\boldsymbol{x}_i, y_i)\}_{i=1}^n$, *where* $\boldsymbol{x}_i = \langle x_{i,1}, \ldots, x_{i,k} \rangle$, *the linear regression surface is given by*

$$\hat{y}(\boldsymbol{x}) = \alpha_0 + \beta_1 x_1 + \beta_2 x_2 + \cdots + \beta_k x_k \qquad \text{(normal form)}$$

$$\hat{y}(\boldsymbol{x}) = \alpha_* + \beta_1 x_1^* + \beta_2 x_2^* + \cdots + \beta_k x_k^* \qquad \text{(centered form)}$$

$$= \alpha_* + \beta_1(x_1 - \overline{x_1}) + \beta_2(x_2 - \overline{x_2}) + \cdots + \beta_k(x_k - \overline{x_k}),$$

where the coefficients α *and* β *minimize the total square distance between observed value* y_i *and predicted value* $\hat{y}(\boldsymbol{x}_i)$.

The method of *matrix regression* is easily generalized to this new setting, and the only difference is that we get a larger X matrix, since we have to make room for all the dimensions of $\boldsymbol{x}$.

Method 3.4.2 Given the data set $\{(\boldsymbol{x}_i, y_i)\}_{i=1}^n = \{(x_{i,1}, x_{i,2}, \ldots, x_{i,k}, y_i)\}_{i=1}^n$, we define the design matrix X and the response vector $\boldsymbol{y}$ as follows:

$$X = \begin{bmatrix} 1 & x_{1,1} & \cdots & x_{1,k} \\ \vdots & \vdots & & \vdots \\ 1 & x_{n,1} & \cdots & x_{n,k} \end{bmatrix} \qquad \boldsymbol{y} = \begin{bmatrix} y_1 \\ \vdots \\ y_n \end{bmatrix}.$$

Then

$$\beta = \begin{bmatrix} \alpha \\ \beta_1 \\ \vdots \\ \beta_k \end{bmatrix} = \left(X^{\mathrm{T}} X \right)^{-1} X^{\mathrm{T}} y,$$

where α and $\beta_1, \ldots, \beta_k$ are the coefficients of the regression surface

$$\hat{y} = \alpha + \beta_1 x_1 + \beta_2 x_2 + \cdots + \beta_k x_k.$$

In normal form, $\alpha = \alpha_0$, whereas in centered form, $\alpha = \alpha_* = \bar{y}$ and the x are x^*.

The centered form is defined relative to the mean of each component of x, so $x_i^* = x_i - \overline{x_i}$.

Notice that the key matrices are

$$X^{\mathrm{T}} X = \begin{bmatrix} n & \Sigma_{x_1} & \Sigma_{x_2} & \cdots \\ \Sigma_{x_1} & \Sigma_{x_1^2} & \Sigma_{x_1 x_2} & \cdots \\ \Sigma_{x_2} & \Sigma_{x_1 x_2} & \Sigma_{x_2^2} & \cdots \\ \vdots & \vdots & \vdots & \ddots \end{bmatrix} \quad \text{and} \quad X^{\mathrm{T}} y = \begin{bmatrix} \Sigma_y \\ \Sigma_{x_1 y} \\ \Sigma_{x_2 y} \\ \vdots \end{bmatrix}$$

so when we are using centered form,

$$X_*^{\mathrm{T}} X_* = \begin{bmatrix} n & 0 & 0 & \cdots \\ 0 & SS_{x_1} & SS_{x_1 x_2} & \cdots \\ 0 & SS_{x_1 x_2} & SS_{x_2} & \cdots \\ \vdots & \vdots & \vdots & \ddots \end{bmatrix}.$$

Example 3.4.3 We are going to find the regression surface for the data in Table 3.4:

Table 3.4

Raw data $\{(x_{i,1}, x_{i,2}, y_i)\}$

	x_1	x_2	y
	4	2	2
	12	2	0
	5	1	1
	7	5	3
Average:	7	2.5	1.5

Centered data $\{(x_{i,1}^*, x_{i,2}^*, y_i)\}$

x_1^*	x_2^*	y
−3	−0.5	2
5	−0.5	0
−2	−1.5	1
0	2.5	3
0	0	1.5

Normal form: We leave the calculation as an exercise for the reader, and just state the result before moving on to show the detailed calculations for the centered form:

$$\hat{y} = \frac{283}{169} - \frac{41}{169}x_1 + \frac{103}{169}x_2.$$

Centered form: The design matrix is

$$X_* = \begin{bmatrix} 1 & -3 & -0.5 \\ 1 & 5 & -0.5 \\ 1 & -2 & -1.5 \\ 1 & 0 & 2.5 \end{bmatrix},$$

which gives us

$$X_*^T X_* = \begin{bmatrix} 4 & 0 & 0 \\ 0 & 38 & 2 \\ 0 & 2 & 9 \end{bmatrix}$$

and

$$\left(X_*^T X_*\right)^{-1} = \begin{bmatrix} \frac{1}{4} & 0 & 0 \\ 0 & \frac{9}{338} & -\frac{1}{169} \\ 0 & -\frac{1}{169} & \frac{19}{169} \end{bmatrix}.$$

Furthermore,

$$y = \begin{bmatrix} 2 \\ 0 \\ 1 \\ 3 \end{bmatrix}$$

so

$$X_*^T y = \begin{bmatrix} 6 \\ -8 \\ 5 \end{bmatrix}$$

then

$$\beta_* = \left(X_*^T X_*\right)^{-1} \times \left(X_*^T y\right) = \begin{bmatrix} \frac{3}{2} \\ -\frac{41}{169} \\ \frac{103}{169} \end{bmatrix}.$$

The linear regression surface is then

$$\hat{y} = \frac{3}{2} - \frac{41}{169}(x_1 - \overline{x_1}) + \frac{103}{169}(x_2 - \overline{x_2}) = \frac{3}{2} - \frac{41}{169}(x_1 - 7) + \frac{103}{169}(x_2 - 2.5).$$

This is the same surface we got when we did our calculations in normal form:

$$\frac{3}{2} - \frac{41}{169}(x_1 - 7) + \frac{103}{169}\left(x_2 - \frac{5}{2}\right) = \frac{283}{169} - \frac{41}{169}x_1 + \frac{103}{169}x_2. \qquad \square$$

3.4.1 Residuals and Standard Errors for Multilinear Regression

Definition 3.4.4 *Given data a set* $\{(x_{i,1}, x_{i,2}, \ldots, x_{i,k}, y_i)\}_{i=1}^n$,

- $\varepsilon_i = y_i - \langle 1, x_{i,1}, x_{i,2}, \ldots, x_{i,k}\rangle \times \boldsymbol{\beta}$ *(error for measurement i)*
- $SS_e = \sum_{i=1}^n \varepsilon_i^2$ *(total squared error)*
- $s_e^2 = \frac{1}{n-(k+1)} SS_e$ *(standard error for linear regression, squared)*.

Just like in the bi-variate case,

Rule 3.4.5 *Given the data set* $\{(x_{i,1}, x_{i,2}, \ldots, x_{i,k}, y_i)\}_{i=1}^n$

$$SS_e = \boldsymbol{y}^T\boldsymbol{y} - \boldsymbol{\beta}^T X^T X \boldsymbol{\beta}.$$

Example 3.4.6 We continue Example 3.4.3, and calculate the following:

(1) The total squared error SS_e
(2) The standard error s_e

Answer:

(1) **Total squared error:** We do the calculation in centered form. First, note that $\vec{\boldsymbol{y}}^T\boldsymbol{y} = |\boldsymbol{y}|^2 = 14$. Next, our calculations give us $(X_*\boldsymbol{\beta}_*)^T = \langle\frac{25}{13}, -\frac{3}{169}, \frac{181}{169}, \frac{511}{169}\rangle$. The transpose of this is $\boldsymbol{\beta}_*^T X_*^T = (X_*\boldsymbol{\beta}_*)^T$, so

$$\boldsymbol{\beta}_*^T X_*^T X_* \boldsymbol{\beta}_* = (X_*\boldsymbol{\beta}_*)^T(X_*\boldsymbol{\beta}_*) = |X_*\boldsymbol{\beta}_*|^2 = \frac{2\,364}{169}$$

$$SS_e = \boldsymbol{y}^T\vec{\boldsymbol{y}} - \boldsymbol{\beta}_*^T X_*^T X_* \boldsymbol{\beta}_*$$

$$= 14 - \frac{2\,364}{169} = \underline{\frac{2}{169}} = 0.011\,834\,3.$$

(2) **Standard error:** $s_e^2 = \frac{1}{4-(2+1)} \times \frac{2}{169} = \frac{2}{169}$, so $s_e = \underline{\underline{\frac{\sqrt{2}}{13}}} \approx 0.108\,786.$ $\qquad \square$

3.5 Exercises

1 **Review:** Read the chapter.
   (a) What is the difference between covariance and correlation?
   (b) What is "centered form"?
   (c) What is SS_e a measure of?

2 You are given the data set $D = \{(x, y)\}_{i \in I} = \{(-1, 3), (0, 5), (3, 9), (5, 7)\}$.
   (a) Set up the data in a table.
   (b) Plot the data in a diagram.
   (c) Calculate the covariance between x and y (both the population and the sample versions).
   (d) Calculate the correlation between x and y.

3 You are given the data set $D = \{(x, y)\}_{i \in I} = \{(-1, 3), (0, 5), (3, 9), (5, 7)\}$. Find the linear regression line.
   (a) Write the data in centered form, $D_* = \{(x^*, y)\}_{i \in I}$.
   (b) Find the regression coefficient β by employing Rule 3.3.5.
   (c) Find the regression line in centered form by employing Method 3.3.3.
   (d) Find the regression line in normal form by employing Method 3.3.3.
   (e) Multivariate calculus:
      i. Let $y_{a,b}(x) = a + bx$, and calculate $f(a, b) = \sum_{i=1}^{4}(y_{a,b}(x_i) - y_i)^2$.
      ii. Find the extremal point (α_0, β) of $f(a, b)$. This is the point where $\frac{\partial}{\partial a}f(a, b) = \frac{\partial}{\partial b}f(a, b) = 0$, and is a minimum.
      iii. Write down the linear regression line $y = a + bx$.
   (f) Check to see if you have the same regression line/coefficients in all the four methods.
   (g) Find the best estimate for the y-value when $x = 10$.
   (h) Draw the regression line in a diagram together with the data points.

4 For the data sets below, calculate the following.
   • The covariances σ_{xy} and s_{xy}.
   • The correlation between x and y.
   • The linear regression line. Use matrix regression.
   • The square of the standard error, s_e^2.
   • Illustrate at least one of them with a regression line and data points.
      (a) $D = \{(85, 221.5), (103, 146.1), (98, 262.6), (94, 139.8), (90, 189), (80, 126), (75, 146.5), (102, 221.4), (107, 252.9), (102, 121.4)\}$.
      (b) $D = \{(5.65, 2.64), (15.62, 9.03), (-2.96, -1.19), (1.29, 2.02), (3.84, -1.82)\}$.
      (c) $D = \{(28, 24), (66, 69), (44, 48), (39, 44), (9, 9), (1, 15), (73, 64), (41, 44)\}$.

5 For the data sets below, calculate the following.

- The linear regression surface. Use matrix regression.
- The square of the standard error, s_e^2.

    (a) $D = \{(42, 79, 1056), (62, 51, 564), (57, 47, 507), (37, 49, 655), (17, 26, 337), (39, 78, 1155), (43, 43, 593), (20, 13, 174), (97, 52, 485), (82, 94, 1158)\}$.

    (b) $D = \{(101.6, 64.8, 91.5), (14.9, 37.6, 66.6), (37.9, 48.9, 84.8), (-23.7, 28.4, 91.4), (8.7, 33.5, 43.6), (43.1, 16.5, 93.3), (44.3, 32.6, 65.)\}$.

    (c) $D = \{(-8.5, 22.8, 73.1), (-10.8, 14.8, 69.3), (-5.8, 24.6, 74.9), (-12.9, 20.8, 71.5), (-10.5, 18.9, 75.3)\}$.

4

Set Theory and Combinatorics

CONTENTS

We may think of a *set* as a container for its *elements*. A box with a deck of cards inside is a physical realization of a set, and the elements are the individual cards in the stack. If we want to illustrate set operations with the set of cards, we can spread out the cards, and draw lines around groups of cards. The groups we can create in this way are the *subsets* of the original set.

4.1 The Set Operation Symbols

- **Element** $\in$: See Figure 4.1. This symbol indicates that something is an *element* of a set. $x \in A$ means that x is an element of the set A. Notice that an element either *is* or is *not* a member of a set; there is no number to "how much" it is a member, so listing an element twice still just means it *is* a member, not that there are two of it in the set.

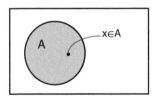

Figure 4.1 The element relation $x \in A$.

- $\{ , | \}$: These four symbols are the ones we use to specify sets. We may write a set in the following two ways.

The Bayesian Way: Introductory Statistics for Economists and Engineers, First Edition.
Svein Olav Nyberg.

1) As a list: $A = \{1, 3, 5, \text{shoe}, \text{Lars}, \text{ace of spades}, \text{Superbowl}\}$. The elements can literally be anything. If we can make a list of it, it is a set! We see that the required building blocks are the parentheses "{" and "}", and the commas. Together with the elements themselves, that is what we need to make a set from a list. The order is irrelevant, so two lists with the same elements in different order will still mean the same set.

2) As a description of the elements: $B = \{x | x \text{ wears green and plays football}\}$. This introduces a new symbol, the vertical line "|"; it means "where (the following condition applies):". We would like to use these descriptions of conditions to define subsets. For instance, if C is a deck of cards: $D = \{x \in C | x \text{ has a value less than nine}\}$, the set D will be the set of all cards in the deck that have a value less than nine.

- **Subset** ⊂, ⊃: See Figure 4.2. If a set B contains *all* elements that are also contained in another set A, and maybe some more as well, we say that A is a *subset* of B. We will write this as $A \subset B$. We can also turn the symbol around and write $B \supset A$ to mean the same. This is also stated as B being a *super*set of A. If you find it hard to remember which is which, recall the inequality signs from mathematics: the opening is always toward the larger one.

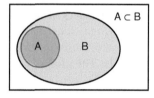

Figure 4.2 Subset $A \subset B$.

- Union and intersection. These are two very useful symbols for building sets.
  - **Union** ∪: See Figure 4.3. Union is inclusive, and $A \cup B$ is the set that contains all elements that are contained in at least one of the sets. In logical terminology $A \cup B$ are the elements that are in A *or* B. Notice that we use the term "or" in the sense of mathematical logic, where it means "at least one of them is the case", as in "I'll make dinner if your father or mother comes for a visit", and not "or" as in "I'll marry Brenda or April when I graduate".

    An example: If $A = \{1, 3, 5, 9\}$ and $B = \{2, 5, 8\}$, then $A \cup B = \{1, 2, 3, 5, 8, 9\}$. Notice that we list elements occurring in both sets only once.

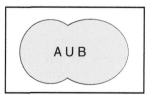

Figure 4.3 Union $A \cup B$.

- **Intersection** ∩: See Figure 4.4. In mathematics, we describe the intersection between the sets A and B like this: $A \cap B$, whereas the more common use in statistics is AB, that is: no symbol between A and B. Intersection is exclusive, so AB contains only those elements that are in *both A and B*. An example: If $A = \{1, 3, 5, 9\}$ and $B = \{2, 5, 8\}$, then $AB = \{5\}$, since the only element present in both sets is 5.

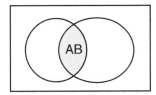

Figure 4.4 Intersection $A \cap B$ or AB.

- If we are taking the union or intersection of many sets, we use a notation of *indexes*, where we enumerate the sets, for instance $A_1, A_2, A_3, \ldots$, so that we can (for instance) write the union of the first thousand sets as follows: $\cup_{k=1}^{1000} A_k$. That is, the set of all elements that are members of *at least one* of those 1000 sets. Or if we wanted the intersection between sets 42 through 513, we would write $\cap_{k=42}^{513} A_k$. That would give us only those elements that were present in *all* of those sets.

  Sometimes enumeration itself can be a bit cumbersome, especially if we have a non-counting labelling system for our sets. In those instances, we make the *index set I*, so that $i \in I$ indexes the set A_i, and the union of all the sets is $\cup_{i \in I} A_i$. The sets could for instance be the list of books read this year by current students at US universities. The set I would be the set of indexes of students, for instance indexing each by their university's name, followed by their last name and a number, so that an index could be "HarveyMudd.Johnson.23". That student's book list would then be the set $A_{HarveyMudd.Johnson.23}$, and the union $\cup_{i \in I} A_i$ would be all books read by at least one current student at a US university. Writing $\cup_{i \in I} A_i$ instead of $\cup_{i=a}^{b} A_i$ can also be justified simply by simplicity, even when both forms are available. The form $\cup_{i=a}^{b} A_i$ is really only necessary when it is important to specify the numbers a and b.

- **Set difference** \: See Figure 4.5. The symbol "\" means *set difference*. $A \backslash B$ indicates the members of A that are left when we have removed the elements that are also members of B. If $A = \{1, 3, 5, 9\}$ and $B = \{2, 5, 8\}$, then $A \backslash B = \{1, 3, 9\}$, whereas $B \backslash A = \{2, 8\}$.

  A concrete example is if A are the drivers through the Lincoln tunnel on a given Friday evening, and B are the sober drivers on the same Friday evening. If the police tested drivers for intoxication by the Lincon tunnel that Friday evening, then $A \backslash B$ would be the drivers they fined or gave other legal reaction.

In some mathematics and statistics texts, you will also come across the symbol $-$ for set difference. That is, $A - B$ instead of $A \backslash B$.

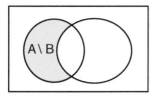

Figure 4.5 Set difference $A \backslash B$.

- **The Universe** Ω: See Figure 4.6. In statistics, we need to specify our universe of discourse. Which objects, events, possibilities, sequences, etc. we are studying. *These, and none other.* We name this set Ω, which is the most common name. You will also find S used for this purpose. In all the set illustrations, Ω is the surrounding rectangle.

    It is helpful to think of Ω as the total space of possible events for our probabilistic studies. In practical calculations, however, Ω rarely has a function as anything but a theoretical entity in the background to ensure the theory stays consistent, and we often employ an Ω that is far larger than the space of possible events.

Figure 4.6 The universe Ω.

- **The empty set** $\varnothing$: See Figure 4.7. The set containing no elements at all is very important in statistics. We write it $\varnothing = \{ \}$. It might seem like a pointless set, but it expresses the important feature of two sets having an empty intersection. That is, no elements in common. We write $AB = \varnothing$ to express that A and B are disjoint.

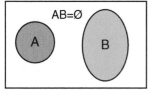

(a) $AB = \varnothing$.   (b) The empty set $\varnothing$.

Figure 4.7 The empty set has no elements; in diagrams, it is a non-existent region.

- **Complement A^c:** See Figure 4.8. This is really a special case of set difference, since technically, $A^c = \Omega \backslash A$. So A^c are all the elements of Ω that are not in A. If Ω is a deck of cards, and A are the face cards, then A^c are the number cards.

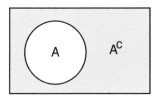

Figure 4.8 The complement A^c.

- **Cardinality $|A|$** (number of elements): We let $|\cdot|$ be the cardinality of a set. If for instance $A = \{\text{Thor, Eric, Gunther, John, Hassan}\}$, then $|A| = 5$.

Sets You Should Know
- $\mathbb{N} = \{1, 2, 3, 4, \ldots\}$, "the natural numbers" are the positive integers.
- $\mathbb{N}_0 = \{0, 1, 2, 3, \ldots\}$ are the positive integers, plus 0.
- $\mathbb{Z} = \{\ldots, -3, -2, -1, 0, 1, 2, 3, \ldots\}$ are all the integers, both positive and negative.
- $\mathbb{Q} = \{\frac{a}{b} | a, b \in \mathbb{Z}, b \neq 0\}$ – all the fractions.
- $\mathbb{R}$ are all the real numbers.

Overview
We employ, and *should* employ, Euler diagrams to illustrate properties of sets. We typically draw a rectangle to signify the universal set Ω, and use other shapes inside of that for sets A and B. This is very helpful for visualizing union, intersection, and number of sets. We sum up and gather all the illustrations in Figure 4.9.

See For Yourself
It is common to list relations between set operations. It is usually a waste of time to memorize these. A far better use of your time is to *draw* them, and see with your own eyes. Try illustrating the list below, and you will remember the relations without even trying!

- $|A \cup B| = |A| + |B| - |AB|$
- $|AB| = |A| + |B| - |A \cup B|$
- $|A \backslash B| = |A| - |AB|$
- $|A^c| = |\Omega| - |A|$
- $A \cup B = B \cup A$
- $AB = BA$
- $A \cup \varnothing = A$ (How do you draw the empty set $\varnothing$? Do you draw it at all?)
- $A\varnothing = \varnothing$
- $(A^c)^c = A$

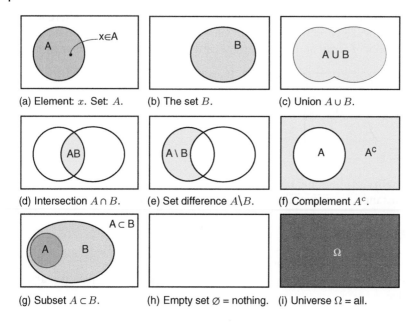

(a) Element: x. Set: A.   (b) The set B.   (c) Union $A \cup B$.

(d) Intersection $A \cap B$.   (e) Set difference $A \backslash B$.   (f) Complement A^c.

(g) Subset $A \subset B$.   (h) Empty set $\varnothing$ = nothing.   (i) Universe Ω = all.

Figure 4.9 Sets and set operations.

4.2 Combinatorics and Product Sets

We are often looking for the probability of a compound event. It then matters that we know how to count. Not as in counting 1-2-3..., but as in being able to find the cardinality of a set. For the simpler operations union, intersection, and set difference, the Euler diagrams are our best friends and helpers for this task.

But we have other ways of making sets from sets: *Product sets!* We are really doing this all the time in everyday life, for instance when we are ordering pizza. What kind of meat are we going to have? Chicken, beef, ham, pepperoni, or none? (5 options) Should we have onion? (2 options) Olives? (2 options) Which of the cheeses they offer should we choose? (maybe 4 different ones, so: 4 options). Each of these choices are independent of the others, which gives us $5 \times 2 \times 2 \times 4$ possible pizzas – 80 different pizzas! And we have yet to enter the dispute over pineapple and more exotic pizza toppings.

The elements of product sets are *sequences*. If $A = \{a_1, a_2, a_3, a_4\}$ and $B = \{b_1, b_2, b_3\}$, then the product set of the two is $A \times B = \{(a, b) \mid a \in A, b \in B\}$. Table 4.1 lists all the elements in the product $A \times B$, multiplication table style.

Table 4.1 Elements of the product set $A \times B$

	a_1	a_2	a_3	a_4
b_1	(a_1, b_1)	(a_2, b_1)	(a_3, b_1)	(a_4, b_1)
b_2	(a_1, b_2)	(a_2, b_2)	(a_3, b_2)	(a_4, b_2)
b_3	(a_1, b_3)	(a_2, b_3)	(a_3, b_3)	(a_4, b_3)

Table 4.2 Brian's table

Brian	walk	cycle
work	(Brian, walk, work)	(Brian, cycle, work)
school	(Brian, walk, school)	(Brian, cycle, school)
pub	(Brian, walk, pub)	(Brian, cycle, pub)

Another example is if $A = \{$Brian, Howard$\}$, while $B = \{$walks, cycles$\}$ and $C = \{$work, school, pub$\}$. How many possible sentences are there, then, saying that "A used means B for getting to C"? We'll list the options to see:

1. Brian walks to work.
2. Brian walks to school.
3. Brian walks to the pub.
4. Brian cycles to work.
5. Brian cycles to school.
6. Brian cycles to the pub.
7. Howard walks to work.
8. Howard walks to school.
9. Howard walks to the pub.
10. Howard cycles to work.
11. Howard cycles to school.
12. Howard cycles to the pub.

This case can also be captured in a table. Or *tables*, since we have a third variable, and must create a separate table for each option of one of the variables. We create a table for Brian (4.2) and a table for Howard (4.3).

Another way to visualize product sets, is by a *tree diagram*. This is more general than just simple products, since the branches can be pruned according to certain rules (as we will see later in Section 4.3, repeated sampling). We illustrate the example in Figure 4.10.

If you prefer a vertical flow, that is of course doable (Figure 4.11).

Let S be all the sentences we can form in the example above. Clearly, $S = A \times B \times C$, and unsurprisingly, $|S| = |A| \times |B| \times |C|$.

Table 4.3 Howard's table

Howard	walk	cycle
work	(Howard, walk, work)	(Howard, cycle, work)
school	(Howard, walk, school)	(Howard, cycle, school)
pub	(Howard, walk, pub)	(Howard, cycle, pub)

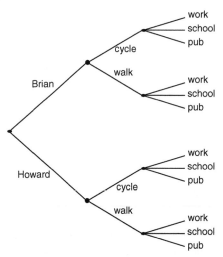

Figure 4.10 A horizontal tree diagram.

If we choose to expand the sentence by specifying a day of the week, we see that we must write each of the sentences in S seven times, once for each of the days of the week in $W = \{$Monday, Tuesday, Wednesday, Thursday, Friday, Saturday, Sunday$\}$. The set T of sentences "A used means B to get to C on day W" is seven times as large as the set of sentences "A used means B to get to C". To calculate, since $T = A \times B \times C \times W$, then $|T| = |A| \times |B| \times |C| \times |W|$.

Stated as a general rule, this is as follows.

Rule 4.2.1   The product rule: *If you have a sequence of n sets $A_1, A_2, \dots, A_n$, and pick one element a_k each from each set A_k, the total possible number of ways of doing this, equal to the total number of possible sequences $(a_1, a_2, \dots, a_n)$, is*

$$|A_1| \times |A_2| \times \cdots \times |A_n|$$

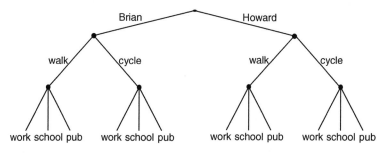

Figure 4.11 A vertical tree diagram.

4.3   Repeated Sampling

Some of the most important product sets are the (repeated) products of a set with itself, or almost itself. For instance: The set of outcomes for a coin toss is $U = \{H, T\}$. The set of possible sequences of three tosses is $S = U^3 = U \times U \times U$, and its size is $|S| = |U|^3 = 2^3 = 8$.

Assignment: A single sports game has three possible outcomes: H (home win), A (away win), and D (draw). If you are going to post a betting slip for a simultaneous bet at 12 simultaneous games, how many such betting slips are possible?

We have four basic types of sampling, determined according to two criteria: Is the sampling done with replacement? Are the results ordered or unordered?

Method 4.3.1   Repeated sampling

1) You first need to determine what type of sampling you are looking at. Ask and answer the following two questions.
   (a) **With/without replacement:** Are the same elements available for the next sampling? This is what *with replacement* means. For dice and coins, this is the default. For sampling from an urn, this is the case only if you *replace* the elements before the next sampling.
      • Yes: "with replacement"
      • No: "without replacement"
   (b) **Ordered/unordered:** Does the order of the results matter?
      • Yes: "ordered" – a sequence, like Heads-Tails-Tails-Heads-Heads
      • No: "unordered" – a combination, like three Heads, two Tails
2) Find the parameters n for k:
   (a) How many elements are available on the first sampling? This is n.
   (b) How many tries? This is k.

We see typical examples from each of the four categories in Table 4.4.

The most important categories are "ordered, with replacement" and "unordered, without replacement".

The formulas: When we perform k samples, with n initial possibilities, the number of possibilities are, depending on replacement and ordering, as shown in Table 4.5.

Table 4.4 Examples from the four categories

	with replacement	without replacement
ordered (sequence)	Daily 4 (Exact/Straight)	Solitaire
unordered (combination)	Daily 4 (Any/Box)	Powerball (white balls)

Table 4.5 The counting formulas for the four categories

	with replacement	without replacement
ordered (sequence)	n^k	$\dfrac{n!}{(n-k)!}$
unordered (combination)	$\dbinom{k+n-1}{k}$	$\dbinom{n}{k}$

This might be your first encounter with the binomial coefficient $\binom{n}{k}$. The binomial coefficient is the following expression: $\binom{n}{k} = \frac{n!}{(n-k)! \times k!}$. We read $\binom{n}{k}$ as "n choose k" or "n over k". Most calculators have the binomial coefficient nCr built-in, often next to the permutation command nPr and factorial (!).

- $\boxed{x!}$ for $n! = 1 \times 2 \times 3 \times \cdots \times (n-1) \times n$
- $\boxed{nPr}$ for $\frac{n!}{(n-r)!}$
- $\boxed{nCr}$ for $\binom{n}{r} = \frac{n!}{r!(n-r)!}$.

Now, why are the sampling formulas as they are? We will illustrate through the following representative examples.

Example-proof 4.3.2 Ordered sampling with replacement: Let us examine the Daily 4 (Exact/Straight).

1) What kind of sampling this is:
   (a) **With/without replacement:** We have four columns, and pick a number from each. The pick is *with* replacement, since number remains eligible for the other rows even after you've chosen it for one of the rows.
   (b) **Ordered/unordered:** It is also *ordered*, since when you play Straight/ Exact, you need the exact sequence to win.
2) The parameters n and k are:
   (a) For each column, you have 10 options, one for each digit 0–9. So $n = 10$.
   (b) You must pick a number four times, once for each column, so $k = 4$.

According to our overview, we are to use the formula n^k. Let us see how this formula arises: with just 1 single column, you have 10 options; one per digit. With 2 columns, your choices correspond to the two-digit numbers $00, 01, \ldots, 98, 99$, for a total of 100. So you had to multiply the original 10 by a factor of 10 (number of options for the second digit). Adding another column means that each of these 100 numbers may be expanded by another digit 0–9 (10 options), so you'd have to multiply by 10 yet again, for a total of 1000, which makes sense, since there are exactly a thousand numbers from $000, 001, \ldots$, up to $998, 999$. We see how the system goes, and multiply by 10 for each new column. For the four columns in Daily 4, that means $10^4 = 10\,000$, which is indeed n^k since the number of possibilities for each column was $n = 10$, and the number of "draws" is the number of columns. That is, $k = 4$. □

Example-proof 4.3.3 Ordered sampling without replacement: Solitaire is a class of card games known to most people, and involves laying out a specified amount of cards on the table, in a specified order. The best known is perhaps *Klondike*, where the starting configuration has seven cards lying face-up, at the top of their respective piles. It matters not only which cards are there, but also which card is on top of which pile. So how many arrangements of seven face-up cards are possible?

1) What kind of sampling this is:
   (a) **With/without replacement:** Once a card is in one of the seven positions, it cannot be in any of the others. So this is *without replacement*.
   (b) **Ordered/unordered:** The placement of the cards – their order – matters, so it is *ordered*.
2) The parameters n and k are:
   (a) You have an initial choice of 52 cards, so $n = 52$.
   (b) There are 7 face-up cards, so $k = 7$.

According to our overview, we are to use the formula $\frac{n!}{(n-k)!}$. Let us see how this formula arises: we see the following.

- The first card is drawn from 52 available cards.
- Once the first card is selected, you have only $52 - 1 = 51$ available options for the second card.
- Similarly, for the third card, you have $51 - 1 = 50$ options.
- … and so on: 49 options for the fourth card, 48 for the fifth, 47 for the sixth, and 46 for the seventh.

The number of possible configurations is thus

$$52 \times 51 \times 50 \times 49 \times 48 \times 47 \times 46$$
$$= \frac{52 \times 51 \times 50 \times 49 \times 48 \times 47 \times 46 \times \cdots \times 3 \times 2 \times 1}{45 \times 44 \times \cdots \times 3 \times 2 \times 1}$$
$$= \frac{52!}{45!} = \frac{52!}{(52 - 7)!}.$$

We notice that regardless of the size of the deck or pile or urn we pick from, the number of available elements will decrease by one for each draw, which means the general formula will always be the one we employed in this example.

In other words, the formula for k ordered draws without replacement from n possibilities, is

$$\frac{n!}{(n-k)!}.$$

In our example, $n = 52$ and $k = 7$, so the total possible number of 7 picked from 52 is $\frac{52!}{(52-7)!} = 674\,274\,182\,400$. $\qquad\qquad\square$

Example-proof 4.3.4 Unordered sampling without replacement: This time we will look at regular five-card hand poker.

1) What kind of sampling this is:
   (a) **With/without replacement:** Once a card is dealt to your hand, it cannot be dealt again, so this is *without replacement*.
   (b) **Ordered/unordered:** In poker, which cards are in your hand matters, but not the order in which you received them, so this is *unordered*.
2) The parameters n and k are:
   (a) You have an initial choice of 52 cards, so $n = 52$.
   (b) You are dealt 5 cards, so $k = 5$.

According to our overview, we are to use the formula $\binom{n}{k}$. Let us see how this formula arises: since the sampling is unordered, different sequences of five cards are *the same* sequence, and this the same hand in poker. So, how many sequences make up the same hand? This is the same as asking how many ways we can order the five cards on our hand. We do it card by card: We have five different positions, and so the first card has all of those five positions to choose from. The second hand has four, since one is taken by the first card. The third card has three, the fourth card two, and then the fifth and last card has to take the one available spot. The total number of orderings is thus $5 \times 4 \times 3 \times 2 \times 1 = 5! = 120$, meaning every poker hand can be realized through $5! = 120$ different sequences of how they were dealt. To find the total number of possible poker hands, divide the possible number of deal sequences by the number of sequences per hand:

$$\frac{\left(\frac{52!}{(52-5)!}\right)}{5!} = \frac{52!}{(52-5)!5!} = \binom{52}{5}.$$

This reasoning applies to the general case of picking k elements from a total of n – unordered and without replacement. So the general formula is

$$\binom{n}{k} = \frac{n!}{(n-k)! \times k!}.$$

(that is: "n choose k", which is named after precisely this kind of sampling).

In our example, $n = 52$ and $k = 5$, so the total number of possibilities is $\binom{52}{5} = 2\,598\,960$. □

Notice that when we draw k elements from a total of n, we are splitting a set of n elements into two sets of respectively k and $n - k$ elements each. So whether you choose to pick the set of k afterwards, or to pick the set of $n - k$, or if you are content with the splitting without picking one afterwards, the formula holds to describe the number of ways to split the set.

Example-proof 4.3.5 Unordered sampling with replacement: How many combinations of results are possible as a result of tossing 13 D_8 dice, one after the other? An example of such a combination is "five 1s, two 3s, one 4, one 5, two 7s, and two 8s".

1) What kind of sampling this is:
   (a) **With/without replacement:** A die cannot "expend" a number by displaying it. It is still available as a possibility for each of the other dice. So repeated die tosses, or tosses with multiple dice, is done *with replacement.*
   (b) **Ordered/unordered:** In this scenario, we only count the 1s, the 2s, etc., and we do not care exactly which dice has which result, as long as the numbers are right. In other words, the order does not matter, so this is *unordered.*
2) The parameters n and k are as follows.
   (a) Each die has 8 possible outcomes, so $n = 8$.
   (b) There are 13 dice to be tossed, so $k = 13$.

According to our overview, we are to use the formula $\binom{k+n-1}{k}$. Let us see how this formula arises: let us pick 13 white dice for our experiment. We will add 7 black dice, for a total of 20 dice. After we have tossed the white dice, we sort the dice into a 20-space row as follows: first, we put all the white dice showing 1. Then a black die. After that, the white 2s. Then a new black die. And so on until, after the last black die, we put the white 8s. If there should happen to be no white 2s, for instance, then the first and second black die will be neighbours. In Figure 4.12, we see possibility, with two 1s, two 2s, no 3s, three 4s, two 5s and one 6, no 7s, and three 8s.

We notice that when we know where in the lineup of 20 the 7 black dice are positioned, we also know the values of the white dice between them. So the results of tossing 13 dice D_8 correspond uniquely to the ways of placing 7 black

Figure 4.12 Thirteen white dice, separated by seven black ones.

dice in 20 spaces. That is, $\binom{20}{7}$. Or, equivalently, the number of ways to position 13 white dice in those 20 spaces: $\binom{20}{13}$.

Going through this again, with the parameters k and n not bound to numbers, we have tossed k dice D_n, which requires $n - 1$ black dice, and $k + n - 1$ spaces for all the dice. The general formula for k dice D_n is then $\binom{k+n-1}{k}$.

In our example, $n = 8$ and $k = 13$, so the total number of possibilities is $\binom{13+8-1}{13} = 77\,520$. □

4.3.1 Multiple Sets

Here, we will make use of the *multinomial formula* from Appendix A.2, a generalization of the following binomial formula.

Definition 4.3.6

$$\binom{n}{k_1, k_2, \ldots, k_m} = \frac{n!}{k_1! k_2! \cdots k_m!}.$$

Rule 4.3.7 *The number of ways to divide a set of n elements into m subsets of $k_1, k_2, \ldots, k_m$ elements each is*

$$\binom{n}{k_1, k_2, \ldots, k_m}.$$

Example-proof 4.3.8 You are going to distribute $n = 14$ one-hour tasks over a $k = 3$ day period, and all that matters is how many hours you will be working each day. You have $k_1 = 4$ hours available on Friday, $k_2 = 7$ on Saturday, and $k_3 = 3$ on Sunday. In how many ways can the 14 tasks be distributed among the three days, when a task has to be finished on the day it was started?

You mark the tasks $A, B, \ldots, N$, and make an array of 14 boxes, one for each hour, and each box gets one task. Task A has 14 boxes to choose from, whereas task B has 13, etc., meaning that you have $14 \times 13 \times \cdots = 14!$ ways to distribute the tasks among the boxes. You then mark the first 4 boxes "Friday", the next 7 "Saturday", and the last 4 "Sunday". Moving tasks within one day is obviously the same distribution of tasks among days. Say Friday has tasks H, N, B, and F. If this is rearranged to B, N, F, H, then Friday still has the same tasks. There are 4! ways to rearrange that Friday. Similarly, Saturday can have its 7 tasks rearranged in 7! ways, and Sunday sees 3! ways to arrange *its* tasks. In total, $(4!)(7!)(3!)$ arrangements are possible without altering which task ends up on which day.

The number of ways to split the tasks between the days is then the total number of ways to arrange the tasks into 14 hours, divided by the number of arrangements that have the same tasks on the same day:

$$\frac{14!}{(4!)(7!)(3!)} = \binom{14}{4,7,3} = \binom{n}{k_1, k_2, k_3}.$$

□

In the converse situation, you are not dividing a set into subsets, but rather combining a specified number of elements from different sets into one single set. How many ways are there of doing that?

Rule 4.3.9 *You are sampling from k sets, $\{A_j\}_{j=1}^k$. From A_j, you pick n_j elements in a specified way (ordered/unordered; with/without replacement). The number of ways to sample n_j elements from set j in the specified way is M_j. The total number of ways you can perform this sampling is then*

$$M = \prod_{j=1}^k M_j = M_1 \times M_2 \times \cdots \times M_k.$$

Example 4.3.10 You have 3 urns; the first has 14 red balls, the second has 12 blue, and the last one has 20 green. You are going to pick 3 red, 7 blue, and 5 green balls. How many ways can you do this?

Answer: From all three urns, you will be sampling without replacement, and the number of ways of doing this is respectively $M_1 = \binom{14}{3}$, $M_1 = \binom{12}{7}$, $M_1 = \binom{20}{5}$. The number of ways to pick 3 red, 7 blue, and 5 green balls is then

$$M = M_1 \times M_2 \times M_3 = \binom{14}{3} \times \binom{12}{7} \times \binom{20}{5} = \underline{4\,469\,617\,152}.$$

□

Example 4.3.11 Your name is Bilbo, and you are on an expedition with the 13 dwarves Thorin, Fili, Kili, Balin, Dwalin, Bifur, Bofur, Bombur, Oin, Gloin, Dori, Nori, and Ori. A sign in the forest you are passing through warns "Do not leave the path", so naturally you just *have to* check out the woods. You can pick 4 dwarves for company. You will each carry *one* weapon: sword, bow, or staff. How many configurations of you are possible for this little excursion?

Answer: The first set, A_1, is the dwarves. They are 13, and you are picking 4; the order in which you pick them does not matter, hence it is unordered. The number of ways to pick companions is thus $M_1 = \binom{13}{4} = 715$.

The other set is the weapons: {sword, bow, staff}. Since you may all choose freely and independently, this is *with replacement*. But it matters who gets which weapon, so the choice of weapons is *ordered*. You are 5, so the number of ways to choose weapons for all 5 is $M_2 = 3^5 = 243$.

The total number of configurations is thus $M = M_1 \times M_2 = 715 \times 243 = 173\,745$. $\qquad\qquad\qquad\qquad\qquad\qquad\qquad\qquad\qquad\qquad\qquad\qquad\qquad$ □

Summary
1) Sequences may be illustrated by drawing tree diagrams, and by counting.
2) The binomial coefficient is $\binom{n}{k} = \frac{n!}{k! \times (n-k)!}$, and is read as "$n$ pick k".
3) Sampling may be *ordered* or *unordered*, and *with* or *without* replacement. This makes for four different types of sampling/draws, each with its own formula.
4) **Sequence** means: *The order matters.* Corresponds to ordered sampling.
5) **Combination** means: *The order is irrelevant.* Corresponds to unordered sampling.

Test Yourself!
The formula for dividing a set of n elements into sets of respectively k and $n - k$ elements is generalized to the *multinomial* formula for the number of ways to divide a set of n into m sets of $k_1, k_2, \ldots, k_{m-1}$ and k_m elements (where $k_1 + \cdots + k_m = n$. The number of ways to do this, is $\frac{n!}{k_1! \times k_2! \cdots k_m!}$. Show how you get this formula, either by a special case, or by demonstrating the general formula directly. Hint: make one division first, and then subdivide one of these afterwards.

4.4 Exercises

Set Theory
1 **Review:** Read the chapter.
   (a) What is a set?
   (b) What is a sequence?
   (c) What is a combination?
   (d) Why does $\binom{n}{k} = \binom{n}{n-k}$?
   (e) What are the connections between binomial and multinomial?

2 $A = \{a, b, c, d, e, h, i, j\}$, $B =$ vowels, $C =$ letters with an even numbered place in the alphabet. Find $A \setminus (B \cap C)$.

3 Kari and Mona are looking at who in their class they have beaten at arm wrestling. Kari has beaten 15, wheras Mona has beaten 13. Of these, 7 have been beaten by both Kari and Mona.

(a) How many have been beaten by at least one of them?

(b) Call the set of classmates beaten by Kari, K, and the set beaten by Mona M. How do you write the set of those who were beaten by both Kari and Mona?

(c) How do you write the set of those who were beaten by at least one of Kari or Mona?

(d) Recall that the cardinality (size) of the set A is $|A|$. Write a formula to link the cardinalities of the sets in the previous two exercises and those of M and K themselves.

4 Palle and Jens are looking at which capitals they have visited. Palle has visited 17 capitals, whereas Jens has visited 23. The number of capitals visited by at least one of them is 33.

(a) How many capitals have been graced by a visit from both?

(b) Call Palle's capitals P, and Jens' capitals J. How do you write "capitals visited by at least one of the two"?

(c) How do you write the set of capitals visited by both?

(d) Recall that the cardinality (size) of the set A is $|A|$. Write a formula to link the cardinalities of the sets in the previous two exercises and those of P and J themselves.

5 In the Scottish village Glenwhisky there are many pubs. Half of the pubs serve Dalwhinnie, and a third serve Laphroig. Only 5% of the pubs serve both Dalwhinnie and Laphroig.

(a) Angus MacAbstainer drinks only these two brands of whisky, and is otherwise a teetotaler. What is the proportion of pubs that serve at least one whisky that Angues drinks?

(b) Call the set of pubs serving Dalwhinnie, D, and the ones serving Laphroig L. For a set of pubs, A, let $p(A)$ be the proportion of pubs that are A. Write your calculation of Angus's pubs as a statement about the relation between $p(D)$, $p(L)$, $P(D \cup L)$, and $p(DL)$.

6 In your church's stock of hymnals, $\frac{3}{5}$ of the hymnals are of the old edition. Half the hymnals have a flyer about your upcoming Christmas concert. Nine out of ten of your church's hymnals are either old or have a flyer about your Christmas concert.

(a) What is the proportion of hymnals that also have a flyer?

(b) Call the set of old edition hymnals E, and the set of hymnals with a flyer in them F. Writing $p(A)$ as the proportion of a set A, write the relation above as a general statement about the proportions $p(S)$, $p(L)$, $p(SL)$, and $p(S \cup L)$.

7 $A = \{$red, orange, green, indigo, violet$\}$, and $B = \{$yellow, green, blue$\}$. Ω consists of all seven colors of the rainbow, and is our universe.
  (a) What is A^c?
  (b) What is $A \cup B$?
  (c) What is $A \cap B$?
  (d) What is $A \setminus B$?
  (e) What is $B \setminus A$?
  (f) Is it true that "green $\in A$"?
  (g) Is it true that "yellow $\in A$"?
  (h) Is it true that "green $\notin B$"?
  (i) Is it true that "yellow $\notin B$"?
  (j) Is it true that "red $\notin (A \cap B)^c$"?

Combinatorics

8 Calculate $\binom{28}{2,3,5,7,11}$

9 Calculate $\binom{231}{4}$

10 The jedi master N's light saber display at a small venue at your university is fully booked, and the arrangement committee decides to expand with an extra show, and to divide the hopeful viewers into two batches. There is only room for 58 at the first show (which is also everyone's primary choice), but a total of 80 hopeful attendees. How many different ways can you divide the 80 so that 58 get to see the first show, whereas the remaining 22 get to attend the second?

11 You are sports dictator in the UK for a day, and have decided that precisely 7 of the next 12 Premier League (European) football matches should be home wins. In how many ways can you pick those 7 games?

12 Your university has a mid-semester break (which everyone knows means a self study week). You are taking 7 subjects this semester, but knowing yourself, you decide to pursue 3 of them over the week. How many different combinations of 3 subjects can you choose, from the 7 you are following?

13 You have a CD collection of 60 CDs, but as you are going to your cabin you find that your bag can hold at most 13. You decide to bring the maximum number; how many different combinations of CDs can you bring?

14 You are playing 5-dice Yatzee. How many different *full houses* are possible?

15 You are playing strip poker. Wearing briefs only, you realize your hand is terrible. You decide to trade in 3 of your cards. Any, since you consider them all equally bad. How many different sets of 3 cards is it possible to discard from your hand of 5?

16 You are deer hunting with 6 buddies. As a deer shows up at the edge of the clearing, you all pull your rifles and shoot. The deer dies instantly. As you later quarter the deer, you find only 3 bullets. How many different combinations of hunters could have contributed to the kill?

17 In Norway, a phone number consists of 8 digits, where the first cannot be 0. A phone number was recently sold for 1 million NOK. It consisted of a number, followed by 6 equal digits different from the first, before the first digit was repeated.
(a) How many such phone numbers are possible?
(b) Can you think of other rare types of phone number?
(c) How many are there of these?

18 **Cinema:** Ewan and Aiden notice that they have both bought tickets to the same movie, and both of them in row 13, which has 25 seats.
(a) How many different ways can Ewan and Aiden be placed in row 13?
(b) How many of these ways put them next to one another?
(c) (Chapter 5) If the choice of tickets is random, what is the probability that they end up next to one another?

19 There are 3 cookies left on the table. You realize that if you are quick, you may get away with grabbing two before someone else grabs the third.
(a) How many pairs of cookies is it possible to pick from the 3 (the order is irrelevant)?
(b) What if the other guy got to grab one cookie, whereupon you grabbed two: how many different ways would there be for him to grab his one cookie?
(c) Why are these two answers the same?

20 You see 8 attractive women on the beach. You know you'll get around to inviting 3 of them for a barbecue party before your friend Bram has invited the remaining 5 to *his* barbecue party.
(a) How many different combinations of your 3 invitations are possible, given the 8 women?
(b) How many different combinations of Bram's 5 invitations are possible, given the 8 women?
(c) Why are these two answers the same?

21 You have decided to divide a deck of cards into two piles; one of 22 cards, and one of 30.

(a) You pick the 30 cards for pile 1, and the remaining 22 are put into pile 2. How many divisions into 2 such piles are possible?

(b) You decide to instead pick 22 cards for pile 1, and let the remaining 30 be put into pile 2. How many divisions into two such piles is possible?

(c) Why are these two answers the same?

22 There are 70 white roses on a rosebush. You are either painting them red, or leaving them white.

(a) You are painting 47 roses red (and leaving the rest white). In how many ways can you do that?

(b) You are leaving 23 white (and painting the rest red). In how many ways can you do that?

(c) Why are these two answers the same?

23 The number of ways to sample s elements from a set of n possible, unordered, and without replacement, has a certain formula.

(a) What is this formula?

(b) What is the formula for sampling $n - s$ elements from a set of n possible?

(c) Why are the answers to these two questions the same?

24 You and your friends Gregory, Bear, and Arne are doing a project together. You have subdivided the project into 34 tasks, and you have decided that you'll do 8 tasks, Gregory will do 5, Bear will do 9, and Arne will do 12. How many ways are there to divide the tasks among you in this fashion?

25 Repeated sampling: In how many ways can you sample 5 elements from a collection of 12, when the sampling is …

(a) ordered, with replacement;

(b) unordered, with replacement;

(c) ordered, without replacement;

(d) unordered, without replacement.

(e) Give a practical example of each of the four kinds of sampling.

5

Probability

CONTENTS

5.1   The Concept of Probability

The concept of probability arose in an exchange between Fermat and Pascal, as they were analyzing how justly to split the pot of a game that had been interrupted before its conclusion. Their solution was to distribute the pot proportional to the *probability* of winning. For instance: say we have a game of dice with \$12 in the pot, and all that remains is a single toss of a D_{20} die. You win if it lands on 4, 7, or 13. Otherwise, someone else wins. How much should you get if the game is interrupted and the D_{20} never is tossed? The answer is that since your position in the game is 3 out of the 20 sides, you should have $\frac{3}{20}$ of the \$12 pot. That is, 1 dollar and 80 cents. The number $\frac{3}{20}$ is the *probability* for you to win the entire \$12 pot.

5.1.1   Kolmogorov's Axioms of Probability

The axioms of probability most commonly referred to were set up by the Russian mathematician Andrej N. Kolmogorov (1903–1987) in his 1933 book *Foundations of the Theory of Probability*. These axioms are best understood by thinking of probabilities as *proportions*.

The Bayesian Way: Introductory Statistics for Economists and Engineers, First Edition.
Svein Olav Nyberg.
© 2019 John Wiley & Sons, Inc. Published 2019 by John Wiley & Sons, Inc.

Definition 5.1.1 (Kolmogorov) *P is a probability if*

(1) $0 \le P(A) \le 1$

(2) $P(\Omega) = 1$

(3) $P\left(\bigcup_{i=1}^{n} A_i\right) = \sum_{i=1}^{n} P(A_i)$ *if* $A_i A_j = \emptyset$ *when* $A_i \ne A_j$,

where A, A_* *are subsets of* Ω.

The last point is often written for pairs of sets. Generalization via induction is then left to the reader. For pairs, axiom (3) is that $P(A \cup B) = P(A) + P(B)$ whenever $AB = \emptyset$.

5.1.2 Probability and Euler Diagrams

When we view probability through the eyes of Kolmogorov, as a kind of proportion, Euler diagrams are natural tools for understanding and calculating simple probabilities. The whole, or "universe" for our probability calculations, is Ω. We often think of Ω as the set containing all possible outcomes, and call it the *sample space*. It can be more fine-grained than that as well, to make room for mathematical techniques like approximations, but for our mental pictures, we should think of it as being precisely the set of all the possible outcomes.

Subsets of the sample space Ω are then best viewed as collections of outcomes. If your Ω is the set of possible outcomes when tossing a D_6 5 times, A can for instance be all outcomes whose sum is between 10 and 15. You may think of the area marking A as a claim on a part of the pot proportional to $P(A)$.

We have everything to small sample spaces like $\Omega = D_6 = \{1, 2, 3, 4, 5, 6\}$ and $\Omega = \{heads, tails\}$, up via $\Omega = \{$all infinite card sequences $card_1, card_2, \dots \}$ or $\Omega = \{$all possible sets of stock price graphs on NASDAQ over the course of one year$\}$, to the Ωs describing large infinite sets in mathematical probability theory, like the possible paths of Brownian motions.

Kolmogorov's axioms lend themselves easily to Euler diagram illustrations: that $P(\Omega) = 1$ is immediate, since the whole's part of the whole – is 1. That $0 \le P(A) \le 1$ is readily seen by just drawing the subset $A \subset \Omega$. The area for A can never be larger than Ω itself (that is: 1), but not smaller than the empty set (that is: 0). The third axiom deserves an explicit illustration (Figure 5.1), and a warning: it matters whether sets overlap or not!

Our illustration contains only a finite number of sets, but the rule applies to any countable collection of sets, $A_1, A_2, \dots$

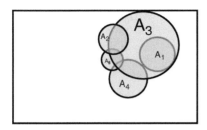 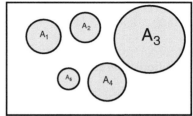

(a) Overlapping sets.   (b) Non-overlapping sets.

Figure 5.1 When the sets don't overlap, the total is larger.

5.1.3 Notation

Which notation should we use for probability? We saw that Kolmogorov's nota-tion for probability was through the function P, where the probability of A is $P(A)$. Some Bayesians, like E. T. Jaynes, point out that this notation omits the context of the probability, *and that there always is a context*. Their notation can be seen as an amendment of Kolmogorov's, where the context is included after "|", a vertical line: the probability of A in context K is $P(A|K)$. Notice that the line "|" is vertical, and not slanted like "/" or the set difference symbol "\". The vertical line "|" is to be read as "given (that)", so the fleshed-out meaning of $P(A|K)$ is "the probability of A given (context) K". As an example, $P(7|D_8)$ is "the probability that the die will show a 7, given that the die is a fair D_8".

What, then, is the probability of striking a 2 when you toss a die? If you write $P(2)$, a fair deal has been taken for granted. Most readers would prob-ably respond right away that $P(2) = \frac{1}{6}$. This means taking for granted that the die is a fair D_6, and not for instance a D_{20}. If it has not been otherwise specified D_6, we should then write $P(2|D_6)$.

But $P(A)$ is a compact and neat notation. Since the context often is given or is fair to assume, this will be our main probability notation. However, when we need to specify the context, we will write $P(A|K)$, or preferably $P_K(A)$ when this makes sense.

This kind of notation will be very useful in Chapter 6, where new obser-vations will narrow our context from the initial Ω_0 of all possibilities, to ever narrowing contexts Ω_1, then Ω_2, ... that are limited to the contexts where these observations are the case. An example of this is if you are analyzing the UEFA cup. Let's say you look at event A = "Bayern Munich wins the UEFA cup final". The context Ω_0 might simply be that the team is in the tournament. Maybe the next context we look at, Ω_1, is that we know they have made it all the way to the finals. A further sharpening of context, Ω_2, could be if in addition goalkeeper Manuel Neuer is injured. For notational simplicity, we put the index on P

rather than on Ω. In other words, we write $P_0(A) = P(A|\Omega_0)$, $P_1(A) = P(A|\Omega_1)$, $P_2(A) = P(A|\Omega_2)$, etc.

5.1.4 Definitions of Probability

We have so far not defined what probability *is*. This is not without reason, for there is no consensus on this issue among statisticians and probabilists.

We have two basic intuitions about probability: symmetry, and relative frequency. Considerations of *symmetry* are the reason we assign a probability of $\frac{1}{30}$ to getting a 7 on a D_{30}, before we have tossed such a die even once. We may also start counting the successes S_n at getting a 7 in n tosses, to get the relative frequency $a_n = \frac{S_n}{n}$. If the D_{30} is fair, a_n will converge to $\frac{1}{30}$ as n grows larger. We will see later, in Rule 9.2.5, that as a general rule, the relative frequency a_n converges to the probability p as $n \to \infty$. This rule holds independently of your definition of probability. But more importantly, as we will see in a little while: one school uses this property as the very definition of probability.

Common to all definitions is that they obey Kolmogorov's axioms.

The first definition corresponds to the following immediate, naive intuition most of us have of probability.

Definition 5.1.2 Probability (subjective): *This probability is personal, and you find your probability of A by establishing what you think is a fair bet for A: if you think it is a fair bet that the price for betting on A is m, while the price for betting against is n, in a gamble where the winner takes the pot of $n + m$, then your personal probability of A is*

$$P(A) = \frac{m}{m + n}.$$

In other words, $P(A)$ is A's fraction of what was put into the pot in a fair bet.

The fair bet is the equilibrium point where you consider both bets equally advantageous. If, at this equilibirum point, the cost of betting *on A* is lowered, or the price of betting *against A* is raised, you would not bet *against A*, but only *on*. Conversely, at the equilibrium point, if the cost of betting on A is raised or the price of betting against A is lowered, you would not find it in your own interest to do anything but bet *against A*.

Some objections to this type of probability are as follows.

- Two people will rarely assign the same probability to an event A, so $P(A)$ does not have a unique, objective, value.
- Subjective (non-objective) probabilities are not applicable in arenas that require objectivity.

- Complex situations do not easily lend themselves to such subjective evaluations, and evaluations may easily turn out to be inconsistent.
- Personal psychology may bias evaluations, especially if you are risk averse or risk seeking.

Definition 5.1.3   Probability (frequentist): *If an experiment can be repeated in an* identical manner, *but still have different outcomes, the outcome is said to be* random. *For this type of experiment, we define the frequentist* probability *of a class of outcomes, A, as*

$$P(A) = \lim_{n \to \infty} \frac{S_n}{n},$$

where S_n is the frequency of outcomes of type A after n repetitions. P(A) is in other words the limit of the relative frequency as the number of trials tends to infinity.

Some objections to the frequentist type of probability are as follows.

- "Randomness": what is considered an "identical" experiment is a subjective opinion.
- Choice of the underlying model is a subjective choice.
- An infinite number of experiments – identical or not – has never been performed (and given what we know of physics: never can be), so there exist no probabilities by this definition.
- The relative frequencies may never approach a limit: $a_n = \frac{S_n}{n}$ may fluctuate between 0 and 1 without ever settling, as n grows to ∞.
- The definition rules out assignment of probabilities to one-time events. That is, it rules out speaking about
  - "the probability that the dinosaurs were wiped out by a comet";
  - "the probability that the USA will get another President Bush";
  - "the probability that she will say *yes* when you propose".

The first two points require an explanation. Many believe that the frequentist school is objective, and a way to "let the data alone speak" without any subjective component whatsoever. But let us examine a typical experiment: tossing a coin. It is a physical experiment where the tacit assumption is that if we do an infinity of identical flips, some coins will land heads, and some will land tails, and that the relative frequency of tails will converge to a limit p as $n \to \infty$. And indeed, in our everyday experience, we have seen that such coin flips land roughly half tails and half heads, so it shouldn't be unreasonable to say that p is somewhere around $\frac{1}{2}$. But are the coin tosses identical experiments?

If we look at the underlying physics, calculating the path of the coin is complicated. But with identical starting points, the laws of mechanics tell us that the coin should follow the same path each and every time, and land at the same spot, with the same side facing up. Nothing random at all! What makes the outcome *seem* random is that the coin flip is sensitive to initial conditions: if the coin is slightly higher forwards, up, rotated ... or if you flick it just a smidgeon harder, these minute differences may be sufficient to cause it to land on the opposite side. But with sufficient precision in all these variables, the result will be the same every time.

So the "randomness" of these seemingly identical experiments is simply the result of these experiments *not* being identical, but that we do not *know* what these minute differences are, or that we have not calculated their effects. This applies to coin tosses, but also to medical experiments or Gallup polls, where the "randomness" is simply a way of saying that we don't know. The randomness states a property of our state of knowledge, i.e. a subjective factor, rather than an objective facet of the experiment itself.

This does not mean that the frequentist approach is useless, but it *does* mean that, when its proponents claim that objectivity requires us to use frequentist methods, they are simply wrong. The frequentist approach is not objective; it has just disguised its subjective components.

There is one school, however, that refers probability wholly to the objective realm.

Definition 5.1.4 Probability (propensity): *Probability is a tendency towards the different outcomes of a system, where the tendencies are inherent in the system itself. P(A) is the degree to which this tendency points to outcome A. It is often called a* propensity.

The main objection to this definition is that it is no definition at all! It turns out to be hard to formalize the propensity probabilities. Yet, despite the difficulties, this is an avenue worth pursuing – not as the *only* concept of probability, but as *a* form of probability applicable to physical systems that behave the way our current theories of physics, say quantum mechanics, work.

This book follows the *classical* school. This school is known in modern times as the *objective Bayesian school.* This as opposed to the first definition, the *subjective Bayesian school.*

Definition 5.1.5 Probability (classical): *Probability is a state of knowledge, or* degree *of knowledge. The basic probabilities are defined from symmetry in our* knowledge *about the system: if A and B are symmetrical (interchangeable) in the model, then P(A) = P(B).*

For example, if a die is symmetrical (like the typical D_n), its sides are interchangeable in our model, so $P(1) = P(2) = \cdots = P(6)$. Since $P(\Omega) = 1$, we have $P(1) + P(2) + \cdots + P(6) = 6 \times P(1) = 1$, meaning $P(1) = \frac{1}{6}$. From this, we calculate all the other probabilities pertaining to the D_n.

Some objections to the classical (objective Bayesian) definition of probability are as follows.

- There is no agreed-upon or objective procedure that can determine symmetry in all possible cases and models.
- If there is too little structure or information to determine any symmetry, there is no useful basis for probabilities for decision making.
- Practical calculations of problems of a certain complexity tend to require more computing power than corresponding frequentist techniques.

Though our favorite is clear, we will refrain from appointing our favorite as the winner. We notice that, with the exception of the propensity definition, all the schools depend on a subjective component – a component of knowledge or evaluation.

The philosophy of probability is an interesting area of study in its own right, but we shall move on to statistical methods. These methods are mostly common to all schools, but when we get to statistical inference, we will focus on the Bayesian kind. Perhaps the main advantage of Bayesian methodology is that it's conceptually whole, and thus easily understood. We will also include the basic frequentist methods in their own chapter, since they have in many ways set the stage for modern statistics, as we believe it is beneficial for students of both schools to know the basics of both methodologies – not just philosophically, but also for a practitioner's practical working understanding of the methods.

We will also see that Bayesians and frequentists often end up with equal or similar answers, even though the theoretical backdrop and the interpretations are different. This facilitates cooperation across schools. We should still be aware of the differences, so as not to mix up evaluations. A classical example illustrates the difference that lies at the core. We will introduce three characters who will be with us throughout the book: Bard the Bayesian, Frederick the frequentist, and their friend Sam.

Example 5.1.6 Frederick notices that there is a coin under a piece of cloth next to him on the table. He can't see which side is up. Sam asks his two friends what the probability is that it is heads up?

- Bard says that there is a 50% probability that it's heads up. "The symmetry," he says, "means our state of knowledge is symmetrical with respect to the two alternatives, heads and tails."
- Frederick, on the other hand, is not interested in Bard's state of knowledge, but asks what would happen if he performed an infinite number of trials.

There is no tossing involved, as that would change the state of the coin, but just observation. "What," he says out loud, "is the relative frequency of heads when I observe the state of this coin repeatedly?" He shows Sam, by lifting the cloth to reveal the coin, putting the cloth back over it, and lifting it again. "Same result every time!" he exclaims, "and think about it, Sam. Didn't we already know that before we lifted the cloth? It would either be heads all the time, in which case the probability of heads would be 100%, or tails all the time, in which case the probability of its being heads was 0%." Sam nods, and Frederick concludes: "In such cases, where the outcome is fixed but unknown, the probability is always either 0 or 100%, *but we don't know which one!*"

Most readers will probably be a bit surprised at Frederick's claim that the probability of heads is an unknown "0 or 100%, but we don't know which", rather than the 50% that Bard talked about. The reason is that Fredrick's concept of probability is different from that of our common intuition, worked up through hard nights of betting and calculating probabilities of poker hands even after the cards had already been dealt. □

Some textbooks use this or similar examples to admonish the student to adhere to the "correct" concept of probability. That is, Frederick's frequentism. If you with Bard answered 50%, you will be told that you are in the wrong. Well, you aren't. In this book, we are going by the opposite premise: we openly state our preference for the Bayesian way, but will not admonish you as being wrong simply for disagreeing with us.

5.2 Basic Probability

Rule 5.2.1 *The following rules are frequently used in probability calculations:*

(a) $P(A \cup B) = P(A) + P(B) - P(AB)$ (d) $P(A^c) = 1 - P(A)$
(b) $P(AB) = P(A) + P(B) - P(A \cup B)$ (e) $P(\Omega) = 1$
(c) $P(A \backslash B) = P(A) - P(AB)$ (f) $P(\varnothing) = 0$.

Proof: You prove these rules by employing Kolmogorov's axioms: divide the universe Ω into four disjoint parts: AB, $A \backslash B$, $B \backslash A$, and $(A \cup B)^c = A^c B^c$, and write out the composition of the sets in Rule 5.2.1 in terms of these four parts. For instance: A consists of AB and $A \backslash B$. The equalities follow by applying Kologorov's third axiom. For example: $P(A) = P(AB) + P(A \backslash B)$.

Verifying these rules is perhaps especially simple by using the Euler diagram in Figure 5.2. □

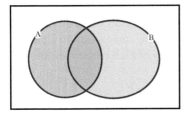
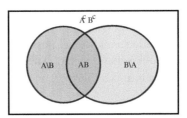

(a) The sets A and B. (b) ... with all four subdivisions marked.

Figure 5.2 Use Euler diagrams to verify Rule 5.2.1 graphically.

Example 5.2.2 It is November, and the meteorologists say there is a 14% probability of rain for tomorrow. What is the probability it will *not* rain tomorrow?

Answer: Here, $A = $ "It will rain tomorrow", and $P(A) = 0.14$. We are calculating $P(A^c)$. Rule 5.2.1.d gives us the answer: $P(A^c) = 1 - P(A) = 1 - 0.14 = \underline{0.86}$, i.e. an 86% probability that it won't rain tomorrow. □

Example 5.2.3 Signal diagram: See Figure 5.3. When a switch is *closed*, it conducts electricity. The probability that switch A is closed is 0.7, the probability that both are closed is 0.3, and the probability that none of them is closed is 0.2. Find the probability that

(a) ... B is closed;
(b) ... the circuit conducts electricity.

Answer: We name the events by their switches, so A means switch A is closed, and B means switch B is closed. The space of possibilities can be decomposed into AB, $A\backslash B$, $B\backslash A$, and A^cB^c. The basic probability rules in Rule 5.2.1 then give us the probabilities we are looking for: $P(B) = P(AB) + P(B\backslash A) = 0.1 + 0.3 = \underline{0.4}$, and $P(\text{circuit conducts electricity}) = P(A \cup B) = 0.4 + 0.3 + 0.1 = \underline{0.8}$. □

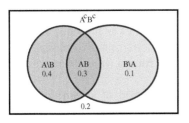
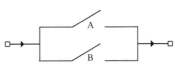

(a) The Euler diagram of the circuit. (b) The circuit diagram.

Figure 5.3 Using an Euler diagram to calculate probabilities for the circuit.

Example 5.2.4 In the following examples, we will find probabilities according to the classical definition (Definition 5.1.5): identify mutually exclusive events $E_1, \ldots, E_k$ that together make up all possibilities. These correspond to a partition of Ω into disjoint sets $A_1, \ldots, A_k$. The probability of each of these events or sets is then the same, and is therefore $\frac{1}{k}$.

(a) What is the probability of getting ball 2 if you draw from a bag of 3 equal balls, numbered 0, 1, 2?
(b) What is the probability of getting 2 heads if you toss an ordinary coin twice?
(c) What is the probability of getting a prime number on a D_{12}?
(d) What is the probability of getting the sum 5 when you throw two D_6?
(e) You mark 1 radian of a tire with white paint, drive for a long drive, and stop. What is the probability that the point on the wheel that points straight down is painted white?
(f) You have lost your goldfish in a $1000 \, \text{m}^3$ pool of muddy water, and pump out $513 \, \text{m}^3$ that you filter through a net. What is the probability that you will find your goldfish in the net?

Answers:

(a) The draw of the balls are symmetric possibilities, so the probability to draw ball 2 is $\underline{1/3}$.
(b) In this instance, the possible numbers of heads are *not* symmetric possibilities, since you can get 1 heads in 2 ways (heads–tails and tails–heads), whereas 0 and 2 heads are realizable through only 1 sequence each. Here, the *sequences* are the symmetric possibilities (see Sections 5.4 and 5.5). Since 2 heads are realized by 1 out of 4 possible sequences, $P(2) = \underline{0.25}$.
(c) The primes less than 12 are 2, 3, 5, 7, 11. On a D_{12}, all possibilities are equal, so $P(\text{prime}) = \underline{5/12}$.
(d) As in the case of the coin, here the *sequences* a–b are symmetrical, where a and b range from 1 to 6. This means there are $6 \times 6 = 36$ possibilities. Five is the sum of these sequences: 1–4, 2–3, 3–2 and 4–1. So $P(5) = \frac{4}{36} = \frac{1}{9}$. Then, $P(\text{non } 5) = 1 - P(5) = \underline{8/9}$.
(e) This is the continuous version of the symmetry method. There are no preferred angles, so the probability of a given interval of angles equals their part of the whole. A wheel – of *any* circumference – spans from 0 to 2π radians, for a total span of 2π. The probability of hitting white paint, which covers 1 radian, is then $P(\text{white}) = 1/2\pi \approx \underline{0.159}$.
(f) Continuous and three-dimensional, but the method of symmetry still works. No cubic meter of water is preferred, so the probability of finding the goldfish equals the proportion of water that was pumped out. In other words, $P(\text{goldfish}) = \frac{513 \, \text{m}^3}{1000 \, \text{m}^3} = \underline{0.513}$. $\square$

Example 5.2.5 In the following examples, we will find probabilities according to the subjective definition (Definition 5.1.2) of probability: by deciding how we would bet.

(a) You have built a bridge of straws, and a friend wants to bet that a five-year old child can blow it down. You consider it a fair bet if you pay $10 and he pays $5, and the winner takes all. What is your probability that a five-year old child will succeed in blowing down the bridge?

(b) The odds at an online gambling site tell you that a $100 bet that the Green Bay Packers will win the Superbowl will pay $350 if you win. You trust these odds to be good. What is your probability that the Green Bay Packers will win the Superbowl?

(c) You consider investing $1000 in a project that will yield $x if successful, but where you will lose your $1000 if it fails. You consider the probability of success to be 80%. What is the smallest payoff x that will justify your investment?

Answers:

(a) $P(\text{succeeds in blowing down}) = \dfrac{\text{bet for "succeeds"}}{\text{total pot}} = \dfrac{5}{10+5} = \underline{\dfrac{1}{3}}.$

(b) Here you know the total pot of $m + n = \$350$, and your required bet of $n = \$100$. That means $P(\text{Green Bay Packers wins the Superbowl}) = \dfrac{\$100}{\$350} = \underline{\dfrac{2}{7}}.$

(c) For a fair bet, you would demand $\dfrac{1000}{x} = 0.8$, in other words a payoff of at least $x = \$1250$ given success. □

We now have two of the four philosophies of probability left: the propensity school, and the frequentist school. The propensity theory is as stated not a well formed theory, which makes calculations difficult. But there are situations in physics (i.e. quantum mechanics) where it can be well argued that mathematical descriptions of physical symmetries partition the set of possibilities into sets of equal probability; if a system has precisely five interchangeable states, then the probability of each state is $\frac{1}{5}$. More complex continuous style calculations can be performed on the basis of the wave function Ψ. So to the extent that propensities may be calculated, they correspond to similar calculations in classical probability. The difference is that in classical probability, the symmetries are among states of knowledge, whereas in the propensity theory, these symmetries are assumed to be physical, or objective.

In the frequentist school, we know a probability only after an infinite sequence of trials. Even a coin that has yielded 1000 tails in a row may turn out to give only 50% tails in the long run, so no finite number of trials – no matter how large – may fill the requirement of the frequentist definition of probability. What is done in the frequentist school is to make an *estimate* on the probability, and then to say you have a certain degree of *confidence* that the probability is within a certain range from the estimated value. What exactly "confidence" is, is hard to say, but it is commonly considered an error to equate it with a probability.

Definition 5.2.6 *An estimate m on a magnitude μ is called* unbiased *if an infinite sequence of estimates $m_1, m_2, m_3, \ldots$, performed in precisely the same manner, would give*

$$\lim_{n\to\infty} \frac{m_1 + m_2 + m_3 + \cdots + m_n}{n} = \mu.$$

Example 5.2.7 In the following examples, we will make simple *estimates* on probabilities according to the frequentist definition (Definition 5.1.3) of probability. A frequentist estimate on a probability should be unbiased. It turns out that the unbiased estimate of $P(A)$ is $\tilde{P}(A) = S_n/n$, where S_n is the number of occurrences of A after n trials.

(a) You have tossed a D_6 die 300 times, and got 45 ones, 51 twos, 46 threes, 52 fours, 58 fives, and 48 sixes. What are the frequentist estimates of $P(1), \ldots, P(6)$?

(b) You have tossed a D_6 die twice, and got one 3 and one 5. Estimate $P(1), \ldots, P(6)$.

Answers:

(a) The estimates are (to three decimal places) $\tilde{P}(1) = \frac{45}{300} = \underline{0.150}$, $\tilde{P}(2) = \frac{51}{300} = \underline{0.170}$, $\tilde{P}(3) = \frac{46}{300} = \underline{0.153}$, $\tilde{P}(4) = \frac{52}{300} = \underline{0.173}$, $\tilde{P}(5) = \frac{58}{300} = \underline{0.193}$, $\tilde{P}(6) = \frac{48}{300} = \underline{0.160}$.

(b) The estimates are (to three decimal places) $\tilde{P}(1) = \frac{0}{2} = \underline{0.000}$, $\tilde{P}(2) = \frac{0}{2} = \underline{0.000}$, $\tilde{P}(3) = \frac{1}{2} = \underline{0.500}$, $\tilde{P}(4) = \frac{0}{2} = \underline{0.000}$, $\tilde{P}(5) = \frac{1}{2} = \underline{0.500}$, $\tilde{P}(6) = \frac{0}{2} = \underline{0.000}$. □

5.2.1 Simple Calculations

We often state simple probabilities in tables, as illustrated by the following example.

Example 5.2.8 Your mobile phone does not react when you press the start button. Your friend Frederick has created a table (Table 5.1) of the probabilities of the causes.

Table 5.1 Causes and probabilities for your dead phone

Cause	Probability
Empty battery	0.65
Pushing the wrong button	0.25
Dead phone	0.07
No battery	0.03

(a) What is the probability that the fault is battery related?
(b) What is the probability that the fault is *not* battery related?

Answers:

(a) P(fault is battery related) $= 0.65 + 0.03 = \underline{0.68}$.

(b) P(fault is not battery related) $= 1 - P$(fault is battery related)
$$= 1 - 0.68 = \underline{0.32}.$$
□

5.3 Conditional Probability

You are on a plane with a friend, heading for your vacation destination, and you recently heard that only 1 in 200 000 flights end in a crash. You estimate the probability of *your* flight ending in a crash as were it randomly picked among 200 000, so you get a very small number indeed. $P(\text{crash}) = \frac{1}{200\,000} = 0.000\,005$. But then your friend asks if you would change your mind if a fire broke out in one of the engines. You have read that one in ten planes with an engine on fire, crashes. You immediately update your probability estimate to $P(\text{crash}) = \frac{1}{10} = 0.1$. Your estimate is *conditioned* on fire in an engine.

We illustrate this in Figure 5.4. To the left, we have the universe Ω of all possible flights, where A are the flights that end in a crash, B are the ones where

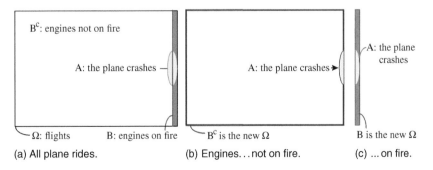

(a) All plane rides. (b) Engines...not on fire. (c) ...on fire.

Figure 5.4 The conditional probability of a plane crash, given that the engines are on fire.

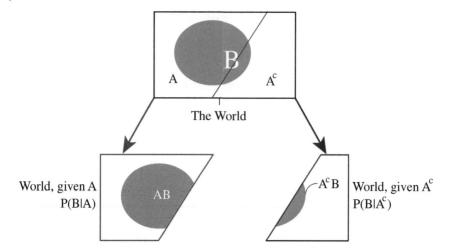

Figure 5.5 $P(B|A)$ is the proportion of AB in A.

an engine catches on fire, and B^c are the ones where the engines *don't* catch on fire. We have inflated B for the sake of illustration. In the right diagram, we see how A's proportion of flights increases dramatically when the engine's on fire, whereas in the middle one we see how A's proportion of the flights remains minuscule when the engines are not on fire.

In general, we do as follows: you want to know the probability of an occurrence A *given* another occurrence B. We call this the probability of A *conditional on B*, or "A given B", and write it $P(A|B)$. The conditional probability $P(\cdot|B)$ is a probability, with all the properties of the full probability $P(\cdot) = P(\cdot|\Omega)$; the only difference is that for $P(\cdot|B)$, the universe is B, not the old Ω.

Just like $P(\cdot) = P(\cdot|\Omega)$ is a proportion (of Ω), so $P(\cdot|A)$ is a proportion of a more limited universe: the universe where A is the case. $P(\cdot|A)$ is a probability where A *is the context*, just like Ω is the context in $P(\cdot) = P(\cdot|\Omega)$.

$P(B|A)$ is the proportion of A that is *also* B. This is, as we see in Figure 5.5, a fraction between $P(AB)$ and $P(B)$. Similarly, $P(B|A^c)$ is the proportion of A^c that is also B, which, as we see, is a totally different number.

This leads us to formal definition of $P(A|B)$, Definition 5.3.1.

Definition 5.3.1 $P(A|B) = \dfrac{P(AB)}{P(B)}$.

We illustrate this in Figure 5.6. The straightforward probability is represented as a solid object, with the understanding that its size is its probability. The conditional probability is represented as a solid object within a larger

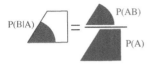 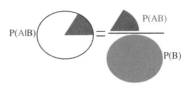

Figure 5.6 $P(B|A) = \frac{P(AB)}{P(A)}$ and $P(A|B) = \frac{P(AB)}{P(B)}$.

object, with the understanding that the ratio of the size of the smaller object to the larger object is the conditional probability.

Strictly speaking, all probabilities are conditional probabilities. This is perhaps particularly important for us Bayesians, where an ever underlying condition to the Ω is our state of knowledge: if I toss a coin and hide from you that it landed heads, then *for you*, $P(\text{tails}) = \frac{1}{2}$, whereas *for me* $P(\text{tails}) = 0$. So strictly speaking, if we flesh out any $P(A)$, we should really be writing $P(A|\text{total state}$ of your knowledge about everything). But this would for one thing lead to very cumbersome notation, so we will leave the greater context as a background and out of our notation, and reserve the conditional notation for conditioning on occurrences within the already set-up model.

It will at times be both notationally and conceptually useful to write $P_B(\cdot)$ instead of $P(\cdot|B)$, and in Chapter 6 on Bayes, that is exactly what we will do!

Example 5.3.2 $P(AB) = 0.5$ and $P(B) = 0.8$. What is $P(A|B)$?

Answer: $P(A|B) = \frac{P(AB)}{P(B)} = \frac{0.5}{0.8} = \underline{0.625}$. $\qquad\qquad\square$

Example 5.3.3 $P(A|B) = 0.7$ and $P(B) = 0.8$. What is $P(AB)$?

Answer: Multiply by $P(B)$ on both sides of Definition 5.3.1, to get $P(AB) = P(A|B) \times P(B)$. In our problem, that gives us $P(AB) = 0.7 \times 0.8 = \underline{0.56}$. $\quad\square$

We sum this up in the generally useful formula for the joint probability $P(AB)$, expressed in Rule 5.3.4, and in Figure 5.7.

Rule 5.3.4 $P(A) \times P(B|A) = P(AB)$.

Figure 5.7 $P(A) \times P(B|A) = P(AB) = P(A|B) \times P(B)$.

Example 5.3.5 You have a bag with one each of the gamer dice D_4, D_6, D_8, D_{10}, D_{12}, D_{20}. You will draw one, and then toss it. What is the probability that you pick a D_8 and then toss a 2?

Answer: Let B = "drawing a D_8", and A = "getting 2 when you toss the die you drew". There are six dice, so the probability of drawing the D_8 is $P(B) = 1/6$. The probability of getting a 2, *given that* you picked the D_8, is of course $P(A|B) = 1/8$. The probability of drawing a D_8 and then throwing a 2 is then

$$P(AB) = 1/8 \times 1/6 = \frac{1}{48}.$$ □

Example 5.3.6 The probability of a serious personnel accident at the Lille-foss Steelworks on a given day is 0.001. The probability that production will be stopped for the rest of the day after a serious personnel accident is 0.73. What, then, is the probability of a personnel accident followed by a stop in production for the rest of the day?

Let A = production stop, and B = serious personnel accident. We have been given that $P(B) = 0.001$, and that $P(A|B) = 0.73$, and have been asked to calculate $P(AB)$. The answer is $P(AB) = P(A|B) \times P(B) = 0.73 \times 0.001 = 0.00073$. □

Example 5.3.7 Mørkvik Marine Solutions construct mini platforms with two load-carrying beams they call the A beam and the B beam. An analysis of 468 mini platforms reveals that 23 of them have a crack in the A beam, 18 have a crack in the B beam, and 15 have cracks in both. Your company owns a Mørkvik mini platform, and you have just discovered a crack in the B beam. Use the relative frequencies of cracks as your probabilities, and answer the following question: What is the probability that the A beam has a crack as well?

Answer: We let the sets A and B designate cracks in the A and B beams, respectively. Our probabilities are $P(A) = \frac{23}{468}$, $P(B) = \frac{18}{468}$, and $P(AB) = \frac{15}{468}$. Our task is to find the probability of a crack in the A beam *given that* there is a crack in the B beam. The probability is

$$P(A|B) = \frac{P(AB)}{P(B)} = \frac{\frac{15}{468}}{\frac{18}{468}} = \frac{5}{6} \approx 0.83.$$

Notice that $P(A)$ does not enter into the calculation of $P(A|B)$. □

Example 5.3.8 We continue Example 5.2.8. Your phone is connected to a charger cable, so it should have worked if the error had been battery related. Make a table of the conditional probabilities for the non-battery-related errors, conditioned on the error not being battery related.

Table 5.2 Causes and conditional probabilities for your dead phone

Table 5.2 Causes and conditional probabilities for your dead phone

Cause	Conditional probability
Pressing the wrong button	$0.25/0.32 = 0.781\,25$
Dead telephone	$0.07/0.32 = 0.218\,75$

Answer: We already know that P(the error is not battery related) $= 0.32$. We get the conditional probabilities by dividing by this number. All are collected in Table 5.2. □

5.3.1 Multiple Conditioning

You might wonder what happens if the conditional probability $P(A|B)$ itself is conditioned on yet another event C. The probability of "A, given B, given C", is it $P((A|B)|C)$? No, fortunately not, as that would have led us down the path to some pretty ugly notation, much along the lines of the old saying *If only we had bacon, we could make eggs&bacon if only we had eggs!* That's just a very cumbersome way to say that we could make eggs&bacon if only we had eggs and bacon. In probability notation, it is the same: "A, given B, given C" is simply "A, given B and C". That means our conditional probability is not $P((A|B)|C)$, but rather $P(A|BC)$.

Notation taken care of, we now expand Rule 5.3.4 to multiple events, in what is known as the *product rule of probability*. Given sets $A_1, \ldots, A_n$, we have

$$P(A_n A_{n-1} \cdots A_2 A_1) = P(A_{n-1} \cdots A_2 A_1) \times P(A_n | A_{n-1} \cdots A_2 A_1).$$

We expand, one set at a time, with the first set first:

$$P(A_{n-1} \cdots A_2 A_1) = P(A_{n-1} | A_{n-2} \cdots A_2 A_1) \times P(A_{n-2} \cdots A_2 A_1).$$

Repeating the process, we get the formula

Rule 5.3.9

$$P\left(\bigcap_{j=1}^{n} A_j \right) = P(A_1) \times \prod_{k=2}^{n} P\left(A_k \,\middle|\, \bigcap_{j=1}^{k-1} A_j \right).$$

Example 5.3.10 In a simple market without any other mechanisms at play, the four insurance giants ShiRe, gRendel, ReRe and VirGo have mutually reassured one another. The probability that VirGo goes bankrupt within a given year is 0.05. If VirGo goes bankrupt, then the probability that ReRe too is going bankrupt in the same year, is 0.31. If both VirGo and ReRe go bankrupt within the same year, then the probability that gRendel will be bankrupt as well that year, is 0.47. If all the three other companies go bankrupt within a given year, the probability that ShiRe too will be bankrupt that year, is 0.38. (Notice that our probabilities are about simultaneity, so the precise order of the bankruptcies do not matter.) What is the probability that all four companies will be bankrupt in a given year?

Answer: Let A_1 be that VirGo goes bankrupt that year. Let further A_2, A_3 and A_4 be the respective bankrupcties of ReRe, gRendel and ShiRe in the same year. We have been given the probabilities

$$P(A_1) = 0.05$$
$$P(A_2|A_1) = 0.31$$
$$P(A_3|A_1A_2) = 0.47$$
$$P(A_4|A_1A_2A_3) = 0.38.$$

The probability that they all will be bankrupt within a given year is then

$$P(A_1A_2A_3A_4) = 0.05 \times 0.31 \times 0.47 \times 0.38 = \underline{0.002\,768\,3} \approx \underline{0.002\,8}. \qquad \square$$

Summary and Rules
- $P(A|B)$ is the probability of A, given that B is the case.
- $P(A|B)$ is a probability. Think of B as the context, and if it helps *write $P(A|B)$ as $P_B(A)$ to accustom yourself to conditional probability being a regular probability.*
- $P(A|B)$ is best illustrated in Euler diagrams as the part of B that is at the same time A.
- Definition: $P(A|B) = \frac{P(AB)}{P(B)}$.
- Important relation: $P(AB) = P(A|B) \times P(B)$.

Test Yourself!
- Show that $P(A|B)$ is a probability.
- Draw Euler diagrams to show: $P(AB) = P(A|B) \times P(B)$.
- Draw one or two diagrams with the sets A and B. Draw A large, and B small, and make them intersect. Use the Euler-diagram(s) to visualize how $P(A|B)$ and $P(B|A)$ differ.

5.4 Independence

Dependence between two events A and B means that the occurrence of the one influences the probability of the occurrence of the other. *Independence* is then the opposite: that the occurrence of one of them does not influence the probability of the occurrence of the other.

Example 5.4.1 You are on an airplane, and then an engine catches fire. *Before* the fire, you estimated the probability of a plane crash to be $\frac{1}{200\,000}$. *After* the fire, you estimate the probability of a crash to be $\frac{1}{10}$. Since the probability changed as a result of the information about the fire, we say that "engine fire" and "plane crash" are *dependent* events. □

Example 5.4.2 You are on an airplane, and then announce that it's chicken for dinner. *Before* the announcement, you estimated the probability of a plane crash to be $\frac{1}{200\,000}$. *After* the announcement, you estimate the probability of a crash to be $\frac{1}{200\,000}$. Since the probability remained unchanged as a result of the announcement, we say that "chicken dinner" and "plane crash" are *independent* events. □

You can come up with a multitude of examples for yourself by simply considering if the occurrence of the one event influences the occurrence of the other. In probability theory, our simplest examples are repeated tosses of a die or coin, where the outcome of one try does not influence the probabilities of the results in subsequent trials. That is: the results of the tosses are independent. We state this mathematically, as the definition of independence:

Definition 5.4.3 *A is independent of B iff $P(A|B) = P(A)$.*

That is, A is independent of B if the proportion that is A remains unchanged when we change context from the entire universe Ω to the sub-universe B. The next rule states a common way to test (and sometimes also define) independence. It follows from our basic definition above, but has the advantage of a neater form in itself and for its generalizations. As a bonus, it also shows that when A is independent of B, then B is independent of A. So independence is a symmetrical property, and we will say that "A and B are independent (of each other)".

Rule 5.4.4 *A and B are independent iff $P(AB) = P(A) \times P(B)$.*

Proof: $P(AB) = P(A|B) \times P(B) = P(A) \times P(B)$. □

A good way to visualize independence, is to think of each set as independently ruling the extension along its own axis. We see this for two sets in Figure 5.8, and for three sets in Figure 5.9.

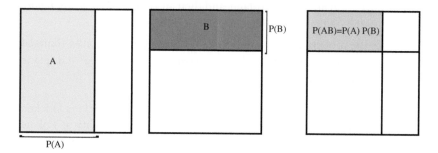

Figure 5.8 Independence for two events.

Example 5.4.5 We have a coin with $P(H) = 0.65$. What is $P(HT)$?

Answer: The coin tosses are independent events: A is "toss one resulted in an H", while B is "toss two resulted in a T". Because of independence,

$$P(HT) = P(H) \times P(T) = 0.65 \times 0.35 = \underline{0.227\,5}.$$ □

Independence is readily generalized to multiple events. For one event being independent of two others, we say that A is independent of both B_1 *and* B_2 when the probability of A remains unchanged regardless of whether B_1 or B_2 are the case. In other words:

$$P(A|B_1 B_2) = P(A|B_1 B_2^c) = P(A|B_1^c B_2) = P(A|B_1^c B_2^c) = P(A).$$

For multiple sets, this generalizes to the following definition.

Definition 5.4.6 *A is independent of* $B_1, \ldots, B_n$ *iff* $P(A|B_1 B_2 \cdots B_n) = P(A)$, *and the inequality holds when you replace one or more* B_k *by* B_k^c.

Independence between multiple sets can be expressed in a myriad ways, but outside of the definition, the most important one is this, a formula that is itself frequently also employed as a definition:

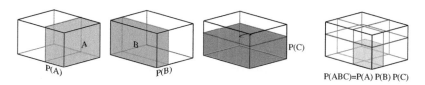

Figure 5.9 Independence for three events.

Rule 5.4.7 $A_1, \ldots, A_n$ *are independent of each other iff*

$$P(A_1 A_2 \cdots A_n) = P(A_1) \times \cdots \times P(A_n)$$

and the inequality holds if you exchange one or more A_k for A_k^c.

Example 5.4.8 A_1, A_2, A_3 are independent, and $P(A_1) = 0.4$, $P(A_2) = 0.6$, $P(A_3) = 0.3$. What is $P(A_1 A_2 A_3)$?

Answer: $P(A_1 A_2 A_3) = P(A_1) \times P(A_2) \times P(A_3) = 0.4 \times 0.6 \times 0.3 = \underline{0.072}$. $\square$

Example 5.4.9 You toss three dice; a D_4, a D_{10}, and a D_{20}. What is the probability that D_4 and D_{20} display prime numbers, but that at the same time, D_{10} doesn't?

Answer: Let A_1, A_2, A_3 be the events that D_4, D_{10}, D_{20}, respectively, displayed primes. We are then calculating the probability of $A_1 A_2^c A_3$. Since each of these three events are about independently tossed dice, the events themselves are independent.

- $P(A_1) = \frac{\text{primes}}{\text{total}} = \frac{n(\{2,3\})}{n(D_4)} = \frac{2}{4} = 0.5$
- $P(A_2) = \frac{\text{primes}}{\text{total}} = \frac{n(\{2,3,5,7\})}{n(D_{10})} = \frac{4}{10} = 0.4$
  - $P(A_2^c) = 1 - P(A_2) = 1 - 0.4 = 0.6$
- $P(A_3) = \frac{\text{primes}}{\text{total}} = \frac{n(\{2,3,5,7,11,13,17,19\})}{n(D_{20})} = \frac{8}{20} = 0.4$
- $P(A_1 A_2^c A_3) = 0.5 \times 0.6 \times 0.4 = \underline{0.12}$ $\square$

Example 5.4.10 A biased coin has probability $p = P(H) = 0.55$ of heads, and $1 - p = P(T) = 0.45$ of tails. What is $P(HHTTTHTHTHT)$?

Answer:
$$P(HHTTTHTHTHT)$$
$$= P(H)P(H)P(T)P(T)P(T)P(H)P(T)P(H)P(T)P(H)P(T)$$
$$= P(H)^5 \times P(T)^6 = p^5 \times (1 - p)^6 = 0.55^5 \times 0.45^6 = \underline{0.000\,417\,916}.$$
$\square$

We see that the probability is independent of the order of the heads and tails, and depended only on the number of each. For a coin with $P(H) = p$, the general formula for the probability of a sequence of k heads and $n - k$ tails is $p^k (1 - p)^{n-k}$.

NOTE: The word "independence" has several uses, and we need to be aware of one particular use that often causes confusion in statistics: In politics and geography we speak of independence as a kind of *separation*, which would correspond to *disjoint* sets. See Figure 5.10 for the Euler diagram.

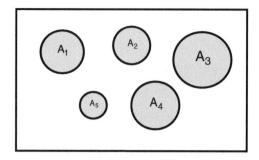

Figure 5.10 Political independence corresponds to disjoint sets.

This is rather the opposite of statistical independence, as depicted in Figures 5.8 and 5.9, where the sets intersect precisely as much with the other sets as with their complements.

We should rather think of independence in a more *moral* or *psychological, decision-making* sense. Imagine you are buying a jacket, and you see a nice mocha coloured one that you'd like to wear. But you have heard that M, a person you are a bit at odds with, has such a jacket. What is then the (morally and psychologically) *independent* decision? (a) Not buying the jacket, *because* M has such a jacket? Or (b) to buy that jacket regardless of what M might own or think about it? And you probably answered correctly: (b) is the independent decision, corresponding to statistical independence.

Assignment: Think about different pairs of sets, events and decisions from life around you, and determine which one are independent, which ones are mutually exclusive, and which ones are neither. It may be useful to do this as a group exercise where one comes up with a pair and the rest assess the pair. Try picking examples to trick each other into the wrong assessment!

5.4.1 Conditional Independence

If A_1 and A_2 are independent, then $P(A_1 A_2) = P(A_1) \times P(A_2)$.

Question: What about conditional probabilities? Is $P(A_1 A_2 | B) = P(A_1 | B) \times P(A_2 | B)$ when A_1 and A_2 are independent?

Answer: That actually depends on the conditioning event B! Recall that the conditional probability $P(\cdot | B) = P_B(\cdot)$ works just like any (other) regular probability, only that the context is B, not Ω. Let us try to device a B such that a pair A_1 and A_2 that are independent in context Ω are *not* so in context B. If you recall our discussion above, that disjointness is the opposite of statistical independence, just let $B = (A_1 A_2) \cup (A_1^c A_2^c)$. Draw an Euler diagram and see that in this case, $P(A_1 A_2 | B) = P(A_1 | B) = P(A_2 | B)$, so that $P(A_1 A_2 | B) \neq P(A_1 | B) \times P(A_2 | B)$ (unless they are all equal to 1).

For a practical example, let us device a B that works much in the same way as our B above: Let A_1 be that Nathan turns up at the party tonight, and A_2 that Beatrice turns up. If Nathan and Beatrice don't know each other, then we may consider A_1 and A_2 to be independent events, where the event that one of them turns up doesn't alter the probability of the other one turning up. But let us introduce event that maybe happened earlier that day: B = "Nathan fell in love with Beatrice, and wants to be at the parties she's at". Will A_1 and A_2 *then* be independent? No, for if B is the case, then Beatrice's presence at the party increases the probability that Nathan wil turn up. So A_1 and A_2 are conditionally *dependent*. A new event, C = "Nathan has gotten over Beatrice", may make the events independent again, and this time conditionally *independent* given C.

When determining if A_1 and A_2 are independent given B, consider what kind of world B creates, and assess how A_1 and A_2 influence each other's probabilities in that world. We defines ordinary independence between A_1 and A_2 by that the probability of A_1 was uninfluenced by whether A_2 was the case or not. Conditional independence is then that the conditional probability of A_1 is uninfluenced by whether A_2 is the case or not. We write this as a definition:

Definition 5.4.11 A_1 *and* A_2 *are conditionally independent given B, if*

$$P(A_1|A_2B) = P(A_1|B).$$

Just as for regular independence, this means

Rule 5.4.12 A_1 *and* A_2 *are conditionally independent given B, if*

$$P(A_1A_2|B) = P(A_1|B) \times P(A_2|B).$$

Example 5.4.13 You have an urn with two red (R) and two white (W) balls. You are going to draw two balls; let A_1 = "the first ball is R", and let A_2 = "the second ball is R". After drawing the first ball, you toss a coin to decide whether to return the ball to the urn. Let B = "you return the ball". Are A_1 and A_2 independent, given B? Are A_1 and A_2 independent?

If B is returned after the first draw, you have four balls at both draws: two W and two R.

- $P(A_1|B) = \frac{reds}{total} = \frac{2}{4}$.
- $P(A_2|B) = \frac{reds}{total} = \frac{2}{4}$.
- A_1A_2 is the set of two draws where the outcome is RR. We may draw a tree diagram and see that 4 out of 16 branches yield RR, meaning $P(A_1A_2|B) = \frac{red-red}{total} = \frac{4}{16}$.

We see that $P(A_1A_2|B) = \frac{4}{16} = \frac{2}{4} \times \frac{2}{4} = P(A_1|B) \times P(A_2|B)$, so A_1 is independent of A_2, given B.

But notice that the independence is conditional on B, return of the ball. If the ball is not returned, corresponding to B^c, the independence vanishes. If the ball is not returned, there are only three balls left for the next draw, so the probability of getting R on the second draw if you had R on the first, is reduced to $\frac{1}{3}$. So while $P(A_1|B^c) = \frac{1}{2}$ and $P(A_2|B^c) = \frac{1}{2}$, we still have $P(A_1A_2|B^c) = \frac{1}{6}$, so $P(A_1|B^c) \times P(A_2|B^c) \neq P(A_1A_2|B^c)$, meaning that A_1 and A_2 are *not* independent, given B^c. □

Conditional independence for multiple sets is captured in the formula

When $A_1, \ldots, A_n$ are conditionally independent of each other, given B, then
$P(A_1 \cdots A_n|B) = P(A_1|B) \cdots P(A_n|B)$.

5.5 Repeated Sampling and Probability

In Chapter 4, we studied how many *ways* there were to sample from a given set, and got different formulas (Method 4.3.1) depending on whether the sampling was ordered or unordered, and whether the sampled unit was replaced. We will now study *probabilities* for repeated samplings. More precisely, we will divide the set to be sampled from into *kinds*, and look at the probabilities of getting certain *sequences* or *combinations* of kinds as we sample.

Definition 5.5.1 *Sequence and combination:*

- A **sequence** *of n samplings from a set is list of the results of the sampling, in the order they occurred. It is* ordered.
- A **combination** *of n samplings from a set is an overview of how many were sampled of each kind. It is* unordered.

Example 5.5.2 Coin toss. A 7 toss *sequence* yields *HTTHHTH*. The corresponding *combination* is 4H and 3T. □

Example 5.5.3 You have an urn with 4 red balls ("R"), 17 green ("G"), and 8 blue ("B"). Sampling 9 balls yields the *sequence RGGGBGGGBG*; the corresponding *combination* is 1R, 6G and 2B. □

We know that when we remove an element from an urn, a deck of cards, or some other finite repository, we thereby alter the proportions of the different kinds of item. If you have a bag of assorted candies with 2 Almond Joy and 1 Twizzler left, and you pick an Almond Joy, the proportion of Almond Joy changes from $\frac{2}{3}$ to $\frac{1}{2}$. But if we replace the Almond Joy, the proportions are unchanged. So when the probability of drawing a given type equals the proportion of that type, then the probability will remain unchanged if you replace after the draw.

When we sample with replacement, the probabilities remain unchanged when the probability of a type equals its proportion. This is then a model for any kind of repeated trial with identical probabilities, like tossing dice or coins, or waiting for a red car to drive by. We will later come to know these as *Bernoulli processes* (Definition 9.2.4). For now, we will use the term *sampling with replacement*.

So our first division line is whether the trial is a "sampling with replacement", or a "sampling without replacement". Given m different possible outcomes of the trial, then

- in sampling *without* replacement, we need to specify the total number of elements, N, the number of types, m, and the total number of elements of each type, $S_1, S_2, \ldots, S_m$;
- in sampling *with* replacement, we need only specify the number of types, m, and the proportions for each type, $p_1, p_2, \ldots, p_m$. If you know S_k and N, then $p_k = S_k/N$.

Since we will be working mainly with cases of only two types, we make two tables: one for the special case of sampling from sets of only two types, and one general table for sampling from sets of m types. For the latter, we will be employing the *multinomial* in Section A.2.

Note 5.5.4 These probabilities resurface in Chapter 9 as probability distributions, respectively in Section 9.4 (unordered without replacement) and in Section 9.3 (unordered with replacement).

Rule 5.5.5 *For n samples with or without replacement, recorded as ordered sequences or as unordered combinations, the formulas are as follows.*

- *We set the parameters to two types, N elements (if applicable), $S = S_1$ elements of the first type (if applicable), and $p = p_1$. Further, k is the number of positives (the first kind), whereas $n - k$ is the number of negatives. The probability of such a result is then given by the following table.*

	With replacement	Without replacement
Sequence (ordered)	$p^k(1-p)^{n-k}$	$\dfrac{\binom{N-n}{S-k}}{\binom{N}{S}}$
Combination (unordered)	$\binom{n}{k}p^k(1-p)^{n-k}$	$\dfrac{\binom{S}{k}\binom{N-S}{n-k}}{\binom{N}{n}} = \dfrac{\binom{N-n}{S-k}\binom{n}{k}}{\binom{N}{S}}$

- We set the parameters to m types, N elements (if applicable), S_j elements of the type j (if applicable), and the proportion of type j is p_j. Further k_j is the number results of the jth kind. The probability of such a result is given by the following table.

	With replacement	Without replacement
Sequence (ordered)	$p_1^{k_1} \cdots p_m^{k_m}$	$\dfrac{\binom{N-n}{S_1-k_1\ldots S_m-k_m}}{\binom{N}{S_1\ldots S_m}}$
Combination (unordered)	$\binom{n}{k_1\ldots k_m}p_1^{k_1} \cdots p_m^{k_m}$	$\binom{n}{k_1\ldots k_m}\dfrac{\binom{N-n}{S_1-k_1\ldots S_m-k_m}}{\binom{N}{S_1\ldots S_m}}$

Note 5.5.6 Approximation: If n, the number of samples, is a lot smaller than the total number of elements, N, the calculated values for sampling with and without replacement will be very similar. In situations where small errors do not matter, it is expedient to use the faster formulas of sampling *with* replacement, even when the sampling is *without*. See Chapter 9 for more details.

Example-proof 5.5.7 A coin has $P(H) = 0.6$ and $P(T) = 0.4$. What is $P(HTTHHHHT)$?

Answer: This is a *sequence*, "with replacement", so the right formula is $p^k(1-p)^{n-k}$. Here, $p = P(H) = 0.6$, $n = 8$, and $k = 5$. But instead of just plugging into the formula, we will look at how it arises in this specific instance.

The coin tosses are independent events: A_1 is "toss 1 landed H", A_2 is "toss 2 landed T", and so on up to A_8 being "toss 8 landed T". Therefore,

$$P(HTTHHHHT) = P(H)P(T)P(T)P(H)P(H)P(H)P(H)P(T)$$
$$= (P(H))^5 \times (P(T))^3 = p^k(1-p)^{n-k},$$

which is the formula; we use that $p = 0.6$, $n = 8$, and $k = 5$, and get

$$= 0.6^5 \times 0.4^3 \approx \underline{0.005\,0}.$$

$\square$

Example-proof 5.5.8 A coin has $P(H) = 0.6$ and $P(T) = 0.4$. What is the probability of precisely 5 heads in 8 tosses?

Answer: This is a *combination*, "with replacement", so the right formula is $\binom{n}{k}p^k(1 - p)^{n-k}$. Here, $p = P(H) = 0.6$, $n = 8$, and $k = 5$. But instead of just plugging into the formula, we will look at how it arises in this specific instance.

In the previous example, we calculated the probability that a given sequence of 5 heads in 8 tosses was $(P(H))^5 \times (P(T))^3$. The probability is the same for all such sequences, so we need to multiply by the number of such sequences. So how many are they? In the example above, we got heads in tosses 1, 4, 5, 6, and 7. This corresponds to marking the numbers 1, 4, 5, 6, 7 in the sequence from 1 to 8. We recall that an unordered sampling of k from a total of n may be done in $\binom{n}{k}$ ways, so the number of different sequences with 5 heads in 8 tosses is $\binom{8}{5}$.

$$P(5 \text{ heads in 8 tosses}) = \binom{8}{5}0.6^k 0.4^{n-k}$$
$$= \binom{n}{k}p^k(1 - p)^{n-k},$$

which is the formula; we use that $p = 0.6$, $n = 8$, and $k = 5$, and get

$$= \binom{8}{5} \times 0.6^5 \times 0.4^3 \approx \underline{0.28}.$$

$\square$

Example-proof 5.5.9 A certain deck of cards has 19 marked cards (M), and 33 unmarked cards (U). You are dealt a hand of 5 cards. What is the probability that precisely 2 of the cards are marked – that is, 2M and 3U?

Answer: This is a *combination*, "without replacement", so the right formula is

$$\frac{\binom{S}{k}\binom{N-S}{n-k}}{\binom{N}{n}}.$$

Here, $N = 52$, $S = 19$, $n = 5$, and $k = 2$. But instead of just plugging into the formula, we will look at how it arises in this specific instance.

This is a simple application of the formula $P = \frac{\text{positive}}{\text{total}}$. From a deck of 52 cards, how many hands of 5 are possible? Combinatorics tells us that the total number of ways to choose n elements from a total of N is $\binom{N}{n}$. So the possible number of hands is $\binom{52}{5}$.

The *positives* are the hands that have 2 of the 19 marked cards, and 3 of the 33 unmarked. So we must see how many ways we can pick the 2 from the 19, and 3 from the 33. This is respectively $\binom{19}{2}$ and $\binom{33}{3}$ ways. By Rule 4.3.9, the number of positive hands is then $\binom{19}{2}\binom{33}{3}$. This means

$$P(2M \text{ and } 3U) = \frac{\text{positive}}{\text{total}}$$

$$= \frac{\binom{19}{2}\binom{33}{3}}{\binom{52}{5}}$$

$$= \frac{\binom{S}{k}\binom{N-S}{n-k}}{\binom{N}{n}},$$

which in this case is

$$= \frac{932\,976}{2\,598\,960} \approx \underline{0.358\,981}. \qquad \square$$

Example-proof 5.5.10 A certain deck of cards has 19 marked cards (M), and 33 unmarked cards (U). You are dealt a hand of 5 cards. What is the probability that you are dealt your hand in the sequence *MUUMU*?

Answer: This is a *sequence*, "without replacement", so the right formula is

$$\frac{\binom{N-n}{S-k}}{\binom{N}{S}}.$$

Here, $N = 52$, $S = 19$, $n = 5$, and $k = 2$. But instead of just plugging into the formula, we will look at how it arises in this specific instance.

Consider all possible sequences of M and U if we put *all* the 52 cards in a row. Each sequence is a unique distribution of $S = 19$ marked cards among the $N = 52$ positions in the sequence. The number of possible sequences is $\binom{N}{S} = \binom{52}{19}$.

How many such sequences start in *MUUMU*? Well, these are the sequences starting in *MUUMU*, and ending in a tail of $19 - 2$ U and $33 - 3$ M. So how many such tails are possible? The tail has $S - k = 19 - 2$ M, out of a total of $N - n = 52 - 5$ cards. The number of sequences starting in *MUUMU* is then $\binom{N-n}{S-k} = \binom{52-5}{19-2}$. Then

$$P(MUUMU) = \frac{\text{positive}}{\text{total}} = \frac{\text{number of sequences starting in } MUUMU}{\text{total possible sequences}}$$

$$= \frac{\binom{52-5}{19-2}}{\binom{52}{19}} = \frac{\binom{N-n}{S-k}}{\binom{N}{S}},$$

which in this case is

$$= \frac{2\,741\,188\,875\,414}{76\,360\,380\,541\,900} \approx \underline{0.035\,898\,1}. \qquad \square$$

Example 5.5.11 We will now consider an example with $m > 2$ alternatives. In all the sub examples, we'll be looking at a bag of chocolates with 4 different types: dark, milk, white, and mint. The bag contains $S_1 = 10$ dark chocolates, $S_2 = 18$ milk chocolates, $S_3 = 6$ white chocolates, and $S_4 = 11$ mint chocolates. A total of $N = 45$ chocolates.

(1) Frederick is curious what's inside the bag, so he picks one, inspects it, and returns it to the bag. He does this 7 times, and gets mint, milk, milk, dark, mint, white, and milk. What is the probability of this exact sequence?

Answer: This is an *ordered* sampling, *with* replacement. We calculate the p values: $p_1 = \frac{S_1}{N} = \frac{10}{45}, p_2 = \frac{S_2}{N} = \frac{18}{45}, p_3 = \frac{S_3}{N} = \frac{6}{45}, p_4 = \frac{S_4}{N} = \frac{11}{45}$. The number of positives for each is $k_1 = 1, k_2 = 3, k_3 = 1, k_4 = 2$. The probability of the sequence is then

$$P = p_1^{k_1} \times p_2^{k_2} \times p_3^{k_3} \times p_4^{k_4} = \left(\frac{10}{45}\right)^1 \times \left(\frac{18}{45}\right)^3 \times \left(\frac{6}{45}\right)^1 \times \left(\frac{11}{45}\right)^2$$

$$= \frac{3\,872}{34\,171\,875} \approx 0.000\,11.$$

(2) The gambler Sam saw what Frederick got, but for his own sake, he is more interested in the probability that such a sampling would yield 1 dark, 3 milk, 1 white, and 2 mint. What is the probability of this result?

Answer: This is unordered sampling with replacement. The formula for this probability is like the sequence in the previous example, but with an added multinomial factor for the number of sequences corresponding to these numbers.

$$P = \binom{7}{1,3,1,2} \times p_1^{k_1} \times p_2^{k_2} \times p_3^{k_3} \times p_4^{k_4} = 420 \times \frac{3\,872}{34\,171\,875}$$

$$= \frac{108\,416}{2\,278\,125} \approx 0.048.$$

(3) Sam now wants to make a bet with the Bayesian Bard about what Bard will get if he samples 7 chocolates. Sam forgets that Bard is a glutton when it comes to chocolate, and will *eat* each piece he picks. That is: his sampling will be *without* replacement. What is the probability that Bard ate mint, milk, milk, dark, mint, white, and milk?

Answer: This is ordered sampling without replacement. The formula requires not just the numbers from the result, but also the number of elements in the bag. The required values are $S_1 = 10, S_2 = 18, S_3 = 6, S_4 = 11$, $N = 45, k_1 = 1, k_2 = 3, k_3 = 1, k_4 = 2, n = 7$.

$$P = \frac{\binom{45-7}{10-1,18-3,6-1,11-2}}{\binom{45}{10,18,6,11}} = \frac{\binom{38}{9,15,5,9}}{\binom{45}{10,18,6,11}} = \frac{68}{481\,299} \approx 0.000\,141\,28.$$

(4) Bard cares more for which pieces he got to eat than the order in which he ate them. What is the probability that Bard ate 1 dark, 3 milk, 1 white, and 2 mint?

Answer: This is unordered sampling without replacement. The formula for this probability is like the sequence in the previous example, but with an added multinomial factor for the number of sequences corresponding to these numbers.

$$P = \binom{7}{1,3,1,2} \times \frac{\binom{45-7}{10-1,18-3,6-1,11-2}}{\binom{45}{10,18,6,11}} = 420 \times \frac{68}{481\,299}$$

$$= \frac{1\,360}{22\,919} \approx 0.059\,339. \qquad \square$$

Summary and Rules
- A and B are independent when $P(AB) = P(A) \times P(B)$, or equivalently when $P(A|B) = P(A)$, which is also equivalent to $P(B|A) = P(B)$.
- Informally, we say A and B are independent of the occurrence if one of them does not influence the probability of the occurrence of the other.
- If A and B are mutually exclusive, then they are *not* independent.

Test Yourself!
- Show, by employing Euler diagrams, that if A and B are mutually exclusive, then this is not true unless the product is zero: $P(AB) = P(A) \times P(B)$.
- Show, either by employing Euler diagrams or by formal proof, that if two of the following are equal, then all three must be: $P(A)$, $P(A|B)$, $P(A|B^c)$.

Example 5.5.12 $P(B|A) = 0.5$, $P(B) = 0.8$, and $P(A) = 0.6$; what is $P(A|B)$?

Answer: $P(A|B) = \frac{P(A) \times P(B|A)}{P(B)} = \frac{0.6 \times 0.5}{0.8} = \underline{0.375}.$ $\qquad \square$

5.6 Exercises

1 **Review:** Read the chapter.
  (a) What is an Euler diagram? What can it teach us about the basic laws of probability?
  (b) What is frequentist probability?
  (c) What is Bayesian probability?
  (d) What does *conditional* probability mean, and how does it differ from regular probability?
  (e) What does statistical *independence* mean?

Exploration

2 Coin flipping: best done in groups, as a competition. Try to control the out-come of a coin flip, flipped from a reasonable (not too far, not too close) distance from a level surface. After a period of practice, set up a compe-tition sequence of 10 coin flips, where you compete on who gets closest to 100% (or 0%) heads. To win, you should try to make the initial condi-tions of your coin flips as consistent as possible. Notice and control height and distance to the surface, as well as where on your hand you place the coin, how it is oriented, and how hard you flip it with your thumb. In your preparatory training, see what makes for the best consistency: a soft or a hard surface; hard or limp slipping; etc.

3 Resistors: pick some resistors with the same nominal resistance. Measure the actual resistance, and compare it to the nominal: is it over or under? Measure in batches of 10. Do you think you can influence the number that are over? Give reasons for why or why not. Then try it out!

Definitions

4 The cooler: *your local corner supermarket* tends to be a bit negligent about checking their stock, and you know from experience that one in five milk cartons is past its *sell-by* date. Earlier today, you bought a carton of milk, and forgot to check the date.

   (a) What is the frequentist probability that your milk is past its sell-by date?

   (b) What is the objective Bayesian probability that your milk is past its sell-by date?

5 The cooler II: your flat mate is going to the local corner supermarket to buy milk, and she never checks the sell-by date.

   (a) What is the frequentist probability that the milk she buys is past its sell-by date?

   (b) What is the objective Bayesian probability that the milk she buys is past its sell-by date?

Basic Probability

6 A given coin has probability $p = 0.37$ of heads. What is the probability of tails?

Conditional Probability

7 $P(A) = 0.5$, $P(B) = 0.25$, $P(AB) = 0.125$.

   (a) What is $P(A \cup B)$?

   (b) What is $P(A|B)$?

   (c) What is $P(B|A)$?

8 $P(A) = 0.5$, $P(B|A) = 0.2$, $P(A|B) = 0.4$. Find $P(B)$.

9 A biased D_6 die has probabilities $P(1) = 0.1$, $P(2) = 0.1$, $P(4) = 0.2$, $P(5) = 0.2$, $P(6) = 0.15$. What is $P(3)$?

10 $P(A) = \frac{1}{3}$, $P(B) = \frac{1}{4}$, and $P(B|A) = \frac{1}{5}$.
(a) What is $P(AB)$?
(b) What is $P(A \cup B)$?
(c) What is $P(A|B)$?

11 $P(A) = 0.3$, $P(B) = 0.2$, and $P(B|A) = 0.25$.
(a) What is $P(AB)$?
(b) What is $P(A \cup B)$?
(c) What is $P(A|B)$?

12 $P(AB) = \frac{1}{4}$ and $P(B) = \frac{3}{4}$. What is $P(A|B)$?

13 $P(AB) = 0.12$ and $P(A) = 0.6$. What is $P(B|A)$?

14 Illustrate the following formulas with Euler diagrams, and give five concrete examples of each:

$$P(A \cup B) = P(A) + P(B) - P(AB)$$

and

$$P(AB) = P(A) + P(B) - P(A \cup B).$$

Repeated Sampling
15 A coin has probability $p = 0.37$ of heads. What is the probability of the sequence $HHTHTHHHTTTHHHHTTHTHTHHT$?

16 A coin has probability $p = 0.37$ of heads. What is the probability of 14 heads in 37 flips?

17 A coin has probability $p = 0.53$ of heads. What is the probability of 21 heads in 47 flips?

18 A coin has probability $p = 0.61$ of heads. What is the probability of 30 heads in 60 flips?

19 *Assorted Candies I:* you are tidying up after a party, and find a bag of assorted candies. The only pieces left in the bag are 7 Almond Joy and 13 Twizzlers.

(a) You pick a candy at random. What is the probability that you get an Almond Joy?

(b) You like neither Twizzlers nor Almond Joy, so you return your piece to the bag to try again. You try a total of five times, and return the piece to the bag every time. What are the probabilities that you got Almond Joy 0, 1, 2, 3, 4, and 5 times, respectively.

(c) Later in the morning, you are so hungry you'll eat even Almond Joy and Twizzlers, so you pick 5 and eat. What are the probabilities that you ate respectively 0, 1, 2, 3, 4, or 5 Almond Joy?

20 You have a box of 40 fuses that look exactly alike except for the colour. There are 4 violet (V), 22 blue (B), 7 red (R), 5 orange (O), and 2 yellow (Y) fuses. Find the probabilities of the following samples.

(a) YBBROBROV (sampling with replacement).

(b) YBBROBROV (sampling without replacement).

(c) 1V, 3B, 2R, 2O, 1Y (sampling with replacement).

(d) 1V, 3B, 2R, 2O, 1Y (sampling without replacement).

21 A bag of Christmas candies contains 20 candies in red foil, 12 in green foil, and 8 in blue foil. If you eat 10 candies at random, what is the probability that you eat 5 red, 3 green, and 2 blue?

22 A bag of 50 identical looking chocolate balls contains 30 chocolates filled with liqueur (L), and 20 filled with marzipan (M). If you eat 10 random chocolates, what is the probability that you eat them in the following sequence: $LMLMLMLMLM$?

23 You have flipped a coin 50 times. The probabilities of heads and tails are equal: $\frac{1}{2}$

(a) Find $P(HTHTTTHHHTHHTTTTHTTHHH$ $HTTTHHHHTTTTHHHTTTTHHHTHTTT)$.

(b) If we rearrange the Hs and the Ts into a different sequence of 27 tails in 50 attempts, what is the probability of this new sequence?

(c) How many different sequences amount to a combination with 27 heads in 50 attempts?

(d) How many different sequences of heads and tail are possible in 50 coin flips?

(e) What is the probability of 27 tails in 50 coin flips?

24 You've been flipping coins again. This time only 8 times. The probabilities of heads and tails are equal: $\frac{1}{2}$.

(a) What is the probability of *TTHTTTHT*?

(b) If we rearrange the *H*s and the *T*s into a different sequence of 5 tails in 8 attempts, what is the probability of this new sequence?

(c) How many different sequences with 5 tails in 8 attempts are there?

(d) How many different sequences of heads and tail are possible in 8 attempts?

(e) What is the probability of 5 tails in 8 coin flips?

25 You've been flipping a coin again, 8 times. But this time the coin is biased, with a probability of heads of $\frac{1}{3}$.

(a) What is the probability of *TTHTTTHT*?

(b) If we rearrange the *H*s and the *T*s into a different sequence of 6 tails in 8 attempts, what is the probability of this new sequence?

(c) Is the probability of *HHHHHHHH* the same as the probability of *TTHTTTHT*?

(d) What determines the probability of a given sequence? Why?

(e) How many different sequences with 6 tails in 8 attempts are there?

(f) What is the probability for 6 tails in 8 coin flips?

26 Gold Digger Airlines run a daily shuttle between San Remo and Dry Creek. They have two aircraft: a two-engine DC-3 that used to belong to a Columbian drug cartel, and an old four-engine DC-6B bought from the US Army. All the six engines, that is, all the engines on both planes, have the same probability p of failing. For both planes: find the probability that half the engines will fail

(a) when $p = 0.001$;

(b) for p in general;

(c) let's assume you hold your life dear, and that a plane will crash if half or more of the engines fail. For which values of p should you prefer the two-engine DC-3 (given that you fly at all)?

Combinatorics

27 Birthdays: we are investigating to figure out how many persons may be in the same room, before the probability that at least 2 of them have the same birthday exceeds $\frac{1}{2}$. Are you able to make a good guess at the answer in advance?

(a) Musa and Ibrahim wonder what the probability that the two are *not* born on the same date happens to be. Calculate this!

(b) Their common friend Issa found the answer, but he wonders what the probability that none of the *three* have the same birthday happens to be.

(c) What is the probability that at least two of them share a birthday.

(d) Math teacher Mohammed thinks the three have an interesting investigation, and wonders what the probability is if they include him as a fourth person. What is the probability that Musa, Ibrahim, Issa, and Mohammed all have different birthdays?

(e) What is the probability that at least two of the four share a birthday?

(f) Their fellow student Aisa tells them to finish their warm-ups, and go for the main question: How many persons can be present in a room before the probability that at least two of them have the same birthday exceeds $\frac{1}{2}$? She hints that the answer is less than 30. What is answer to Aisa's question?

28 For each of the official hands in regular five-card hand poker, calculate the number of ways to realise such a hand. (Warning: calculations of the simplest hands are somewhat demanding! Hint: Start out with the rare hands, and work your way down.)

6

Bayes' Theorem

CONTENTS

6.1 Bayes' Formula

We have seen that $P(A|B)$, which is A's proportion of B, very rarely equals $P(B|A)$, that is, B's proportion of A. But are the two still related? They are, and this relation is important enough to have its own name: *Bayes' formula*.

> **Rule 6.1.1** Bayes' formula $P(A|B) = \dfrac{P(A) \cdot P(B|A)}{P(B)}$

Proof: Use $P(AB) = P(A) \cdot P(B|A)$ (5.3.4) for $P(AB)$ in $P(A|B) = \frac{P(AB)}{P(B)}$ (5.3.1), as illustrated in Figure 6.1. □

Example 6.1.2   Mammography:[1] Knowing that roughly one in ten women will get breast cancer at some point in life, you want to check yourself regularly

---

[1] The numbers vary, so we have employed a simplified version of numbers from https://www.cancer.gov and Professor Odd Aalen.

The Bayesian Way: Introductory Statistics for Economists and Engineers, First Edition.
Svein Olav Nyberg.
© 2019 John Wiley & Sons, Inc. Published 2019 by John Wiley & Sons, Inc.

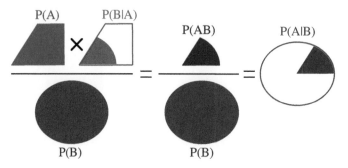

P(A)   P(B|A)   P(AB)   P(A|B)

P(B)   P(B)

Figure 6.1 $\frac{P(A) \cdot P(B|A)}{P(B)} = \frac{P(AB)}{P(B)} = P(A|B)$.

so that you catch the disease early. Like most women, you take the test several times, so the probability that you have cancer as you go to your next check-up is 5‰. You also know that 10% of all mammograms are positive, and also that about 80% of those who actually have cancer get a positive mammogram. So now you have been tested, and your mammogram is positive. Does this mean that the probability that you have breast cancer is 80%?

Answer: We translate our question into mathematical notation, and let $A =$ {has cancer} and $B =$ {has positive mammogram}. What we know thus far, is that $P(B|A) = 80\%$. But what you are looking for is $P(A|B)$, the conditional probability that you have cancer if your mammogram is positive. To find this latter quantity, *Bayes' formula* tells us that we need $P(A)$ and $P(B)$ as well. We read above that $P(B) = 10\%$, and that $P(A) = 5‰$. Then

$$P(A|B) = \frac{P(A) \cdot P(B|A)}{P(B)} = \frac{0.005 \cdot 0.8}{0.1} = \underline{0.04 = 4\%}. \qquad \Box$$

You may get even more accurate results if you narrow down to the numbers for your demographic, but the take-home message of this example is clear: a positive mammogram does not by itself mean you have cancer. It just means you need a follow-up test. For we see that the two conditional probabilities $P(A|B)$ and $P(B|A)$ are very different. This is illustrated in Figure 6.2.

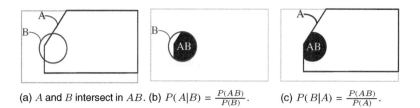

(a) A and B intersect in AB. (b) $P(A|B) = \frac{P(AB)}{P(B)}$. (c) $P(B|A) = \frac{P(AB)}{P(A)}$.

Figure 6.2 A's share of B may be large, while at the same time B's share of A is small.

Bayes' formula is important, since we need to know both the difference and the relation between $P(A|B)$ and $P(B|A)$. Many confusions and much abuse of probabilities hinge on confusing $P(A|B)$ and $P(B|A)$. By demonstrating the difference, we can clear up confusion and unmask abuses.

6.2 The Probability of an Observation

The machinery we will soon get to know as *Bayes' theorem* begins with Bayes' formula, and with another important rule, the rule of *total probability*. This is expressed in Rule 6.2.1 and in Figure 6.3.

Rule 6.2.1 Total probability: *If A_j are disjoint sets, and $\cup_j A_j = \Omega$, then we have for any B that*

$$P(B) = \sum_j P(A_j B) = \sum_j P(A_j) \cdot P(B|A_j).$$

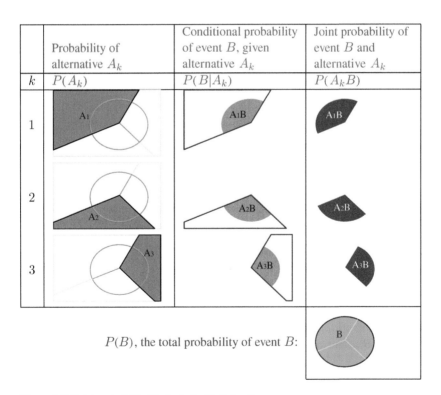

Figure 6.3 Total probability illustrated with Euler diagrams.

Figure 6.4 The possibilities and their probabilities.

Proof: The first equality follows from Kolmogorov's 2nd axiom: let B be a disjoint union of sets $A_k B$; that is, let $B = A_1 B \cup \cdots \cup A_n B$. Then $P(B) = P(A_1 B) + \cdots + P(A_n B)$. The second equality is Rule 5.3.4. □

Example 6.2.2 A purely numerical example.

			Joint probability = Product
k	$P(A_k)$	$P(B\|A_k)$	$P(A_k) \cdot P(B\|A_k) = P(A_k B)$
1	0.6	0.3	$0.6 \cdot 0.3 = 0.18$
2	0.3	0.9	$0.3 \cdot 0.9 = 0.27$
3	0.1	0.5	$0.1 \cdot 0.5 = 0.05$
Total probability = Sum:			$P(B) = 0.18 + 0.27 + 0.05 = \underline{0.5}$

□

Example 6.2.3 Gamesmaster has six dice in a box: D_4, D_6, D_8, D_{10}, D_{12}, and D_{20}, all of them having four red sides and the remainder white. He picks a die at random, and tosses it, all out of view of the gamers and us. He will tell us if it landed red or white. What is the probability that it landed white?

Answer: We will go through the set-up and the calculations in detail. The answer consists of three equally important phases: the preparatory phase, where we *identify* the alternatives A_k and the (potential) observation B. The second phase is to *quantify* the probabilities of A_k and B, and the final phase is the *calculation* itself, where we fill in the table and calculate the probability $P(B)$. The possibilities and their probabilities are illustrated in Figure 6.4.

Identification: Here, we identify the alternatives A_k, and the (potential) observation B.

- The alternatives A_k are the mutually exclusive events that our universe is divided into. Here, A_k is "we drew D_k", for $k = 4, 6, 8, 10, 12, 20$.
- B is the (potential) observation of which we are calculating the probability. Here, B = "Gamesmaster's die landed on *white*".

Quantification: We first quantify the probabilities of each alternative A_k, and then the conditional probabilities of B given A_k for each A_k.

- Since there is no preferred die, the principle of symmetry says they are equally probable. Since there are 6 dice, $P(A_k) = \frac{\text{prositive}}{\text{total}} = \frac{1}{6}$ for all k.
- $P(B|A_k)$ is the probability of *white* given the die is D_k. The principle of symmetry says this probability is given by the number of sides on the die, so $P(B|A_k) = 1 - \frac{4}{k}$.

Calculation: We enter the probabilities in the table. We get that $P(B) = \frac{29}{60}$:

k	$P(A_k)$	$P(B|A_k)$	$P(A_k) \times P(B|A_k) = P(A_k B)$
4	(1/6)	0/4	$(1/6) \times 0 = 0$
6	(1/6)	(2/6)	$(1/6) \times (2/6) = 1/18$
8	(1/6)	(4/8)	$(1/6) \times (4/8) = 1/12$
10	(1/6)	(6/10)	$(1/6) \times (6/10) = 1/10$
12	(1/6)	(8/12)	$(1/6) \times (8/12) = 1/9$
20	(1/6)	(16/20)	$(1/6) \times (16/20) = 2/15$
			$P(B) = $ Sum $= 29/60$

□

6.3 Bayes' Theorem

Bayes' theorem is by many considered the crown jewel of statistics. So what does it do? Bayes' theorem works like a machine taking old probabilities and new data as its inputs, and whose outputs are new, updated probabilities. Bayes' theorem is stated in Rule 6.3.1, and shown graphically in Figure 6.5.

Rule 6.3.1 Bayes' theorem: *If we divide Ω into n disjoint (mutually exclusive) sets $A_1, A_2, \dots A_n$, then*

$$P(A_k|B) = \frac{P(A_k) \times P(B|A_k)}{\sum_j P(A_j) \times P(B|A_j)}.$$

Proof: Use the expression in Rule 6.2.1 for $P(B)$ in the denominator in Bayes' formula (Rule 6.1.1). □

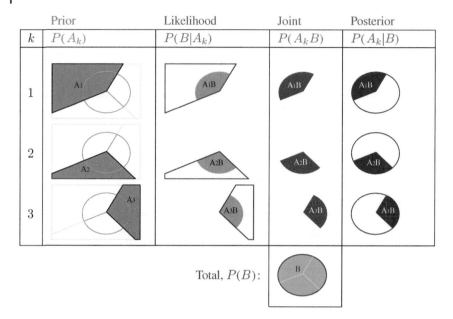

	Prior	Likelihood	Joint	Posterior
k	$P(A_k)$	$P(B\|A_k)$	$P(A_kB)$	$P(A_k\|B)$

Total, $P(B)$:

Figure 6.5 Bayes' theorem illustrated with Euler diagrams.

Bayes' theorem is easily calculated in a table, both on paper and in a spreadsheet. We start out with the following numerical example.

Example 6.3.2 We have three alternatives A_1, A_2, A_3, and an observation B. The probabilities are $P(A_1|\Omega) = 0.7$, $P(B|A_1) = 0.3$, $P(A_2|\Omega) = 0.1$, $P(B|A_2) = 0.6$, $P(A_3|\Omega) = 0.2$, and $P(B|A_3) = 0.8$. Find $P(A_2|B)$.

Answer: Using the plain formula, we get

$$P(A_2|B) = \frac{0.1 \times 0.6}{0.7 \times 0.3 + 0.1 \times 0.6 + 0.2 \times 0.8} \approx \underline{\underline{0.140}}.$$

We then perform the calculations in a table. The table we employ for calculating Bayes' theorem is the same that we used to calculate total probability, expanded with an additional column for $P(A_k|B)$, which we get by dividing the joint probabilities by the total probability:

k	$P_0(A_k)$	$P_0(B\|A_k)$	$P_0(A_kB)$	$P_0(A_k\|B) = P_1(A_k)$
1	0.7	0.3	$0.7 \times 0.3 = 0.21$	$0.21/0.43 = 0.488$
2	0.1	0.6	$0.1 \times 0.6 = 0.06$	$0.06/0.43 = 0.140$
3	0.2	0.8	$0.2 \times 0.8 = 0.16$	$0.16/0.43 = 0.372$
			Sum $= 0.43$	

Notice that we enumerated the probabilities in the example above. Probability 0 is the base probability $P_0(A) = P(A) = P(A|\Omega)$, conditioned on only the fundamental context Ω. Probability 1 is in addition conditioned on B being the case. We may include B in the context rather than making it explicit in the notation, so we write $P_1(A) = P(A|B\Omega)$. In the next section, we will update one more time, with observation B_2. This gives us the probability $P_2(A) = P(A|B_1B_2)$.

Note on notation and columns
The table for Bayes' theorem is an expanded version of the table for total probability, but notice that we are naming of the columns slightly differently.

- The initial probability, $P_0(A_k)$, is the *prior* probability.
- $P_0(B|A_k)$ is a *likelihood*.
- The product $P_0(A_k) \times P_0(B|A_k)$ is the *joint probability* $P_0(BA_k)$.
- The sum of the joint probabilities $P_0(BA_k)$ is the *total probability* $P_0(B)$.
- The final column, $P_1(A_k) = P_0(A_k|B)$, is the *posterior* probability.

It may feel strange to think that knowledge can update a probability. But recall that, classically understood, probability is degree of knowledge. Probability is a truth with fractional value. But to aid this understanding, we need a logical example.

Example 6.3.3 Hassan and Rizwan are among the five qualified applicants for an engineering position with GrenDrill Ltd in the North Sea. They meet at their mailboxes, and they both have a letter from GrenDrill. What is the probability that Hassan got the job? Since there are five applicants, the symmetry criterion tells us that it is $\frac{1}{5}$.

Rizwan opens his envelope first; *he got the job!* What is *now* the probability that Hassan got the job? Logic tells us that since the job is now taken, Hassan can't get the job. So we see that the knowledge that Rizwan got the job forced us to update the probability that Hassan got the job, from $\frac{1}{5}$ to 0. □

Now another example, but with a few more fractions.

Example 6.3.4 Hassan and Rizwan are among the five qualified applicants for two engineering position with GrenDrill Ltd in the North Sea. They meet at their mailboxes, and they both have a letter from GrenDrill. What is the probability that Hassan got the job? Since there are five applicants and two jobs, the symmetry criterion tells us that it is $\frac{2}{5} = 40\%$.

Rizwan opens his envelope first; *he got one of the two positions GrenDrill!* What is *now* the probability that Hassan got a job with GrenDrill? The knowledge that Rizwan got one of the jobs makes us realize that there are four applicants left, and one job. Logic then dictates that the probability that Hassan

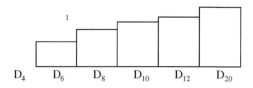

Figure 6.6 The world, when $B = white$.

$$D_4 \quad D_6 \quad D_8 \quad D_{10} \quad D_{12} \quad D_{20}$$

got this other job is $\frac{1}{4} = 25\%$ – down from the 40% that it was before Rizwan opened his envelope. □

Example 6.3.5 This is a continuation of Example 6.2.3, where we looked at Gamesmaster's dice.

Gamesmaster has six dice in a box: D_4, D_6, D_8, D_{10}, D_{12}, and D_{20}, all of them having four red sides and the remainder white. He picks a die at random, and tosses it, all out of view of the gamers and us. He announces that it landed white. (That is, the event is $B_1 = W$, "white".) For all the k, find the updated (posterior) probabilities $P_1(A_k) = P_0(A_k|B_1)$ that he had picked the D_k die. In Figure 6.6, we follow up Figure 6.4, by conditioning on knowing that $B = white$.

Answer: We expand the table from Example 6.2.3 by one column, and find:

k	Prior $P_0(A_k)$	Likelihood $P_0(B_1|A_k)$	Joint $P_0(B_1 A_k)$	Posterior $P_1(A_k) = P_0(B_1 A_k)/P_0(B_1)$
4	1/6	0	0	0
6	1/6	1/3	1/18	10/87
8	1/6	1/2	1/12	5/29
10	1/6	3/5	1/10	6/29
12	1/6	2/3	1/9	20/87
20	1/6	4/5	2/15	8/29
			$P_0(B_1) = 29/60$	

□

6.4 Next Observation and Update

What happens, and what should we do, if we run more rounds of observations and updated probabilities? This is often the case, so it is expedient to employ a tidy form of notation.

Our rounds of updates will then progress as follows.

- The basic probability is P_0, where the probability of A is $P_0(A)$.
- After observing B_1, we update the probabilities:
  - the new probability of A is $P_1(A) = P_0(A|B_1)$.
- After the next observation, B_2, we update the probabilities again:
  - the probability of A is now $P_2(A) = P_1(A|B_2) = P_0(A|B_1 B_2)$.
- This continues after new observations $B_3, B_4, \ldots$

Even though there is nothing wrong with the conditional probability notation $P(A|B_1B_2 \ldots B_n)$, it is far neater and efficient to write $P_n(A)$.

We illustrate this with Euler diagrams in Figure 6.7, and name our observations B, C, and D for visual clarity. Notice how A_2's proportion of the totality changes as the totality itself changes upon new information.

> **Rule 6.4.1 Next observation:** *Given a (posterior) probability P_m, the probability that the next observation is C is*
>
> $$P_m(C) = \sum_j P_m(A_jC) = \sum_j P_m(C|A_j) \times P_m(A_j).$$

We recognize this as Rule 6.2.1 (*total probability*) applied to P_m. We find $P_m(C)$ as before, in the Sum cell in the table when we fill in $P_m(A_k)$ in the first column, and $P_m(C|A_k)$ in the second.

Example 6.4.2 We continue Example 6.3.5 by looking at the probability that Gamesmaster gets *RRR*, three reds, in the next three tosses.

Gamesmaster has previously tossed a die he picked at random, and he got W. The exploration is continuing with the same, unknown, die. Let B be the event "his first toss landed W". Let further C be the event that his next three tosses land *RRR*. We are going to find the probability of C. In Figure 6.8, we follow up Figure 6.6 by showing the new event $C = RRR$.

Answer: We have already done the identification stage, so now we must quantify the probabilities. We already know the likelihoods $P(\text{red}|A_k) = \frac{4}{k}$, so

$$P(RRR|A_k) = \left(\frac{4}{k}\right)^3.$$

But what are the (*prior*) probabilities of the A_k? A common beginner's mistake is to employ $P_0(A_k)$ as the prior also at this stage. But an error? Why is it an error? It is an error because we are then discarding the information and the conclusion from the previous round. Let us see what goes wrong:

k	Wrong Prior $P_0(A_k)$	Likelihood $P_0(C|A_k)$	Joint probability	
4	1/6	1	1/6	
6	1/6	8/27	4/81	
8	1/6	1/8	1/48	
10	1/6	8/125	4/375	
12	1/6	1/27	1/162	
20	1/6	1/125	1/750	
			$P_0(C) \approx 0.255\,056$	**Wrong probability**

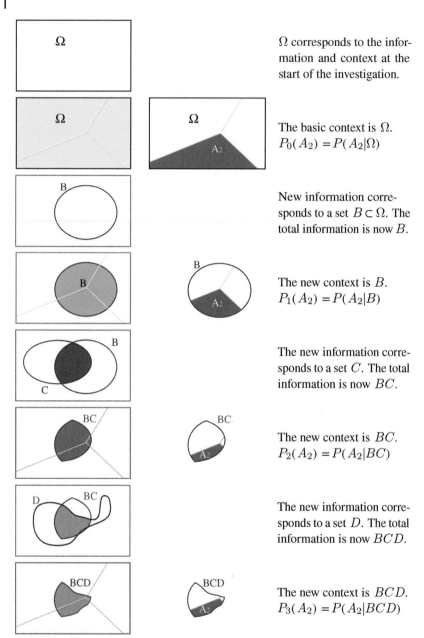

The following annotations appear alongside the diagrams:

Ω corresponds to the information and context at the start of the investigation.

The basic context is Ω.
$P_0(A_2) = P(A_2|\Omega)$

New information corresponds to a set $B \subset \Omega$. The total information is now B.

The new context is B.
$P_1(A_2) = P(A_2|B)$

The new information corresponds to a set C. The total information is now BC.

The new context is BC.
$P_2(A_2) = P(A_2|BC)$

The new information corresponds to a set D. The total information is now BCD.

The new context is BCD.
$P_3(A_2) = P(A_2|BCD)$

Figure 6.7 Progression of updates illustrated with Euler diagrams.

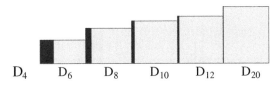

Figure 6.8 Observation
$C = RRR$, given $B = white$.

D$_4$ D$_6$ D$_8$ D$_{10}$ D$_{12}$ D$_{20}$

We see that we have a contribution from the D_4, a die we already *know* is out of the picture. The contribution in the joint probability column says that the probability of "The die is the D_4, and we will get *RRR*" is $\frac{1}{6}$. But when D_4 is out of the picture, the probability of this event is not $\frac{1}{6}$, but 0. So to get it right, we need to bring with us the information from the previous round, which excludes this die and alters the probabilities of the rest. We carry that information with us by using the old *posterior* probabilities $P_1(A_k) = P_0(A_k|B)$ as our new *prior* probabilities for the new round.

	Prior	Likelihood	Joint probability
k	$P_1(A_k)$	$P_1(C|A_k)$	$P_1(A_kC)$
4	0	0	0
6	10/87	8/27	0.034 057
8	5/29	1/8	0.021 551 7
10	6/29	8/125	0.013 241 4
12	20/87	1/27	0.008 514 26
20	8/29	1/125	0.002 206 9
			0.079 571 3

So the right probability of getting *RRR* is $P_1(C) = P(C|B) = 0.079\,571\,3$, and not $P_0(C) = 0.255\,056$.   □

We now look at the *posterior* probability if Gamesmaster's die actually landing *RRR*.

Example 6.4.3 (Continuation of Example 6.4.2) Gamesmaster announces that he has tossed the die three more times. He got *RRR*. What are the new and updated probabilities of which die he picked? We expand the table from Example 6.4.2 by a *posterior* column, and find:

	Prior	Likelihood	Joint	Posterior
k	$P_1(A_k)$	$P_1(C|A_k)$	$P_1(A_kC)$	$P_2(A_k)$
4	0	1	0	0
6	10/87	8/27	0.034 057	0.428 007
8	5/29	1/8	0.021 551 7	0.270 848
10	6/29	8/125	0.013 241 4	0.166 409
12	20/87	1/27	0.008 514 26	0.107 002
20	8/29	1/125	0.002 206 9	0.027 734 8
			0.079 571 3	

□

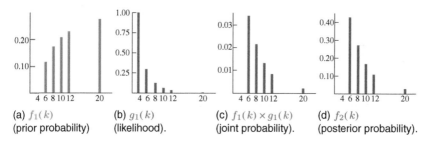

(a) $f_1(k)$ (prior probability) (b) $g_1(k)$ (likelihood). (c) $f_1(k) \times g_1(k)$ (joint probability). (d) $f_2(k)$ (posterior probability).

Figure 6.9 Bayes' theorem with discrete functions.

We may also write the probabilities $P_n(A_k)$ and $P_n(B_{n+1}|A_k)$ as functions $f_n(k)$ and $g_n(k)$ for more efficient notation when the number of alternatives is large, and the values are reasonably easy to express by formulas. We will introduce probabilities as functions in the next chapter, and will generalize Bayes' theorem to probability functions in Chapter 12. So we let this be a taster and an illustration for now. f_n and g_n are discrete functions of the variable k, and we plot the graph of functions in the example above in Figure 6.9.

We see clearly how $f_2(k)$ is a scaled version of $f_1(k) \times g_1(k)$.

Now, what would have happened if we had updated directly from P_0 to P_2 by gathering the observations in Examples 6.4.3 and 6.4.2 into one observation, $WRRR$? The calculation is left as an exercise to the reader (Exercise 11), where we find that the resulting *posterior* is identical to the posterior from the two-step updating – as it should be. We capture this observation as the following rule.

Rule 6.4.4 *Assume we have an initial probability P_0, and update it in n steps, first applying Bayes' theorem to observation B_1 for the first update to P_1, and then to B_2 to P_2, ... and so on all the way to P_n. If we update P_0 in one go, by applying Bayes' theorem once, to the totality of the observations, $B = B_1 B_2 ... B_n$, we get the same posterior probability P_n.*

6.5 Updating, When the Probability of B_{n+1} Depends on $B_1, ..., B_n$

In the example of Gamesmaster and the dice, the outcome of the first toss of the die did not affect the probabilities of the subsequent tosses. This is an example of a type of exploration where the probability of B_{n+1} is indepen-dent of the previous observations $B_1, ..., B_n$. This frequently corresponds to sampling with replacement. The most common type of experiment where the probability of B_{n+1} *does* depend on the previous observations $B_1, \ldots, B_n$ corresponds to sampling *without* replacement. We will look at an example.

Example 6.5.1 We have 3 urns containing 50 balls each. The first contains 40 red (R) and 10 yellow (Y) balls, the second contains 25 red and 25 yellow balls, and the third and last contains 10 red and 40 yellow balls. We mix the urns, and pick one at random, not knowing which of the three it is. By the symmetry criterion, it is then equally probable that we picked any of them. We have 3 possibilities:

- A_1: The urn we picked has 40 red and 10 yellow balls. $f_0(1) = P_0(A_1) = \frac{1}{3}$
- A_2: The urn we picked has 25 red and 25 yellow balls. $f_0(2) = P_0(A_2) = \frac{1}{3}$
- A_3: The urn we picked has 10 red and 40 yellow balls. $f_0(3) = P_0(A_3) = \frac{1}{3}$.

We then sample 5 balls from the urn, without replacing them. Our first observation, B_1, is $RYRRY$. Common for all 3 urns is that $N = 50$ (total number of balls in the urn). For each urn, we have S_k red balls, $n = 5$ sampled balls, and $s = 3$ sampled reds. Then $g(k) = P(B_1|A_k) = \binom{N-n}{S_k-s}/\binom{N}{S_k}$, giving us

- A_1: $S_1 = 40$, so $g_0(1) = P_0(B_1|A_1) = \dfrac{\binom{50-5}{40-3}}{\binom{50}{10}} = \dfrac{2\,223}{105\,938}$

- A_2: $S_2 = 25$, so $g_0(2) = P_0(B_1|A_2) = \dfrac{\binom{50-5}{25-3}}{\binom{50}{25}} = \dfrac{75}{2\,303}$

- A_3: $S_3 = 10$, so $g_0(3) = P_0(B_1|A_3) = \dfrac{\binom{50-5}{10-3}}{\binom{50}{10}} = \dfrac{234}{52\,969}$.

Then,

k	Prior	Likelihood	Joint	Posterior
1	1/3	2 223/105 938	741/105 938	741/2 047 ≈ 0.361 993
2	1/3	75/2 303	25/2 303	50/89 ≈ 0.561 798
3	1/3	234/52 969	78/52 969	156/2 047 ≈ 0.076 209 1
			$\Sigma = 89/4\,606$	

Figure 6.10 visualizes the table as functions. Again, we see that $f_1(k)$ is a scaled version of $f_0(k) \times g_0(k)$. It always is. □

Example 6.5.2 We sample more balls to get a closer estimate of which urn we picked. Without replacement this time as well. This time, we get $B_2 = RRYYYRYR$. What are the new probabilities that we picked urns 1, 2, and 3?

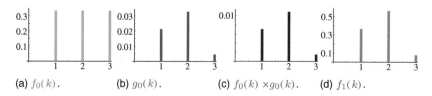

(a) $f_0(k)$.   (b) $g_0(k)$.   (c) $f_0(k) \times g_0(k)$.   (d) $f_1(k)$.

Figure 6.10 Visualized as functions.

Answer: The new prior probabilities equal the old posterior probabilities:

- $f_1(1) = P_1(A_1|B_1) = \dfrac{741}{2\,047}$
- $f_1(2) = P_1(A_2|B_1) = \dfrac{50}{89}$
- $f_1(3) = P_1(A_3|B_1) = \dfrac{156}{2\,047}$.

The likelihoods:

- $P(B_2|A_1)$: Here, $N = 50$, $S = 40$, …

Wait a minute! The urn no longer has 50 balls, since we removed 5 in the first round. There are now $N = 50 - 5 = 45$ left. The number of reds has changed as well; diminished by 3. So the likelihoods are not simply $P(C|A_1)$, that is, the probability of *RRYYYRYR* given $A_1 =$ "sampling from urn 1", but rather $P(C|A_1B_1)$, that is, the probability of sampling *RRYYYRYR* given $A_1B_1 =$ "sampling from urn 1, where we removed 5 balls, whereof 3 were red". So for the correct likelihoods, $N = 45$, $n = 8$, and $s = 4$, and

- A_1: $S_1 = 37$, so $g_1(1) = P_1(B_2|A_1) = P_0(B_2|A_1B_1) = \dfrac{\binom{45-8}{37-4}}{\binom{45}{37}} = \dfrac{4\,403}{14\,370\,213}$
- A_2: $S_2 = 22$, so $g_1(2) = P_1(B_2|A_2) = P_0(B_2|A_2B_1) = \dfrac{\binom{45-8}{22-4}}{\binom{45}{22}} = \dfrac{1\,771}{412\,542}$
- A_3: $S_3 = 7$, so $g_1(3) = P_1(B_2|A_3) = P_0(B_2|A_3B_1) = \dfrac{\binom{45-8}{7-4}}{\binom{45}{7}} = \dfrac{259}{1\,512\,654}$.

The fractions are getting large, so we switch to decimals. Then,

	Prior	Likelihood	Joint	Posterior
1	0.361 993	0.000 306 398	0.000 110 914	0.043 740 9
2	0.561 798	0.004 292 9	0.002 411 74	0.951 113
3	0.076 209 1	0.000 171 222	0.000 013 048 7	0.005 145 99
			$\Sigma = 0.002\,535\,7$	

Figure 6.11 visualizes the table as functions.   □

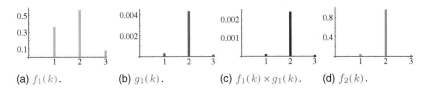

(a) $f_1(k)$. (b) $g_1(k)$. (c) $f_1(k) \times g_1(k)$. (d) $f_2(k)$.

Figure 6.11 Visualized as functions.

In Example 6.5.2 we updated the posterior probabilities from Example 6.5.1 by means of new data. Would we have gotten the same results if we instead had updated with all the data at once? The answer is again *yes*, but let us first look at the example, and then explain why.

Example 6.5.3 We are going to do a one-stop update of the same updating that we did in two steps through Examples 6.5.1 and 6.5.2: The total observation is $B = B_1 B_2 = RYRRYRRYYYRYR$. This is a sampling without replacement, with $N = 50$, $n = 13$, and $s = 7$, and

- $A_1: S_1 = 40$, so $g(1) = P(B|A_1) = \dfrac{\binom{50-13}{40-7}}{\binom{50}{40}} = \dfrac{1\,887}{293\,493\,662} \approx 6.429 \times 10^{-6}$

- $A_2: S_2 = 25$, so $g(2) = P(B|A_2) = \dfrac{\binom{50-13}{25-7}}{\binom{50}{25}} = \dfrac{6\,325}{45\,242\,106} \approx 1.398 \times 10^{-4}$

- $A_3: S_3 = 10$, so $g(3) = P(B|A_3) = \dfrac{\binom{50-13}{10-7}}{\binom{50}{10}} = \dfrac{111}{146\,746\,831} \approx 7.564 \times 10^{-7}$.

The table is then

k	Prior	Likelihood	Joint	Posterior
1	1/3	6.429×10^{-6}	2.143×10^{-6}	0.043 740 9
2	1/3	1.398×10^{-4}	4.660×10^{-5}	0.951 113
3	1/3	7.564×10^{-7}	2.521×10^{-7}	0.005 145 99
			4.900×10^{-5}	

We see that it's the same *posterior* probability that we got after the second step of the two-step updating. □

So we have shown by example that it does not matter if you update all at once, or in steps. The theoretical side is simple: just notice that $f_2 \propto f_1 g_1$ (where $\propto$ means *is proportional to*), since $f_2 = f_1 g_1 / S$. In the same way, $f_1 \propto f_0 g_0$. This means $f_2 \propto f_0 g_0 g_1$, corresponding to a direct update from f_0 to f_2, with $g = g_0 g_1$ as the likelihood. In other words: the likelihood of the one-stop update equals the product of the likelihoods of the many-step update. We show this, that g is the right likelihood for observation $B = B_1 B_2$, by a little bit of formal manipulation:

$$g(k) = g_0(k)g_1(k) = P(B_1|A_k)P(B_2|A_kB_1) = \frac{P(B_1A_k)}{P(A_k)} \times \frac{P(B_1A_kB_2)}{P(A_kB_1)}$$

$$= \frac{P(B_1A_kB_2)}{P(A_k)} = P(B_1B_2|A_k) = P(B|A_k).$$

6.6 Applied Examples

Both frequentists and Bayesians make use of Bayes' theorem, and both do so when they can construct a *prior* probability within the framework of their theories of probability. There is a difference, however. For a Bayesian, the prior reflects their state of knowledge, and may therefore in principle always be found. The frequentist, on the other hand, must use some form of empirical estimate of what an infinite sequence of experiments would have yielded, and this may not always be obtained, or be obtained with sufficient confidence.

How important is the prior? Some might say that it does not matter at all, and that "observations are to speak for themselves"; they will employ the *likelihood* to do the *posterior*'s job. So to answer this anti-prior sentiment, we will show an example where the use of a prior makes all the difference, and justly so. (Notice that this is an example that would cause no disagreement between Bayesians and frequentists; it is not a matter of schools.)

Example 6.6.1   The cab accident: One night, a man was run down by a cab, and the cab flew the crime. The town court, not knowing which precise driver commited the crime, is now going to decide which taxi company is responsible. The town has three taxi companies: *Blue Taxi*, *Zebra*, and *EcoCab*. The court has summoned the only witness in the case, 89 year old Olga Johnson. Olga swears under oath that the taxi was green. Zebra and Blue Taxi's lawyers thank Olga, and argue that the case is now closed, since the only taxi company in town with green cabs is EcoCab. Blue Taxi's cabs are blue, and Zebra's have black and white stripes.

But EcoCab's lawyer asks the court to hear him out before passing judgement. He has examined Olga and her ability to recognize colors and patterns in low light conditions, and it turns out that

- 100% of the time, she sees green as green,
- 0% of the time, she sees zebra stripes as green,
- 20% of the time, she sees blue as green.

Upon seeing this, the judge sends Zebra's lawyer home. Blue Taxi's lawyer asks why he was not dispensed as well. For surely the company he represents "is five times less likely to be the culprit than is EcoCab."

So why should not EcoCab be judged to be the most likely offender?

It turns out that Blue Taxi is the big cab company in town, with 25 cabs, whereas EcoCab has only 2. So if we were to pick a non-striped taxi at random, the probabilities are $P(\text{blue}) = 25/27$, whereas $P(\text{green}) = 2/27$. These numbers are the *prior* probabilities when we are going to calculate the probabilities of which cab was responsible for the hit-and-run. We write this in a table:

	Prior	Likelihood	Joint	Posterior
Blue	25/27	20%	5/27	$5/7 \approx 71\%$
Green	2/27	100%	2/27	$2/7 \approx 29\%$
			7/27	

The probability that the responsible taxi was an EcoCab is thus no more than 29%. ☐

What we saw in this example was that the underlying shares of the cab park, in other words the *prior* probabilities, played a major role in determining the final probability of guilt. The likelihood by itself seemed to point in one direction, but all facts were not on the table until the prior was accounted for, and then the summing up of the knowledge from both prior and likelihood in the *posterior* turned around the verdict that we had expected from the likelihood alone.

The underlying proportion plays an important role in medical diagnosis as well, and in that domain it is known as the *base rate*. The language used when employing Bayes' theorem in the diagnosis of disease is slightly different even though it means the same. A table for diagnosing a disease in a population will typically look like Table 6.1.

The numbers a, b, c, and d may be the numbers in each category after a major survey, or they may be exact or estimated percentages of the population, where $a\% + b\% + c\% + d\% = 100\%$. Specificity is the share of the healthy who get

Table 6.1 Sensitivity–specificity table for medical diagnosis

	Diseased	Healthy	
Positive test	a	b	$a + b$
Negative test	c	d	$c + d$
Base rate	$\frac{a+c}{T}$	$\frac{b+d}{T}$	$T = a + b + c + d$
Sensitivity	$\frac{a}{a+c}$	$\frac{d}{b+d}$	Specificity

a negative test result, whereas sensitivity is the share of the diseased who get a positive test result.

So we have two concerns to balance: sensitivity is about the percentage of the diseased whose disease is revealed by the test, and we naturally want as many of those cases found as we can. But false diagnosis may also be a problem, and we want the share of healthy people who get a negative test result to be as high as possible as well. This is specificity. While the ideal test is 100% specific and 100% sensitive, a higher number in one tends to mean a lower number in the other, so we have to do with less, and try to find the optimal balance.

Example 6.6.2 (Continuation of Example 6.1.2) We lay out the numbers concerning the test results in more detail. We will calculate the probability that you have breast cancer if your mammogram came back positive, both by looking at the sensitivity–specificity table, and by using Bayes' theorem the way we are used to using it. Let $A = \{$has breast cancer$\}$ and $B = \{$has positive mammogram$\}$. We are told that

- $P(A) = 5‰$ (so $P(A^c) = 99.5\%$)
- $P(B|A) = 80\%$ and $P(B^c|A^c) = 95\%$.

Sensitivity–specificity table: We fill in the information as percentages:

	Diseased	Healthy	
Positive test	a	b	$a + b$
Negative test	c	d	$c + d$
Base rate	$a + c = 0.5\%$	$99.5\% = b + d$	
Sensitivity	$\frac{a}{a+c} = 80\%$	$95\% = \frac{d}{b+d}$	Specificity

We resolve the values: $a = \frac{a}{a+c} \cdot (a + c) = 0.8 \cdot 0.005 = 0.004$, $b = (a + c) - a = 0.005 - 0.004 = 0.001$, $d = \frac{d}{b+d} \cdot (b + d) = 0.95 \cdot 0.995 = 0.945\,25$, $b = (b + d) - d = 0.995 - 0.945\,25 = 0.049\,75$, and enter them into the table:

	Diseased	Healthy	
Positive test	0.4%	4.975%	5.375%
Negative test	0.1%	94.525%	94.625%
Baserate	0.5%	99.5%	
Sensitivity	80%	95%	Specificity

Bayes' theorem: We start from the information of a positive mammogram (B), and know that $P(B|A) = 0.8$. We then find that $P(B|A^c) = 1 - P(B^c|A^c) = 1 - 0.95 = 0.05$. Further, $P(A) = 0.005$ and $P(A^c) = 0.995$, so the table for Bayes' theorem becomes

	Prior	Likelihood	Joint	Posterior
Diseased	0.005	0.8	0.004	$\approx 7.4\%$
Healthy	0.995	0.05	0.049 75	$\approx 92.6\%$
			0.053 75	

$\square$

6.7 Bayesian Updating in the Long Run

We make two main uses of Bayes' theorem:

- circling inwards in ever better guesses at the underlying alternatives;
- making predictions of future observations through increasingly improved probabilities.

Sometimes, only the first point is of interest. In Example 6.6.2, we are only interested in knowing if we have breast cancer or not. If we have a battery of tests, we would like our probabilities to be as close to 0 or 100% as possible. We are really not interested in the mammograms themselves, and finding the probabilities of results of future mammograms tells us nothing more about whether we have that cancer or not. The same is the case in Example 6.6.1, the cab accident: we want to establish responsibility for *this* accident, not predict how well Olga will report the next accident she sees.

But at other times, the underlying alternatives are of mere transient interest, as tools to make guesses about the future. This is the mindset of a gambler of any stripe, be it at the casino or in more serious businesses like insurance or military modelling. In the case of gambling, we are interested in the probability of future results, for that is how we decide how to bet.

Let us have a closer look at what a multiply repeated updating by Bayes' theorem would look like: both for the *circling in* on the alternatives, and for the predictions of one or of multiple future observations. We continue the example of Gamesmaster and his dice, as previously studied in Examples 6.2.3, 6.3.5, 6.4.2, and 6.4.3.

Example 6.7.1 Let us just change perspective for a moment. Instead of being one of the players, we are now Gamesmaster. We picked a new die this time, and since we have Gamesmaster's privileged viewpoint, we know it is a D_{10}. With its 4 red and 6 white sides, $P(R) = 0.4$ and $P(W) = 0.6$, and the probability of a given sequence S of k reds and l whites is $P(S) = 0.4^k 0.6^l$.

We toss the die ... first 3 times, then 67 more for a total of 70, and then finally we add enough tosses for a total of 500. The result of the first 3 tosses was RRW. The result of the 70 first (including the first 3), is 31 red and 39 white. The score for all the 500 tosses was a total of 193 red and 307 white.

So let us give this information to the players, who only get to know the results but not the actual die, and watch their estimates. They will look at three things:

- the probability of the alternatives: which die we had picked ($P_n(D_k)$);
- the probability of red ($P_n(\bullet)$) and of white ($P_n(\bigcirc)$) in the next toss;
- the probability of k red in the next five tosses ($P_n(k$ red in five tosses)).

But let us start at the very beginning, in Figure 6.12, before we have made any tosses at all, and the players only have the prior probabilities to go by.

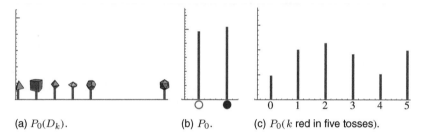

(a) $P_0(D_k)$.   (b) P_0.   (c) $P_0(k$ red in five tosses).

Figure 6.12 The prior probabilities.

Then we start, in Figure 6.13, and look at their estimates after a small number of trials (the first three).

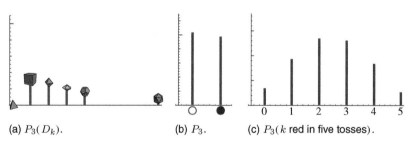

(a) $P_3(D_k)$.   (b) P_3.   (c) $P_3(k$ red in five tosses).

Figure 6.13 The posterior probabilities after three trials.

Then we move on to a medium number of trials (70) in Figure 6.14.

Finally, we look at their estimates after many trials (all the 500) in Figure 6.15.

As a postscript, let us look at our *own* estimates in Figure 6.16, from our privileged position as Gamesmaster; let us call our probabilities P_G. Then $P_G(D_{10}) = 1$, since *we know* we picked a D_{10}.

We see that the players had converged pretty well on our privileged point of view after 500 tosses, and have a very high probability that the right die is the

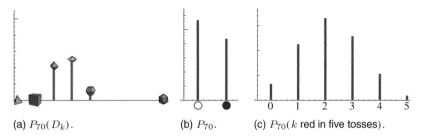

(a) $P_{70}(D_k)$. (b) P_{70}. (c) $P_{70}(k$ red in five tosses$)$.

Figure 6.14 The posterior probabilities after 70 trials.

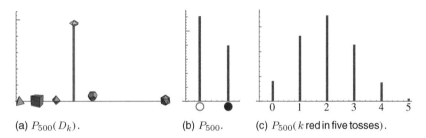

(a) $P_{500}(D_k)$. (b) P_{500}. (c) $P_{500}(k$ red in five tosses$)$.

Figure 6.15 The posterior probabilities after 500 trials.

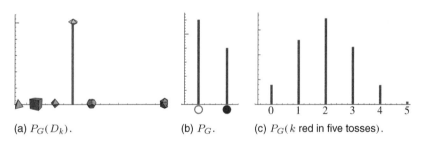

(a) $P_G(D_k)$. (b) P_G. (c) $P_G(k$ red in five tosses$)$.

Figure 6.16 The probabilities from Gamesmaster's perspective.

D_{10}. Their probabilities for the different results are a very good match for ours in both 1 and 5 tosses. A deeper result of the theory of probability tells us that this was no odd statistical fluke, but rather what we should expect will happen in the long run. With probability 1, the players' $P_n(D_{10})$ will tend to 1 as the number of observations, n grows large, and the probabilities of the results of the next tosses will converge to the probabilities of the privileged viewpoint of we who *know* that it was the D_{10} that was picked. □

Note 6.7.2 Keen students might be reading this book a second time, or they have looked ahead at the graphs for the binomial probability distribution ("repeated sampling with replacement") in Section 9.3. If so, they will probably find the graph of $P_n(k$ red in 5 tosses) for $n = 0$ or 3 a bit odd compared to those of the binomial distribution. The graphs for the binomial distribution have a pleasant dromedary shape, whereas the graphs we refer to here have a rare rise at the far right end that can't be explained through a binomial distribution.

Maybe, these students may ask themselves, the graph should have followed the distribution of repeated tosses of a single die whose probability of white is $p = P(\text{red})$. That is, a "$D_{4/p}$" die. The short answer to that is *no*. The somewhat longer answer is to ask these students to calculate for themselves, in Exercises 12–14, and see how the predictive probability is a weighted average of the probabilities for the different dice, and that this is different from the results of some "average die".

We saw, in Examples 6.6.1 and 6.6.2, that the *posterior* probability depends on the *prior* probability. Or the *base rate*, in medical diagnosis. We speak of this as *sensitivity to the prior*. But at the other end of this sensitivity: what happens when our data set is massive?

Example 6.7.3 We revisit Gamesmaster and the dice, but this time with a non-uniform prior. Let $P(D_4) = P(D_6) = 0.4$, and $P(D_k) = 0.05$ for the rest. For a realistic reason for such a skew prior, imagine a scenario where there are far more D_4 and D_6 than any other type of die. We see the results in Figure 6.17. □

We see that the *posterior* is close to indistinguishable from the posterior in Example 6.7.1. The tendency of the data to assert themselves over the prior when they come in large numbers is something we speak of as the *dominance of the data*. It means that in the long run, after many trials, the weight of the information from the *prior* becomes insignificant compared to the weight of the information from the gathered data.

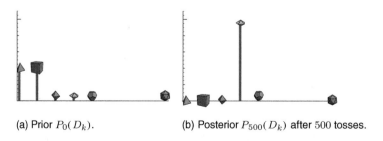

(a) Prior $P_0(D_k)$. (b) Posterior $P_{500}(D_k)$ after 500 tosses.

Figure 6.17 The long-term posterior (500 tosses) with different prior probabilities.

Summary

(1) When we get new data B, this will change the probabilities of other events and occurrences A_k, and we are able to calculate the new probabilities with Bayes' theorem.

(2) The probability before the event of the data B is called the *prior*, and we write $P_0(A_k)$, or $P(A_k|\Omega)$, or simply $P(A_k)$ if Ω is understood. The probablity after the event of the data B is called the *posterior* and we write $P(A_k|B)$ or $P_1(A_k)$.

(3) Posterior probabilities are probabilities conditioned on the data B (and on the initial context, Ω).

(4) The "reverse" probabilities, $P(B|A_k)$ are called *likelihoods*, and are used in the context of Bayes' theorem to calculate *posterior* probabilities from *prior* probabilities.

(5) Bayes' formula and Bayes' theorem are not exclusive to Bayesians, but are also used by for instance frequentists, since these formulas may be used whenever we have conditional probabilities.

(6) The difference between Bayesians and frequentists stems from their different concepts of probability. Expressed in terms of conditional probabilities, the Bayesian look at both $P(\text{event}|\text{model})$ and $P(\text{model}|\text{event})$ as conditional probabilities, whereas a frequentist will as a rule only accept the former.

6.8 Exercises

1 Read the chapter.
  (a) Why was EcoCab not found guilty in Example 6.6.1, even though the eye witness stated that the offending taxi was green (EcoCab's color)?
  (b) Is it possible to use Bayes' theorem when sampling *without* replacement?
  (c) As a follow-up to the previous question: "Yes, but … ?" – yes, but *what* do we need to be careful to do?
  (d) What is sensitivity?
  (e) What is specificity?
  (f) What happens to the probabilities of the alternatives under Bayes' theorem in the long run, after many trials?

2 … **the a-ha version:** Your best friend is holding a black and a white marble, and asks you to open your hands and close your eyes.
  (a) He deposits one marble in each of your hands, and asks you to close your hands before you open your eyes. What is the probability that the marble in your right hand is black?
  (b) He asks you to open your left hand, and there you see a black marble. What is *now* the probability that the marble in your right hand is black?

3 **Just calculation:** Do all of the exercises in a Bayes' theorem table. Do in addition try out at least one by using the formula directly, to see for yourself that the two methods give the same result.

   i. Calculate $P(B)$.

   ii. Calculate all posterior probabilities $P(A_k|B)$.

     (a) Prior: $P(A_1) = \frac{1}{2}, P(A_2) = \frac{1}{2}$

       Likelihood: $P(B|A_1) = 0.7, P(B|A_2) = 0.3$

     (b) Prior: $P(A_1) = \frac{1}{2}, P(A_2) = \frac{1}{2}$

       Likelihood: $P(B|A_1) = 0.07, P(B|A_2) = 0.03$

     (c) Prior: $P(A_1) = \frac{1}{3}, P(A_2) = \frac{1}{3}, P(A_3) = \frac{1}{3}$

       Likelihood: $P(B|A_1) = 0.12, P(B|A_2) = 0.87, P(B|A_3) = 0.01$

     (d) Prior: $P(A_1) = \frac{1}{3}, P(A_2) = \frac{1}{3}, P(A_3) = \frac{1}{3}$

       Likelihood: $P(B|A_1) = 0.0012, \quad P(B|A_2) = 0.0087, \quad P(B|A_3) = 0.0001$

     (e) Prior: $P(A_1) = 0.3, P(A_2) = 0.7$

       Likelihood: $P(B|A_1) = 0.001, P(B|A_2) = 0.011$

     (f) Prior: $P(A_1) = 0.1, P(A_2) = 0.9$

       Likelihood: $P(B|A_1) = \frac{2}{3}, P(B|A_2) = \frac{3}{11}$

     (g) Prior: $P(A_1) = \frac{1}{11}, P(A_2) = \frac{8}{11}, P(A_3) = \frac{2}{11}$

       Likelihood: $P(B|A_1) = 0.012, P(B|A_2) = 0.001, P(B|A_3) = 0.7$

     (h) Prior: $P(A_1) = 0.237\,1, P(A_2) = 0.455\,4, P(A_3) = 0.307\,5$

       Likelihood: $P(B|A_1) = 0.8, P(B|A_2) = 0.07, P(B|A_3) = 0.2$

     (i) Prior: $P(A_1) = 0.07, P(A_2) = 0.07, P(A_3) = 0.07, P(A_4) = 0.79$

       Likelihood: $P(B|A_1) = 0.8, \quad P(B|A_2) = 0.07, \quad P(B|A_3) = 0.007, P(B|A_4) = 0.000\,7$

     (j) Prior: $P(A_1) = \frac{1}{55}, P(A_2) = \frac{2}{55}, P(A_3) = \frac{3}{55}, P(A_4) = \frac{4}{55}, P(A_5) = \frac{5}{55}, P(A_6) = \frac{6}{55}, P(A_7) = \frac{7}{55}, P(A_8) = \frac{8}{55}, P(A_9) = \frac{9}{55}, P(A_{10}) = \frac{10}{55}$

       Likelihood: $P(B|A_1) = 0.1, \quad P(B|A_2) = 0.2, \quad P(B|A_3) = 0.3, P(B|A_4) = 0.4, \quad P(B|A_5) = 0.5, \quad P(B|A_6) = 0.6, \quad P(B|A_7) = 0.7, P(B|A_8) = 0.8, P(B|A_9) = 0.9, P(B|A_{10}) = 1.0$

     (k) Prior: $P(A_1) = \frac{1}{55}, P(A_2) = \frac{2}{55}, P(A_3) = \frac{3}{55}, P(A_4) = \frac{4}{55}, P(A_5) = \frac{5}{55}, P(A_6) = \frac{6}{55}, P(A_7) = \frac{7}{55}, P(A_8) = \frac{8}{55}, P(A_9) = \frac{9}{55}, P(A_{10}) = \frac{10}{55}$

       Likelihood: $P(B|A_1) = 0.1, \quad P(B|A_2) = 0.09, \quad P(B|A_3) = 0.08, P(B|A_4) = 0.07, P(B|A_5) = 0.06, P(B|A_6) = 0.05, P(B|A_7) = 0.04, P(B|A_8) = 0.03, P(B|A_9) = 0.02, P(B|A_{10}) = 0.01$

     (l) Prior: $P(A_1) = \frac{1}{55}, P(A_2) = \frac{2}{55}, P(A_3) = \frac{3}{55}, P(A_4) = \frac{4}{55}, P(A_5) = \frac{5}{55}, P(A_6) = \frac{6}{55}, P(A_7) = \frac{7}{55}, P(A_8) = \frac{8}{55}, P(A_9) = \frac{9}{55}, P(A_{10}) = \frac{10}{55}$

       Likelihood: $P(B|A_1) = 0.1, \quad P(B|A_2) = 0.081, \quad P(B|A_3) = 0.064, P(B|A_4) = 0.049, \quad P(B|A_5) = 0.036, \quad P(B|A_6) = 0.025, \quad P(B|A_7) = 0.016, P(B|A_8) = 0.009, P(B|A_9) = 0.004, P(B|A_{10}) = 0.001$

(m) Create your own problems to solve and to share with your friends for comparison, like the following.

  i. Choose the number of alternatives n.

  ii. Choose $P(A_k)$ for all the n alternatives. Make sure that $P(A_k) \geq 0$ and $\sum_k P(A_k) = 1$.

  iii. Choose likelihoods $P(B|A_k) \in [0,1]$ (their sum does not need to be 1).

4 You have 3 urns. In the first urn, there are 91 red and 34 blue balls. The second contains 14 red and 25 blue balls, and the third has 40 red and 25 blue. You pick an urn that random.

(a) What are A_1, A_2, and A_3?

(b) What are the *prior* probabilities $P(A_1)$, $P(A_2)$, and $P(A_3)$?

(c) What are the probabilities $P(R|A_k)$ of drawing a red ball from each urn, and what are the probabilities of blue $P(B|A_k)$?

(d) You now sample balls from urn 30, with replacement. Your sequence of results (we will call it S) had 20 red and 10 blue balls. Calculate the *likelihoods* $P(S|A_k)$.

(e) Calculate the *posterior* probabilities $P(A_k|S)$.

(f) Your pocket contains two coins, and you pick one of them. Let A_1 be that you picked coin 1, and A_2 that you picked coin 2.

  i. What are $P(A_1)$ and $P(A_2)$ if none of the coins are preferred?

  ii. Coin 1 is fair. What is $P(H|A_1)$ and $P(T|A_1)$?

  iii. Coin 2 is biased, with $P(H|A_2) = 0.407$. What is $P(T|A_2)$?

  iv. (Bayes' theorem) You flip the coin twice, and get tails both times. Calculate $P(TT|A_1)$ and $P(TT|A_2)$.

  v. Calculate the *posterior* probabilities $P(A_1|TT)$ and $P(A_2|TT)$ in a table.

  vi. You flip the coin 3 more times, and get HTH. Set up a new table, using the *posterior* from the previous exercise as your new *prior*. Find the new likelihoods, and calculate the new posterior probabilities.

5 **Text exercises** For each exercise:

(a) identify the alternatives A_k;

(b) identify the pivotal event B;

(c) calculate the prior probabilities $P(A_k)$ and the likelihoods $P(B|A_k)$;

(d) calculate the posterior probabilities $P(A_k|B)$.

  i. You have 6 dice, D_1, D_2, D_3, D_4, D_5, and D_6. A friend of yours picks one of them at random. He tosses it, and gets 3. For each k, find the probability that the die he picked was the D_k.

  ii. Some students have a biased D_{20}. For this die:

  • the probability of an even number is $\frac{9}{15}$;

  • the probability that the outcome is in the top six is $\frac{4}{5}$;

- the conditional probability of an even number, *given that* the outcome is in the top six, is $\frac{2}{3}$.

  If you know the die has landed on an even number, what is the probability that it is in the top six?

iii. The proud nation of Molvania has 2 weightlifting teams. On one team ("the steroid team"), 70% use steroids, whereas on the other team ("the clean team"), 10% use steroids.

- What is the probability that a randomly picked lifter uses steroids? (The teams are equally large, so each lifter has the same chance of being picked.)
- To decide which team to send to the Olympics, Molvania's minister of sports tosses a die, and the probability that he sends the steroid team is $\frac{2}{3}$. Now, we read in the newspaper that a Molvanian lifter was randomly picked for a steroid test during the Olympics, and his test came back negative. Use this information to update the probability that the steroid team was sent to the Olympics.

iv. *Genius Consulting* are testing out 3 different statistics software bundles: *Freequen*, *Gnormal* and *Halfling*. When they test the software with a problem with 100 variables, Freequen crashes 10% of the time, Gnormal crashes 20% of the time, and Halfling crashes 30% of the time.

  Sophie is one of Genius Consulting's ten software testers. Six testers are assigned to Freequen, three to Gnormal, and the last one to Halfling. Sophie's program crashes when she tests it with a 100-variable problem. What is probability that Sophie was assigned to Halfling?

6 A certain Mr. Claus has mixed up his Scandinavian gift bags this Christmas, and he does not know which one goes to Denmark, which one goes to Sweden, and which one goes to Norway. The Danish bag has 70% soft gifts, the Norwegian one has 40%, whereas the Swedish bag has 20%. Mr. Claus has now put one of the bags in his sleigh, and it's your job as an engineering elf to find the probabilities of which country it is from. You sample 8 gifts, and by squeezing carefully, you determine that 4 of them are soft.

(a) Identify the alternatives A_k.

(b) What are the values $P(A_k)$?

(c) Identify the pivotal event B.

(d) Calculate $P(B|A_k)$. Note: the bags contain millions of gifts each, so the difference between sampling with and without replacement is negligible.

(e) Calculate the posterior probabilities $P(A_k|B)$.

(f) Say in your own words what you just found out.

(g) You sample 42 more gifts, to get a sharper guess before Mr. Claus takes off. Seventeen of the gifts are soft. What are your new probabilities for which bag is on the sleigh?

7 Sinterklaas Inc. make two kinds of advent calendar. Type A contains 8 pieces of marzipan and 16 pieces of chocolate. Type G contains 16 pieces of marzipan and 8 pieces of chocolate. They make 3 times as many A calendars as G calendars. You have bought one of their advent calendars, but do not know which.

(a) What are the probabilities that you bought the respective calendars? That is, what are $P(A)$ and $P(G)$?

(b) On December 2nd you have opened 2 hatches, and you got 2 marzipan. Use this to update the probabilities that you bought the respective calendars.

(c) What is the probability that you will get marzipan in your calendar on December 3rd?

8 It is rumoured that, every year, the Alaskan Easter Bunny brings all children in his state an easter egg filled with 30 chocolates. The eggs are usually filled with 22 milk chocolates and 8 white chocolates, but this year the Mr. Easter has hired some new bunny ladies in the chocolate kitchen, and so this year there are three different types of egg:
- 10 000 eggs are like the classical ones described above;
- 4 000 eggs have milk chocolate only;
- 6 000 eggs have 15 of each kind.

Charlie Chocoholic wakes up when Mr. Easter arrives, so he tries to make Charlie fall asleep by telling him about the bunny ladies and their new types of egg. But then Charlie just *has to check* what kind of egg he just got. Mr. Easter tells him he is allowed 5 chocolates, but no more; the rest must wait until daybreak.

(a) What are your prior probabilities for the three egg types?

(b) Charlie picks 5 chocolates and eats: milk, milk, white, white, and milk. What are the likelihoods for the 3 egg types?

(c) What are your updated, *posterior* probabilities for the three egg types?

9 A confidential poll by the polling company Giddyup showed that in a certain area
- 1 in 4 lawyers belong to a secret society;
- 2 out of 3 lawyers who are members of secret societies had lied to protect a client, whereas:
- half of all lawyers in general had lied to protect a client.

Which proportion of the lawyers who had lied to protect a client are also members of some secret society?

10 *Assorted candies II*: You have 4 bags of mixed candies on the table. They are leftovers from a party, so naturally there are only Almond Joys (A) and Twizzlers (T) left. The contents are as follows:
- bag 1: 5 Almond Joys and 3 Twizzlers
- bag 2: 7 Almond Joys and 8 Twizzlers
- bag 3: 2 Almond Joys and 9 Twizzlers
- bag 4: 11 Almond Joys and 13 Twizzlers.

   a. A bit tired from tidying up, you zone out and pick a bag at random. You start picking out candies, looking at them, then returning them to their bag. You repeat this 7 times, and observe *ATTATAA*.

     i. What are the (prior) probabilities that you had picked the different bags?

     ii. What is the probability of your observation, given the different bags.

     iii. Now find the posterior probabilities for the bags.

   b. You pick three more pieces and return them, this time getting *LKL*. Update the probabilities for the bags on the basis of this new information.

   c. A bit later you put the bag back, and you are joined by your brother. Like you, he picks a bag at random, and then he picks out a few pieces. But unlike to you, he eats them. He eats *TTA*. Find, for each of the bags, the updated probability that it was picked by your brother.

11 Calculate the *posterior* probabilities for the dice in Examples 6.4.3 and 6.4.2 when you collect all the observations into one big observation for your one update, i.e. *WRRR*. Compare your one-step updated probability – let's call it P_1' – to P_2 from Example 6.4.3. What do you see?

12 You have two D_6 dice. Die 1 is all white, whereas die 2 is all black. Gamesmaster picks one of them. Let A_k be that he picked die k.

   (a) What are the prior probabilities $P_0(A_1)$ and $P_0(A_2)$?

   (b) What is the probability that his first toss lands white, $p = P(\text{white})$?

   (c) What is the probability that his tosses of the die are white first, then black?

   (d) Let die 3 be "the average die" between the all-white die 1 and the all-black die 2, that is, a die with 3 white sides and 3 black. What is $p = P(W)$ for this die, and what is $P(WB)$?

   (e) Is repeated tosses with die 3 equivalent to randomly picking either die 1 or die 2, and then tossing the picked die repeatedly?

13 Details for Example 6.7.1. You got *RRW*.

   (a) Calculate the *posterior* probabilities P_1 for each of the dice.

   (b) Find $p = P_1(R)$, the probability that the next observation is R.

(c) If you have *sampling with replacement*, where the alternatives are red (R) and white (W), and $p = P_1(\text{red})$ from the above, what is then $P(RRRRR)$?

(d) Find $p = P_1(RRRRR)$, the probability that the next five observations are $RRRRR$.

(e) Why are the probabilities in the two die questions different?

14 [Use a decent tool for calculating] You are selling a car brand that has 2 factories. The factories' output is equal. Lately, 2.5% of the cars from Factory A have had problems with their brakes, whereas only 0.5% of the cars from Factory B have had such problems. The manufacturer never reveals which factory has produced which cars, and you now have a big new shipment of cars from one of these factories, but you don't know which. You test 150 of these cars, and find brake problems in only one of them.

(a) What is the *prior* probability that the shipment is from Factory B?

(b) What is the *posterior* probability that the shipment is from Factory B?

(c) What is the probability p that there is a problem with the brakes of the next car?

(d) What is the probability that there is a problem with the brakes of 50 or more of the next 3000 cars?

(e) What error could you have committed in the previous calculation?

7

Stochastic Variables on $\mathbb{R}$

In Chapter 2, we mentioned that the values of a stock on a historical chart are *data* about that stock, whereas the projected (and unknown) *future* values are *stochastic variables.*

Norges Bank's *key policy rate* is the Norwegian central bank's equivalent of the US Federal *funds rate.* In Figure 7.1, we see their data and their projections from 2008, a very interesting year. In the figure, they use color coding to indicate how likely it is that the interest rate ends up within such and such bounds at a given date. The policy rate for January 2009, for instance, was in October 2008 a *stochastic variable* with a 15% probability of being below 4.75, and a 70% probability of being between 4.75 and 7.00.

7.1 Real Stochastic Variables

How do we indicate a magnitude we don't know the value of? The simple answer is: as probabilities of being so and so large. And this is also the complex answer. But precisely how do we implement this, mathematically? To implement these probabilities, we start out with a set Ω. The set Ω *may* be something as simple as the set of possible values, but we get a much more generally useful theory if we allow Ω to be uncoupled from the specific values, just the way in a car we

The Bayesian Way: Introductory Statistics for Economists and Engineers, First Edition.
Svein Olav Nyberg.
© 2019 John Wiley & Sons, Inc. Published 2019 by John Wiley & Sons, Inc.

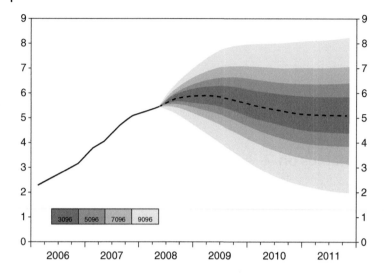

Figure 7.1 Key policy rate; the future in blue. Source: Norges Bank 2008.

uncouple the engine and the wheels, leaving us with the option of reconnecting them through a transmission system with gears and all kinds of controls.

So we start out with a for now unspecified set Ω, and will explore possible specifications below. The minimum requirement is that this set must contain subsets $A \subset \Omega$ and a probability function P allowing us to find probabilities $P(A)$. We will further require Ω to be at least large enough that we can map it onto the possible values of the magnitude we want to study. And it is this map that is our first concern. We call it a *stochastic variable*.

Definition 7.1.1 *A* real *stochastic variable* is a function $X : \Omega \to \mathbb{R}$. The probability P on Ω then becomes a probability on the possible values of X in the following way: for an interval $I = (a, b)$, let $C \subset \Omega$ be the set of elements for which X has values in I. Then*

$$P(\{\omega \mid X(\omega) \in I\}) = P(C).$$

For the rest of the book, we will use the shorthand $P(X \in I)$ for this probability.

This may look a bit "mathematically complicated" at first glance, so let us make Ω and X concrete through a few examples.

Example 7.1.2 Let Ω be all the fish in the salmon fishing river *Loppa*; that is: each ω is a particular fish. $X(\omega)$ is the weight of fish ω, and $P(X \in (2, 3))$ is the probability that a randomly sampled fish in Loppa weights between 2 and 3 kg. □

This simple image is really all you need. Many fish – many individual ω – may have the same weight value, but every possibly sampleable weight has at least one fish whose weight is of that particular value. But as you may already have guessed, we will often do well to let Ω be a much larger set than even the possible weighable candidates – even all the way up to an infinitely large set, even though our entire population may be a rather small and finite set. This applies in particular to unknown magnitudes, where we really do not have much knowledge to go on. Indeed, even the fish in the example above would be best studied with an infinite Ω, as in the following example.

Example 7.1.3 Let Ω be all the fish in the river Loppa, *at a specified time t* this month. That is, each ω is a given fish *at a given time*. $X(\omega)$ is then the weight of that fish *at that given time*, and $P(X \in (2, 3))$ is the probability that a salmon in Loppa, randomly sampled at a random time, weighs between 2 and 3 kg. □

And this is just how it begins ...

When we proceed to processes of not just one sample but many, in the course of doing inference, the set Ω needs to take into consideration these factors as well, and the description becomes a bit more complex.

Example 7.1.4 In Example 6.7.1, we update probabilities. Here, each ω is not just which die was possibly picked, but also each possible infinite sequence of test results from the die tosses we did or could have performed in order to determine the probabilities for each die. So, for instance, a single ω could be $\omega = \text{``}D_8, WRRW \ldots\text{''}$.

This also highlights another useful reason for employing a large Ω rather than just the possibilities we want to know the probability of: on a single Ω we may have many different stochastic variables. We do have the X, which is simply $X = \text{``which } k \text{ is our } D_k\text{''}$. But we also have others, like $Y = \text{``the number of } R \text{ in the next five tosses''}$, or $Z = \text{``the number of tries before our first successive 10 } R\text{s in a row''}$.

This Ω also illustrates conditional probability, as follows.

- $P_{D_4}(R) = P(R|D_4)$ is the probability of R (red) *given that* we had picked a D_4. For this conditional probability we restrict our attention to the $\omega \in \Omega$ beginning in D_4, and look at the relative probability of having R first in the sequence of tosses.
- $P_{RWR}(D_{12})$ is the probability that we had picked D_{12} *given that* the first three tosses yielded RWR. For this conditional probability we restrict our attention to the $\omega \in \Omega$ whose tossing sequence start RWR, and look at the relative probability of starting D_4.
- Some of the most important probabilities stem from updating probabilities by Bayes' theorem; that is, P_0, P_1, P_2, etc.

We are still talking about the same stochastic variable X, but this time as *conditioned*, since we are now looking at its action on a subset of Ω. □

But once it is introduced and explained, most of us can leave Ω behind, and concentrate on the properties of the stochastic variables, like X. If you need to think about Ω and are not a theoretical statistician or mathematician, just think "salmon in Loppa", and you will have all that you need. Notation-wise, we will forget the little ω as well, and simply write X instead of $X(\omega)$.

The key property of X is the probability of its values, which is where our investigation started out. We usually indicate these probabilities by the *cumulative probability distribution*. All stochastic variables X have one, and we write $F_X(x) = P(X \leq x)$.

Definition 7.1.5 *F is a* cumulative probability distribution *for X iff*

- *when $x < y$, then $F(x) \leq F(y)$;*
- *$F(-\infty) = 0$ and $F(\infty) = 1$;*
- *$F(x) = P(X \leq x)$.*

We will look at three types of stochastic variables on ℝ: discrete, continuous, and mixed, the latter in Chapter 8.2. Figure 7.2 shows us the three types.

Purely discrete and purely continuous stochastic variables are each characterized by their type of probability distribution $f(x)$. We will investigate these in the coming sections. Not all stochastic variables have a probability distribution even though they all have a *cumulative probability distribution* $F(x)$. We will look at mixed stochastic variables, which are among those that do not have a probability distribution function, in Chapter 8.2. But first, we discuss the two types of stochastic variable that are recognized by their specific type of probability distribution.

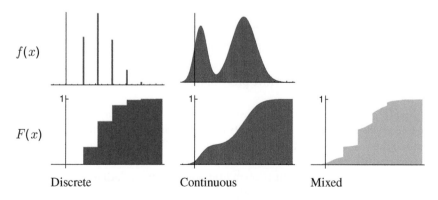

Figure 7.2 The three different types of stochastic variables we will be looking at.

7.2 Discrete Probability Distributions on ℝ

Just as we made tables and diagrams of proportions for data, we can make tables and diagrams of probabilities for stochastic variables. The table and diagram are ways of specifying a discrete function f by connecting the argument (outcome) x to the function's value (probability) $f(x)$. A stochastic variable whose probabilities we may specify by a discrete function is called a *discrete stochastic variable*. We define this formally as follows.

Definition 7.2.1 *A function f on a finite or countable set U ("sample space") is a* discrete probability distribution *for X if*

(1) $f(x) \geq 0$ *for all $x \in U$;*

(2) $\displaystyle\sum_{x \in U} f(x) = 1;$

(3) $f(x) = P(X = x)$ *for all $x \in U$.*

We write $X \sim f$, which we read as "X has probability distribution f". We will also find it useful to speak of the point probability $p_x = f(x)$.

Example 7.2.2 You have found five old radio tubes at the storage at your new job, and have been told that there is a 50% probability that a given tube functions. Let the stochastic variable X be the number of functioning tubes among the ones you have found. We consider each tube to be an independent trial, so we calculate the probabilities for X by means of the formula for unordered sampling with replacement. The parameters are $p = 0.5$ and $n = 5$, so $f(k) = P(X = k) = \binom{5}{k} 0.5^5$. We then have that $X \sim f$, and we specify the values of f in a probability table, and then plot the graph of the discrete function f. We do this in Figure 7.3. □

Outcome k	$f(k)$
0	0.03125
1	0.15625
2	0.3125
3	0.3125
4	0.15625
5	0.03125
SUM	1

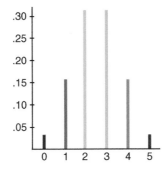

Figure 7.3 This is a *binomial distribution*. More about binomial distributions in Section 9.3.

Table 7.1 The skippings with their probabilities

$x \in U$	−0.2	−0.1	0	0.1	0.2	0.3	0.4	0.5	0.6
$f(x) = p_x$	0.11	0.03	0.18	0.21	0.16	0.01	0.08	0.19	0.03

Rule 7.2.3 *If X has a discrete probability distribution f, we find the probability that X ∈ A by summing the point probabilities of the elements in A:*

$$P(X \in A) = \sum_{x \in A} p_x = \sum_{x \in A} f(x).$$

Example 7.2.4 You have an old milling machine that is stepwise adjustable in tenths of a millimeter. It has become a bit loose, so it sometimes skips to a different step, closer to you or away from you. Let the stochastic variable X be how many millimeters it has skipped away from you. Your colleague in charge of the milling machine tells you that, by long and careful observation, they have arrived at the probabilities of the different values of X shown in Table 7.1 and Figure 7.4.

1) Show that f is a discrete probability distribution.
2) $A = \{-0.2, 0, 0.2, 0.3\}$. What is $P(X \in A)$? (See Figure 7.5.)
3) What is $P(X < 0)$? (See Figure 7.6.)

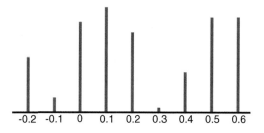

Figure 7.4 The graph of the skipping probabilities.

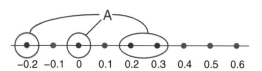

Figure 7.5 The set A.

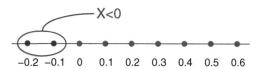

Figure 7.6 The set of x where x < 0.

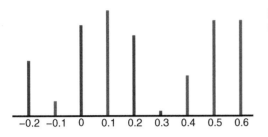

Figure 7.7 $P(X \in A) = \sum_{x \in A} p_x = 0.11 + 0.18 + 0.16 + 0.01 = \underline{0.46}.$

Answer:

(1) We check the criteria from Definition 7.2.1:
   (a) is $f(x) \geq 0$ for all $x \in U$? Yes;
   (b) does $\sum_{x \in U} f(x) = \sum_{x \in U} p_x = 1$.
   Both criteria hold, so f is a discrete probability distribution.
(2) We see the answer in Figure 7.7.
(3) We see the answer in Figure 7.8.

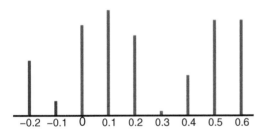

Figure 7.8 $P(X < 0) = p_{-0.2} + p_{-0.1} = 0.11 + 0.03 = \underline{0.14}.$

The sample space U does not need to be a finite set. It may also be infinite, as long as you can enumerate the possible values: first value, second value, third...and by that cover all the values. Not all sets are countable; it is, for instance, impossible to enumerate the elements of $[0, 1]$.

Example 7.2.5 You return to the tubes, but this time you find yourself in tube heaven! The five tubes in Example 7.2.2 came from a storage that held thousands of tubes! You decide to test tubes until you have found a functioning one. The probability that a given tube functions is still $\frac{1}{2}$ for each individual tube. Let X be the number of tubes you test before finding finally one that works. Include only the non-functioning tubes. Then

$$X \sim f(n) = \frac{1}{2^{n+1}}.$$

(1) What is the sample space U?
(2) Show that f is a discrete probability distribution.
(3) $A = \{1, 4, 5\}$. What is $P(X \in A)$?
(4) What is $P(X > 2)$?

Figure 7.9 $P(X > 2) = P(X \geq 3)$
$$= \sum_{n=3}^{\infty} \left(\frac{1}{2}\right)^{n+1} = \underline{\frac{1}{8}}.$$

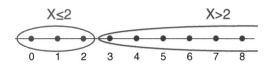

Figure 7.10 $P(X > 2) = 1 - P(X \leq 2)$
$$= 1 - \left(\sum_{n=1}^{2} \left(\frac{1}{2}\right)^{n+1}\right) = \underline{\frac{1}{8}}.$$

Answer: We notice that $f(n) = \left(\frac{1}{2}\right)^{n+1}$, so we may employ the formula for geometric series in Appendix A.2.

(1) We may find the functioning tube after having gone through 0, 1, 2, ... non-functioning tubes, so $U = \{0, 1, 2, \ldots\} = \mathbb{N}_0$.

(2) We check the two criteria:

  (a) is $f(n) \geq 0$ for all $n \in U$? Yes, since $f(n) = \frac{1}{2^{n+1}} > 0$ for all $n \in \mathbb{N}_0$;

  (b) $\sum_{n=0}^{\infty} f(n) = \sum_{n=0}^{\infty} \left(\frac{1}{2}\right)^{n+1} = \sum_{n=0}^{\infty} \frac{1}{2} \cdot \left(\frac{1}{2}\right)^{n} = \frac{1}{2} \cdot \frac{1}{1-\frac{1}{2}} = 1$.

  Both criteria hold, so f is a discrete probability distribution.

(3) $P(X \in A) = p_1 + p_4 + p_5 = \frac{1}{2^2} + \frac{1}{2^5} + \frac{1}{2^6} = \underline{\frac{19}{64}}$.

(4) We may solve this in two ways. We show by both direct calculation (Figure 7.9) and by complementary probability (Figure 7.10).

Notice that since the values are discrete, $X > 2$ is the same as $X \geq 3$, whereas the complimentary event $X \leq 2$ may be written $X < 3$. The neighboring sets $X \leq 2$ and $X > 2$ do, despite being immediate neighbors, have a gap of one unit between them. □

This is a *geometric distribution*. More about these in Section 9.5.

7.2.1 Cumulative Discrete Distribution

We will now revisit the cumulative probability distribution, but now from the perspective of already having a discrete probability distribution function. We will in other words look at the relation between cumulative probability and discrete probability distributions.

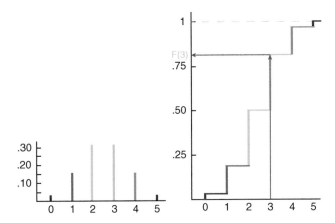

Figure 7.11 We read from the cumulative graph that $F(3) = \underline{0.8125}$.

Rule 7.2.6 *If $X \sim f$, that is if X is a discrete stochastic variable with probability distribution $f(k) = P(X = k)$, then the* cumulative *distribution is given by*

$$F(k) = P(X \leq k) = \sum_{j \leq k} f(j).$$

Example 7.2.7 (Continuation of Example 7.2.2) X is the number of functioning tubes when you have 5 tubes that each have a 50% probability of working. What is the probability that 3 or fewer are functioning? We solve the problem graphically in Figure 7.11, and numerically in Table 7.2. ☐

Table 7.2 The probability that fewer or fewer tubes work, is $F(3) = \underline{0.8125}$

Outcome k	probability $f(k)$		cumulative probability $F(k) = \sum_{j \leq k} f(j)$	0 1 2 3 4
0	0.03125		= 0.03125	)● ● ● ● ●
1	0.15625	+0.03125	= 0.18750	◉) ● ● ● ●
2	0.31250	+0.18750	= 0.50000	● ◉) ● ● ●
3	0.31250	+0.50000	= 0.81250	● ● ◉) ● ●
4	0.15625	+0.81250	= 0.96875	● ● ● ◉) ●
5	0.03125	+0.96875	= 1.00000	● ● ● ● ◉)

Example 7.2.8 We continue Example 7.2.4 of the loose milling machine.

(1) Find the cumulative distribution F.
(2) What is the probability that is skips less than 0.2 mm away from you?
(3) Use F to find the probability that it skips more than 0 but no more than 0.4 mm away from you.

Answer:

(1) We may calculate individual values by the summing formula, like for instance

$$F(0.1) = \sum_{x \le 0.1} f(x) = f(-0.2) + f(-0.1) + f(0) + f(0.1) = 0.53.$$

or we can do one better by extending the probability table by an extra row for F, like in Example 7.2.7, by remembering that for discrete distributions, $F(x)$ equals the previous F value plus $f(x)$. We do this in Table 7.3.
(2) This is $P(X < 0.2) = P(X \le 0.1) = F(0.1) = 0.53$
(3) This is $P(0 < X \le 0.4) = F(0.4) - F(0) = 0.78 - 0.32 = 0.46$ □

Since $F(x) = P(X \le x)$, we may find the probability that X is in the interval $\langle a, b]$ like this:

$$P(X \in \langle a, b]) = F(b) - F(a)$$

Example 7.2.9 We continue Example 7.2.5, where we found that $X \sim f(n) = \frac{1}{2^{n+1}}$ for $n \in \mathbb{N}_0$

(1) Find the cumulative probability distribution F.
(2) Use F to find the probability that you must go through 3 non-working tubes before you find one that works.

Answer:

(1) The cumulative probability distribution is

$$F(n) = \sum_{k \le n} f(k) = \sum_{k=0}^{n} \frac{1}{2^{k+1}} = \sum_{k=0}^{n} \frac{1}{2} \cdot \left(\frac{1}{2}\right)^k$$

$$= \frac{1}{2} \cdot \frac{1 - (1/2)^{n+1}}{1 - 1/2} = 1 - (1/2)^{n+1}$$

(2) This is $P(X > 3) = 1 - P(X \le 3) = 1 - F(3) = 1 - (1 - (1/2)^{3+1}) = \underline{\underline{1/16}}$ □

Table 7.3 Cumulative probability distribution for the skipping of the milling machine

x	−0.2	−0.1	0	0.1	0.2	0.3	0.4	0.5	0.6
$f(x)$	0.11	0.03	0.18	0.21	0.16	0.01	0.08	0.19	0.03
$F(x)$	0.11	0.14	0.32	0.53	0.69	0.7	0.78	0.97	1

Table 7.4 The original probabilities of the balls

x	1	3	4	7	9	Two-digit ball
probabilities	0.13	0.04	0.21	0.14	0.12	0.36

Table 7.5 The conditional probabilities of the single-digit balls

x	1	3	4	7	9
Probabilities	0.203 125	0.062 5	0.328 125	0.218 75	0.187 5

Example 7.2.10 We have an urn with enumerated balls. But someone has removed the two-digit balls, and only those. The probabilities for each type of ball before the two-digit balls were removed are given in Table 7.4. Find the new probabilities for the remaining balls.

Answer: This is a conditional probability, conditioned on the alternative "single-digit ball", which has a total of P(single-digit ball) $= 0.64$, so we divide the stated probabilities by the probability of the total, 0.64, and get the conditional probabilities, in Table 7.5. □

7.3 Continuous Probability Distributions on ℝ

For a continuous stochastic variable, single values have probability zero. Positive probabilities are reserved for *sets* of positive extension. Variables that are able to take any value in an interval are typically continuous, whereas variables that are limited to a finite set of values are discrete. Our possible measurement values of the height of a human being are discrete. Most of the time, we measure in centimeters: 181, 182, 183 cm, etc., but sometimes we refine it to millimeters. But typically, we will have a limit to our accuracy. The height itself, on the other hand, is a continuous variable capable of taking any value within the cm or mm intervals. The probability that we *measure* a person to be 182 cm is positive, whereas the probability that he *is* 182.000 … cm is zero.

We will look at an example before formally defining continuous stochastic variables.

Example 7.3.1 Let X be the point on your right front tire pointing straight down right now. What is the probability that X is in any of the following areas.

(1) A: from 10 to 130° counterclockwise from the valve?
(2) B: from 220 to 240° counterclockwise from the valve?
(3) C: precisely at 300° counterclockwise from the valve?

Answer: We calculate the probability on the assumption that it is spread uniformly over the circumference of the tire, meaning that the probability of it

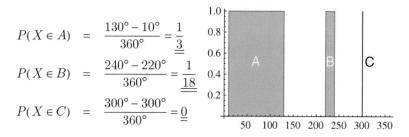

$$P(X \in A) = \frac{130° - 10°}{360°} = \frac{1}{3}$$

$$P(X \in B) = \frac{240° - 220°}{360°} = \frac{1}{18}$$

$$P(X \in C) = \frac{300° - 300°}{360°} = 0$$

Figure 7.12 Probabilities of hitting certain segments of a tire.

being in a given segment equals that segment's proportion of the circumference. See Figure 7.12 for the illustrated calculations. □

The continuous stochastic variable X above assigns positive probability to intervals, whereas its point probabilities all are zero.

Definition 7.3.2 *The function f on the interval (A, B) is a* continuous probability distribution *for X if*

(1) $f(x) \geq 0$;
(2) $\int_A^B f(x)\, dx = 1$;
(3) $P(a < X < b) = \int_a^b f(x)\, dx$.

If this holds, X is furthermore a continuous stochastic variable. *We write $X \sim f$, and read that "X has probability distribution f."*

Probability in the continuous case is not like in the discrete case given by the sum of probabilities for individual outcomes, with a discrete graph of point probabilities. In the continuous case, probability is rather given as the area under the graph of a function F defined on an entire interval, in other words the integral of f. The continuous probability distribution f is often called a probability *density*.

Note that $P(X \in \langle a, b \rangle) = P(X \in [a, b])$, since the probability of X taking an endpoint value is zero. Indeed, any finite collection of points has probability zero for a continuous stochastic variable. Working with continuous distributions, we may then without any loss write our intervals (a, b), leaving it indeterminate whether the endpoints are included or not. For discrete and mixed distributions, though, (end)points may have positive probability, and thus we need to include them.

Example 7.3.3 $X \sim f$, where

$$f(x) = \begin{cases} 2x & 0 \leq x \leq 1 \\ 0 & \text{otherwise.} \end{cases}$$

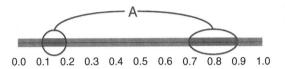

Figure 7.13 *A is the union of two intervals.*

(1) Show that X is a continuous stochastic variable.
(2) $A = (0.1, 0.2) \cup (0.7, 0.9)$. What is $P(X \in A)$? (see Figure 7.13).

Answer:

(1) We check the criteria:
   (a) $f(x) \geq 0$ for all x;
   (b) $\int_{-\infty}^{\infty} f(x)\, dx = \int_0^1 2x\, dx = 1$.
   which means that f is a continuous probability distribution.
(2) We calculate by dividing into intervals, integrating over each of them, and then summing. The total is the sum of the areas under the graph in the intervals, as illustrated in Figure 7.14.

$$P(X \in A) = P(X \in (0.1, 0.2) \cup (0.7, 0.9))$$
$$= P(X \in (0.1, 0.2)) + P(X \in (0.7, 0.9))$$
$$= \int_{0.1}^{0.2} 2x\, dx + \int_{0.7}^{0.9} 2x\, dx = 0.35.$$

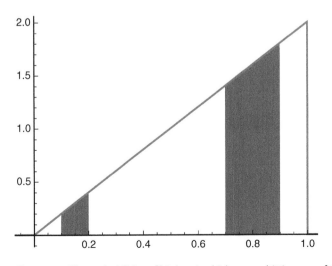

Figure 7.14 The probabilities of hitting A, which means hitting any of the two intervals. ☐

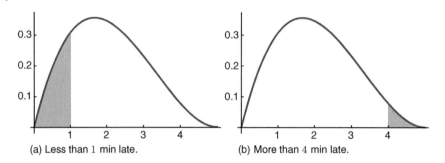

(a) Less than 1 min late.   (b) More than 4 min late.

Figure 7.15 Lecturer N's delay in Example 7.3.4.

Example 7.3.4 Lecturer N always comes late for lectures after lunch, but he is never more than five minutes late. The delay is a continuous stochastic variable X with probability density $f(t) = \frac{12}{625}t(5-t)^2$ in the interval $[0, 5]$. (Figure 7.15.)

(a) What is the probability that N is less than 1 minute late? (Figure 7.15a.)
(b) What is the probability that N is 4 or more minutes late? (Figure 7.15b.)

Answer:

(a) $P(X \le 1) = \int_0^1 f(t)\, dt = \int_0^1 \frac{12}{625}t(5-t)^2\, dt \approx 0.18.$
(b) We may solve this in two ways. We show by both direct calculation (Figure 7.16) and by complementary probability (Figure 7.17).

Notice that since these values are continuous, then $P(X > 4)$ is the same as $P(X \ge 4)$, whereas the complimentary probability $P(X \le 4)$ may be written $P(X < 4)$. The neighbors $X < 4$ and $X > 4$ are flush with one another; the gap between them is 0 units wide.

Figure 7.16 $P(X > 4) =$
$P(X \ge 4) = \int_4^5 f(t)\, dt =$
$\int_4^5 \frac{12}{625}t(5-t)^2\, dt \approx 0.02.$

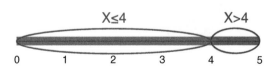

Figure 7.17 $P(X > 4) =$
$1 - P(X \le 4) =$
$1 - \int_0^4 \frac{12}{625}t(5-t)^2\, dt \approx 0.02.$

□

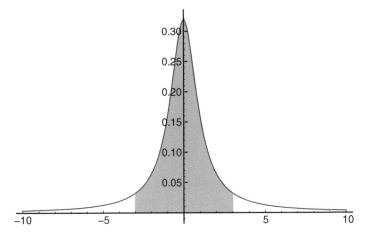

Figure 7.18 $P(|X| < 3)$.

Continuous distributions may be limited to an interval, or they may span over all real numbers from minus to plus infinity, like the following distributions.

Example 7.3.5 $X \sim f(x) = \frac{1/\pi}{1+x^2}$. What is the probability that $|X| < 3$? (See Figure 7.18.)

Answer: $P(|X| < 3) = \int_{-3}^{3} \frac{1/\pi}{1+x^2} \, dx = \left[\frac{1}{\pi} \cdot \tan^{-1}(x) \right]_{-3}^{3} \approx \underline{\underline{0.795}}$. □

Example 7.3.6 The height of waves right outside Bolgevik harbor, measured in meters, is $Z \sim f(z) = 2ze^{-z^2}$. What is the probability that a given wave exceeds 2 m in height? (See Figure 7.19.)

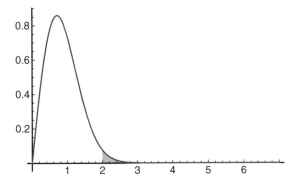

Figure 7.19 $P(Z > 2)$.

Answer:

$$P(Z \le 2) = \int_0^2 2ze^{-z^2}\, dz \text{ (we substitute } u = z^2, \text{ so } \tfrac{du}{dz} = 2z)$$

$$= \int_{z=0}^{z=2} \frac{du}{dz} e^{-u}\, dz = \int_{z=0}^{z=2} e^{-u}\, du$$

$$= \left[-e^{-u}\right]_{z=0}^{z=2} = \left[-e^{-z^2}\right]_0^2$$

$$= -e^{-2^2} - (-e^{-0^2}) = 1 - e^{-4}$$

$$P(Z > 2) = 1 - P(Z \le 2) = \underline{e^{-4} \approx 0.018\,3}. \qquad \square$$

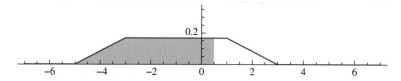

Figure 7.20 $P(-7 \le X \le 0)$.

Example 7.3.7 (Numerical example using a piecewise defined distribution function)

$$X \sim f(x) = \begin{cases} \frac{x+5}{12} & x \in [-5, -3) \\ \frac{1}{6} & x \in [-3, 1) \\ \frac{-x+3}{12} & x \in [1, 3) \\ 0 & \text{otherwise.} \end{cases}$$

Find $P(-7 \le X \le 0)$. (See Figure 7.20.)

Answer:

$$P(-7 \le X \le 0) = \int_{-7}^{0} f(x)\, dx = \int_{-7}^{-5} f(x)\, dx + \int_{-5}^{-3} f(x)\, dx + \int_{-3}^{0} f(x)\, dx$$

$$= \int_{-7}^{-5} 0\, dx + \int_{-5}^{-3} \frac{x+5}{12}\, dx + \int_{-3}^{0} \frac{1}{6}\, dx = 0 + \frac{1}{6} + \frac{1}{2} = \underline{\frac{2}{3}}. \qquad \square$$

7.3.1 Cumulative Continuous Distribution

For a continuous stochastic variable, the link between the probability distribution and the cumulative distribution is one of the fundamental theorems of calculus – see the following definition.

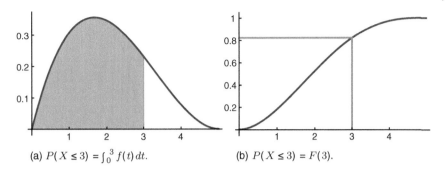

(a) $P(X \leq 3) = \int_0^3 f(t)\,dt.$ (b) $P(X \leq 3) = F(3).$

Figure 7.21 The two ways of finding $P(X \leq 3)$ are essentially the same.

Definition 7.3.8 *If $X \sim f$, that is, if X is a continuous stochastic variable on (A, B) with probability distribution f, the* cumulative *probability distribution is given by*

$$F(x) = P(X \leq x) = \int_A^x f(t)\,dt.$$

Conversely,

$$f(x) = F'(x).$$

Example 7.3.9 We continue Example 7.3.4, where the probability density of lecturer N's delay is given by $f(t) = \frac{12}{625}t(5 - t)^2$. Sonia is 3 minutes late for the lecture, and as she enters the lecture hall she wonders what the probability is that N has arrived.

Answer: Graphically, see Figure 7.21. The calculations go like this:

$$F(t) = \int_0^t f(x)\,dx = \int_0^t \frac{12}{625}x(5 - x)^2\,dx = \frac{1}{625}(150t^2 - 40t^3 + 3t^4)$$

$$P(X \leq 3) = F(3) = \int_0^3 f(t)\,dt = \frac{1}{625}(150 \times 3^2 - 40 \times 3^3 + 3 \times 3^4) \approx \underline{0.82}.$$

□

Example 7.3.10 (Continuation of Example 7.3.5) $X \sim f(x) = \frac{1/\pi}{1+x^2}$.

(1) What is $F(x)$?
(2) Use F to find $P(|X| < 3)$.

We see this illustrated in Figure 7.22.

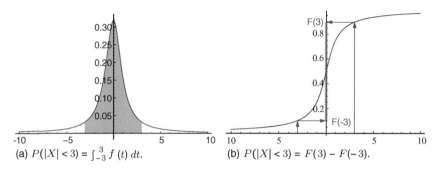

(a) $P(|X| < 3) = \int_{-3}^{3} f(t) \, dt.$ (b) $P(|X| < 3) = F(3) - F(-3).$

Figure 7.22 The two ways of finding $P(|X| < 3)$ are essentially the same.

Answer:

(1) $F(x) = \displaystyle\int_{-\infty}^{x} \frac{1/\pi}{1 + t^2} \, dt = \left[\frac{1}{\pi} \cdot \tan^{-1}(t) \right]_{-\infty}^{x}$

$= \dfrac{1}{\pi} \left(\tan^{-1}(x) - \left(-\dfrac{\pi}{2} \right) \right) = \underline{\underline{\dfrac{1}{2} + \dfrac{1}{\pi} \tan^{-1}(x)}}.$

(2) $P(|X| < 3) = P(-3 < X < 3) = F(3) - F(-3) = \underline{\underline{\dfrac{2}{\pi} \tan^{-1}(3)}}.$ □

Example 7.3.11 (Continuation of Example 7.3.6) The height of waves right outside Bolgevik harbor. Find the cumulative distribution $F(z)$, and use this to find the probability that a given wave is between 0.3 and 0.5 meters tall.

Answer: We use the integration from Example 7.3.6, and get

$$F_Z(z) = P(Z \le z) = \int_0^z 2z e^{-z^2} \, dz = 1 - e^{-z^2}$$

$$P(Z \in (0.3, 0.5)) = F_Z(0.5) - F_Z(0.3) = e^{-0.09} - e^{-0.25} \approx \underline{0.135}.$$ □

Example 7.3.12 (Continuation of Example 7.3.7) Find the cumulative probability distribution for

$$X \sim f(x) = \begin{cases} \frac{x+5}{12} & x \in [-5, -3) \\[6pt] \frac{1}{6} & x \in [-3, 1) \\[6pt] \frac{-x+3}{12} & x \in [1, 3) \\[6pt] 0 & \text{otherwise.} \end{cases}$$

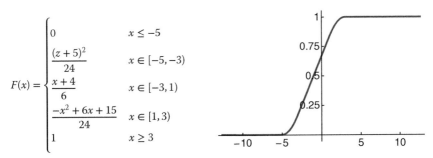

$$F(x) = \begin{cases} 0 & x \le -5 \\ \dfrac{(z+5)^2}{24} & x \in [-5, -3) \\ \dfrac{x+4}{6} & x \in [-3, 1) \\ \dfrac{-x^2 + 6x + 15}{24} & x \in [1, 3) \\ 1 & x \ge 3 \end{cases}$$

Figure 7.23 The conclusion.

Answer: This has to be done piecewise, interval by interval, remembering that the "area under graph" accumulates as we move from left to right in the series of intervals. In general for $x \in \langle a, b]$, we have $F(x) = \int_{-\infty}^{x} f(z)\,dz = \int_{-\infty}^{a} f(z)\,dz + \int_{a}^{x} f(z)\,dz = F(a) + \int_{a}^{x} f(z)\,dz$. We see the conclusion with Figure 7.23. The details are as follows:

$$\langle -\infty, -5] : F(x) = \int_{-\infty}^{x} 0 \, dx = 0$$

$$\langle -5, -3] : F(x) = F(-5) + \int_{-5}^{x} \frac{z+5}{12}\,dz = 0 + \frac{(x+5)^2}{24} = \frac{(x+5)^2}{24}$$

$$\langle -3, 1] : F(x) = F(-3) + \int_{-3}^{x} \frac{1}{6}\,dz = \frac{1}{6} + \frac{x+3}{6} = \frac{x+4}{6}$$

$$\langle 1, 3] : F(x) = F(1) + \int_{1}^{x} \frac{-z+3}{12}\,dz = \frac{5}{6} + \frac{-x^2 + 6x - 5}{24} = \frac{-x^2 + 6x + 15}{24}$$

$$\langle 3, \infty \rangle : F(x) = F(3) + \int_{3}^{x} 0 \, dz = 1 + 0 = 1. \qquad \square$$

7.4 Percentile and Inverse Cumulative Distribution

Just as interesting as the probability that $X \le x$ is the complementary question: "At which value x is the probability that $X \le x$ first equal to or larger than $p = P\% = \frac{P}{100}$?"

Definition 7.4.1 *Let $X \sim f$, and let F be the cumulative distribution. Then the Pth percentile P_P of X is the smallest value of x such that*

$$F(x) \ge \frac{P}{100}.$$

For continuous distributions, this is the solution x of the equation

$$F(x) = \frac{P}{100}.$$

The median is $m_X = P_{50}$.

Example 7.4.2 (Discrete case) X is the number of heads after five tosses with a fair die.

(1) What is the 86th percentile of X?
(2) What is the median of X?

Answer: This continues Example 7.2.2, and we make use of the table there.

(1) We search the $F(k)$ column until we find a number exceeding $\frac{86}{100}$. We see that $F(3) = 0.8125 < 0.86 \leq 0.96875 = F(4)$, so $F(4)$ is the first number equal to or in excess of $\frac{86}{100}$. This means $P_{86} = 4$.

   We may also solve this graphically: we find P_{86} at $\frac{86}{100}$ on the y-axis, going horizontally until we hit the graph, and then vertically until we hit the x-axis and may read the percentile value.

(2) $m_X = P_{50}$, so we find the smallest number in excess of (or equal to) 0.5 in the right column; that number is 0.5 itself. Then we read the entry in the "outcome" column of the same row, which says 2, so $m_X = \underline{\underline{2}}$. □

For continuous probability distributions, the percentile goes by a different name: *the cumulative inverse*, iCDF for short.

> **Rule 7.4.3** *Let $X \sim f$, and let F be the cumulative distribution. If $F(x) < F(y)$ when $x < y$, then* the cumulative inverse, $F^{-1}(p)$, exists, and
>
> $$F^{-1}(p) = P_p,$$
>
> *where $P = 100p$.*

Example 7.4.4 (Continuous case) We continue Example 7.3.4 about N's delay, $X \sim f(t) = \frac{12}{625}t(5 - t)^2$.

(1) Dan T. is a pragmatist, and wants optimal use of his time. He wonders how late he himself may be if he wants a 50% probability that N is there before him.
(2) Ahmad would like to present the class with an overview, with illustrations of P_{10}, P_{40}, P_{65}, and P_{90}.

Answer: This is a continuation of Example 7.3.9, where we found that $F(t) = \frac{1}{625}(150t^2 - 40t^3 + 3t^4)$. We use this to answer the following questions.

(1) Dan T. wants to find P_{50}, which is also the median, m_X. It is *also* $F^{-1}(0.5)$. He must then solve

$$F(t) = \frac{50}{100} = 0.5, \text{ which is the equation } \frac{1}{625}(150t^2 - 40t^3 + 3t^4) = 0.5.$$

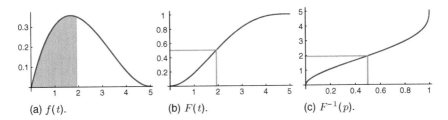

(a) $f(t)$.   (b) $F(t)$.   (c) $F^{-1}(p)$.

Figure 7.24 Three graphical solutions, which are essentially identical.

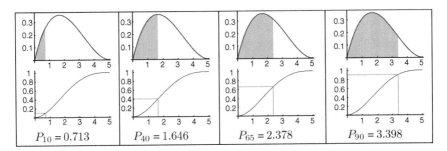

$P_{10} = 0.713$ $P_{40} = 1.646$ $P_{65} = 2.378$ $P_{90} = 3.398$

Figure 7.25 Ahmed's project, using the (cumulative) distribution, $f(t)$ and $F(t)$.

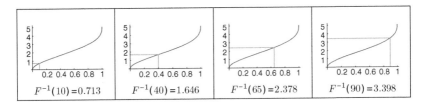

$F^{-1}(10) = 0.713$ $F^{-1}(40) = 1.646$ $F^{-1}(65) = 2.378$ $F^{-1}(90) = 3.398$

Figure 7.26 Ahmed's project, using the inverse cumulative distribution $F^{-1}(p)$.

This is a fourth degree polynomial equation, so Dan T. solves it in Wolfram Alpha, and gets

$$F^{-1}(0.5) = P_{50} = 1.929 = \underline{1 \text{ minute and 56 seconds.}}$$

The different graphical solutions are shown in Figure 7.24.

(2) Ahmed's project is answered in the same way as Dan's. His solutions are shown in Figure 7.25.

Ahmed's questions may also be solved by using the inverse cumulative distribution, as shown in Figure 7.26.   □

7.4.1 Some Easy Functions to Calculate

For the examples below, we urge the reader to integrate $f(t)$ and verify that it is indeed $F(t)$, and to differentiate $F(t)$ and verify that it is indeed $f(t)$. Furthermore, verify that $F^{-1}(p)$ is indeed the inverse of $F(x)$, either by solving $F(x) = p$, or by calculating and finding that $F\left(F^{-1}(p)\right) = p$ and $F^{-1}(F(x)) = x$.

Example 7.4.5 A uniform probability distribution on $[3, 8]$:

- $f(t) = 0.2$
- $F(t) = 0.2t - 0.6$
- $F^{-1}(p) = 5p + 3$.

In Figure 7.27, we see that $P(X \leq 5) = F(5) = 0.4$.

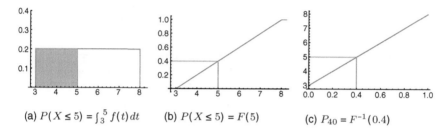

(a) $P(X \leq 5) = \int_3^5 f(t)\,dt$ (b) $P(X \leq 5) = F(5)$ (c) $P_{40} = F^{-1}(0.4)$

Figure 7.27 $P(X \leq 5) = F(5) = \int_3^5 f(t)\,dt = 0.4$ – equivalently $P_{40} = F^{-1}(0.4) = 5$. □

Example 7.4.6 Linearly rising probability on $[0, 10]$:

- $f(t) = \dfrac{t}{50}$
- $F(t) = \dfrac{t^2}{100}$
- $F^{-1}(p) = 10\sqrt{p}$.

In Figure 7.28, we see that $P(X \leq 6) = F(6) = 0.36$.

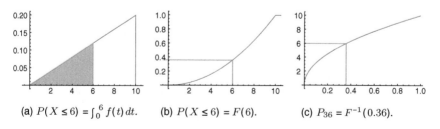

(a) $P(X \leq 6) = \int_0^6 f(t)\,dt$. (b) $P(X \leq 6) = F(6)$. (c) $P_{36} = F^{-1}(0.36)$.

Figure 7.28 $P(X \leq 6) = F(6) = \int_0^6 f(t)\,dt = 0.36$ – equivalently $P_{36} = F^{-1}(0.36) = 6$. □

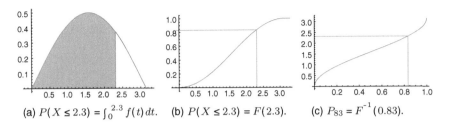

(a) $P(X \le 2.3) = \int_0^{2.3} f(t) \, dt.$ (b) $P(X \le 2.3) = F(2.3).$ (c) $P_{83} = F^{-1}(0.83).$

Figure 7.29 $P(X \le 2.3) = F(2.3) = 0.83$ – equivalently $P_{83} = F^{-1}(0.83) = 2.3.$

Example 7.4.7 A half sine wave on $[0, \pi]$:

- $f(t) = \frac{1}{2}\sin(t)$
- $F(t) = \frac{1}{2}\sin^2\left(\frac{t}{2}\right)$
- $F^{-1}(p) = 2\sin^{-1}\left(\sqrt{p}\right).$

□

In Figure 7.29, we see that $P(X \le 2.3) = F(2.3) = 0.83.$

Example 7.4.8 A compound sine wave on $[0, \sqrt{\pi}]$:

- $f(t) = t\sin(t^2)$
- $F(t) = \sin^2\left(\frac{t^2}{2}\right)$
- $F^{-1}(p) = \sqrt{2\sin^{-1}\left(\sqrt{p}\right)}.$

□

In Figure 7.30, we see that $P(X \le 1.1) = F(1.1) = 0.32.$

Example 7.4.9 Negative exponential function on $[0, \infty]$:

- $f(t) = e^{-t}$
- $F(t) = 1 - e^{-t}$
- $F^{-1}(p) = -\ln(1 - p).$

In Figure 7.31, we see that $P(X \le 1.5) = F(1.5) = 0.78.$

□

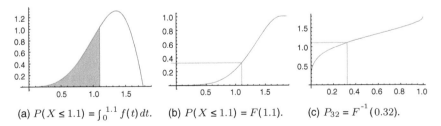

(a) $P(X \le 1.1) = \int_0^{1.1} f(t) \, dt.$ (b) $P(X \le 1.1) = F(1.1).$ (c) $P_{32} = F^{-1}(0.32).$

Figure 7.30 $P(X \le 1.1) = F(1.1) = 0.32$ – equivalently $P_{32} = F^{-1}(0.32) = 1.1.$

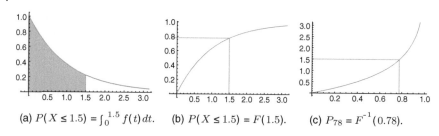

(a) $P(X \leq 1.5) = \int_0^{1.5} f(t)\,dt.$ (b) $P(X \leq 1.5) = F(1.5).$ (c) $P_{78} = F^{-1}(0.78).$

Figure 7.31 $P(X \leq 1.5) = F(1.5) = 0.78$ – equivalently $P_{78} = F^{-1}(0.78) = 1.5.$

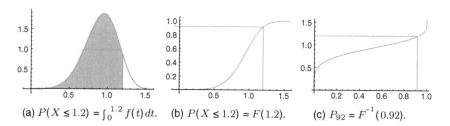

(a) $P(X \leq 1.2) = \int_0^{1.2} f(t)\,dt.$ (b) $P(X \leq 1.2) = F(1.2).$ (c) $P_{92} = F^{-1}(0.92).$

Figure 7.32 $P(X \leq 1.2) = F(1.2) = 0.92$ – equivalently $P_{92} = F^{-1}(0.92) = 1.2.$

Example 7.4.10   A compound function on $[0, \infty]$:

- $f(t) = 5t^4 e^{-t^5}$
- $F(t) = 1 - e^{-t^5}$
- $F^{-1}(p) = \sqrt[5]{-\ln(1-p)}.$

In Figure 7.32, we see that $P(X \leq 1.2) = F(1.2) = 0.92.$ □

7.5   Expected Value

We found that the average of a data set may be written $\bar{x} = \sum_{all\ x} x \cdot p_x$, where p_x is the *proportion* of the data having value x. The proportion formula for the average $\bar{x}$ may be generalized to probabilities, since probability is a kind of proportion. Just let p_x be the point probability of the value x. You then get the *expected value* of the discrete stochastic variable X, which is written $E[X]$ or μ_X.

Definition 7.5.1   *For a discrete stochastic variable X with point probabilities* $P(X = x) = p_x$,

$$\mu_X = E[X] = \sum_{all\ x} x \times p_x.$$

Example 7.5.2 Sam is more interested in what the future may bring than in past history. He has been invited to play a game with a single toss of a D_4. He has to bet £3, and will be paid the amount shown by the die. Sam thinks this is an excellent game, since he gets a prize no matter what! But he's lost in gambling before, so he asks his good friend Bard if he really should join the game, and if he should perhaps negotiate his bet. The die D_4 has, as we remember, uniform probabilities. That is,

$$P(X = k) = 0.25 \text{ for all } x \in U = \{1, 2, 3, 4\}.$$

Bard tells Sam that he can expect to lose in the long run, since his bet is larger than the expected payoff of the game. Bard asks Sam to calculate the expected value of the game for himself. Sam looks at the D_4, and calculates:

$$\mu_X = E[X] = \sum_{k=1}^{4} k \cdot p_k = 1 \cdot 0.25 + 2 \cdot 0.25 + 3 \cdot 0.25 + 4 \cdot 0.25 = \underline{\underline{2.5}}.$$

The expected payoff of £2.5 is *less* than Sam's bet. Sam decides not to join the game, and decides to look for another one instead. □

This formula transfers easily to continuous variables. We just need to replace the point probabilities p_x by the probability density $f(x)$, and replace the summation sign by an integral, and we get the formula

$$\mu_X = \int_{-\infty}^{\infty} x \cdot f(x) \, dx. \tag{7.1}$$

The integration limits may be narrowed to fit the range of the stochastic variable X.

Example 7.5.3 We revisit the example of N's delay, $X \sim f(t) = \frac{12}{625} t(5 - t)^2$, and wonder: What is N's *expected* delay? Figure 7.33 has the answer. □

Example 7.5.4 This continues Example 7.3.7, where

$$X \sim f(x) = \begin{cases} \frac{x+5}{12} & x \in [-5, -3) \\ \frac{1}{6} & x \in [-3, 1) \\ \frac{-x+3}{12} & x \in [1, 3) \\ 0 & \text{otherwise.} \end{cases}$$

We will now find $E[X]$:

$$E[X] = \int_{-\infty}^{\infty} x \cdot f(x)\, dx$$

$$= \int_{-5}^{-3} x \cdot \frac{x+5}{12}\, dx + \int_{-3}^{1} x \cdot \frac{1}{6}\, dx + \int_{1}^{3} x \cdot \frac{-x+3}{12}\, dx = \underline{\underline{-1}}. \qquad \square$$

$$\mu_X = E[X]$$

$$= \int_{0}^{5} t \times f(t)\, dt$$

$$= \int_{0}^{5} t \times \frac{12}{625} t(5-t)^2\, dt$$

$$= 2.$$

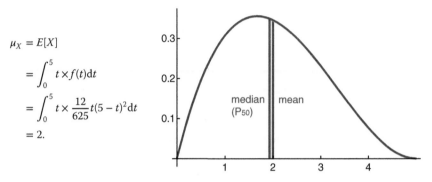

Figure 7.33 The answer to example 7.5.3.

Example 7.5.5 It's Saturday evening, and Sam is sitting on his couch expecting to become a millionaire. He and Bard decided, as an experiment, to fill out a 12-game betting slip using a D_3 to decide whether to bet H (home), D (draw), or A (away) for each game. The betting slip is currently in Bard's office. After Liverpool FC's totally unexpected 5–1 win over Manchester United, the payouts are as follows:

- 12 correct: £500 000
- 11 correct: £8 947
- 10 correct: £52.

What is Sams and Bards expected payout from their betting slip?

Answer:

- The probability of 12 correct is $\left(\frac{1}{3}\right)^{12} \approx 1.88 \cdot 10^{-6}$.

- The probability of 11 correct is $\binom{12}{1} \cdot \frac{2}{3} \cdot \left(\frac{1}{3}\right)^{11} \approx 0.000\,045$.

- The probability of 10 correct is $\binom{12}{2} \cdot \left(\frac{2}{3}\right)^2 \left(\frac{1}{3}\right)^{10} \approx 0.000\,50$.

The expected payout is then

$$1.88 \cdot 10^{-6} \cdot 500\,000 + 0.000\,045 \cdot 8\,947 + 0.000\,50 \cdot 52 = 1.37.$$

The price of the betting slip was just £1, so this bet has an expected positive net result of £0.37. It's the kind of bet you would like to be in. But when Sam asks Bard if they should repeat the experiment next week, Bard declines, since the expected payout on a normal week is 50 pence. ☐

Example 7.5.6 Bard, Svein, and Thor are at the airport with their friend Sam, who is going to attend a poker tournament in Las Vegas. The four friends are making time pass by playing a long-lasting mediaeval card game with open cards called Rondebleu, and they have roughly half an hour left of the game when boarding for Sam's flight is announced. There is £1120 in the pot, and they would like to split it fairly before they part.

Bard instructs that the fair share would be the price they would be willing to pay to resume the game with the same cards and the same pot.

"How much would that be?" Svein asks.

Bard replies that they should not choose to participate if the new bet exceeded the expected payout. So the limiting price to participate would be each player's probability to win, times the total pot.

"Ah OK, just the pot times my probability to win!" exclaims Svein. "That's an easy one, for since I'm in the lead, I have a 100% probability to win!"

Bard then offers Svein to play him afterwards, each keeping their current cards. Svein will have to pay £1120, and Bard 0. Svein doesn't think much of that proposal, he says, since all he gets out of it is to have his money back at the end of the game. So then Bard offers a different deal: Svein pays £1000 to join the game, and gets paid £1120 if and when he wins.

Svein realizes that even this is a poor bet, since there is still a significant probability that one of the others will win, and then he'd lose £1000 on a risky wager. Meekly, Svein asks Bard to calculate the actual probabilities to win, and distribute the pot accordingly.

The Rondebleu probabilities are easy to calculate: they are 60% for Svein to win, 30% for Sam, and 10% for Thor. Bard lost early on, so he's not in.

(1) Expected payout for Svein is $1120 \cdot 0.6 = 672$, so Svein gets £672.
(2) Expected payout for Sam is $1120 \cdot 0.3 = 336$, so Sam gets £336 to bring to Vegas.
(3) Expected payout for Thor is $1120 \cdot 0.1 = 112$, so he gets £112.

The players agree that this is fair, and reason as follows: if Bard put each of their cards in sealed envelopes with the payouts as price tags, any of them would think the naming price fair for any of the envelopes, even if it wasn't their own.

They wave farewell to Sam, and agree to visit Frederick to continue the game. Thor pays £672 to take over Svein's old hand, since he's tired of being the underdog, whereas Bard buys Thor's old hand for £112. Svein buys Sam's hand for £336. Frederick just shakes his head at his crazy gambler friends, and goes to put on some hot water to make them all a nice cup of tea. ☐

Example 7.5.7 It is "student surprise" day at your university, and Tonya, Samuel, and Guillaume have been told that the surprise lecturer is either Svein Nyberg, Bill Gates, or Larry Ellison, and that the probabilities that it is just that person are, respectively 0.8, 0.15, and 0.05. Samuel has decided to ask the lecturer to contribute to the students' education fund, with a promise to do one push-up for every hundred dollars donated.

Samuel has heard that Nyberg is a tightwad, but will give $100 just to see Samuel do a push-up, whereas Ellison is known to give 1000, and Gates is said to give ten thousand if asked. Tonya says she'll give a dollar to charity for every push-up Samuel does, if Guillaume matches her. Guillaume starts worrying that the surprise lecturer might be Bill Gates, and asks Tonya if he can prepay the expected amount instead, "you know, just as in a bet". Tonya says that's fair. How much must Guillaume pay?

Answer: Remember that the probabilities are 0.8, 0.15, and 0.05. Tonya's expected payment is then

$$E[X] = 0.8 \cdot 1 + 0.15 \cdot 10 + 0.05 \cdot 100 = 7.3$$

so Guillaume has to pay 7 dollars and 30 cents. Guillaume doesn't have any cent coins on him, but since he is a generous man, he donates $8 to the fund. □

7.6 Variance, Standard Deviation, and Precision

The most common measures of spread for both stochastic variables and populations/data, are *variance* and *standard deviation*. We will now study these measures for stochastic variables. For data, see Chapter 2.

Definition 7.6.1 *The* variance *of a stochastic variable X is the mean square distance from the expected value* μ_X:

$$Var(X) = E[(X - \mu_X)^2].$$

Rule 7.6.2 *A simpler calculation than the definition, is the identity*

$$Var(X) = E[(X - \mu_X)^2] = E[X^2] - \mu_X^2.$$

• *For discrete X, it then follows that* $Var(X) = \left(\sum_{all\ x} x^2 \cdot p_x \right) - \mu_X^2.$

• *For continuous variables,* $Var(X) = \left(\int_{-\infty}^{\infty} x^2 \cdot p(x)\, dx \right) - \mu_X^2.$

If X is measured in units of unit u, then so is the mean, but the variance goes in units of u^2. To get a measure of spread that is commensurate with the X and with the expectation, we must use the *standard deviation* – see the following definition.

Definition 7.6.3 *The* standard deviation *of a stochastic variable X is*

$$\sigma_X = \sqrt{Var(X)}.$$

Example 7.6.4 We continue Example 7.5.2 concerning Sam's D_4. Recall that $P(X = k) = 0.25$ for all $x \in U = \{1, 2, 3, 4\}$, and that the expected value was $\mu_X = 2.5$. What is the standard deviation of X?

Answer: We find $E[X^2]$, $Var(X)$, and σ_X in turn:

$$E[X^2] = \sum_{\text{all } k} k^2 \times p_k = \sum_{k=1}^{4} k^2 \times 0.25 = 7.5$$

$$Var(X) = E[X^2] - \mu_X^2 = 7.5 - 2.5^2 = 1.25$$

$$\sigma_X = \sqrt{Var(X)} = \sqrt{1.25} \approx \underline{1.118}.$$ $\square$

Example 7.6.5 We revisit the situation in Example 7.3.4 about N's delay $X \sim f(t) = \frac{12}{625}t(5 - t)^2$. We calculated (Example 7.5.3) that his expected delay was 2 minutes. But what is the spread of these delays? We want to answer it in terms of the standard deviation.

Answer: We find $E[X^2]$, $Var(X)$, and σ_X in turn:

$$E[X^2] = \int_0^5 t^2 \cdot f(t)\, dt = \int_0^5 t^2 \cdot \frac{12}{625}t(5 - t)^2\, dt = 5$$

$$Var(X) = E[X^2] - \mu_X^2 = 5 - 2^2 = 1$$

$$\sigma_X = \sqrt{Var(X)} = \sqrt{1} = 1.$$

Illustration: We often consider values inside one standard deviation of the mean as "normal", so we have marked N's normal delays in Figure 7.34. That is, the delays within 1 standard deviation σ_X on either side of μ_X. $\square$

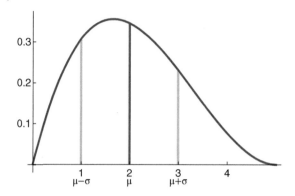

Figure 7.34 The mean μ plus/minus one standard deviation σ.

Example 7.6.6 We continue Examples 7.3.7 and 7.5.4, where

$$X \sim f(x) = \begin{cases} \frac{x+5}{12} & x \in [-5, -3) \\ \frac{1}{6} & x \in [-3, 1) \\ \frac{-x+3}{12} & x \in [1, 3) \\ 0 & \text{otherwise.} \end{cases}$$

We will calculate $Var(X)$ and σ_X, and recall that $\mu_X = E[X] = -1$:

$$E[X^2] = \int_{-\infty}^{\infty} x^2 \cdot f(x)\, dx$$

$$= \int_{-5}^{-3} x^2 \cdot \frac{x+5}{12}\, dx + \int_{-3}^{1} x^2 \cdot \frac{1}{6}\, dx + \int_{1}^{3} x^2 \cdot \frac{-x+3}{12}\, dx = \frac{13}{3}$$

$$Var(X) = E[X^2] - \mu_X^2 = \frac{13}{3} - (-1)^2 = \underline{\underline{\frac{10}{3}}}$$

$$\sigma_X = \sqrt{Var(X)} = \sqrt{\frac{10}{3}}.$$

$\square$

You have probably pondered a bit, and wondered why we have chosen such a seemingly cumbersome measure of spread as σ_X, where we have to subtract, square, sum, and then take the square root again. Surely, it would be a lot easier to just sum $|x - \mu_X| \cdot p_x$ and be done with it! Well, seemingly it would. In Section 8.3 we will see that mean and variance are very well behaved when we add stochastic variables: $Z = X + Y$. The formulas are nice, tidy, and neat. Corresponding formulas if we decided to use $\sum |x - \mu_X| \cdot p_x$ as our measure of spread, on the other hand, are not user friendly at all. And since statistics relies heavily on adding stochastic variables, we need our measure of spread to behave well under addition.

7.6.1 Precision

We will make much use of the inverse of the variance of a stochastic variable, known as the *precision*.

Definition 7.6.7 *The* precision *of a stochastic variable X is*

$$\tau_X = \frac{1}{\sigma_X^2}.$$

Example 7.6.8 X has standard deviation $\sigma_X = 5$. What is its precision τ_X?

Answer: $\tau_X = \frac{1}{\sigma_X^2} = \frac{1}{5^2} = 0.04$. ☐

7.7 Exercises

1 Read the chapter.
   (a) What are the smallest and largest values possible for a discrete distribution function $f(x)$?
   (b) What are the smallest and largest values possible for a continuous distribution function $f(x)$?
   (c) What are the smallest and largest values possible for a cumulative distribution function $F(x)$?
   (d) What *is* $f(x)$, and what is $F(x)$? What is the difference between the two?
   (e) What does discrete probability have in common with data analysis? Which common measures do we have?
   (f) How do you calculate $P(a < X \le b)$ from $F(x)$?
   (g) What is the relation between $P(X > a)$ and $P(X \le a)$?

Discrete Stochastic Variables

2 In the exercises below, you are given concrete stochastic variables X.
   - Make a table of the probabilities larger than 0.
   - Draw a probability diagram from the table.
   - Find $E[X]$.
   - Find $Var(X)$.
   - Find σ_X.
   - Find τ_X.

(a) X is the outcome of flipping a fair coin with "0" on one side and "1" on the other.

(b) X is the sum of the outcomes of *two* fair coins with "0" on one side and "1" on the other.

(c) X is the sum of the outcomes of *three* fair coins with "0" on the one side, and "1" on the other.

(d) X is the outcome of tossing a fair D_4.

(e) X is the sum of *two* fair D_4.

(f) X is the outcome of tossing a fair D_6.

(g) X is the sum of *two* fair D_6.

3 For the following sets M and functions f, do the following.
- Make a table and diagram for f.
- Determine if f is a discrete probability distribution. If it is, let $X \sim f$ and do the following.
  i. Find μ_X.
  ii. Find $Var(X)$.
  iii. Find σ_X.
  iv. Find τ_X.
  v. Find $P(X \in M)$.

(a) $M = [-3, 2]$ and $f(x) = \begin{cases} \frac{x}{4} & \text{if } x = 1, 2, 3, 4 \\ 0 & \text{otherwise.} \end{cases}$

(b) $M = \{1, 4\}$ and $f(x) = \begin{cases} \frac{x}{10} & \text{if } x = 1, 2, 3, 4 \\ 0 & \text{otherwise.} \end{cases}$

(c) $M = \{2, 4, 6\}$ and $f(x) = \begin{cases} \frac{x}{k} & \text{if } x = 1, 2, 3, 4, 5, 6 \\ 0 & \text{otherwise.} \end{cases}$

Answer the assignment for the value of k that makes $f(x)$ a discrete probability distribution.

4 We create a family of distributions, determined by a parameter r. For each r, let

$$h_r(n) = \begin{cases} \frac{k}{(n-1)!} & \text{if } n = 1, 2, \dots, r \\ 0 & \text{otherwise} \end{cases}$$

with a value of k so that h_r is a probability distribution.
(a) Calculate the value of k for $r = 4$.
(b) Let $Y \sim h_4$. Write down the probability table for Y.
(c) Calculate μ_Y.
(d) Calculate $E[Y^5]$.

Continuous Stochastic Variables

5 Which of the following functions are continuous probability distributions?

- $f(x) = \begin{cases} 1/40 & x \in [-10, 0] \\ 3/40 & x \in [0, 10] \\ 0 & x \notin [-10, 10] \end{cases}$

- $f(x) = \begin{cases} 1/2 & x \in [2, 6] \\ -1 & x \in [6, 7] \\ 0 & x \notin [4, 7] \end{cases}$

- $f(x) = \begin{cases} -\frac{1}{2}\sin(x) & x \in [\pi, 2\pi] \\ 0 & x \notin [\pi, 2\pi] \end{cases}$

- $f(x) = \begin{cases} 1/6 & x = 4 \\ 1/3 & x = 5 \\ 1/2 & x = 6 \\ 0 & x \notin \{5, 6, 7\}. \end{cases}$

6 For the functions $f(x)$ above that are continuous probability distributions: let $X \sim f(x)$, and find μ_X, σ_X^2, and $P(X \geq 5)$.

7 For the following sets M and functions f, do the following in (a)–(d).
- Graph f, and mark $[a, b]$ along the horizontal axis.
- Determine whether $f(x)$ is a continuous probability distribution. If it is, let $X \sim f$ and do the following.
  - **i.** Find μ_X.
  - **ii.** Find $Var(X)$.
  - **iii.** Find σ_X.
  - **iv.** Find τ_X.
  - **v.** Find $P(X \in M)$.

(a) $M = [0.5,\ 1.5]$ and function $f(x) = \begin{cases} 2x & x \in [0, 1] \\ 0 & \text{otherwise.} \end{cases}$

(b) $M = [0,\ 5]$ and function $f(x) = \begin{cases} \cos(x) & x \in [-\frac{\pi}{2}, \frac{\pi}{2}] \\ 0 & \text{otherwise.} \end{cases}$

(c) $M = \{0,\ 5\}$ and function $f(x) = \begin{cases} \sin(x) & x \in [-\frac{\pi}{2}, \frac{\pi}{2}] \\ 0 & \text{otherwise.} \end{cases}$

(d) $M = [0,\ 0.5]$ and function $f(x) = \begin{cases} 3x^2 & x \in [0, 1] \\ 0 & \text{otherwise.} \end{cases}$

(e) The *Christmas tree distribution* looks like its name. Its distribution function is

$$f(x) = \begin{cases} x + 1 & \text{for } x \in [-1, 0] \\ 1 - x & \text{for } x \in [0, 1] \\ 0 & \text{for all other values of } x. \end{cases}$$

i. What is μ_X?

ii. What is σ_X?

iii. What is $P(X \in (0.5, 1.5))$?

(f) The Mohawk proto distribution is given by the continuous function

$$\bar{f}(x) = \begin{cases} 1 - 10x^2 - 24x^3 - 15x^4 & \text{if } -1 \leq x < 0 \\ 1 - 10x^2 + 24x^3 - 15x^4 & \text{if } 0 \leq x \leq 1 \\ 0 & \text{if } |x| > 1. \end{cases}$$

i. Why is $\bar{f}$ not a probability distribution?

ii. Find the constant k that makes $f(x) = k \cdot \bar{f}(x)$ a probability distribution.

iii. The stochastic variable X has probability distribution f. Find

A. μ_X.

B. $P(X \in [-2, -0.5] \cup [0.5, 2])$.

8

Stochastic Variables II

CONTENTS

In Chapter 5, we looked at stochastic variables X of just one variable $x \in \mathbb{R}$. Mixed distributions have, in addition to the input variable x, a choice between different distributions. In the flow diagram in Figure 8.1, the choice is between distribution 1 and distribution 2. The extra dimension is $\{1, 2\}$ – the choice between 1 and 2.

Given n distributions, the extra dimension is $\{1, 2, \ldots, n\}$. In a limiting case, illustrated in Figure 8.2, the choice dimension gets infinitely large, maybe even as large as $\mathbb{R}$, which moves us on to two-variable stochastic variables, with probability distributions on $\mathbb{R} \times \mathbb{R}$, that is: $\mathbb{R}^2$.

8.1 Mixed Distributions*

Mixed distributions between a continuous and a discrete distribution, as illustrated in Figure 8.3, can be very useful. Purely discrete and purely continuous stochastic variables each have their type of f probability distribution function. Such probability distribution functions are very useful. However, discrete–continuous mixed distributions have cumulative distributions F only.

An elevator provides examples of both mixed waiting time and mixed state.

Example 8.1.1 Waiting time: When you arrive at work, you run up the stairs 60% of the time, and ride the elevator in the remaining 40%.

The Bayesian Way: Introductory Statistics for Economists and Engineers, First Edition.
Svein Olav Nyberg.
© 2019 John Wiley & Sons, Inc. Published 2019 by John Wiley & Sons, Inc.

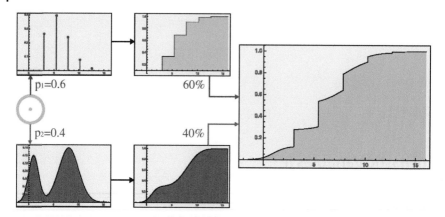

Figure 8.1 A mixed probability distribution.

The time it takes you to run up the stairs, X_t, has a continuous distribution $f(t) = \frac{1}{32\,000}(-3600 + 240t - 3t^2)$ for $20 \le t \le 60$. The time it takes to take the elevator, on the other hand, X_h, has a discrete distribution, with point probabilities $p_{25} = 0.5$, $p_{33} = 0.25$, $p_{41} = 0.125$, $p_{49} = 0.062\,5$, and $p_{57} = 0.062\,5$.

The time to get to your floor is therefore a discrete–continuous mixed stochastic variable X which is a mix between X_t and X_h with probabilities $w_t = 0.6$ and $w_h = 0.4$. □

Example 8.1.2   State of the elevator: You want to know where the elevator might be. Some of the time, it stands still on one floor. The rest of the time, it is moving between floors. Its position, given that it stands still, is a discrete stochastic variable. Its position, given that it is moving, is a continuous variable. The unconditioned total picture of where the elevator is at any given time is therefore a discrete–continuous mixed stochastic variable. □

Example 8.1.3   The weather: The amount of rain on a rainy day follows a continuous distribution. When it doesn't rain, the amount of rain is a discrete

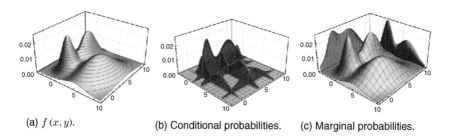

(a) $f(x, y)$.

(b) Conditional probabilities.

(c) Marginal probabilities.

Figure 8.2 We will be studying two-variable distributions in Section 8.2.

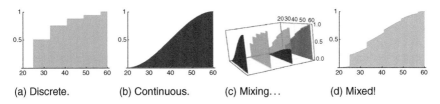

(a) Discrete.   (b) Continuous.   (c) Mixing...   (d) Mixed!

Figure 8.3 Discrete–continuous mixed cumulative distribution.

stochastic variable whose value always is 0. So the amount of rain on any given day follows a discrete–continuous mixed distribution. □

Example 8.1.4 Water flow: Some watercourses have both dry and wet periods. The water flow of these is a discrete–continuous mixed stochastic variable. □

So now that we know mixed stochastic variables exist and are relevant, how do we calculate them? The answer is to divide the universe of the problem into mutually exclusive conditions $\theta_1, \dots, \theta_n$. For each condition θ_k, we find the stochastic variable X_k, its cumulative distribution function F_k, and a weight $w_k = P(\theta_k)$ which is the probability that the condition is realized.

If X is the height of a randomly chosen Norwegian, the conditions may be M(ale) and F(emale), making X a mixture of X_M and X_F, with weights w_M and w_F.

If X is a discrete–continuous mix, the primary conditioning is discrete (with stochastic variable X_d) and continuous (with stochastic variable X_c).

Rule 8.1.5 *If X is a weighted mix of stochastic variables $X_1, \dots, X_n$, with weights $w_1, \dots, w_n$, then*

$$F_X(x) = \sum_{k=1}^{n} w_k \cdot F_k(x)$$

$$f_X(x) = \sum_{k=1}^{n} w_k \cdot f_k(x) \text{ (iff all the } X_k \text{ are of the same type)}$$

$$E[X^m] = \sum_{k=1}^{n} w_k \cdot E[X_k^m]$$

$$\text{Var}(X) = \sum_{k=1}^{n} w_k \cdot \text{Var}(X_k) + \sum_{k=1}^{n} w_k(\mu_k - \mu)^2.$$

For 2 variables X_1, X_2, we may use the simplified formula

$$\text{Var}(X) = w_1 \cdot \text{Var}(X_1) + w_2 \cdot \text{Var}(X_2) + w_1 w_2(\mu_1 - \mu_2)^2.$$

Example 8.1.6 For the elevator example (Example 8.1.1), find

1) $P(X \le 40)$;
2) $E[X]$;
3) $\text{Var}(X)$.

Answer: We answer the corresponding questions for the two components, X_d and X_c, and combine according to Rule 8.1.5:

1)
$$F_c(x) = \int_{20}^{x} f(t)\, dt = \frac{1}{32\,000}(-x^3 + 120x^2 - 3\,600x + 32\,000)$$

$$F_c(40) = \frac{1}{2}$$

$$F_d(40) = \sum_{x \le 40} f_d(x) = p_{25} + p_{33} = 0.75$$

$$P(X \le 40) = F(40) = 0.6 \cdot F_c(40) + 0.4 \cdot F_d(40) = \underline{0.6}.$$

2)
$$E[X_c] = \int_{20}^{60} x \cdot \frac{1}{32\,000}(-3\,600 + 240x - 3x^2)\, dx = 40$$

$$E[X_d] = \sum_{\text{all } x} x \cdot f_d(x) = 32.5$$

$$E[X] = F(40) = 0.6 \cdot E[X_c] + 0.4 \cdot E[X_d] = \underline{37}.$$

3)
$$E[X_c^2] = \int_{20}^{60} x^2 \cdot \frac{1}{32\,000}(-3600 + 240x - 3x^2)\, dx = 1680$$

$$\text{Var}(X_c) = E[X_c^2] - E[X_c]^2 = 80$$

$$E[X_d^2] = \sum_{\text{all } x} x^2 \cdot f_d(x) = 1148$$

$$\text{Var}(X_d) = E[X_d^2] - E[X_d]^2 = 91.75$$

$$\text{Var}(X) = 0.6 \cdot 80 + 0.4 \cdot 91.75 + 0.6 \cdot 0.4(40 - 32.5)^2 = \underline{98.2}. \qquad \square$$

We do, however, seldom find the stochastic variable readily decomposed for us, so have to do that job ourselves. We decompose into discrete and continuous as follows.

- The decomposition is $X = w_d \cdot X_d + w_c \cdot X_c$.
- Find the points x_k with positive point probability p_k. Let $p = \sum_k p_k$. Then, $X_d \sim f_d$, where $f_d(x_k) = \frac{p_k}{p}$, with weight $w_d = p$.
- The continuous part X_c has a continuous (proto) probability distribution $f(x)$ on an interval I. Then, $X_c \sim f_c$, where $f_c(x) = \frac{1}{\int_I f(x)} f(x)$, with weight $w_c = 1 - w_d$.

Example 8.1.7 You are sponsoring a charitable sack run, and have pledged 1 Norwegian crown per meter, but have promised to pay a minimum of 200 crowns, but a maximum of 1000 crowns. The length of the sack run, in meters, has a probability distribution $f(x) = 7.5 \cdot 10^{-13} x(2000 - x)^2$ in the interval $I = [0, 2000]$. Decompose your pledged amount X into a discrete part X_d and a continuous part X_c, and find the weights w_d and w_c. Then, find your expected sponsorship amount.

Answer: For 200 meters or less, you pay 200 crowns, and the probability of that is $p_{200} = \int_0^{200} f(x)\,dx = 0.0523$. For 1000 meters or more, you pay 1000 crowns, and the probability of that is $p_{1000} = \int_{1000}^{2000} f(x)\,dx = 0.3125$. These are the two points of our discrete component X_d, and the total probability of that condition is $w_d = p_{200} + p_{1000} = 0.0523 + 0.3125 = 0.3648$. Then, $w_c = 1 - w_d = 0.6352$, and

$$X_d \sim f_d(x) = \begin{cases} \dfrac{0.0523}{0.3648} = 0.143366 & \text{for } x = 200 \\[2mm] \dfrac{0.3125}{0.3648} = 0.856634 & \text{for } x = 1\,000 \end{cases}$$

$$E[X_d] = 0.143366 \cdot 200 + 0.856634 \cdot 1\,000 = 885.307$$

$$X_c \sim f_c(x) = \frac{f(x)}{w_c} = 1.1807 \cdot 10^{-12} x(2\,000 - x)^2 \text{ for } 200 < x < 1\,000$$

$$E[X_c] = \int_{200}^{1\,000} x \cdot f(x)\,dx = 618.942$$

$$E[X] = 0.3648 \cdot 885.307 + 0.6352 \cdot 618.942 = \underline{716.112}. \qquad \square$$

Mixed distributions may also be mixes where all components are continuous or all components are discrete. In such cases, we are able to work with a probability distribution f, and not just the cumulative distribution F.

Example 8.1.8 The Loppen river is known for good fishing of Salmon, char, and trout. The weights of the fish of each species are Normally distributed (see Section 10.1) with each their mean weights μ and standard deviations σ. The catches are distributed between the species as follows:

- 63.7% salmon, with $\mu = 4.2$ and $\sigma = 1.2$;
- 22.4% char, with $\mu = 1.7$ and $\sigma = 0.5$;
- 13.9% trout, with $\mu = 0.95$ and $\sigma = 0.1$.

Let X be the weight of a random catch in Loppen, and let the components X_s, X_c, X_t be the weights given that the catch is respectively a salmon, a char, and a trout. You can easily calculate and find that $\mu_X = 3.18825$ and that $\text{Var}(X) = 2.81922$. We are going to find the probability density f of X. This is illustrated

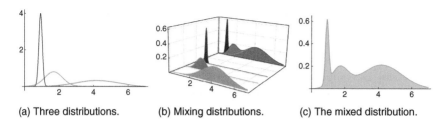

(a) Three distributions. (b) Mixing distributions. (c) The mixed distribution.

Figure 8.4 A mix of three continuous distributions of fish weights.

in Figure 8.4. For the calculation, we have to look a bit ahead to Section 10.1 to find the formula for the Normal distribution, and we get

$$f(x) = w_s f_s(x) + w_c f_c(x) + w_t f_t(x)$$

$$= \frac{0.637}{1.2\sqrt{2\pi}} e^{-\frac{(x-4.2)^2}{2 \cdot 1.2^2}} + \frac{0.224}{0.5\sqrt{2\pi}} e^{-\frac{(x-1.7)^2}{2 \cdot 0.5^2}} + \frac{0.139}{0.1\sqrt{2\pi}} e^{-\frac{(x-0.95)^2}{2 \cdot 0.1^2}}.$$

□

We conclude our investigation of mixed distributions by looking at mixes with not just 2 or 3, or even countably many components, but mixes with a continuum of components. Then, the point weights w_k are replaced by probability densities $w(k)$, and the sums are replaced by integrals, since this is a transition from discrete to continuous stochastic variables.

Rule 8.1.9 *Let X be a mix of stochastic variables X_y of the same type, where the weights are given by a continuous probability distribution $w(y)$. Then,*

$$F_X(x) = \int_{y=-\infty}^{\infty} w(y) \cdot F_y(x)\, dy$$

$$f_X(x) = \int_{y=-\infty}^{\infty} w(y) \cdot f_y(x)\, dy$$

$$E[X^m] = \int_{y=-\infty}^{\infty} w(y) \cdot E[X_k^m]\, dy$$

$$\text{Var}(X) = \int_{y=-\infty}^{\infty} w(y) \cdot \left(\text{Var}(X_y) + (\mu_y - \mu)^2\right) dy.$$

Example 8.1.10 Θ_t, the position of a fault in a flywheel, given that the fault arose at time t, has distribution $f_t(\theta) = \frac{1}{1\pi}(1 + \cos(\theta - t))$. The time that the fault arises, t, has distribution $g(t) = e^{-t}$ for $t \geq 0$. Then Θ, the position of the fault, is a mixed distribution with continuously distributed weights $w_t = g(t)$ and individual components Θ_t. Find the distribution of Θ.

Answer:

$$f_\Theta(\theta) = \int_{t=0}^{\infty} e^{-t} \cdot \frac{1}{1\pi}(1 + \cos(\theta - t))\,dt$$

$$= \frac{\sin(\theta) + \cos(\theta) + 2}{4\pi} = \frac{2 + \sqrt{2}\cos\left(\theta - \frac{\pi}{4}\right)}{4\pi}.$$

$\square$

When we are looking at mixed probability distributions, we are essentially looking at two variables: the index of the weights, w, and the variable y. It is then natural to proceed to the full-blooded theory of two- and multi-variable probability distributions.

8.2 Two- and Multi-variable Probability Distributions*

Definition 8.2.1 *Z is a two-dimensional stochastic variable when its sample space is $U = \mathbb{R}^2$. We may then write Z decomposed into its two dimensions as $Z = \langle X, Y \rangle$. The cumulative probability distribution is F_Z, with*

$$F_Z(x, y) = P(\{X \le x\} \cap \{Y \le y\}).$$

We call $f_Z(x, y) = f_{XY}(x, y)$ the *simultaneous distribution* of X and Y. The relation between the distribution f and the cumulative distribution F has already been explored in Section 8.2.1 for discrete X and Y, and Section 8.2.2 for continuous X and Y.

The variable X has a *marginal* probability distribution, which is simply its probability distribution when we factor out Y and look at X alone. The marginal *cumulative* probability distribution of X is $F_X(x) = P(X \le x) = F_Z(x, \infty)$. We find the marginal probability distribution f_X from F_X, by taking the differences (discrete) or by differentiation (continuous). The same goes for Y.

For a given value x of X, we have the *conditional* stochastic variable $Y_{|X=x}$, or $Y_{|x}$ for short. It has the *conditional* distribution $f_{|x}(y) = f_Z(x, y)/f_X(x)$.

8.2.1 Discrete X and Y

Rule 8.2.2 *If (X, Y) is a pair of discrete stochastic variables with joint point probabilities $p_{ab} = f_{XY}(a, b)$,*

$$p_{ab} = P(X = a, Y = b)$$

$$P((x, y) \in C) = \sum_{(x,y) \in C} p_{xy},$$

then the marginal probability distributions are

$$f_X(a) = \sum_y f_{XY}(a, y)$$

$$f_Y(b) = \sum_x f_{XY}(x, b),$$

whereas the conditional probability distributions are

$$f_{|Y=y}(x) = \frac{f_{XY}(x, y)}{f_Y(y)}$$

$$f_{|X=x}(y) = \frac{f_{XY}(x, y)}{f_X(x)}.$$

Example 8.2.3 Sam flips four coins. Bard wins the ones that land heads. He then flips the remaining coins, and Frederick wins the heads of this new round. Let X and Y be the respective number of coins Bard and Frederick win.

(1) Make a table of the values of f_{XY}, and include the marginal probabilities f_X and f_Y.
(2) Find the conditional probabilities $f_{|Y=2}(x)$; write them up in a table.
(3) Find the probability that Bard and Frederick won one coin each.
(4) Find the probability that Bard and Frederick won equally many coins.
(5) Find the probability that Bard and Frederick won three coins in total.

Answer:

(1) We first need to find the values $f_{XY}(a, b)$, and will here show how to calculate one such value, $f_{XY}(2, 1)$: the probability of two heads in four tosses is $\binom{4}{2} \times 0.5^3 = 0.375$. After Bard has won two coins, there are $4 - 2 = 2$ coins left. The probability of one heads in two tosses is $\binom{2}{1} \times 0.5^2 = 0.5$. This means $f_{XY}(2, 1) = P(X = 2, Y = 1) = 0.375 \times 0.5 = 0.187\,5$.

Corresponding calculations for other values of a and b first give the probability that $X = a$, that a of the five tosses land heads: $\binom{4}{a} \times 0.5^4$. We then find the probability that $Y = b$, that b of the next $4 - a$ tosses land heads: $\binom{4-a}{b} \times 0.5^{4-a}$. The formula to fill the distribution table for f_{XY} is thus

$$f_{XY}(a, b) = \binom{4}{a} \times 0.5^4 \times \binom{4-a}{b} \times 0.5^{4-a}.$$

f_{XY}	$Y = 0$	$Y = 1$	$Y = 2$	$Y = 3$	$Y = 4$	f_X
$X = 0$	0.003 906 25	0.015 625	0.023 437 5	0.015 625	0.003 906 25	0.0625
$X = 1$	0.031 25	0.093 75	0.093 75	0.031 25	0	0.25
$X = 2$	0.093 75	0.187 5	0.093 75	0	0	0.375
$X = 3$	0.125	0.125	0	0	0	0.25
$X = 4$	0.0625	0	0	0	0	0.0625
f_Y	0.316 406	0.421 875	0.210 938	0.046 875	0.003 906 25	

(2) Let us take a closer look at f_{XY} for $Y = 2$, and divide by $f_Y(2)$ to get $f_{|Y=2}(x)$:

x	$f_{XY}(x, 2)$
0	0.023 4375
1	0.093 75
2	0.093 75
3	0
4	0
$f_Y(2)$	0.210 938

x	$f_{|Y=2}(x)$
0	$0.023\,4375/0.210\,938 = 1/9$
1	$0.093\,75/0.210\,938 = 4/9$
2	$0.093\,75/0.210\,938 = 4/9$
3	$0/0.210\,938 = 0$
4	$0/0.210\,938 = 0$
$\sum$	1

(3) The probability that Bard and Frederick each got one coin is $P(X = 1, Y = 1) = f_{XY}(1, 1)$. We read this probability in the first table, where is says that $f_{XY}(1, 1) = \underline{0.093\,75}$.

(4) The probability that they got the same number of coins, is

$$P(X = Y) = f_{XY}(0, 0) + f_{XY}(1, 1) + f_{XY}(2, 2) + f_{XY}(3, 3) + f_{XY}(4, 4)$$
$$= 0.003\,906\,25 + 0.093\,75 + 0.093\,75 + 0 + 0 = \underline{0.191\,406}.$$

(5) The probability that they got three coins in total is

$$P(X + Y = 3) = f_{XY}(0, 3) + f_{XY}(1, 2) + f_{XY}(2, 1) + f_{XY}(0, 3)$$
$$= 0.125 + 0.1875 + 0.093\,75 + 0.015\,625 = \underline{0.421\,875}.$$

$\square$

Rule 8.2.4   Some handy rules for calculations

- Given the point probabilities $f_{XY}(x, y)$, we find the cumulative probability $F_{XY}(x, y)$ as follows:

$$F_{XY}(i, j) = \sum_{\substack{a \le i \\ b \le j}} f_{XY}(a, b).$$

In a table, this means that $F_{XY}(i, j)$ is the sum of all the values in the rectangle spanned by (i, j) and the upper left corner of the f_{XY} table, as shown in Figure 8.5.

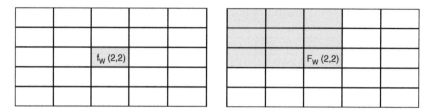

Figure 8.5 Finding the cumulative distribution.

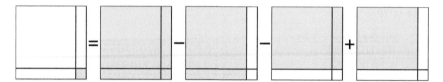

Figure 8.6 Finding the probability distribution.

- Conversely, we may find the point probabilities from the cumulative probabilities by calculating differences, as shown in Figure 8.6.
- Finally, the following rule helps us calculate the probabilities of general rectangles when we know the cumulative distribution F_{XY}:

$$P(a < X < A, b < Y < B) = F_{XY}(A, B) - (F_X(a, B)$$
$$+ F_Y(A, b) - F_{XY}(a, b)).$$

If there is no upper limit A or B, use ∞ as the upper limit, and that $F_Z(x, \infty) = F_X(x)$ and $F_Z(\infty, y) = F_Y(y)$.

Example 8.2.5 (Continuation of Example 8.2.3)

(1) Make a table of the cumulative values F_{XY}, F_X, and F_Y.
(2) Find the probability that Bard and Frederick won at most one coin each.
(3) Find the probability that Bard won at least one coin and that Frederick won at least two.

(1) The cumulative table is then

F_{XY}	$Y = 0$	$Y = 1$	$Y = 2$	$Y = 3$	$Y = 4$	F_X
$X = 0$	0.003 906 25	0.019 5313	0.042 9688	0.058 5938	0.0625	0.0625
$X = 1$	0.035 1563	0.144 531	0.261 719	0.308 594	0.3125	0.3125
$X = 2$	0.128 906	0.425 781	0.636 719	0.683 594	0.6875	0.6875
$X = 3$	0.253 906	0.675 781	0.886 719	0.933 594	0.9375	0.9375
$X = 4$	0.316 406	0.738 281	0.949 219	0.996 094	1.	1.
F_Y	0.316 406	0.738 281	0.949 219	0.996 094	1.	

(2) The probability that Bard and Frederick won *at most* one coin each is $P(X \leq 1, Y \leq 1) = F_{XY}(1,1)$, which we find in the cumulative table: $F_{XY}(1,1) = \underline{0.144\,531}$.

(3) Find the probability that Bard won at least one coin and that Frederick won at least two. We illustrate this by coloring the f_{XY} table:

f_{XY}	$Y = 0$	$Y = 1$	$Y = 2$	$Y = 3$	$Y = 4$
$X = 0$	0.003\,906\,25	0.015\,625	0.023\,437\,5	0.015\,625	0.003\,906\,25
$X = 1$	0.031\,25	0.093\,75	0.093\,75	0.031\,25	0.
$X = 2$	0.093\,75	0.187\,5	0.093\,75	0.	0.
$X = 3$	0.125	0.125	0.	0.	0.
$X = 4$	0.062\,5	0.	0.	0.	0.

In this particular case, summing the colored rectangles is an easy job, since only three of them are non-zero. But we will employ the general technique in Rule 8.2.4, and then we get

$$P(X \geq 1, Y \geq 2) = P(X > 0, Y > 1)$$
$$= 1 - \left(F_X(0) + F_Y(1) - F_{XY}(0,1)\right) = \underline{0.218\,75}. \quad \square$$

8.2.2 Continuous *X* and *Y*

Wind speed is a typical example of a continuous stochastic variable that has both an X and a Y component. If we regard wind as a three-dimensional phenomenon, it also has a Z component. Wind speed has this in common with any other geometry related stochastic variable: it may be decomposed into components by coordinates. Conversely, any pair (X, Y) of continuous stochastic variables has a joint cumulative probability distribution F_{XY}, and a joint probability density f_{XY}.

Rule 8.2.6 *If (X, Y) is a pair of continuous variables, with probability densities f and cumulative densities F, then*

$$F_{XY}(A, B) = P(X \leq A \text{ and } Y \leq B) = \int_{-\infty}^{A} \int_{-\infty}^{B} f_{XY}(x, y)$$
$$F_X(x) = F_{XY}(x, \infty) \quad \text{and} \quad F_Y(y) = F_{XY}(\infty, y).$$

Notice that it does not matter if we include the end points or not, since for continuous stochastic variables all point probabilities are zero. The marginal probability distributions are

$$f_X(x) = \int_{-\infty}^{\infty} f_{XY}(x, y) \, dy = \frac{d}{dx} F_X(x)$$
$$f_Y(y) = \int_{-\infty}^{\infty} f_{XY}(x, y) \, dx = \frac{d}{dy} F_Y(y)$$

whereas the conditional probability distributions are

$$f_{|Y=y}(x) = \frac{f_{XY}(x, y)}{f_Y(y)}$$

$$f_{|X=x}(y) = \frac{f_{XY}(x, y)}{f_X(x)}$$

when the denominator differs from 0.

See Figure 8.7 for different illustrations of bivariate probability distributions. Before we proceed, we will write up the rules for two continuous stochastic variables, analogous to Rule 8.2.4:

Rule 8.2.7 *If* (X, Y) *is a pair of continuous stochastic variables, then*

$$P(X \le a, Y \le b) = F_{XY}(a, b) = \int_{-\infty}^{b} \int_{-\infty}^{a} f_{XY}(x, y) \, dx \, dy$$

$$P(X > a, Y > b) = 1 - F_X(a) - F_Y(b) + F_{XY}(a, b)$$

$$P(W \in \langle a, b] \times \langle c, d]) = F_{XY}(b, d) - F_{XY}(a, d) - F_{XY}(b, c) - F_{XY}(a, c).$$

For general areas, the rule is stated in Rule 8.2.8. Notice that this rule requires multivariate integration.

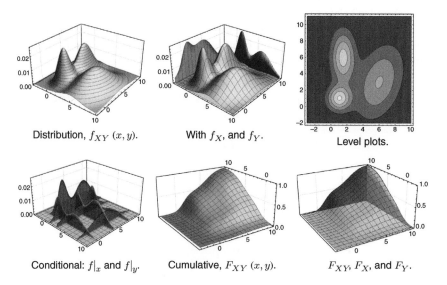

Distribution, $f_{XY}(x, y)$.　　With f_X, and f_Y.　　Level plots.

Conditional: $f|_x$ and $f|_y$.　　Cumulative, $F_{XY}(x, y)$.　　F_{XY}, F_X, and F_Y.

Figure 8.7 Illustrations of bivariate probability distributions.

Rule 8.2.8* *If $D \subset \mathbb{R}^2$, then*

$$P((X, Y) \in D) = \iint_D f_{XY}(x, y) \, dA.$$

Example 8.2.9 Loading the HTML code of Sondromatic Ltd's web pages takes between 0 and 1 milliseconds. The logo is loaded from a different server, and this loading also takes between 0 and 1 milliseconds. The loading times are a stochastic variable pair, X and Y, with joint probability distribution $f_{XY}(x, y) = 6(y - x)^2$ in the unit square $[0, 1] \times [0, 1]$.

1) Find the cumulative probability distribution F_{XY}.
2) Find the probability that the HTML code loads in less than 0.5 milliseconds *and* that the logo loads in less than 0.7 milliseconds.
3) Find the probability that the sum of the HTML code and logo loading times is less than 1 millisecond.
4) Find the marginal cumulative probability distributions F_X and F_Y.
5) Find the marginal probability distributions f_X and f_Y.

Answers:

(1) $F_{XY}(x, y) = \displaystyle\int_{-\infty}^{y} \int_{-\infty}^{x} f_{XY}(z, w) \, dz \, dw$

$\qquad\qquad = \displaystyle\int_{0}^{y} \int_{0}^{x} 6(w - z)^2 \, dz \, dw$ (since $X, Y \geq 0$)

$\qquad\qquad = \underline{\underline{2x^3y - 3x^2y^2 + 2xy^3}}.$

(2) We are finding the probability that $0 \leq X \leq 0.5$ and $0 \leq Y \leq 0.7$. See Figure 8.8. The simplest method is using F_{XY}.

$$P(X \leq 0.5, Y \leq 0.7) = F_{XY}(0.5, 0.7) = \underline{\underline{0.1505}}.$$

(3) Here, we want to find the probability that $X + Y \leq 1$. See Figure 8.9. Since this area is not rectangular, we can't use the cumulative distribution F_{XY}. We recall that $0 \leq X, Y \leq 1$, and integrate f_{XY} over indicated area.

$$P(X + Y \leq 1) = \int_{0}^{1} \int_{0}^{1-x} f_{XY} \, dy \, dx$$

$$= \int_{0}^{1} \int_{0}^{1-x} 6(y - x)^2 \, dy \, dx = \underline{\underline{\tfrac{1}{2}}}.$$

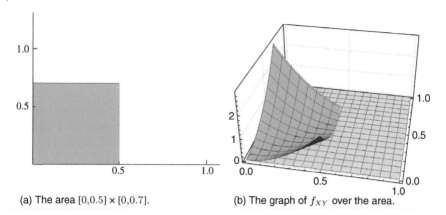

(a) The area $[0,0.5] \times [0,0.7]$.

(b) The graph of f_{XY} over the area.

Figure 8.8 f_{XY} over a rectangular area.

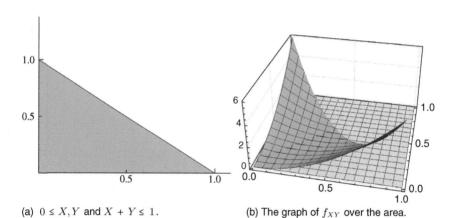

(a) $0 \leq X, Y$ and $X + Y \leq 1$.

(b) The graph of f_{XY} over the area.

Figure 8.9 f_{XY} over a triangular area.

(4) $F_X(x) = F_{XY}(x, \infty) = F_{XY}(x, 1)$ (since $Y \leq 1$)

$\quad = 2x^3 \cdot 1 - 3x^2 \cdot 1^2 + 2x \cdot 1^3 = \underline{\underline{2x^3 - 3x^2 + 2x}}$

$F_Y(y) = F_{XY}(\infty, y) = F_{XY}(1, y)$ (since $X \leq 1$)

$\quad = 2 \cdot 1^3 y - 3 \cdot 1^2 y^2 + 2 \cdot 1 y^3 = \underline{\underline{2y - 3y^2 + 2y^3}}.$

(5) $f_X(x) = \dfrac{\partial}{\partial x} F_X(x) = \dfrac{\partial(2x^3 - 3x^2 + 2x)}{\partial x} = \underline{\underline{6x^2 - 6x + 2}}$

$f_Y(y) = \dfrac{\partial}{\partial y} F_Y(y) = \dfrac{\partial 2y - 3y^2 + 2y^3}{\partial y} = \underline{\underline{2 - 6y + 6y^2}}.$

$\square$

Example 8.2.10   In Windy Bay, the wind always comes from some southeastern direction. Let $Z = (X, Y)$, where X is the eastern component, and Y is the southern. The joint probability density is $f_Z(x, y) = 4xye^{-(x^2+y^2)}$ for $x, y \in [0, \infty)$.

(1) Find the cumulative probability density $F_Z(x, y)$.
(2) Find the cumulative marginal probability densities $F_X(x)$ and $F_Y(y)$.
(3) Find the marginal probability densities $f_X(x)$ and $f_Y(y)$.
(4) Find the probability that the absolute wind speed is less than r.

Answers:

(1) $F_Z(x, y) = \displaystyle\int_{-\infty}^{x} \int_{-\infty}^{y} f_Z(z, w) \, dw \, dz$

$\qquad = \displaystyle\int_{0}^{x} \int_{0}^{y} f_Z(z, w) \, dw \, dz$ (since $f_Z(x, y) = 0$ when $x, y < 0$)

$\qquad = \displaystyle\int_{0}^{x} \int_{0}^{y} 4ze^{-z^2} we^{-w^2} \, dw \, dz = 4 \int_{0}^{x} ze^{-z^2} \int_{0}^{y} we^{-w^2} \, dw \, dz$

$\qquad = 4 \displaystyle\int_{0}^{x} ze^{-z^2} \, dz \times \int_{0}^{y} we^{-w^2} \, dw$ (see the calculation in 7.3.6)

$\qquad = 4 \times \dfrac{1 - e^{-x^2}}{2} \times \dfrac{1 - e^{-y^2}}{2} = \underline{\underline{\left(1 - e^{-x^2}\right)\left(1 - e^{-y^2}\right)}}.$

(2) $F_X(x) = F_Z(x, \infty) = 1 - e^{-x^2}.$
$\quad\; F_Y(y) = F_Z(\infty, y) = 1 - e^{-y^2}.$
(3) We may find these magnitudes in either of two ways, and will show both.
  • Method 1, finding f_X and f_Y by differentiating F_X and F_Y:

$$f_X(x) = \tfrac{\partial}{\partial x} F_X(x) = \tfrac{\partial}{\partial x}\left(1 - e^{-x^2}\right) = \underline{\underline{2xe^{-x^2}}}.$$
$$f_Y(y) = \tfrac{\partial}{\partial y} F_Y(y) = \tfrac{\partial}{\partial y}\left(1 - e^{-y^2}\right) = \underline{\underline{2ye^{-y^2}}}.$$

  • Method 2, finding f_X and f_Y by integrating f_{XY}:

$$f_X(x) = \int_{-\infty}^{\infty} f_Z(x, y) \, dy = \int_{0}^{\infty} 4xye^{-(x^2+y^2)} \, dy.$$
$$= 4xe^{-x^2} \int_{0}^{\infty} ye^{-y^2} \, dy = 4xe^{-x^2} \times \tfrac{1}{2} = \underline{\underline{2xe^{-x^2}}}$$
$$f_X(x) = \int_{-\infty}^{\infty} f_Z(x, y) \, dx = \underline{\underline{2ye^{-y^2}}}.$$

(4) Here, we employ Rule (8.2.8) and integration from multivariate calculus:

$$P(|Z| \le r) = P(X^2 + Y^2 \le r^2) = \iint_{x^2+y^2 \le r} f_Z(x, y) \, dA$$

$$= \int_0^r \int_0^{\sqrt{r^2-x^2}} 4xye^{-(x^2+y^2)} \, dy \, dx$$

$$= \int_0^r 2xe^{-x^2} \left(\int_0^{\sqrt{r^2-x^2}} 2ye^{-y^2} \, dy \right) dx$$

$$= \int_0^r 2xe^{-x^2} \left(1 - e^{x^2-r^2} \right) dx = \int_0^r 2x \left(e^{-x^2} - e^{-r^2} \right) dx$$

$$= \underline{\underline{1 - (1 + r^2)e^{-r^2}}}.$$

□

8.2.3 Covariance and Correlation

Covariance and correlation are as important to studying pairs of stochastic variables as they are to the treatment of pairwise data. They tell us how tight and close the two magnitudes X and Y are.

Definition 8.2.11 *The* covariance *("co-variance")*

$$\sigma_{XY} = E[(X - \mu_X)(Y - \mu_Y)].$$

As is so often the case, the following formula for practical calculation differs from the formula used for the definition, and is simpler.

Rule 8.2.12 $\sigma_{XY} = E[XY] - E[X] \cdot E[Y]$, *where*

$$E[XY] = \begin{cases} \int_{\mathbb{R}^2} xyf_{XY}(x, y) \, dA & \text{(continuous)} \\ \sum_{x,y} xyf_{XY}(x, y) & \text{(discrete).} \end{cases}$$

We will first look at the following discrete example.

Example 8.2.13 (Continuation of Examples 8.2.3 and 8.2.5)

$$\mu_X = \sum_x xf_X(x)$$

$$= 0 \times 0.062\,5 + 1 \times 0.25 + 2 \times 0.375 + 3 \times 0.25 + 4 \times 0.062\,5 = 2$$

$$\mu_Y = \sum_y yf_Y(y) = 0 \times 0.316\,406 + 1 \times 0.421\,875 + 2 \times 0.210\,938$$

$$+ 3 \times 0.046\,875 + 4 \times 0.003\,906\,25 = 1$$

$$E[XY] = \sum_{x,y} xyf_{XY}(x,y) = \sum_x \left(x \times \sum_y yf_{XY}(x,y) \right) = 0 \times \left(\sum_y yf_{XY}(0,y) \right)$$

$$+ 1 \times (0 \times 0.031\,25 + 1 \times 0.093\,75 + 2 \times 0.093\,75 + 3 \times 0.031\,25 + 4 \times 0)$$

$$+ 2 \times (0 \times 0.093\,75 + 1 \times 0.187\,5 + 2 \times 0.093\,75 + 3 \times 0 + 4 \times 0)$$

$$+ 3 \times (0 \times 0.125 + 1 \times 0.125 + 2 \times 0 + 3 \times 0 + 4 \times 0)$$

$$+ 4 \times (0 \times 0.062\,5 + 1 \times 0 + 2 \times 0 + 3 \times 0 + 4 \times 0)$$

$$= 1.5$$

$$\sigma_{XY} = E[XY] - \mu_X\mu_Y = 1.5 - 2 \times 1 = \underline{\underline{-0.5}}. \qquad \square$$

Example 8.2.14 (Continuation of Example 8.2.9) What is the covariance between the load time of the HTML code and of the logo of Sondromatics' front page?

Answer: We calculate μ_X, μ_Y, and $E[XY]$, and insert the numbers into formula (8.2.12) for σ_{XY}:

$$\mu_X = E[X] = \int_{-\infty}^{\infty} xf_X(x)\,dx = \int_0^1 x(6x^2 - 6x + 2)\,dx = \frac{1}{2}$$

$$\mu_Y = E[Y] = \int_{-\infty}^{\infty} yf_Y(y)\,dx = \int_0^1 y(2 - 6y + 6y^2)\,dy = \frac{1}{2}$$

$$E[XY] = \iint_{\mathbb{R}} xyf_{XY}(x,y)\,dA = \int_0^1 \int_0^1 6xy(y-x)^2\,dx\,dy = \frac{1}{6}$$

$$\sigma_{XY} = E[XY] - \mu_X\mu_Y = \frac{1}{6} - \frac{1}{2} \cdot \frac{1}{2} = \underline{\underline{-\frac{1}{12}}}. \qquad \square$$

Example 8.2.15 (Continuation of Example 8.2.10) What is the covariance between the southern and the eastern wind components?

Answer: We will make use of the fact that $\int_0^{\infty} z^2 e^{-z^2}\,dz = \frac{\sqrt{\pi}}{4}$. This is a tough integral, so do the calculation by means of a calculator or other such tool. We used Mathematica®, to obtain an exact answer.

$$\mu_X = \int_0^{\infty} xf_X(x)\,dx = \int_0^{\infty} x \cdot 2xe^{-x^2}\,dx = \frac{\sqrt{\pi}}{2}$$

$$\mu_Y = \int_0^{\infty} yf_Y(y)\,dy = \int_0^{\infty} y \cdot 2ye^{-y^2}\,dy = \frac{\sqrt{\pi}}{2}$$

$$E[XY] = \int_0^\infty \int_0^\infty xy f_{XY}(x,y)\, dx\, dy = \int_0^\infty \int_0^\infty xy \cdot 4xye^{-(x^2+y^2)}\, dx\, dy$$

$$= 4 \int_0^\infty x^2 e^{-x^2}\, dx \cdot \int_0^\infty y^2 e^{-y^2}\, dy$$

$$= 4 \cdot \frac{\sqrt{\pi}}{4} \cdot \frac{\sqrt{\pi}}{4} = \frac{\pi}{4}$$

$$\sigma_{XY} = E[XY] - \mu_X \mu_Y = \frac{\pi}{4} - \frac{\sqrt{\pi}}{2} \cdot \frac{\sqrt{\pi}}{2} = \underline{\underline{0}}.$$ □

The covariance tells us how the two variables are related, but its primary function is instrumental rather than being an end measure in itself. If we were to scale the variables X and Y by a factor of r, we do that by scaling σ_{XY} by a factor of r^2. For an and-in-itself measure of relation, we would like it to be invariant when we scale the variables, or convert from (say) millimeters to inches. The *correlation* ρ_{XY} is such a measure.

Definition 8.2.16   *The* correlation ρ_{XY} *is given by*

$$\rho_{XY} = \frac{\sigma_{XY}}{\sigma_X \sigma_Y}$$

and is a number between −1 *and* 1 *expressing the strength of the linear relation between X and Y.*

In addition to our previous calculations, we need the standard deviations σ_X and σ_Y.

Example 8.2.17   (Continuation of Example 8.2.14) What is the correlation between the loading times for the HTML code and for the logo at Sondromatics' front page?

Answer: We calculate the variance and standard deviations of X and Y, and insert into formula (8.2.16) for ρ_{XY}:

$$E[X^2] = \int_{-\infty}^\infty x^2 f_X(x)\, dx = \int_0^1 x^2(6x^2 - 6x + 2)\, dx = \frac{11}{30}$$

$$\mathrm{Var}(X) = E[X^2] - \mu_X^2 = \frac{11}{30} - \left(\frac{1}{2}\right)^2 = \frac{7}{60}$$

$$\sigma_X = \sqrt{\mathrm{Var}(X)} = \sqrt{\frac{7}{60}}$$

$$E[Y^2] = \int_{-\infty}^\infty y^2 f_Y(y)\, dy = \int_0^1 y^2(2 - 6y + 6y^2)\, dx = \frac{11}{30}$$

$$\mathrm{Var}(Y) = E[Y^2] - \mu_Y^2 = \frac{11}{30} - \left(\frac{1}{2}\right)^2 = \frac{7}{60}$$

$$\sigma_Y = \sqrt{\text{Var}(Y)} = \sqrt{\frac{7}{60}}$$

$$\rho_{XY} = \frac{\sigma_{XY}}{\sigma_X \sigma_Y} = \frac{-\frac{1}{12}}{\sqrt{\frac{7}{60}} \cdot \sqrt{\frac{7}{60}}} = \underline{\underline{-\frac{5}{7}}}.$$

$\square$

We conclude this section with the following definition.

Definition 8.2.18 *Z is an n-dimensional stochastic variable when its sample space is $U = \mathbb{R}^n$. We write $Z = \langle Z_1, \ldots, Z_n \rangle$, where $Z \sim F$ and $Z_i \sim F_i$ for each of $i = 1, \ldots, n$.*

8.2.4 Independence

The concepts *conditional probability* and *independent sets* translate seamlessly into the context of stochastic variables. Note that even though the definition of independence is derived from the concept of independence for sets, (5.4.3) and (5.4.4), we prefer the following characterization, which we will let be our working definition of independence for stochastic variables.

Definition 8.2.19 *If X and Y are independent stochastic variables, and $Z = (X, Y)$, then $F_z(x, y) = F_x(x) \cdot F_y(y)$ for all x and y.*

Rule 8.2.20 *If X and Y have probability distributions respectively f_x and f_y, we may also characterize independence between X and Y by the fact that $f_z(x, y) = f_x(x) \cdot f_y(y)$, or by the fact that $f_{|x}(y) = f_Y(y)$.*

This "product check" for independence applies to independence between multiple stochastic variables as well.

Example 8.2.21 (X, Y) is given by the following table. Determine whether X and Y are independent.

y \ x	1	2
1	0.5	0
2	0	0.5

Answer: We see that $f_X(x) = 0.5$ for $x = 1, 2$, and that $f_Y(y) = 0.5$ for $x = 1, 2$. The product then becomes 0.25 in all four rectangles, and thus different from $f_{XY}(x, y)$. Since, then, $f_X \cdot f_Y \neq f_{XY}$, we conclude that X and Y are *not* independent. Or more briefly: that they are dependent. $\square$

Example 8.2.22 (X, Y, Z), and $f_{XYZ}(x, y, z) = e^{-(x+y+z)}$ for $x, y, z \in [0, \infty)$. Are $X, Y,$ and Z independent?

Answer: We have that

$$f_X(x) = \int_0^\infty \int_0^\infty e^{-(x+y+z)} \, dy \, dz = e^{-x}$$

and, correspondingly, $f_Y(y) = e^{-y}$ and $f_Z(z) = e^{-z}$, which means that $f_{XYZ} = f_X \cdot f_Y \cdot f_Z$, which implies that $X, Y,$ and Z are independent. $\square$

An implication worth noting is the following rule.

> **Rule 8.2.23** *If X and Y are independent, then $\sigma_{XY} = 0$. Conversely, if $\sigma_{XY} \neq 0$, then X and Y are dependent.*

Note that the implication is one-way, and that you cannot conclude from $\sigma_{XY} = 0$ that X and Y must be independent. See assignment 11.

8.3   The Sum of Independent Stochastic Variables

Sometimes, we are interested in more than a single stochastic variable X_i. In hypothesis testing (Chapter 14), we need to look at the difference between two stochastic variables X and Y to calculate the probability that $X > Y$.

In the game of Monopoly, we use two dice. What interests us is not the value of each individual die, but the sum. In Yatzee "chance", we similarly care about the sum of 5 or 6 dice, but not about outcomes of the individual dice. How are these sums distributed? That is: what are the probabilities of the different sums? As opposed to the individual dice, whose values are uniformly distributed, the sums of n dice are concentrated around the expected value of the sum.

> **Rule 8.3.1** *If X and $X_1, \ldots, X_n$ are stochastic variables with expected values respectively μ_X and $\mu_1, \ldots, \mu_n$, and $X = a_1 \cdot X_1 + \cdots + a_n \cdot X_n$, then*
>
> $$\mu_X = a_1 \mu_1 + \cdots + a_n \mu_n.$$
>
> *If the variables $X_1, \ldots, X_n$ are independent, and the variances are σ_X^2 and $\sigma_1^2, \ldots, \sigma_n^2$, then*
>
> $$\sigma_X^2 = a_1^2 \sigma_1^2 + \cdots + a_n^2 \sigma_n^2.$$

How to remember this formula? *Think Pythagoras!*

Example 8.3.2 A toss of a D_4 die has expected value $\mu = 5/2$ and standard deviation $\sigma = \sqrt{5/4}$. A toss of a D_6 die has expected value $\mu = 7/2$ and $\sigma = \sqrt{35/12}$. What is the expected value and the standard deviation of the sum, if we toss one D_4 and one D_6?

Answer: The two tosses are independent stochastic variables X_1 and X_2, with $\mu_1 = 5/2$ and $\mu_2 = 7/2$, and $\sigma_1 = \sqrt{5/4}$ and $\sigma_2 = \sqrt{35/12}$. The outcome of the two tosses is the stochastic variable $X = X_1 + X_2$. We employ the formulas above, and get

$$\mu = \mu_1 + \mu_2 = \frac{5}{2} + \frac{7}{2} = 6$$

$$\sigma^2 = \sigma_1^2 + \sigma_2^2 = \frac{5}{4} + \frac{35}{12} = \frac{25}{6}$$

$$\sigma = \sqrt{\sigma^2} = \frac{5}{\sqrt{6}} = \frac{5\sqrt{6}}{6}.$$ □

Rule 8.3.3 *Let $X_1, \ldots, X_n$ be independent stochastic variables with expected values $\mu_1, \ldots, \mu_n$ and common standard deviation σ. The mean value of the variables is $\overline{X} = (X_1 + \cdots + X_n)/n$. Then*

$$\mu_{\overline{X}} = \frac{\mu_1 + \cdots + \mu_n}{n} \quad and \quad \sigma_{\overline{X}} = \frac{\sigma}{\sqrt{n}}.$$

8.4 The Law of Large Numbers*

In this section, we will present a few probability theoretical theorems that provide a theoretical background for later and more advanced topics in statistics, like those in the inference in this book. This theoretical aspect is singled out in this one section, so that students who plan to take statistics further and deeper can pause here to go into proofs, whereas the students who aim more for application without care for deeper mathematical proof may just keep the results here "for reference".

Rule 8.4.1 Markov's inequality: *If $X \geq 0$, then*

$$P(X \geq a) \leq \frac{E[X]}{a}.$$

Proof: Define Y to be 0 when $0 \leq X < a$, and a when $X \geq a$. Since $X \geq Y$ then $E[X] \geq E[Y]$. This means that

$$E[X] \geq E[Y] = P(Y = 0) \cdot 0 + P(Y = a) \cdot a = P(Y = a) \cdot a = P(X \geq a) \cdot a.$$

Divide both ends by a, and you get Markov's inequality. □

Rule 8.4.2 Chebyshev's inequality: *For a stochastic variable X with expected value $E[X] = \mu$ and variance $E[|X - \mu|^2] = \sigma^2$,*

$$P[|X - \mu| \geq k\sigma] \leq \frac{1}{k^2}.$$

Proof: Define the positive stochastic variable $Y = |X - \mu|^2$. It has expected value $E[Y] = E[|X - \mu|^2] = \sigma^2$. Markov's inequality tells us that

$$P[Y \geq \varepsilon^2] \leq \frac{E[Y]}{\varepsilon^2} = \frac{\sigma^2}{\varepsilon^2}.$$

Since $x^2 > y^2$ when $x > y > 0$, it then follows that

$$P[|X - \mu| \geq \varepsilon] = P[|X - \mu|^2 \geq \varepsilon^2] \leq \frac{\sigma^2}{\varepsilon^2}.$$

Let $\varepsilon = k\sigma$. Then

$$P[|X - \mu| \geq k\sigma] \leq \frac{\sigma^2}{(k\sigma)^2} = \frac{1}{k^2}.$$

$\square$

This result – or more precisely the middle result $P[|X - \mu| \geq \varepsilon] \leq \sigma^2/\varepsilon^2$ from the proof – is handy in proving *The Law of Large Numbers*.

Rule 8.4.3 *Let $X_1, X_2, \ldots, X_n, \ldots$ be independent stochastic variables with the same probability distribution with expected value $\mu < \infty$ and variance $\sigma^2 < \infty$, and let $\overline{X}_n = \frac{1}{n} \sum_{k=1}^{n} X_k$. Then*

$$\lim_{n \to \infty} P(|\overline{X}_n - \mu| \geq \varepsilon) = 0 \text{ for all } \varepsilon > 0.$$

Proof: According to Rule 8.3.1, we get that $\mu_{\overline{X}_n} = \mu$ and that $\sigma^2_{\overline{X}_n} = \frac{1}{n}\sigma^2$. Then

$$P(|\overline{X}_n - \mu| \geq \varepsilon) \leq \frac{1}{n} \cdot \frac{\sigma^2}{\varepsilon^2},$$

which goes to 0 when $n \to \infty$.

$\square$

The Law of Large Numbers exists in many forms, many of them stronger than the version we just proved. For those who plan to take statistics further, the following formulations are useful. We list them without proof.

Rule 8.4.4 *Let $X_1, X_2, \ldots, X_n, \ldots$ be independent stochastic variables with the same probability distribution with expected value $\mu_X < \infty$. Let $\overline{X_n} = \frac{1}{n}\sum_{k=1}^{n} X_k$. Then*

$$\lim_{n \to \infty} P(|\overline{X_n} - \mu_X| \geq \varepsilon) = 0 \quad \text{for all } \varepsilon > 0 \text{ (Weak Law of Large Numbers)}$$
$$P(\lim_{n \to \infty} \overline{X_n} = \mu_X) = 1 \qquad \text{(Strong Law of Large Numbers)}.$$

8.5 Exercises

Mixed Distributions

1 In the subproblems below, X is a mixed distribution with k components X_k, with respective weights w_k. For each subproblem, find μ_X, $Var(X)$, and $P(X \geq 0)$.

  (a) 2 components. $w_1 = 0.2$, $w_2 = 0.8$. $\mu_1 = -3$, $\mu_2 = 5$, $Var(X_1) = 9$, $Var(X_2) = 16$, $P(X_1 \geq 0) = 0.2$, $P(X_2 \geq 0) = 0.9$.

  (b) 2 components. $w_1 = 0.9$, $w_2 = 0.1$. $\mu_1 = 7$, $\mu_2 = 11$, $Var(X_1) = 50$, $Var(X_2) = 50$, $P(X_1 \geq 0) = 0.11$, $P(X_2 \geq 0) = 0.11$.

  (c) 2 components. $w_1 = 0.5$, $w_2 = 0.5$. $\mu_1 = 5$, $\mu_2 = -5$, $Var(X_1) = 25$, $Var(X_2) = 25$, $P(X_1 \geq 0) = 1$, $P(X_2 \geq 0) = 0$.

  (d) 3 components. $w_1 = 0.1$, $w_2 = 0.6$, $w_3 = 0.3$. $\mu_1 = 2$, $\mu_2 = 4$, $\mu_3 = 8$, $Var(X_1) = 4$, $Var(X_2) = 8$, $Var(X_3) = 16$, $P(X_1 \geq 0) = 0.841\,345$, $P(X_2 \geq 0) = 0.921\,35$, $P(X_3 \geq 0) = 0.977\,25$.

  (e) 3 components. $w_1 = 0.9$, $w_2 = 0.09$, $w_3 = 0.01$. $\mu_1 = 1$, $\mu_2 = 2$, $\mu_3 = 100$, $Var(X_1) = 1$, $Var(X_2) = 1$, $Var(X_3) = 1$, $P(X_1 \geq 0) = 0.841\,345$, $P(X_2 \geq 0) = 0.977\,25$, $P(X_3 \geq 0) = 1$.

2 In the subproblems below, there is one discontinuity. Graph F_c, F_d, and F_X, and find μ_X and $Var(X)$.

  (a) Continuous: $w_c = 0.7$, $X_c \sim f(x) = 0.1$ for $0 \leq x \leq 10$.
  Discrete: $w_d = 0.3$, and $p_4 = 1$.

  (b) Continuous: $w_c = 0.15$, $X_c \sim f(x) = 2x$ for $0 \leq x \leq 1$.
  Discrete: $w_d = 0.85$, and $p_0 = 1$.

  (c) Continuous: $w_c = 0.6$, $X_c \sim f(x) = 0.5\sin(x)$ for $0 \leq x \leq \pi$.
  Discrete: $w_d = 0.4$, and $p_\pi = 1$.

3 We have a mixed distribution with two discontinuities. The continous part is given by $w_c = 0.2$, and $X_c \sim f(x) = 0.1$ for $0 \leq x \leq 10$. The discrete part is given by $w_d = 0.8$, and the two discontinuities are $p_3 = 0.375$ and $p_9 = 0.625$.
  Graph F_c, F_d, and F, and find μ_X and $Var(X)$.

4 The artist Sandra has insured her concert tour. Insurance agents Beowulf have calculated with a 12% probability that the tour will be in the red, and that, in such a case, the loss will be distributed $g(x) = 5 \times 10^{-7} e^{-5 \times 10^{-7} x}$ (in Section 10.3 we will get to know this as the *exponential distribution* with $\lambda = 5 \times 10^{-7}$). The insurance covers the entire loss. Let X be the payment from Beowulf to Sandra. Beowulf charges their policies at a price of 1.7 times the expected payment, plus 0.1 times the standard deviation. How much are they charging Sandra for her policy?

5 Find the charge for Sandra's policy, should the payment be capped at 5 000 000.

Two- and Multi-variable Probability Distributions

6 For $Z = (X, Y)$ given by following table, find the marginal probabilities, $P(X + Y = 4)$, and the correlation ρ_{xy}:

	$X = 1$	$X = 2$	$X = 3$
$Y = 1$	0.05	0.05	0.3
$Y = 2$	0.05	0.25	0.05
$Y = 3$	0.2	0.05	0

7 Soren K. has studied previous exams in German philosophy, and has discovered that the lecturer seems to love the two obscure philosophers Max Stirner and Karl Werder. Soren sets up a table he believes expresses the joint probabilities for the respective number of questions on Stirner (X) and on Werder (Y):

	$X = 1$	$X = 2$	$X = 3$
$Y = 2$	0	0.1	0.2
$Y = 3$	0.05	0.15	0.1
$Y = 4$	0.15	0.2	0.05

(a) Write up the table for the cumulative probability F_{XY}.

(b) Find the probability that Soren gets a total of 5 questions about these two philosophers.

(c) Find the conditional probabilities $P(X = 1 | X + Y = 5)$, $P(X = 2 | X + Y = 5)$, and $P(X = 3 | X + Y = 5)$, and write them into a table.

(d) What do you find, when you calculate $P(X = a | X + Y = 5)$? (Textual answer; not calculation.)

8 You have a bag of one each of the dice $D_4, D_6, D_8, D_{10}, D_{12}, D_{20}$. You sample a die, and toss it. Let X be the number of faces on the die, and let Y be the value of the toss. Are X and Y independent?

9 $Z = (X, Y) \sim f_Z(x, y) = \frac{4}{5}\left(2 - x - y^3\right)$ for $x, y \in [0, 1]$. Find the marginal probabilities f_x and f_y, the covariance σ_{xy}, and determine whether X and Y are independent.

10 $Z = (X, Y) \sim f_Z(x, y)$, which is $\frac{1}{\pi}$ inside the circle $x^2 + y^2 = 1$, and 0 otherwise (where $\pi = 3.14159\ldots$).
   (a) Find the marginal probability distributions f_x and f_y.
   (b) Are X and Y independent?
   (c) Find the conditional probability distributions $f_{|Y=y}(x)$ and $f_{|X=x}(y)$.

11 Let X be a stochastic variable taking the values $-2, -1, 0, 1, 2$ with probability $\frac{1}{5}$ each, and let $Y = X^2$. Are X and Y independent? What is ρ_{XY}?

The Sum of Independent Stochastic Variables

12 Find μ_Z and σ_Z, when ...
   (a) $Z = X - Y$, $\mu_X = 4$, $\sigma_X = 3$, $\mu_Y = -3$, $\sigma_Y = 2$;
   (b) $Z = X_1 + X_2 + X_3$, and $\mu_{X_1} = 1$, $\mu_{X_2} = 2$, $\mu_{X_3} = 3$, whereas $\sigma_{X_1} = 3$, $\sigma_{X_2} = 4$, $\sigma_{X_3} = 12$;
   (c) $Z = \sum_{k=1}^{6} X_k$ and $\mu_{X_k} = k$ and $\sigma_{X_k} = \sqrt{k}$;
   (d) $Z = \frac{1}{2}X$ and $\mu_X = 14$ and $\sigma_X = 4$;
   (e) $Z = kX$ and $\mu_X = \mu$ and $\sigma_X = \sigma$;
   (f) $Z = \sum_{k=1}^{6} \frac{1}{k}X_k$ and $\mu_{X_k} = k$ and $\sigma_{X_k} = \sqrt{k}$.

13 Let X, Y be independent stochastic variables with standard deviation respectively σ_X and σ_Y, and let $Z = aX + (1 - a)Y$. Let σ_Z be Z's standard deviation. For which value(s) of a is σ_Z the smallest?

9

Discrete Distributions

CONTENTS

In this chapter, we will look at some key discrete probability distributions. With the exception of the uniform distribution, they are built from *binary* events, which are events with only two possible outcomes: $\top$ (positive outcome) and $\bot$ (negative outcome). These distributions then state the probability of the *number* of positive or negative outcomes under different counting scenarios.

9.1　How to Read the Overview

Each distribution begins with a box with the most commonly used properties of the distribution. If the distribution has its own symbol ψ, we will use that to designate it, and the uppercase version of the symbol, Ψ, for the cumulative distribution function. If not, we will use the *name* and *NAME* in these roles. By means of the distribution's *parameters* $p_1, p_2, \ldots, p_k$, we may then describe

- the probability distribution (pdf) $\psi_{(p_1,\ldots,p_k)}(x)$ or $name_{(p_1,\ldots,p_k)}(x)$
- the cumulative probability distribution (CDF) $\Psi_{(p_1,\ldots,p_k)}(x)$ or $NAME_{(p_1,\ldots,p_k)}(x)$
- the inverse cumulative distribution (iCDF) $\Psi^{-1}_{(p_1,\ldots,p_k)}(x)$ or $NAME^{-1}_{(p_1,\ldots,p_k)}(x)$

The Bayesian Way: Introductory Statistics for Economists and Engineers, First Edition.
Svein Olav Nyberg.
© 2019 John Wiley & Sons, Inc. Published 2019 by John Wiley & Sons, Inc.

- the expected value $\mu_X = E[X]$, together with a formula for calulating it from the parameters $p_1, \ldots, p_k$
- the variance σ_X^2, together with a formula for calulating it from the parameters $p_1, \ldots, p_k$.

We also show how to calculate the properties of the distribution in Mathematica®. Where possible we will also show how to calculate the cumulative probability distribution in some higher-end popular graphic calculators from CASIO®, TI®, and HP®.

In Mathematica, the distributions are invoked by the name of the distribution, followed by a parenthetical list of its parameters:

Mathematica: *NameOfDistribution*$[p_1, \ldots, p_k]$.

The key commands for calculating the distribution in Mathematica are:

- PDF[*NameOfDistribution*$[p_1, \ldots, p_k], x$] gives the probability distribution $f(x)$.
- CDF[*NameOfDistribution*$[p_1, \ldots, p_k], x$] gives the cumulative probability distribution $F(x)$, which equals $P(X \leq x)$.
- InverseCDF[*NameOfDistribution*$[p_1, \ldots, p_k], x$] gives $F^{-1}(x)$.
- Probability[*condition on x*, $x \approx$ *NameOfDistribution*$[p_1, \ldots, p_k]$] gives $P(X \in \{x | condition on x\})$; the condition on x may for instance be $x < a$ or $a \leq x < b$. To get the $\approx$ sign, type (esc)dist(esc).
- Mean[*NameOfDistribution*$[p_1, \ldots, p_k]$] gives μ_X, also known as $E[X]$.
- Variance[*NameOfDistribution*$[p_1, \ldots, p_k]$] gives σ_X^2.
- StandardDeviation[*NameOfDistribution*$[p_1, \ldots, p_k]$] gives σ_X.
- RandomVariate[*NameOfDistribution*$[p_1, \ldots, p_k]$] gives a random number, sampled according to the probabilities indicated by the distribution.

For a general exposition of discrete probability distributions, see Section 7.2.

Sometimes, direct calculations on distributions can be very demanding, or even too demanding, given the tools at hand. For that reason, we often employ approximations to simplify calculations. Even advanced tools like Mathematica or R have their limitations: binomial expressions like $\sum_{k=0}^{n} \binom{n}{k} \cdot f(k)$ are not too difficult to handle as long as n is reasonable, but Mathematica may take several hours to complete if $n = 1\,000\,000$. So *do* make use of the approximations where this is expedient. You'll get there faster, and with sufficient precision.

Whilst there are many excellent packages available,[1] we will focus on Mathematica as our advanced tool of choice, as it has a simple and freely available interface online at

http://www.wolframalpha.com

---

[1] For the advanced user, we recommend getting acquainted with the freely available tool *R*.

This book has web support, which includes calculator guides, at
http://bayesians.net

9.2   Bernoulli Distribution, *bern_p*

Bernoulli distribution: A Bernoulli trial X with parameter p is a single trial whose outcome is either success $\top$ or failure $\bot$, where $P(\top) = p$. Then, $X \sim bern_p$, a probability distribution with a single parameter: p.

pdf: $bern_p(x) = \begin{cases} 1-p & x = 0 \\ p & x = 1 \\ 0 & \text{else} \end{cases}$

CDF: $BERN_p(x) = \begin{cases} 0 & x < 0 \\ 1-p & 0 \leq x < 1 \\ 1 & p \geq 1 \end{cases}$

Expected value: $\mu_X = p$ **Variance:** $\sigma_X^2 = p(1-p)$
Binomial: $bern_p = bin_{(1,p)}$
Mathematica: BernoulliDistribution[p]

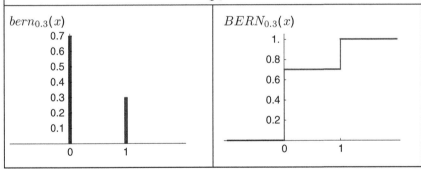

9.2.1   Fair Odds

When you play for money, the essence of the game is a bet. So what is the fair price for joining a bet where the probability of winning £K is p? And what are *odds*?

Definition 9.2.1   Odds *is the proportion of the probability of winning to the probability of losing:*

$$Odds\ for = o_f = P(\top)/P(\bot) = p/(1-p)$$
$$Odds\ against = o_a = P(\bot)/P(\top) = (1-p)/p.$$

Calculate backwards, and the odds will give you back the probability of winning. Let $o_f = a : b$. Then

$$p = 1/(o_a + 1) = 1 - 1/(o_f + 1) = b/(a + b).$$

Odds is a traditional way of stating betting chances by a number, and predated Pascal and Fermat's invention of probability by roughly a hundred years.

It is customary to write the odds ratio a/b as $a : b$ or as $a - b$. If possible, one chooses low integers for a and b, or chooses one of the numbers to be 1. So we would not write odds as 3.25 : 2.5, but would rather multiply both sides by 4, and write the odds as 13 : 10. Alternatively, we would divide both sides by b, and write 1.3 : 1.

A bookmaker's odds tells you how many times your bet gets paid out if you win. If the odds are $a : b$ (against; it is usually "odds against" if nothing else is specified), and you have bet £K, your winnings will be £$\frac{K}{b}(a + b)$. So if the odds are *even*, that is, equal probability of winning and losing, 1 : 1, the pay-out for a £100 bet is $\frac{100}{1}(1 + 1) =$ £200.

Note 1   Continental European bookmakers do not use odds ratios in the way of the anglophone world, but prefer to state the winnings as a multiple of the bet. Their odds is this multiple. This number equals $o_a + 1$, and thus equals $1/p$.

Note 2   The bookmaker is himself no gambler, but a businessman striving to maintain a steady living. His odds are therefore not stipulated on what he believes to be the probabilities of the different outcomes. Instead, the bookmaker will base his odds on how many people make the bets, so that what he gets in exceeds what he pays out, no matter what the outcome. This means that if you *do* know the true probability that the horse *Silky White* will win the next horse race better than anyone else does, your expected payout is positive. But before you wager your student loan on a horse, remember the thousands of broke gamblers who were assured they knew the outcomes of past horse races. Never gamble what you can't afford to lose!

The bookmaker ideally sets his odds in the same way Black and Scholes' 1973 formula dictates the right price for financial derivatives.

Rule 9.2.2   *The fair price for a gamble with payoff G, and probability p to win, is $p \cdot G$.*

Example 9.2.3 The fair odds on *Silky White* is 3 : 2 against.

1) Find o_a as a decimal number.
2) Find o_f as a decimal number.
3) Find the continental European odds.
4) Find the probability p that *Silky White* wins.
5) What is a fair price for the bet where you get £100 if *Silky White* wins?
6) If you bet £100 with these odds, and *Silky White* wins, how large is your payout?
7) A British bookmaker offers odds of 7 : 4 against. Does it pay to place a bet on *Silky White* with him?

Answers:

1) $o_a = 3/2 = 1.5$.
2) $o_f = 1/o_a = 0.667$.
3) Continental European odds $= o_a + 1 = 1.5 + 1 = 2.5$.
4) $p = b/(a + b) = 2/(3 + 2) = 0.4$.
5) $p \cdot G = 0.4 \cdot 100 = £40$.
6) $100/0.4 = £250$.
7) With this bookmaker, you are paid $\frac{7+4}{4} = 2.75$ times your bet if you win, which is in excess of the 2.5 multiplier of a fair bet, meaning this bet pays better for you to play than a fair bet. □

9.2.2 Bernoulli Process

The Bernoulli distribution for a single trial is rarely seen in practical application. We are more interested in repeated trials, in what we call a *Bernoulli process*. We define the following.

Definition 9.2.4 *A Bernoulli process with parameter p is a sequence of trials where*

1) each trial has exactly 2 possible outcomes: $\top$ and $\bot$;
2) $P(\top) = p$ for each trial;
3) the outcomes of the trials are independent.

A single Bernoulli trial is simply a trial of whether a certain event occurs or not. The typical examples are from the gaming world in the tossing of a coin or betting on a roulette colour. But daily life furnishes us with equally good examples: checking whether a light bulb is dead is a Bernoulli trial, asking a girl out for a date is a Bernoulli trial, applying for a scholarship is a Bernoulli trial. We have a Bernoulli *process* of n trials when we perform n identical but independent trials. Examples are five coin flips, or applying for three scholarships. The context for the trials must be the same, yielding $P(\mathsf{T}) = p$ for all trials, independently of the outcomes of the other trials.

Bernoulli trials are in themselves very easy to comprehend, but still the theory built on their backs will have far-reaching consequences. The Law of Large Numbers (8.4.3 and 8.4.4) for Bernoulli trials are central to our understanding of probability, and lie at the base of the frequentist *definition* of probability.

Rule 9.2.5 *Let $X_1, X_2, \ldots, X_n, \ldots$ be independent Bernoulli trials with probability p of success, and let $Z_n = \frac{1}{n} \sum_{k=1}^{n} X_k$. The stochastic variable Z_n is then the proportion of successes after n Bernoulli trials.*

$\lim_{n\to\infty} P(|Z_n - p| \geq \varepsilon) = 0$ *when $\varepsilon > 0$ (Bernoulli's theorem)*

$P(\lim_{n\to\infty} Z_n = p) = 1$ *(Borel's Law of Large Numbers).*

9.3 Binomial Distribution, $bin_{(n,p)}$

Binomial distribution: Given a Bernoulli process with parameter p (probability of success), let the stochastic variable X be the number of T in n trials. Then, $X \sim bin_{(n,p)}$; this means X follows a *binomial probability distribution* $bin_{(n,p)}$ with parameters n and p.

pdf: $bin_{(n,p)}(x) = \binom{n}{x} p^x (1-p)^{n-x}$	CDF: $BIN_{(n,p)}(x)$
CASIO: BinomialPD(x, n, p) **HP:** binomial(n, x, p) **TI:** binompdf(n, p, x)	BinomialCD(x, n, p) binomial_cdf(n, p, x) binomcdf(n, p, x)

Mathematica: BinomialDistribution[n, p]

Expected value: $\mu_X = np$ **Variance:** $\sigma_X^2 = np(1-p)$
Poisson approximation: $pois_{np}$ (9.6) is good when $n^{0.31} p < 0.47$
Normal approximation: $\phi_{(\mu_X, \sigma_X)}$ (10.1.5) is good when $np(1-p) > 5$

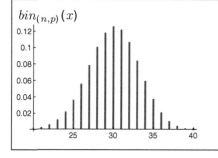

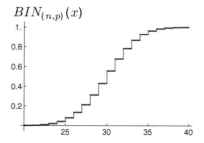

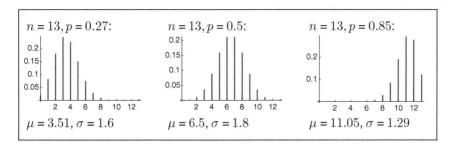

$n = 13, p = 0.27$: $n = 13, p = 0.5$: $n = 13, p = 0.85$:

$\mu = 3.51, \sigma = 1.6$ $\mu = 6.5, \sigma = 1.8$ $\mu = 11.05, \sigma = 1.29$

9.3.1 Application: Repeated Sampling With Replacement

The binomial distribution describes the probabilities of *repeated sampling with replacement* from Section 5.5, and we see that the distribution formula $bin_{(n,p)}(x)$ is the formula for the probabilities for repeated sampling with replacement of two kinds in Rule 5.5.5. Typical situations where this is useful are as follows.

- **Industrial:** The rate of critical failure in the cars you produce is 0.8%. What is the probability that more than 1% of 20 thousand new cars have critical failures?
- **Sports:** Bayern Munich wins 83% of their Bundesliga games. What is the probability that they win 9 out of the next 10?
- **Medicine:** If a disease is misdiagnosed as something else 10% of the time, how probable is it that a doctor will misdiagnose exactly 2 out of his next 9 cases of this disease?
- **Economics:** Your company has submitted proposals for 14 different projects this month, and they tend to win 40% of them. What is the probability that your company will win 5 projects this month?

9.3.2 Calculating Binomial Distributions

Example 9.3.1 $X \sim bin_{(7,0.34)}$, binomially distributed with parameters $n = 7$ and $p = 0.34$. Find

1) μ_X
2) σ_X
3) $P(X \in \{3, 4, 5\})$
4) $P(X \leq 2)$.

Answers:

1) $\mu_X = np = 7 \cdot 0.34 = 2.38$
2) $\sigma_X = \sqrt{np(1 - p)} = \sqrt{7 \cdot 0.34 \cdot (1 - 0.34)} = 1.253$
3) $P(X \in \{3, 4, 5\}) = BIN_{(7,0.34)}(5) - BIN_{(7,0.34)}(2) = 0.437\,054$
4) $P(X \leq 2) = BIN_{(7,0.34)}(2) = f(2) + f(1) + f(0) = 0.555\,284$. $\square$

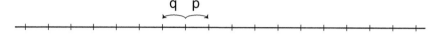

Figure 9.1 Steps for a random walk.

Calculating binomial distributions is generally a straightforward application for the formulas. There is no shortcut for exact calculations of cumulative probability, but you may often end up doing less work if you remember that $P(X < a) + P(X = a) + P(X > a) = 1$.

Example 9.3.2 X is binomially distributed with $p = 0.9$ and $n = 20$. What is $P(X \leq 19)$? Here, we can either sum the point probabilities from p_0 up to and including p_{19}, or we can simply note that

$$P(X \leq 19) = 1 - P(X > 19) = 1 - P(X = 20) = 1 - 0.9^{20} = \underline{0.878\,423}. \qquad \square$$

A common error is to forget to include $P(X = 0)$ in the sum.

9.3.3 Application: Random Walk

Simple random walk is a process on the integers, with probability p of the next step being the current position $+1$, and probability $q = 1 - p$ for -1, as illustrated in Figure 9.1. This is a close relative of the Bernoulli process, with alternatives -1 and $+1$ rather than 0 and 1. A popular illustration is that of a drunken Irishman's pub crawl, where a skew coin determines which pub he visits next: the one up the street, or the one down the street.

The probability distribution of where the Irishman is after n steps with $p = 0.55$ is illustrated in Figure 9.2 for $n = 1, \ldots, 8$.

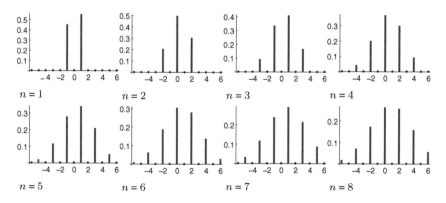

Figure 9.2 Probability distributions after n steps of the random walk.

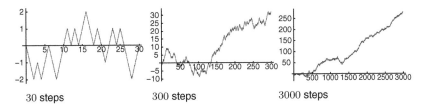

Figure 9.3 The path of the random walk after a given number of steps.

The probability distribution for simple random walk is almost the same as for the binomial distribution. The distribution formula for a random walk X_n with parameter p (= the probability of +1), is then

$$P(X_n = x) = bin_{(n,p)}\left(\frac{x+n}{2}\right).$$

The expected value is $\mu_n = E[X_n] = (2p - 1)n$, and the variance is $\sigma_n^2 = 4np(1 - p)$.

Sample random walks are usually illustrated by using the horizontal axis for time and the vertical axis for position/value. Such an illustration may then look like Figure 9.3 ($p = 0.55$).

Random walk and its continuous counterpart *Brownian motion* are used to model everything from the motion of pollen in liquids to the movements of stock markets.

Example 9.3.3 The stock of Goldum Ltd is in a market where it trades once per day, and the share price follows a random walk with probability $p = 0.51$ of a $1 increase, and a probability $q = 0.49$ of a $1 decrease. The price of one share starts at $105. What is the probability that the price is less than $105 dollars after 20 days have passed? What is the expected share price, and what is the variance?

Answer: The share price in dollars equals $Y_n = X_n + 105$ for a random walk X starting at 0 with parameter $p = 0.51$, so we deduct 105 on both sides, and ask the equivalent question of the probability that $P(X_{20} < 0)$. The calculation:

$$P(X_{20} < 0) = \sum_{x<0} P(X_{20} = x) = \sum_{x<0} bin_{(20,\,0.51)}\left(\frac{x+20}{2}\right)$$

$$= \sum_{x<20} bin_{(20,\,0.51)}\left(\frac{x}{2}\right) = \sum_{x<10} bin_{(20,\,0.51)}(x)$$

$$= \sum_{x=0}^{9} bin_{(20,\,0.51)}(x) = 0.377\,056 \approx 38\%.$$

The expected share price after 20 days is

$$E[Y_{20}] = E[105 + X_{20}] = 105 + \mu_{20} = 105 + (2 \cdot 0.51 - 1) \cdot 20 = 105.4.$$

The variance of Y_{20} equals that of X_{20}, and is $\sigma_{20}^2 = 4 \cdot 20 \cdot 0.51 \cdot 0.49 = 19.992$. □

9.3.4 The Normal Approximation

The binomial distribution is the discrete counterpart of the continuous distribution called the *Normal distribution*, which we shall look at in Section 10.1. If we scale the binomial distribution along the horizontal axis (that is, "in the x-direction"), so that the gaps between the bars are no longer 1 unit, but $n^{-1/2}$, it will approximate the Normal distribution. We illustrate the progression with $p = 0.6$ in Figure 9.4.

Example 9.3.4 X follows the binomial distribution f with parameters $n = 90$ and $p = 0.4$. Find $P(X \in \{30, \ldots, 39\})$ by using the Normal approximation.

Solution: $\mu_X = 90 \cdot 0.4 = 36$ and $\sigma_X = \sqrt{90 \cdot 0.4(1 - 0.4)} = 4.65$, so the Normal approximation is $\phi_{(36,4.65)}$. Rule 10.1.13 then says that

$$P(29 < X \le 39) = \Phi_{(36,\,4.65)}\left(39 + \frac{1}{2}\right) - \Phi_{(36,\,4.65)}\left(29 + \frac{1}{2}\right) = \underline{0.693}.$$

By comparison, direct calculations give $P(29 < X \le 39) = 0.695$. □

Poisson Approximation

The binomial distribution may also be approximated by the Poisson distribution, that we will be visiting in Section 9.6. In Figure 9.5, we see how binomial distributions with parameters n and p converge to the Poisson distribution with parameter $np = \lambda$ when n grows while $p = \frac{\lambda}{n}$:

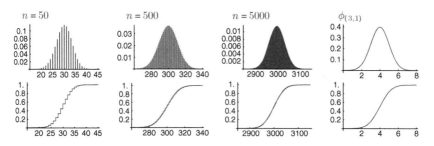

Figure 9.4 Approximating the Normal distribution.

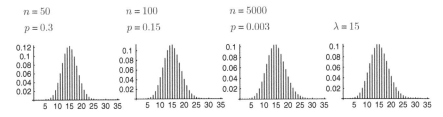

Figure 9.5 Approximating the Poisson distribution.

Example 9.3.5 $X \sim bin_{(223,\,0.07)}$. Find $P(X = 14)$.

Solution: We employ the Poisson approximation with $\lambda = np = 223 \times 0.07 = 15.61$, and get

$$P(X = 14) \approx pois_{15.61}(14) = \frac{15.61^{14}}{14!} \times e^{-15.61} = 0.097\,25.$$

By comparison, direct calculations give $bin_{(223,\,0.07)}(14) = 0.099\,8$. □

9.4 Hypergeometric Distribution, $hyp_{(n,S,N)}$

Hypergeometric distribution: Given n samples *without* replacement from a population of N elements, whereof S have the property $\top$ whereas the rest have the property $\bot$, let the stochastic variable X be the number of $\top$ sampled. Then, $X \sim hyp_{(n,S,N)}$; that is, X follows a *hypergeometric probability distribution* $hyp_{(n,S,N)}$ with parameters n, S, and N. Let $p = \frac{S}{N}$.

pdf: $hyp_{(n,S,N)}(x) = \frac{\binom{S}{x}\binom{N-S}{n-x}}{\binom{N}{n}}$	CDF: $HYP_{(n,S,N)}(x)$
CASIO: HypergeoPD(x, n, S, N)	HypergeoCD(x, n, S, N)
Mathematica: HypergeometricDistribution[n, S, N]	

Expected value: $\mu_X = np$ **Variance:** $\sigma_X^2 = np(1-p) \times \frac{N-n}{N-1}$

Binomial approximation: $bin_{(n,p)}$ (9.3) is good when $n < \frac{N}{10}$

Poisson approximation: $pois_{np}$ (9.6) is good when $\begin{cases} 100 < n < \frac{N}{10} \\ n^{0.31}p < 0.47 \end{cases}$

Normal approximation: $\phi_{(\mu_X,\sigma_X)}$ (10.1.5) is good when $\begin{cases} n < \frac{N}{10} \\ np(1-p) > 5 \end{cases}$

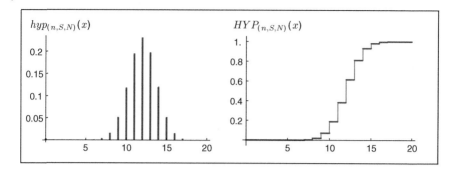

9.4.1 Application: Sampling Without Replacement

The hypergeometric distribution describes the probabilities of *unordered sampling without replacement* in Section 5.5, and we see that the distribution formula $hyp_{(n,S,N)}(x)$ is the formula for the probabilities for repeated sampling without replacement of two kinds in rule 5.5.5.

1) **Industry:** Adobe Illustrator is installed on 8 of your company's 53 laptops. Your team bring 7 random laptops for a mission. Once boarded on your flight, you realise you need at least 2 laptops with AI installed. What is the probability of this happening?
2) **Sports:** Bayern Munich's first team has 32 players, whereof 16 are Germans. Out of the 32, 4 are goalkeepers; 3 of these 4 are German. If a team of 11 is randomly assembled, what is the probability of picking precisely 8 German players?
3) **Medicine:** You work in the cancer ward, and today 5 of its 23 patients will receive bad news on the doctor's round. You are assigned 8 random patients. What is the probability that you will be giving exactly 3 of them bad news?

9.4.2 Calculations on Hypergeometric Distributions

Example 9.4.1 $X \sim hyp_{(6,21,34)}$, following a hypergeometric distribution with parameters $n = 6$, $S = 21$, and $N = 34$. Find

1) μ_X
2) σ_X
3) $P(X \in \{2, 4\})$
4) $P(X > 2)$.

Answers:

1) $\mu_X = np = n \times \frac{S}{N} = 6 \times \frac{21}{34} = 3.705\,88$.

2) $\sigma_X = \sqrt{np(1-p) \times \dfrac{N-n}{N-1}} = \sqrt{n \times \dfrac{S}{N}\left(1-\dfrac{S}{N}\right) \times \dfrac{N-n}{N-1}}$

$\quad = \sqrt{6 \times \dfrac{21}{34}\left(1-\dfrac{21}{34}\right) \times \dfrac{34-6}{34-1}} = 1.096\,48.$

3) $P(X \in \{2,4\}) = P(X=2) + P(X=4) = f(2) + f(4)$

$$= \frac{\binom{21}{2}\binom{34-21}{6-2}}{\binom{34}{6}} + \frac{\binom{21}{4}\binom{34-21}{6-4}}{\binom{34}{6}}$$

$$= 0.111\,644 + 0.347\,11 = 0.458\,754.$$

4) $P(X > 2) = 1 - P(X \le 2) = 1 - (f(2) + f(1) + f(0)) = 0.866\,985.$ □

Calculation of hypergeometric distributions is straightforward, just as for binomial distributions, but it is still the most time consuming of the common probability distributions. We will therefore be moved to employ approximations to other distributions quite frequently, and, in particular, approximation to the binomial distribution.

Binomial Approximation

For fixed n, we see that $hyp_{(n,S,N)}(x)$ becomes increasingly more similar to $bin_{(n,S/N)}(x)$ the larger N becomes. In Figure 9.6, $p = \dfrac{S}{N} = 0.6$ and $n = 5$, and N is increasing.

Example 9.4.2 $X \sim hyp_{(8,\,98,\,213)}$. Find $P(X \in \{5,6,7\})$.

Solution: We employ the binomial approximation, with parameter $p = \dfrac{98}{213} = 0.460\,1$:

$$P(X \in \{5,6,7\}) \approx BIN_{(8,\,0.460\,1)}(7) - BIN_{(8,\,0.460\,1)}(4) = \underline{0.278\,0}.$$

By comparison, direct calculation on the hypergeometric distribution gives 0.274 6. □

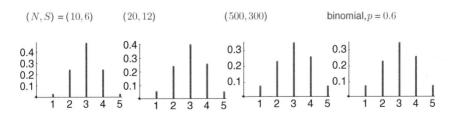

Figure 9.6 Approximating the binomial distribution.

Poisson Approximation

The hypergeometric distribution inherits the Poisson approximation from the binomial distribution.

Example 9.4.3 $X \sim hyp_{(200,150,5000)}$. Find $P(X \in \{3,4,5\})$.

Solution: We employ the Poisson approximation, with parameter $\lambda = \frac{nS}{N} = \frac{200 \times 150}{5000} = 6$:

$$P(X \in \{3,4,5\}) = POIS_6(5) - POIS_6(2) = \underline{0.383\,7}.$$

By comparison, direct calculation on the hypergeometric distribution gives 0.384 2. □

Normal Approximation

The hypergeometric distribution inherits the Normal approximation from the binomial distribution.

Example 9.4.4 $X \sim hyp_{(50,320,800)}$. Find $P(20 < X \leq 30)$.

Solution: $\mu_X = 50 \times \frac{320}{800} = 20$, and $\sigma_X = \sqrt{50 \times 0.4 \times 0.6 \times \frac{800-50}{800-1}} = 3.356\,2$, so our Normal approximation is $\phi_{(20,\,3.356\,2)}$, and so

$$P(20 < X \leq 30) \approx \Phi_{(20,\,3.356\,2)}\left(30 + \frac{1}{2}\right) - \Phi_{(20,\,3.356\,2)}\left(20 + \frac{1}{2}\right) = \underline{0.441}.$$

By comparison, direct calculation on the hypergeometric distribution gives 0.436 5. □

9.5 Geometric and Negative Binomial Distributions, $nb_{(k,p)}$

Negative binomial distribution: Let L be the number of failures ($\perp$) from a Bernoulli process with parameter $p = P(\mathsf{T})$ that is terminated upon getting a total of k successes (T). Then $L \sim nb_{(k,p)}$; that is, L follows a *negative binomial probability distribution* $nb_{(k,p)}$ with parameters k and p.	
pdf: $nb_{(k,p)}(x) = \begin{pmatrix} x+k-1 \\ k-1 \end{pmatrix} p^k (1-p)^x$ $= p \times bin_{(x+k-1,p)}(k-1)$	CDF: $NB_{(k,p)}(x) = I_{(k,x+1)}(p)$ $\qquad\qquad = 1 - BIN_{(x+k,p)}(k-1)$ (formulas A.2.17 and A.4.3)

CASIO:	
$p \cdot \text{BinomialPD}(k-1, x+k-1, p)$	$1\text{-BinomialCD}(k-1, x+k, p)$
HP: $p \cdot \text{binomial}(x+k-1, k-1, p)$	$1\text{-binomial\_cdf}(x+k, p, k-1)$
TI: $p \cdot \text{binompdf}(x+k-1, p, k-1)$	$1\text{-binomcdf}(x+k, p, k-1)$

Mathematica: NegativeBinomialDistribution$[k, p]$

Expected value: $\mu_X = k(1-p)/p$ **Variance:** $\sigma_X^2 = k(1-p)/p^2$

Other distributions: The negative binomial distribution is the discrete counterpart to the Erlang and Gamma distributions (Section 10.3).

There exists a variant of the negative binomial distribution that instead of counting failures counts the total number of trials, but both terminate at k successes. In Mathematica, this variant is called PascalDistribution$[k, p]$.

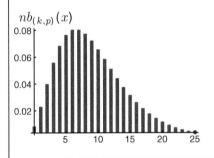
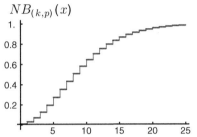

9.5.1 Application: Waiting for Success

Whereas the binomial distribution indicates the probability of x successes in n independent Bernoulli trials with parameter p, the *negative binomial* distribution indicates the probability of x failures when the Bernoulli trials are terminated at k successes: it looks at the number of actual frogs you end up kissing before you have kissed k princes.

- **Industrial:** Your signal has 80% probability of successful transmission, and if the transmission fails, you just resend. How many failed transmissions will you make before the event of a successful transmission?
- **Sports:** You are playing miniature golf on a court where the ball returns to its starting point if you miss the hole. How many misses will you go through before finally putting the ball into the hole?
- **Medicine:** A certain chemotherapy has a 70% chance of success, independently of previous treatments. How many failed attempts will the patient go through before he is cured?

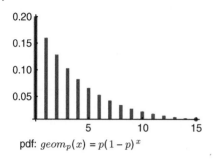

pdf: $geom_p(x) = p(1-p)^x$

CDF: $GEOM_p(x) = 1 - (1-p)^{x+1}$

Figure 9.7 A geometric distribution.

- **Economics:** Your company tends to win 40% of their submitted proposals. How many proposals must they submit to have at least a 50% probability of winning 8 of them?

9.5.2 Geometric Distribution

A special case of the negative binomial distribution is the *geometric* distribution, $geom_p$, which indicates the probability of x failures before a success. $geom_p(x) = nb_{(1,p)}(x)$. Since a Bernoulli process consists of independent trials, and thus can be said to be *memoryless*, the number of failures before k successes equals the sum of k waiting times under a geometric distribution. Formally: if $Y \sim nb_{(n,p)}$, then $Y = X_1 + \cdots + X_n$, where $X_k \sim geom_p$. In Figure 9.7, we see the pdf and the CDF of a typical geometric distribution.

9.5.3 Calculating Geometric Distributions

Example 9.5.1 $L \sim geom_{0.25}(x)$, that is, a geometric distribution with parameter $p = 0.25$. Find: (1) μ_L, (2) σ_L^2, (3) $P(L \in \{3,4\})$, and (4) $P(L > 1)$.

Solution:

(1) $$\mu_L = \frac{1-p}{p} = \frac{1-0.25}{0.25} = 3.$$

(2) $$\sigma_L^2 = \frac{1-p}{p^2} = \frac{1-0.25}{0.25^2} = 12.$$

(3) $P(L \in \{3,4\}) = P(L = 3) + P(L = 4) = f(3) + f(4)$

$$= 0.25 \times 0.75^3 + 0.25 \times 0.75^4 = 0.184\,57.$$

(4) $P(L > 1) = 1 - P(L \le 1) = 1 - F(1)$

$$= 1 - [1 - (1 - 0.25)^{1+1}] = 0.75^2 = 0.562\,5. \qquad \square$$

9.5.4 Calculating Negative Binomial Distributions

Example 9.5.2 $L \sim nb_{(7, 0.25)}$, that is, a negative binomial distribution with parameters $k = 7$ and $p = 0.25$. Find

(1) μ_L.

(2) σ_L^2.

(3) $P(L \in \{4, 5\})$.

(4) $P(L > 0)$.

Answer:

(1) $\mu_L = \frac{k(1-p)}{p} = \frac{7(1-0.25)}{0.25} = 21$.

(2) $\sigma_L^2 = \frac{k(1-p)}{p^2} = \frac{7(1-0.25)}{0.25^2} = 84$.

(3) $P(L \in \{4, 5\}) = f(4) + f(5)$

$= \binom{10}{6} 0.25^7 \times 0.75^4 + \binom{11}{6} 0.25^7 \times 0.75^5$

$= 0.004\,055\,5 + 0.006\,691\,58 = 0.010\,747\,1$.

(4) We employ the principle of complementary probability:

$$P(L > 0) = 1 - P(L = 0) = 1 - \binom{6}{6} 0.25^7 \times 0.75^0 = 0.999\,939.$$

$\square$

9.6 Poisson Distribution, *pois*$_\lambda$

Poisson distribution: The Poisson distribution counts occurrences in trials where the occurrences appear independently and according to a given *rate* λ. Let X be the number of occurrences. Then $X \sim pois_\lambda$; that is, X follows a *Poisson distribution* with parameter λ.	
pdf: $pois_\lambda(x) = \dfrac{\lambda^x}{x!} \times e^{-\lambda}$	CDF: $POIS_\lambda(x)$
CASIO: PoissonPD(x, λ) **HP:** poisson(λ, x) **TI:** poissonpdf(λ, x)	PoissonCD(x, λ) poisson_cdf(λ, x) poissoncdf(λ, x)
Mathematica: PoissonDistribution$[\lambda]$	
Expected value: $\mu_X = \lambda$ **Variance:** $\sigma_X^2 = \lambda$ **Normal approximation:** $\phi_{(\mu_X, \sigma_X)}$ (see Section 10.1.5) is good when $\lambda > 10$.	

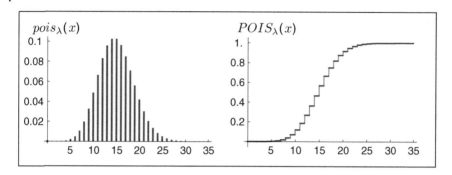

9.6.1 Application: Poisson Processes, and Many Trials with Rare Occurrences

The Poisson distribution is intimately connected to the Poisson *process* (Section A.4.2). We may well view the Poisson process as a continuous counterpart to the Bernoulli process, or as a limiting case of a Bernoulli process where, instead of probing at every integer point in time, we probe at every half, quarter, eighth, …, or $\frac{1}{n}$th unit, but with the probability p of success at each trial going down such that pn remains constant.

That constant is the rate $\lambda = np$, and the Poisson process proper is when we let $n \to \infty$. However, for small p and large n, the Poisson process is an excellent approximation to the Bernoulli process.

In its own right, the Poisson process describes waiting for an occurrence of T when the probability of a T occurring in an interval $[a, b]$ is proportional to the size of the interval, and independent of occurrences outside the interval. The Poisson *distribution* measures the number of occurrences in a fixed interval. Examples are the number of calls to a call center in a given 15 minute time span, the number of smiles from a stranger in the streets in a given hour, the number of shooting stars in the sky in a given year. Other examples are as follows.

- **Industrial:** Blow-outs per day for oil wells in the golden age for oil in Texas.
- **Sports:** Goals during a football (soccer) match.
- **Medicine:** Bone fractures in the ER during a given weekend.
- **Economics:** Unsolicited job applications to your company within a given month.

A Poisson process may also model occurrences in domains other than time. Distance, area, volume, and terabytes of data are examples of such domains. The number of hitch-hikers along a three mile stretch of a uniform road is most likely Poisson distributed. The number of sharks in a cubic mile of water in the Mexican Gulf is also Poisson distributed, as is the number of true transvestites at a showing of *Rocky Horror*. Other examples are as follows.

1) Typos on a given page of a newspaper.
2) Algae in a pint of water from a lake.
3) Number of supernovas in the Andromeda galaxy.
4) Number of bit errors in 10 petabytes of data.

The **Erlang** distribution measures the complementary property of the Poisson process: Whereas the **Poisson** distribution measures the number of occurrences x in an interval of width t, the **Erlang** distribution measures the time t it takes until x occurrences. See Section 10.3.

9.6.2 Calculating the Poisson Distribution

Example 9.6.1 $X \sim pois_{(8)}$, that is, it is Poisson distributed with parameter $\lambda = 8$. Find

1) μ_X
2) σ_X
3) $P(X \in \{7, 8, 9\})$
4) $P(X > 0)$.

Answer:

1) $\mu_X = \lambda = 8$
2) $\sigma_X = \sqrt{\lambda} = \sqrt{8} = 2.828\,43$
3) $P(X \in \{7, 8, 9\}) = f(7) + f(8) + f(9) = \dfrac{8^7}{7!} \times e^{-8} + \dfrac{8^8}{8!} \times e^{-8} + \dfrac{8^9}{9!} \times e^{-8}$

$$= 0.139\,587 + 0.139\,587 + 0.124\,077 = 0.403\,25$$

4) $P(X > 0) = 1 - P(X = 0) = 1 - f(0) = 0.999\,665$. □

An example of how the Poisson distribution is a limiting case of the binomial distribution with large n and small p is this: look at the atoms in a lump of radioactive material. A good and vivid image of it is that each atom has a 10^{12}-sided die, a $D_{1\,000\,000\,000\,000}$, and tosses it in wait for the single event (perhaps a 1) signalling that the atom should split spontaneously. The die may be a bit smaller or a bit bigger, depening on which kind of atom we are looking at. But given $n = 1\,000\,000\,000\,000$ atoms, how many will spontaneously undergo fission during one round of die tossing?

We could, in principle, calculate this by means of the binomial distribution, but in practice, $n! = 1\,000\,000\,000\,000!$ is a bit much to handle, even for advanced computing equipment. Luckily, if we go even further, by increasing n to infinity, and at the same time diminishing p such that np stays constant ($\lambda = np$), our calculations suddenly turn very tractable: we get the Poisson distribution with parameter $\lambda = np$. And the larger the value of n, for the same value of λ, the better an approximation the Poisson distribution is.

Example 9.6.2 A test of Crazy Mint's hard candies has shown that there are on average 3.7 cracked candies per 100 bags. Find the probability that next time you buy 100 Crazy Mint hard candies, you will find exactly 1 cracked candy. Then find the probability that you will find 3. Finally, find the probability that you will find 5, 6, or 7 cracked candies.

Answer: Cracked candies are rare,[2] so we consider their occurrences to be Poisson distributed with some parameter λ.

Since $\mu = \lambda$ for Poisson distributions, then $\lambda = \mu$, and we already know that $\mu = 3.7$. The number of cracked candies, X, is then Poisson distributed with parameter $\lambda = 3.7$, and so

$$f(x) = pois_{3.7}(x) = \frac{3.7^x}{x!} \cdot e^{-3.7}.$$

Then

$$P(X = 1) = f(1) = \frac{3.7^1}{1!} \cdot e^{-3.7} \approx 0.091$$

$$P(X = 3) = f(3) = \frac{3.7^3}{3!} \cdot e^{-3.7} \approx 0.208$$

$$P(X \in \{5, 6, 7\}) = POIS_{3.7}(7) - POIS_{3.7}(4) = \underline{0.278}. \qquad \square$$

Normal Approximation

Example 9.6.3 $X \sim pois_{100}$. Find $P(X \in \{88, 89, \dots, 102, 103\})$.

Answer: $\mu_X = \lambda = 100$ and $\sigma_X = \sqrt{\lambda} = \sqrt{100} = 10$ give Normal approximation $\phi_{(100,10)}$. Rule 10.1.13 then says that

$$P(87 < X \leq 103) \approx \Phi_{(100, 10)}\left(103 + \frac{1}{2}\right) - \Phi_{(100, 10)}\left(87 + \frac{1}{2}\right) = \underline{0.531}.$$

By comparison, direct calculation on the Poisson distribution gives 0.539. $\square$

9.7 Discrete Distributions: Overview

Which distributions do we use when? And when may we use an approximation? The section for each distribution gives a numerical guideline for when to use a simpler distribution as an approximation. Outside of this, some handy rules for deciding which distribution to use are as follows.

- You have a finite population where each element has an equal chance of being chosen: *Uniform distribution.*

---

[2] Technically, we have satisfied the rule of thumb for when the Poisson approximation to the binomial distribution may be applied, since $n^{0.31}p < 0.47$.

- The number of trials is fixed, and you measure the probabilities of the number of successes.
  - For sampling from a small population where the element you sample disappears or is otherwise excluded from further sampling after having been chosen once: *Hypergeometric distribution.*
  - For Bernoulli processes, or by sampling from medium-sized populations where the outcomes so far have negligible effect on the next sampling probability: *Binomial distribution.*
  - For Poisson processes or sampling from large populations where the probability of success is low: *Poisson distribution.*
- The number of successes is fixed, and you measure the probabilities of the number of trials before the required number of successes is achieved.
  - You are waiting for 1 success: *Geometric distribution.*
  - You are waiting for more than 1 success: *Negative binomial distribution.*

9.8 Exercises

9.8.1 Discrete Uniform Probability Distribution

The discrete uniform distribution has been omitted as a section so that the student may have at hand a tractable probability distribution to build up and study its properties. The first assignment is the key.

1 X follows a discrete uniform distribution over the n numbers $\{a, a + 1, \ldots, b - 1, b\}$ if $P(X = c) = 1/n$ whenever c is in the list, and 0 otherwise. Use the rules from chap. 7 to answer the problems below.
  (a) Find μ_X.
  (b) Find σ_X^2.
  (c) Sketch the probability distribution function (pdf) $f(x)$, and the cumulative probability distribution function (CDF) $F(x)$ for the uniform distribution.

2 The stochastic variable X follows a uniform probability distribution over the integers $\{0, \ldots, 99\}$. What, then, is $P(X \in \{2, 3, 5, 7, 11, 13, 17, 19\})$?

3 You have a fair D_{100} die with the numbers $0 \ldots 99$ printed on each face. What is the probability that a toss yields a prime number below 20?

4 For the two questions above: what are the values of μ_X and of σ_X? (You must have solved problem 1 to answer this.)

5 You have a fair D_{60} die with the numbers $0 \ldots 59$ printed on each face. Find the expected value and standard deviation of the outcomes from a single toss, and find the probability that 4 is one of its digits. (You must have solved problem 1 to answer this.)

9.8.2 Bernoulli Distribution, $bern_p$

6 What is the Bernoulli distribution, and what is its use?

7 For which value of the parameter p does a Bernoulli distributed stochastic variable X have the largest expected value μ_X?

8 For which value of the parameter p does a Bernoulli distributed stochastic variable X have the largest variance σ_X^2?

9 What is the maximal value of the precision τ_X for a Bernoulli distributed stochastic variable X? Which values of p give the largest values of the precision?

10 Odds: the probability that Bayern Munich wins the Bundesliga next year is 81.25%.
(a) Find o_f as a decimal number.
(b) Find o_f as an integer fraction $a : b$.
(c) Find o_a.
(d) Find the continental European odds.
(e) What is the fair price for a bet that gives you a return of 100 euro if Bayern Munich wins the Bundesliga next year?
(f) If you bet 100 euro with the odds given above, and Bayern Munich wins the Bundesliga, how much do you win?
(g) A British bookmaker offers you odds of 1 : 10 against. Would it pay to place the bet that Bayern Munich wins the Bundesliga with him?

9.8.3 Binomial Distribution, $bin_{(n,p)}$

11 The stochastic variable X is the number of T from five trials in a Bernoulli process with parameter $p = 0.55$. Find the distribution function f of X, and calculate its expected value and standard deviation. What is $P(X \in \{0, 1, 2\})$?

12 You have a motorized lawn mower whose probability of starting the first time you pull the cord is $p = 0.55$. If you try this on five separate occasions: what is the probability that, on two or fewer of those occasions, it would start on the first pull of the cord?

13 The stochastic variable Y is the number of T from 14 trials in a Bernoulli process with parameter $p = 0.4$. Find the distribution function g of Y, and calculate its expected value and standard deviation. What is $P(Y \in \{3, 4, 5\})$?

14 In a delayed food shipment, $p = 40\%$ have passed their expiry date. If you pick 14 random food items, what is the probability that 3, 4, or 5 have passed their expiry date?

15 The stochastic variable Z is the number of T from 20 trials in a Bernoulli process with parameter $p = 0.8$. Find the distribution function h of Z, and calculate its expected value and standard deviation. What is $P(Z \in \{0, 1, 2, 3, 4, 5\})$?

16 The craftsmen where you live on average keep 80% of their deadlines. You hire 20 craftsmen for an expansion of your home. What is the probability that no more than 5 go past their deadlines?

17 Assume that precisely 30% of the population will be voting Liberal Democrat at the next election. You are conducting a poll. What is the probability that between 27 and 33% of a random sample will tell you that they are going to vote Liberal Democrat, if ...
(a) you poll 10 people;
(b) you poll 100 people;
(c) you poll 1 000 people;
(d) you poll 10 000 people?

18 DiscCo's latest 8TB hard drive has an error rate given by the fact that the probability of error in a given bit within a given hour is $p = 50 \times 10^{-15}$. (We use the fact that $8\text{TB} = 8 \times 8 \times 1024^4 = 70\,368\,744\,177\,664$ bits.)
(a) What is the expected number of errors in 1 hour?
(b) What is the variance in the number of errors in 1 hour?
(c) What is the probability of at most 5 errors within 1 hour?

19 A repair protein by the name of *helicase* moves along a strand of DNA in a random walk driven by water molecules. If we assume an upwards movement with probability $p = 0.51$, and then down with probability $q = 0.49$, ...
(a) what are the possible positions relative to the starting point after 10 steps;
(b) what is the probability of having moved at least 4 units up in 10 steps;

(c) what is the expected number of units moved upwards after 10 000 steps;

(d) what is the variance in the number of units moved upwards after 10 000 steps;

(e) what is the probability of having moved at least 100 units upwards after 10 000 steps?

9.8.4 Hypergeometric Distribution, $hyp_{(n,S,N)}$

20 A stochastic variable X follows a hypergeometric distribution with parameters $N = 20$, $S = 13$, and $n = 5$. Find the distribution function f of X, and calculate its expected value and standard deviation. What is $P(X \in \{0, 1, 2\})$?

21 In a bag of assorted candies, there are 13 Almond Joys and 7 Twizzlers left. You sample 5 candies at random. What is the probability that you get 2 or fewer Almond Joys?

22 A stochastic variable X follows a hypergeometric distribution with parameters $N = 30$, $S = 7$, and $n = 9$. Find the distribution function f of X, and calculate its expected value and standard deviation. What is $P(X \in \{0, 1\})$?

23 A bag contains 23 copper rings and 7 gold rings. You reach into the bag and grab 9 rings. What is the probability that at most 1 of them is a gold ring?

24 You have a hand in 5-card poker with 2 kings, 1 ace, and 2 numbered cards. You consider trading in the two numbered cards and the ace for three new cards. What is the probability that at least one of your three new cards is an ace?[3]

9.8.5 Poisson Distribution, $pois_\lambda$

25 X is Poisson distributed and $\lambda = 2.8$. Calculate μ_X, σ_X, $P(X = 1)$, and $P(X \neq 1)$.

26 X is Poisson distributed and $\mu_X = 3.5$. Calculate σ_X, $P(X = 5)$, and $P(X > 0)$.

27 X is Poisson distributed and $\sigma_X = 5$. Calculate μ_X, $P(X = 7)$, and $P(X \leq 2)$.

---

[3] Since all the other cards are unknown to you, the remainder is all 47 cards, including those of the other players. The distribution of the number of aces is thus hypergeometric with parameters $N = 47$, $S = 3$, and $n = 3$, and your job is to find $f(1) + f(2) + f(3)$ – or, equivalently, $1 - f(0)$.

28 X is Poisson distributed and $P(X = 0) = 0.1$. Calculate λ, μ_X, σ_X, and $P(X = 1)$.

29 The stochastic variable X is Poisson distributed with parameter $\lambda = 2.37$. Find the distribution function f for X, and calculate the expected value and standard deviation. What is $P(X \in \{0, 1\})$?

30 You have tested a lump of Cryptonite with a Geiger counter, and found that it gives an average of 2.37 clicks on the counter per minute. Find the probability of 1 click or less in the next minute.

31 The stochastic variable X is Poisson distributed with parameter $\lambda = 1$. Find the distribution function f of X, and calculate the expected value and standard deviation. What is $P(X \in \{0\})$?

32 The gambling joint *Backstab Saloon* by Crater Lake in Oregon evict an average of one cheat per hour. You are on your way to Seattle to talk to "Bill" about a lucrative job, but have an hour to spare at Backstab. You position yourself outside the joint in hopes of seeing how they evict a cheat. What is the probabilty that you won't get to see any eviction?

33 X, the number of foxes the Easter Bunny sees in the forest on a given day, is Poisson distributed. During the seven days of Week 5, Mr Bunny saw respectively $1, 5, 4, 7, 3, 2$, and 8 foxes. By an odd coincidence, μ_X and the average number of foxes in Week 5 are exactly the same. Calculate μ_X, and use that to answer the following questions.
   (a) What is λ?
   (b) What is σ_X?
   (c) What is the probability that the Easter Bunny will see precisely five foxes today?

9.8.6 Overview and General

34 The difference between the binomial and uniform distributions.
   (a) You toss a fair D_{12} whose faces have values $1, \ldots, 12$. What is the probability of an outcome of 3 or less?
   (b) You toss a fair coin 12 times. What is the probability of 3 or fewer heads? Why does this answer differ from the problem above?

35 **The PirateBay problem:** Find the distribution for yourself! Is it uniform, binomial, hypergeometric, or Poisson? For some of the problems, you should use an approximation rather than try to calculate exactly. You may indicate the exact distribution, but use the approximation for the calculation.

One the one side, we have five activists against music and video piracy: Martha, Orrin, Tamara, and Otto. Their job is to gather the IP numbers of pirates. On the opposite side, we have four independent pirates: the students Ringo Smith, John Washington, Paul Dibley, and George Atwick.

(a) One evening, 28 people downloaded an album by the Norwegian band Ragnarök from PirateBay. Among them were our four students. Martha's quota that evening was 1 single IP-address from that torrent. What is the probability that her catch was one of the four students? Which distribution do you use for your calculation?

(b) Tamara followed a torrent at PirateBay, where 37 people downloaded John Denver's last album. Among them were three of our students. Tamara's quota was 5 IP addresses. What is the probability that at least 2 of these 5 were from our students? Which distribution do you use for your calculation?

(c) Orrin watched a torrent at a Norwegian tracker, where 10 000 eager young boys downloaded Mira Craig's latest album. All our 4 students were among them. Orrin's target was 300 IP addresses. What is the probability that his catch will include precisely 2 of our students? Which distribution do you use for your calculation?

(d) Otto followed at Swedish porn torrent with 16 384 participants (in the torrent, not the film!). He sampled all IP addresses ending in 133. Assume that these three digits are evenly distributed from 000 to 255. What is the expected number of IP addresses Otto will gather? What is the probability that his catch will be exactly 15 pirates? Which distribution do you use for your calculation?

36 **The Autopol problem:** There is a new toll ring around the town Autopol. There are some initial hiccups, so the company running the toll booths only get to check on one in every four cars passing through the toll gates. But the cars that have not paid when checked get a major fine. The cruisers Hans Rust-Holch and Fritz Königsegg want to test the new system, and discuss the probabilities of getting caught dodging the toll payments.

(a) What is the probability that they will be caught dodging if they do it once (1 time)?

(b) What is the probability that they will be caught dodging twice if they dodge 5 times?

(c) What is the smallest number of times they must dodge the toll payment for the expected number of times to get caught to be at least 2? What is the standard deviation?

(d) What is the largest number of times they can dodge the payment before the probability of getting caught at least once exceeds 50%?

(e) After 12 dodgings, Hans and Fritz have been caught 4 times. Seven of the 12 times they dodged the payment, Hans sat in the rear, reading

"Wunderbaum Heute". What is the probability that he read that magazine exactly 3 of the times they were caught?

(f) On average, every 1024th car passing the toll ring around Autopool has "Wunderbaum Heute" lying in the rear seats. The magazine "Wunderbaum Heute" has no impact on the probability of getting caught, so WH readers are caught just the same as everyone else. In the week Hans and Fritz tested out the dodging, 16 384 drivers were caught. What is the probability that exactly 15 of these had "Wunderbaum Heute" in the backseats?

10

Continuous Distributions

CONTENTS

In this chapter, we will look at some key discrete probability distributions. Each distribution begins with a box with the most commonly used properties of the distribution. If the distribution has its own symbol ψ, we will use that to designate it, and the uppercase version of the symbol, Ψ, for the cumulative distribution function. If not, we will use the *name* and *NAME* in these roles. By means of the distribution's *parameters* $p_1, p_2, \ldots, p_k$, we may then describe the ...

- probability distribution (pdf) $\psi_{(p_1,\ldots,p_k)}(x)$ or $name_{(p_1,\ldots,p_k)}(x)$;
- cumulative probability distribution (CDF) $\Psi_{(p_1,\ldots,p_k)}(x)$ or $NAME_{(p_1,\ldots,p_k)}(x)$;
- inverse cumulative distribution (iCDF) $\Psi^{-1}_{(p_1,\ldots,p_k)}(x)$ or $NAME^{-1}_{(p_1,\ldots,p_k)}(x)$;
- expected value $\mu_X = E[X]$, together with a formula for calulating it from the parameters $p_1, \ldots, p_k$;
- variance σ_X^2, together with a formula for calulating it from the parameters $p_1, \ldots, p_k$.

You will also be shown how to perform calculations on these distributions and their properties – primarily in Mathematica, but where possible also on CASIO, TI, and HP calculators. For details, see the beginning of Chapter 9.

For a general exposition of continuous probability distributions, see Section 7.3.

The Bayesian Way: Introductory Statistics for Economists and Engineers, First Edition.
Svein Olav Nyberg.
© 2019 John Wiley & Sons, Inc. Published 2019 by John Wiley & Sons, Inc.

10.1 Normal Distribution, $\phi_{(\mu,\sigma)}$

Normal distribution: $X \sim \phi_{(\mu,\sigma)}$ is a continuous stochastic variable taking values in $\mathbb{R}$; it is Normally distributed with parameters $\mu \in \mathbb{R}$, and $\sigma > 0$.

pdf: $\phi_{(\mu,\sigma)}(x) = k \times e^{-(x-\mu)^2/2\sigma^2}$, where $k = 1/\sqrt{2\pi}\sigma$

CDF: $\Phi_{(\mu,\sigma)}(x)$	iCDF: $\Phi^{-1}_{(\mu,\sigma)}(p)$
CASIO: NormCD$(-10^{99}, x, \sigma, \mu)$ **HP:** normald_cdf(μ, σ, x) **TI:** normalcdf$(-10^{99}, x, \mu, \sigma)$	InvNormCD$(-1, p, \sigma, \mu)$ normald_icdf(μ, σ, p) invNorm(p, μ, σ)

Wolfram: NormalDistribution$[\mu, \sigma]$

You may also use $\phi(x) = \phi_{(0,1)}(x)$, the *standard Normal distribution*, and use tables for the values of $\Phi(x)$. Then $\Phi_{(\mu,\sigma)}(x) = \Phi\left(\dfrac{x-\mu}{\sigma}\right)$.

Expected value: $\mu_X = \mu$	Variance: $\sigma_X^2 = \sigma^2$

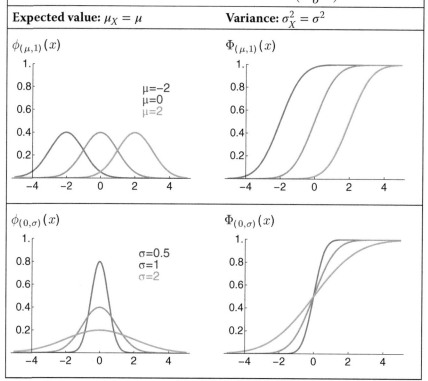

10.1.1 Applications: Nearly Everything!

Show a person in the street the curve of the Normal distribution, and the immediate response will be *statistics!* No other function graph is as characteristic of statistics as precisely the Normal distribution. The Normal distribution appears so often in statistical analysis that its name stems from precisely that: it is the *Normal* distribution. The Normal distribution is completely characterized once you know its means μ and standard deviation σ, and therefore the probability distribution normally uses precisely these two values as its parameters: $\phi_{(\mu, \sigma)}$. Variants are using the variance, writing $\phi_{(\mu, \sigma^2)}$, while a small minority prefer the precision $\tau = 1/\sigma^2$, writing $\phi_{(\mu, \tau)}$. As far as the curve is concerned, μ indicates its middle, where its maximum is located, whereas σ indicates how it is spread, and is (as stated) its standard deviation. It is more common to write ϕ as N, but since the use of Φ for the cumulative distribution is universal, we will stick with consistency and use ϕ for the distribution function.

The name "Normal distribution" is easily seen to be true when we see its many areas of application. Typical domains covered by the Normal distribution are the following.

- The weight of animals and humans often follow a Normal distribution around the average.
- Performance: how fast 16 year olds at a school will complete a 100 yard dash will often follow a Normal distribution.
- Scores on proper IQ tests are Normally distributed around a middle of 100.
- Product quality: the length of the firewood you buy will be Normally distributed around the average length, and the standard deviation will be smaller the more experienced the woodcutter.
- Scientific measurements: different measurements of one and the same magnitude may give different results for different measuring tools or even different measurings with the same tool. The results will follow a Normal distribution, and the standard deviation indicates *measurement uncertainty*.

Another important use of the Normal distribution is as an approximation to other distributions, through the so-called *Normal approximation*. This does in particular hold for sums of identically distributed, independent stochastic variables, since such sums tend to a Normal distribution when the number of variables increases. This has its own subsection at the end of this section on the Normal distribution.

10.1.2 Calculating the Normal Distribution

The values of Φ may be found in several ways.

Example 10.1.1 $X \sim \phi_{(3.1, 5.7)}$. What is $P(X \leq 12)$?

Answer: $P(X \leq 12) = \Phi_{(3.1, 5.7)}(12)$. We calculate the answer in several ways.

1) We use the fact that $\Phi_{(\mu, \sigma)}(x) = \Phi((x - \mu)/\sigma)$, and find the values for Φ in Table C.6.

$$\Phi_{(3.1, 5.7)}(12) = \Phi\left(\frac{12 - 3.1}{5.7}\right) = \Phi(1.561\,4) \approx \Phi(1.56) = \underline{0.940}.$$

2) Mathematica: CDF[NormalDistribution[3.1, 5.7], 12] gives <u>0.940 786</u>.

3) CASIO: NormCD(-10^{99}, 12, 5.7, 3.1) gives <u>0.940 785</u>.

4) TI: normalcdf(-10^{99}, 12, 3.1, 5.7) gives <u>0.940 785</u>.

5) HP: normald_cdf(3.1, 5.7, 12) gives <u>0.940 785</u>. ☐

Example 10.1.2 Abe has waded into the middle of his favourite salmon river *Flomma*. The weight of the next salmon he catches in Flomma is Normally distributed with mean $\mu = 3.2$ kg and standard deviation 1.3 kg. What is the probability that the next salmon Abe catches weighs between 2 and 5 kg?

Answer: $\mu = 3.2$ and $\sigma = 1.3$, and we need to find $P(X \in (2, 5))$, so

$$P(X \in (2, 5)) = \Phi_{(3.2,\, 1.3)}(5) - \Phi_{(3.2,\, 1.3)}(2) = 0.738\,931.$$

The probability that the next salmon Abe catches is between 2 and 5 kg, is 74%. We find the probability by means of calculation tools, thus:

- Mathematica: Probability[$2 < x < 5$, $x \approx$ NormalDistribution[3.2,1.3]];
- CASIO: NormCD(2, 5, 1.3, 3.2);
- TI: normalcdf(2, 5, 3.2, 1.3);
- HP: normald_cdf(3.2, 1.3, 5) $-$ normald_cdf(3.2, 1.3, 2). ☐

10.1.3 z_α, the Inverse of Φ

The most common calculation for a Normally distributed variable $X \sim \phi_{(\mu,\sigma)}$ is to find $P(X \leq a)$. The second most common is the inverse question: given a probability p, for which value of x is $P(X \leq x) = p$? As we saw in Section 7.4, the answer is in the iCDF, so $x = \Phi^{-1}_{(\mu,\sigma)}(p)$. For both theoretical and practical pre-calculator reasons, this inverse is often calculated via the inverse of the cumulative standard Normal distribution, $\Phi^{-1} = \Phi^{-1}_{(0,1)}$, giving us $x = \Phi^{-1}_{(\mu,\sigma)}(p) = \mu + \sigma \times \Phi^{-1}(p)$. This method is so common that we give $\Phi^{-1}(\alpha)$ its own name: z_α. $\Phi(x)$ and $z_\alpha = \Phi^{-1}(\alpha)$ are illustrated in Figure 10.1.

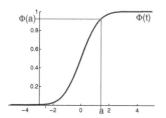

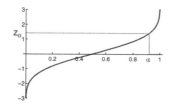

(a) The cumulative standard Normal distribution Φ.

(b) The inverse of Φ, the function $z_\alpha = \Phi^{-1}(\alpha)$.

Figure 10.1 The cumulative standard normal distribution Φ and its inverse.

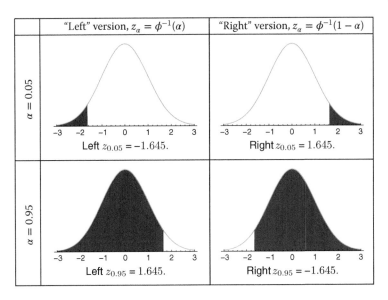

Figure 10.2 Left and right versions of z_α for $\alpha = 0.05$ and $\alpha = 0.95$.

Definition 10.1.3 $z_\alpha = \Phi^{-1}(\alpha)$

It is more common to use z_α to designate the *right* tail (that is, that z_α is the solution to $P(X \geq z) = \alpha$) rather than the left tail of the straight inverse (that is, that z_α is the solution to $P(X \leq z) = \alpha$). But they are easy to use around one another, since the "right" z_α equals the "left" $z_{1-\alpha}$, and conversely. We will, as we said, stick with the "left" z_α. See Figure 10.2 for illustration of the difference and complementarity.

The symmetry between left and right is succinctly expressed in the following rule.

Rule 10.1.4 $z_p = -z_{1-p}$.

Example 10.1.5 $z_{0.8} = -z_{0.2} = -(-0.841\,621) = 0.841\,621.$ □

Rule 10.1.6 Let $X \sim \phi_{(\mu,\sigma)}$. Then the solution x to $P(X \leq x) = \alpha$ is given by the fact that $x = \Phi^{-1}_{(\mu,\sigma)}(\alpha) = \mu + \sigma \cdot z_\alpha$.

We may calculate the values of the iCDF of the Normal distribution, $\Phi^{-1}_{(\mu,\sigma)}$, as follows.

1) Standard Normal: Find z_α, and use that $\Phi^{-1}_{(\mu,\sigma)}(\alpha) = \mu + \sigma \cdot z_\alpha$.
2) Mathematica: $\Phi^{-1}_{(\mu,\sigma)}(p)$ is InverseCDF[NormalDistribution[μ, σ],p].
3) CASIO: InvNormCD(p, σ, μ).
4) TI: invNorm(p, μ, σ).
5) HP: normald_icdf(μ, σ, p).

Example 10.1.7 Abe is fishing in Flomma (Example 10.1.2) and wonders how heavy the 90% heaviest salmon there are.

Answer: We understand that Abe wants to find an x such that $P(X > x) = 0.9$, that is, $P(X \le x) = 0.1$. Since $X \sim \phi_{(3.2,\, 1.3)}(x)$, then $x = \Phi^{-1}_{(3.2,\, 1.3)}(0.1)$. We calculate the value as follows.

- The table gives us that $z_{0.1} = -1.281\,6$, so

$$x = \mu + z_{0.1}\sigma = 3.2 - 1.281\,6 \cdot 1.3 = 1.533\,92.$$

- Mathematica: InverseCDF[NormalDistribution[3.2, 1.3],0.1] gives 1.533 98.
- CASIO: InvNormCD(0.1, 1.3, 3.2) gives 1.533 98.
- TI: invNorm(0.1, 3.2, 1.3) gives 1.533 98.
- HP: normald_icdf(μ, σ, p).

So 90% of the salmon in his river are heavier than 1.53 kg. □

Example 10.1.8 (Continued from Example 10.1.7) Abe also wants to know how heavy the 90% *lightest* salmon are.

Answer: Calculation tools do this in the same way as above, but with new numbers, but most tables stop at $p = 0.1$ or 0.2. To get past this, we make use of Rule 10.1.4:

$$x = \mu + z_{0.9}\sigma = \mu - z_{0.1}\sigma = 3.2 - (-1.2816) \times 1.3 = 4.866\,08.$$

So 90% of the salmon weigh less than 4.87 kg. □

Example 10.1.9 In the mass production of vehicles for land based transport (cars, trains, buses, etc.), the main producer will set certain performance standards for the production of parts: for the measured specs, the mean ± four standard deviations should be within some set toleration limits. The underlying assumption is that these properties are Normally distributed. A car consists of 15 000 parts.

(a) What is the probability that a car has components outside of the tolerance limits?

(b) If we assume that all deviations outside of $\pm$ six sigma will lead to complaints, how many complaints may be expected on a production of one million cars?

Answer: This will at the outset seem to be a problem solely about the Normal distribution, but it illustrates how the distributions work together. We calculate the following.

(a) The probability that a given component lies with the $\pm 4\sigma$ limit is

$$p = \Phi(4) - \Phi(-4) = 0.999\,936\,66.$$

The probability that all 15 000 components of a car are within the limits is then $p^{15\,000} = 0.386\,7$. The probability of one or more components being outside the limits is then

$$1 - p^{15\,000} = 1 - 0.386\,7 = 0.613\,3.$$

(b) The probability that a given component lies within the $\pm 6\sigma$ limit, is

$$r = \Phi(6) - \Phi(-6) = 0.999\,999\,998\,026\,824\,7.$$

The probability that all 15 000 components of a car are within the limits is then $r^{15\,000} = 0.999\,970\,402\,8$. The number of complaints will be binomially distributed as $bin_{1\,000\,000,1-r}$. The expected number of complaints is then

$$n \cdot (1 - r) = 100\,000\,0 \times 0.000\,029\,60 = 29.6. \qquad \square$$

10.1.4   The Sum of Independent Normal Distributions

The sum of independent Normally distributed variables will itself follow a Normal distribution. The most common sums are the sum or difference of *two* variables. We write the special case of two components first, and the general case after that.

Rule 10.1.10   *For $Z = X + Y$, we have*

$$\mu_Z = \mu_X + \mu_Y \qquad \sigma_Z^2 = \sigma_X^2 + \sigma_Y^2,$$

whereas for $W = X - Y$, we have

$$\mu_W = \mu_X - \mu_Y \qquad \sigma_W^2 = \sigma_X^2 + \sigma_Y^2.$$

The general form is weighted sums of stochastic variables. The rule then modifies to the following.

Rule 10.1.11   *Let $X_k \sim \phi_{(\mu_k,\sigma_k)}$ and $b_k \in \mathbb{R}$, and let $Y = r + \sum_{k=1}^{n} b_n X_n$. Then $Y \sim \phi_{(\mu,\sigma)}$, where*

$$\mu = r + \sum_{k=1}^{n} b_n \mu_n \qquad and \qquad \sigma^2 = \sum_{k=1}^{n} b_n^2 \sigma_n^2.$$

Important special cases are as follows:

- if $X \sim \phi_{(\mu,\sigma)}$, and $k \in \mathbb{R}$, and $Y = kX$, then $Y \sim \phi_{(k\mu,|k|\sigma)}$;
- if $X \sim \phi_{(\mu,\sigma)}$, and $r \in \mathbb{R}$, and $Z = X + r$, then $Z \sim \phi_{(\mu+r,\sigma)}$;

and one special case merits its own separate rule, as follows.

Rule 10.1.12 *Given n stochastic variables $X_k \sim \phi_{(\mu,\sigma)}$, with mean $\bar{X} = (X_1 + \cdots + X_n)/n$, we have $\bar{X} \sim \phi_{(\mu,\,\sigma/\sqrt{n})}$.*

10.1.5 Normal Approximation

A central reason why the Normal distribution is the most normal probability distribution is that the sum of many small, random variations add up to something looking a lot like the Normal distribution. Not only that, but many of the other probability distributions become increasingly similar to the Normal distribution when their parameters grow large. In these cases, it is often expedient to replace the exact calculation by a Normal approximation.

Rule 10.1.13 *The Normal approximation to the (distribution of the) stochastic variable X is*

$$X \approx \phi_{(\mu_X,\sigma_X)}(x),$$

where we use $\approx$ to indicate "has approximate distribution". For continuous X,

$$P(X \le n) = \Phi_{(\mu_X,\sigma_X)}(n).$$

For discrete X, the formula for a point probability is (as illustrated in Figure 10.3)

$$P(X = n) \approx \Phi_{(\mu_X,\sigma_X)}\left(n + \frac{1}{2}\right) - \Phi_{(\mu_X,\sigma_X)}\left(n - \frac{1}{2}\right).$$

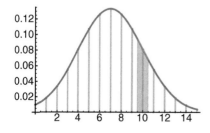

Figure 10.3 Point probability $\Rightarrow$ interval probability.

That is, the discrete value n corresponds to the interval $n \pm \frac{1}{2}$, and so for the calculation of cumulative discrete distributions, we add the so-called *continuity correction*:

$$P(X \leq n) = \Phi_{(\mu_X, \sigma_X)}\left(n + \frac{1}{2}\right).$$

Example 10.1.14 X is a discrete stochastic variable with $\mu_X = 14.7$ and $\sigma_X = 3.4$. Employ the Normal approximation to calculate $P = P(X \in \{14, 15, 16, 17, 18, 19\})$.

Solution: $P = \Phi_{(14.7, 3.4)}(19.5) - \Phi_{(14.7, 3.4)}(13.5) = \underline{0.558\,924}$.   □

Example 10.1.15 Y er is a continuous stochastic variable with $\mu_Y = 42$ and $\sigma_Y = 15$. Employ the Normal approximation to calculate $P(Y \in (20, 40))$.

Solution: $P(Y \in (20, 40)) = \Phi_{(42, 15)}(40) - \Phi_{(42, 15)}(20) = \underline{0.375\,732}$.   □

Precisely when you should use the Normal approximation instead of exact calculation depends on the balance between error tolerance in your calculations and your need for simplification and speed of calculation. In these two chapters about the different probability distributions, we provide rules of thumb for each distribution for when a Normal approximation is acceptable. But keep in mind that these are rules of thumb, and not universally valid criteria regardless of your actual needs.

We round off this section with a known theorem that is closely related to the Normal approximation, and that is often invoked to explain the ubiquitousness of the Normal distribution mathematically.

Rule 10.1.16   Central limit theorem: *Let X_k be a sequence of independent stochastic variables sharing the same $E[X_k] = \mu$ and $Var(X_k) = \sigma^2$, let $\bar{X}_n$ be the average of the first n variables, and $Z_n = (\bar{X} - \mu)/(\sigma/\sqrt{n})$. Then $\lim\limits_{n\to\infty} Z_n = Z \sim \phi_{(0,1)}(x)$.*

10.1.6   In Brief

We illustrate the anatomy of the Normal distribution in Figure 10.4 by looking at a representative Normal distribution.

We notice that

- the peak lies over $x = \mu$;
- $\phi(x)$ is symmetrical around μ;
- $\phi_{(\mu,\sigma)}(\mu) = \dfrac{1}{\sqrt{2\pi\sigma^2}}$;

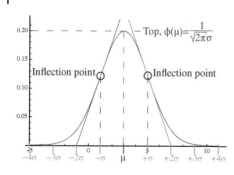

- inflection points at $x = \mu \pm \sigma$;
- a rough graph of the curve will seem to hit the x-axis on each side at between $\mu \pm 3\sigma$ and $\mu \pm 4\sigma$.

10.2 Binormal Distribution, $\phi_{(\mu,\Sigma)}$*

Binormal distribution: If $Z = \begin{bmatrix} X \\ Y \end{bmatrix}$ follows a binormal distribution, it takes values in $\mathbb{R}^2$, and its probability distribution is

$$f_Z(z) = N_{(\mu,\Sigma)}(z) = k \times e^{-\frac{1}{2}(z-\mu)^{\mathrm{T}}\Sigma^{-1}(z-\mu_z)}, \quad \text{where } k = \frac{1}{\sqrt{(2\pi)^2|\Sigma|}}$$

- **Expected value:** $\mu = \mu_z = \begin{bmatrix} \mu_x \\ \mu_y \end{bmatrix} \in \mathbb{R}^2$

- **Covariance matrix:** $\Sigma = \Sigma_z = \begin{bmatrix} \sigma_x^2 & \sigma_{xy} \\ \sigma_{xy} & \sigma_y^2 \end{bmatrix} \in \mathbb{R}^2 \times \mathbb{R}^2$

Mathematica: MultinormalDistribution $\left[\{\mu_x, \mu_y\}, \begin{pmatrix} \sigma_x^2 & \sigma_{xy} \\ \sigma_{xy} & \sigma_y^2 \end{pmatrix} \right]$

3D plot of $f_z(x, y)$

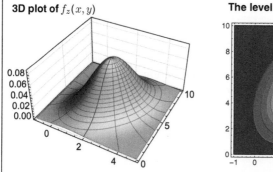

The level curves of $f_z(x, y)$

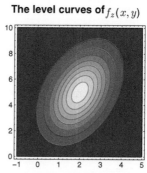

Rule 10.2.1 Conditional and marginal: *The* conditional *variable* $X|_y$ *follows the probability distribution* $f_{x|y}(x) = \phi_{(\mu_{X|y}, \sigma^2_{X|y})}$, *where*

$$\mu_{X|y} = \mu_x + \frac{\sigma_{xy}}{\sigma^2_y}(y - \mu_y)$$

$$\sigma^2_{X|y} = \sigma^2_y(1 - \rho_{xy})^2 \quad \text{where } \rho_{xy} = \sigma_{xy}/\sqrt{\sigma^2_x \sigma^2_y}.$$

The marginal *probability distribution for X is*

$$f_x(x) = \phi_{(\mu_x, \sigma_x)}(x).$$

Correspondingly for Y, when we switch the roles of x and y.

These properties are illustrated in Figure 10.5.

10.2.1 Special Case: X and Y Independent

The simplest kind of binormal distribution is when X and Y are independent. We then know that $f_{x|y} = f_x$, $f_{y|x} = f_y$, and that $f_z(x, y) = f_x(x) \times f_y(y)$ (Rule 8.2.19). Since $\sigma_{xy} = 0$, then Σ is a diagonal matrix:

$$\Sigma = \begin{bmatrix} \sigma^2_x & 0 \\ 0 & \sigma^2_y \end{bmatrix} \quad \text{and} \quad \phi_{(\mu,\Sigma)} = \phi_{(\mu_x, \sigma_x)} \times \phi_{(\mu_y, \sigma_y)}.$$

The level curves of Z are now ellipses with the x- and y-axes as its principal axes. See Figure 10.6 for examples.

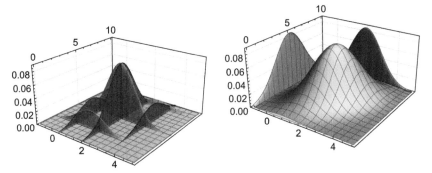

(a) The cross sections are proportional to the conditional distributions.

(b) The marginal distributions, scaled, in blue and green on the side planes.

Figure 10.5 Conditional and marginal distributions for a binormal distribution.

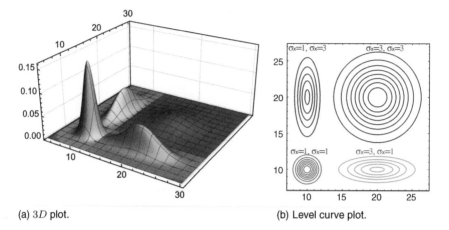

(a) $3D$ plot.   (b) Level curve plot.

Figure 10.6 The function stretches along each axis proportional to the σ for that axis.

10.2.2 The General Case as a Rotation of the Special Case

The general covariance matrix is

$$\hat{\Sigma} = \begin{bmatrix} \hat{\sigma}_x^2 & \hat{\sigma}_{xy} \\ \hat{\sigma}_{xy} & \hat{\sigma}_y^2 \end{bmatrix} = \begin{bmatrix} \hat{\sigma}_x^2 & \rho\hat{\sigma}_x\hat{\sigma}_y \\ \rho\hat{\sigma}_x\hat{\sigma}_y & \hat{\sigma}_y^2 \end{bmatrix},$$

where ρ is the correlation coefficient between the x- and the y-components. For all covariance matrixes $\hat{\Sigma}$, we may find another coordinate system by a rotation with angle θ, so that the covariance matrix is a diagonal matrix Σ in the rotated coordinate system. In brief, we say that we rotate the distribution, covariance matrix, and vector of expected values. From linear algebra, we have that the rotation matrix for rotating a vector $x \in \mathbb{R}^2$ by an angle θ counterclockwise is

$$R_\theta = \begin{bmatrix} \cos\theta & -\sin\theta \\ \sin\theta & \cos\theta \end{bmatrix}.$$

We see an example of a rotated (and shifted) binormal distribution in Figure 10.7.

When we rotate an independent binormal distribution with expectation vector μ and covariance matrix Σ by an angle θ counterclockwise, we get the dependent binormal distribution $\phi_{(\hat{\mu}, \hat{\Sigma})}$, where

$$\hat{\mu} = R_\theta\mu = \begin{bmatrix} \sigma_x\cos\theta - \sigma_y\sin\theta \\ \sigma_y\cos\theta + \sigma_x\sin\theta \end{bmatrix}$$

$$\hat{\Sigma} = R_\theta\Sigma R_{-\theta} = \begin{bmatrix} \sigma_x^2\cos^2\theta + \sigma_y^2\sin^2\theta & (\sigma_x^2 - \sigma_y^2)\cos\theta\sin\theta \\ (\sigma_x^2 - \sigma_y^2)\cos\theta\sin\theta & \sigma_x^2\cos^2\theta + \sigma_y^2\sin^2\theta \end{bmatrix}.$$

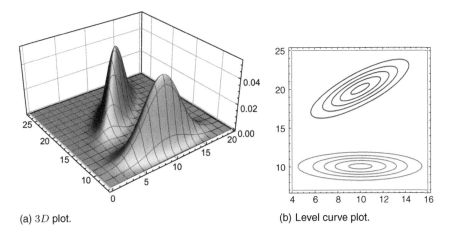

(a) $3D$ plot.   (b) Level curve plot.

Figure 10.7 An independent distribution (orange), and a 40° rotation of this (blue).

But what about the converse case, where we start with a non-independent covariance matrix $\hat{\Sigma}$? How do we find the independent covariance matrix Σ that it is the rotation of? The solution is to start by finding the angle θ, and then use the rotation formula above in reverse:

$$\theta = \frac{1}{2}\tan^{-1}\left(\frac{2\hat{\sigma}_{xy}}{\hat{\sigma}_x^2 - \hat{\sigma}_y^2}\right) \qquad \mu = R_{-\theta}\,\hat{\mu} \qquad \Sigma = R_{-\theta}\,\hat{\Sigma}R_{\theta}.$$

Example 10.2.2 The wind speed at East Cape (in m/s) is a stochastic variable $Z = (\hat{X}, \hat{Y}) \sim \phi_{(\hat{\mu},\hat{\Sigma})}$, where

$$\hat{\mu} = \begin{bmatrix} -1 \\ 3 \end{bmatrix}, \qquad \hat{\Sigma} = \begin{bmatrix} 12.1 & 4.7 \\ 4.7 & 9.7 \end{bmatrix},$$

and X is the eastward wind speed, whereas Y is the northward wind speed. If you want to rotate the coordinate system so that $Z = (X, Y)$ is independent,

1) what is the angle of rotation, θ;
2) what is the rotated coordinate of a vector x;
3) what is μ;
4) what is the diagonal covariance matrix Σ?

Answer:

1) $\theta = \frac{1}{2}\tan^{-1}\left(\frac{2\times4.7}{12.1-9.7}\right) = \underline{0.660\,409\,\text{rad}} = \underline{37.838\,6°}.$

2) The coordinate comes through a rotation with an angle $-\theta$, so the new coordinates are

$$R_{-\theta}\boldsymbol{x} = R_{-0.660\,4}\begin{bmatrix} x \\ y \end{bmatrix} = \begin{bmatrix} 0.789\,7x + 0.613\,4y \\ 0.789\,7y - 0.613\,4x \end{bmatrix}.$$

3) $\boldsymbol{\mu}$ is a rotation of $\hat{\boldsymbol{\mu}} = \begin{bmatrix} -1 \\ 3 \end{bmatrix}$. The formula for the general $\boldsymbol{x}$ gives

$$\boldsymbol{\mu} = R_{-0.660\,4}\hat{\boldsymbol{\mu}} = \begin{bmatrix} 0.789\,7 \times (-1) + 0.613\,4 \times 3 \\ 0.789\,7 \times 3 - 0.613\,4 \times (-1) \end{bmatrix} = \begin{bmatrix} 1.050\,5 \\ 2.982\,5 \end{bmatrix}.$$

4) $\Sigma = R_{-0.660\,4}\Sigma R_{0.660\,4}$. Use a calculation tool. In Mathematica:

RotationMatrix$[-0.660\,4]$.$\begin{bmatrix} 12.1 & 4.7 \\ 4.7 & 9.7 \end{bmatrix}$. RotationMatrix$[0.660\,4]$//MatrixForm.

We get

$$\Sigma = \begin{bmatrix} 15.750\,8 & 0.000\,085\,884\,6 \\ 0.000\,085\,884\,6 & 6.049\,23 \end{bmatrix}.$$

We *should* have had 0 outside of the diagonal, but got 0.000 085 884 6. This is due to numerical inaccuracy. Since statistics is the art of being approximate in an accurate way, and you know that the values outside of the diagonal really are 0, double check your setup, and if it still looks good, round off these values to 0:

$$\Sigma = \begin{bmatrix} 15.75 & 0 \\ 0 & 6.049 \end{bmatrix}.$$

$\square$

10.2.3 Multinormal Distribution

The generalization of a binormal distribution is a *multinormal* distribution. When $X = (X_1, \ldots, X_n)$ is a multinormal distribution over $\mathbb{R}^n$, then

$$f(\boldsymbol{x}) = N_{(\boldsymbol{\mu}, \Sigma)}(\boldsymbol{x}) = k \times e^{-\frac{1}{2}(\boldsymbol{x}-\boldsymbol{\mu})^T\Sigma^{-1}(\boldsymbol{x}-\boldsymbol{\mu})} \quad \text{where } k = \frac{1}{\sqrt{(2\pi)^n|\Sigma|}}.$$

Expected value and position parameter: $\boldsymbol{\mu} = \begin{bmatrix} \mu_1 \\ \cdots \\ \mu_n \end{bmatrix}$.

Covariance matrix and variance parameters: $\Sigma = \begin{bmatrix} \sigma_1^2 & \cdots & \sigma_{1n} \\ \vdots & \ddots & \vdots \\ \sigma_{n1} & \cdots & \sigma_n^2 \end{bmatrix}$.

Mathematica: MultinormalDistribution$\left[\{\mu_1, \ldots, \mu_n\}, \begin{bmatrix} \sigma_1^2 & \cdots & \sigma_{1n} \\ \vdots & \ddots & \vdots \\ \sigma_{n1} & \cdots & \sigma_n^2 \end{bmatrix} \right].$

As with the binormal distributions (two-dimensional multinormal), the three- and higher-dimensional multinormal distributions may be rotated into a coordinate system where the components are independent, and Σ is a diagonal matrix. Graphic rendering is hard in higher dimensions, though. In Figure 10.8, we see the level curves of a multinormal distribution in three dimensions.

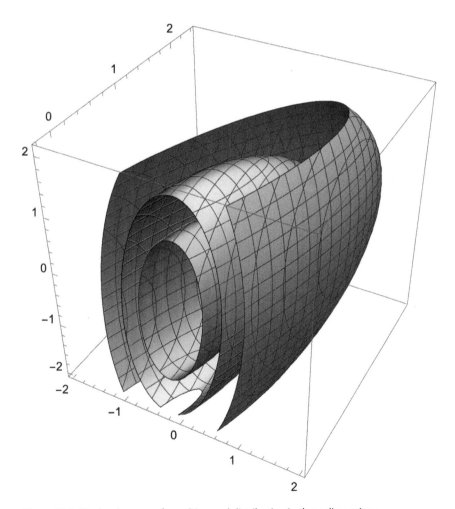

Figure 10.8 The level curves of a multinormal distribution in three dimensions.

10.3 Gamma Distribution, $\gamma_{(k,\lambda)}$ – With Family

Gamma distribution $T \sim \gamma_{(k,\lambda)}$ is a continuous stochastic variable taking values in $[0, \infty)$; we say that it follows the gamma distribution with parameters $k, \lambda > 0$.

pdf: $\gamma_{(k,\lambda)}(t) = \frac{\lambda^k}{\Gamma(k)} t^{k-1} e^{-\lambda t}$ where $\Gamma(k)$ is the Γ *function* (A.2.10)

The γ *distribution* derives its name from the Γ *function*. Below, we use χ^2 (see Section 10.3.1) to calculate CDF and iCDF, due to its uniquitousness in calculators.

CDF: $\Gamma_{(k,\lambda)}(t)$ (when $2k \in \mathbb{N}$)	iCDF: $\Gamma^{-1}_{(k,\lambda)}(p)$ (when $2k \in \mathbb{N}$)
CASIO: ChiCD$(0, 2\lambda t, 2k)$ **HP:** chisquare_cdf$(2k, 2\lambda t)$ **TI:** χ^2cdf$(0, 2\lambda t, 2k)$	InvChiCD$(1 - p, 2k)/(2\lambda)$ chisquare_icdf$(2k, p)/(2\lambda)$ **CX:** Inv$\chi^2(p, 2k)/2\lambda$ **83+:** Solver: χ^2cdf$(0, 2\lambda x, 2k) - p = 0 \gg x = 1$

Wolfram: GammaDistribution$[k, \frac{1}{\lambda}]$

Expected value: $\mu_T = \frac{k}{\lambda}$ **Variance:** $\sigma_T^2 = \frac{k}{\lambda^2}$

Normal approximation: $\phi_{(\mu_T, \sigma_T)}$ is a good approximation when $k > 30$

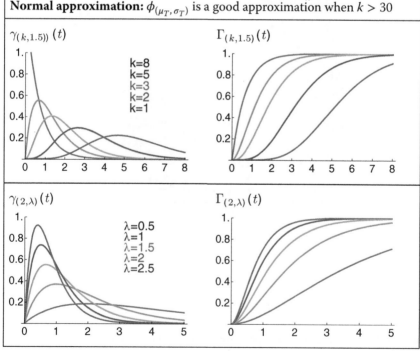

10.3.1 The Family

The gamma distribution is the most general form in a family of related distributions. The other distributions are special cases of the γ distribution, each in their own way, sometimes with a twist in how the parameters are set up. The closest cousin is the Erlang distribution, which differs only in the requirement that k be an integer. We have:

$$
\begin{array}{llll}
\gamma_{(k,\lambda)}(x) = \gamma_{(k,\lambda)}(x) & k, \lambda > 0 & \text{GammaDistribution}[k, \frac{1}{\lambda}] \\
erl_{(k,\lambda)}(x) = \gamma_{(k,\lambda)}(x) & k \in \mathbb{N}, \lambda > 0 & \text{ErlangDistribution}[k, \lambda] \\
exp_{\lambda}(x) = \gamma_{(1,\lambda)}(x) & \lambda > 0 & \text{ExponentialDistribution}[\lambda] \\
\chi^2_{v}(x) = \gamma_{\left(\frac{v}{2}, \frac{1}{2}\right)}(x) & v \in \mathbb{N} & \text{ChiSquareDistribution}[v] \\
S^{-1}\chi^2_{v}(x) = \gamma_{\left(\frac{v}{2}, \frac{S}{2}\right)}(x) & v \in \mathbb{N} & \text{``}S^{-1}\text{ times chi squared''}.
\end{array}
$$

The χ^2 parameter v is traditionally called the "degrees of freedom", and is sometimes written df.

In statistics, calculating inverse cumulative probabilities is as important as calculating the probabilities themselves. This goes for the γ family as well. In Mathematica, you find the inverse by simply invoking InverseCDF for the distribution, but when you are limited to the repertoir of a desktop calculator, you need to rewrite the expression to get what you want. Let $\mathbb{X}^2$ be the cumulative χ^2 distribution, and $\mathbb{X}^{-2}$ its inverse. Then,

- basic identity: $\Gamma^{-1}_{(k,\lambda)}(p) = \dfrac{\mathbb{X}^{-2}_{2k}(p)}{2\lambda}$;
- CASIO: InvChiCD$(1 - p, 2k)/(2\lambda)$;
- TI CX: Inv$\chi^2(p, 2k)/2\lambda$
- TI 83+: Solver: $\chi^2\text{cdf}(0, 2\lambda x, 2k) - p = 0 \gg x = 1$
- HP: chisquare_icdf$(2k, p)/(2\lambda)$.

Let us take a closer look at the family members.

10.3.2 Application of the Exponential and Erlang Distributions: Waiting Times

Historically, the Erlang distribution was invented by the Dane A. K. Erlang to analyse problems in telecommunications, since the incoming calls for a large company may be modelled as a Poisson process (Section A.4.2). In a *Poisson process*, the Poisson *distribution* (9.6) describes the number of occurrences within a fixed time t. The Erlang distribution does the opposite, and describes the waiting time t for a fixed number of occurrences k.

The Poisson process is what we call "memoryless", so the probability of a new occurrence within the next second, minute or hour is independent of the number of occurrences thus far, and how long the process has been going on. An important special case of the Erlang distribution is the waiting time for a single occurrence: $exp_\lambda(t) = erl_{(1,\lambda)}(t)$. The waiting time for one occurrence has its own name and symbol: the *exponential distribution, exp_λ*.

Calculating the Erlang distribution

Calculating the Erlang and exponential distributions is straightforward, especially for low values of k. Since k is an integer, the pdf and the CDF have the simpler form

$$\text{pdf: } erl_{(k,\lambda)}(x) = \frac{\lambda^k}{(k-1)!}x^{k-1}e^{-\lambda x}$$

$$\text{CDF: } ERL_{(k,\lambda)}(x) = 1 - \sum_{n=0}^{k-1}\left(\frac{(\lambda x)^n}{n!}e^{-\lambda x}\right).$$

In Appendix A.4, the identity A.4.6 may make for quicker calculations on some calculators.

Example 10.3.1 A Poisson process has rate $\lambda = 4.4$. Let T be the waiting time for $k = 13$ occurrences. What are μ_T, σ_T, and $P(T > 2)$ for this waiting time?

Answer: $\mu_T = \frac{13}{4.4} = 2.954\,55$, and $\sigma_T = \frac{\sqrt{13}}{4.4} = 0.819\,443$.

$$P(T > 2) = 1 - P(T \le 2) = \sum_{n=0}^{13-1}\left(\frac{(4.4 \cdot 2)^n}{n!}e^{-4.4\cdot 2}\right) = 0.889\,838.$$

Using identity A.4.6, we may also calculate the probability thus:

$$P(T > 2) = 1 - ERL_{(13,4.4)}(2) = POIS_{4.4 \cdot 2}(13 - 1) = 0.889\,838. \qquad \square$$

Example 10.3.2 Your wife works at a call center, and her day is over when she has filled a certain quota of calls. The arrival of these calls is described by a Poisson process. You know neither the rate λ nor how many calls k your wife must complete before she is done, but she said she believed it would take $\mu_T = \frac{1}{3}$ hours, but that the uncertainty was $\sigma_T = \frac{1}{6}$ hours. As you sit there pondering why your wife, who is so proficient at statistics, works at a call center, you decide to estimate the parameters λ and k. What are λ and k?

Answer: $\mu_T = \frac{k}{\lambda}$ and $\sigma_T = \frac{\sqrt{k}}{\lambda}$, which means we must solve the following system of equations:

$$k/\lambda = 1/3$$
$$\sqrt{k}/\lambda = 1/6.$$

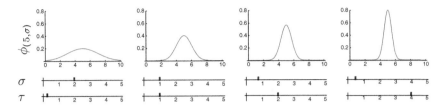

Figure 10.9 Spread, σ and τ itself for different values of τ.

We divide the first equation by the second, and get

$$\frac{k/\lambda}{\sqrt{k}/\lambda} = \sqrt{k} = \frac{1/3}{1/6} = 2.$$

So $\sqrt{k} = 2$, which gives $k = 4$. Inserting that into the first equation, we get

$$4/\lambda = 1/3$$
$$\lambda = 12.$$

$\square$

10.3.3 Application of the γ and χ^2 Distributions: Precision

Probability distributions γ and χ^2 describe the *precision* τ. Recall from Subsection 7.6.1 that the relationship between precision and variance is that $\tau = 1/\sigma^2$. Note that σ^2 has its own probability distributions (inverse γ and inverse χ^2), but that we will stick with the probability distributions for τ, since they are the most tractable. Everything we need to know about σ can be obtained from querying about τ.

In the illustrations in Figure 10.9 and summed up in Figure 10.10a, we see a normal distribution centered around $\mu = 5$, with different values for the spread (standard deviation) parameter σ (and the precision τ). When we do not know

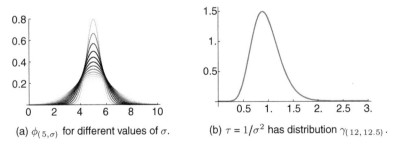

(a) $\phi_{(5,\sigma)}$ for different values of σ.

(b) $\tau = 1/\sigma^2$ has distribution $\gamma_{(12,\,12.5)}$.

Figure 10.10 The possible shapes of the normal distribution for different σ, when $\tau = 1/\sigma^2$ follows a γ distribution.

the precise measure of spread and precision, we indicate it by a probability distribution describing how likely it is that they have such-and-such values. This distribution is usually a gamma distribution, like the one in 10.10b.

Theory tells us that if $X_1, \ldots, X_n \sim \phi_{(0,1)}$ are independent, and $Z = \sum_{k=1}^{n} X_k^2$, then $Z \sim \chi_n^2$. Since variance is calculated from the squares, this means that χ^2 and thereby the γ family are intimately tied to the study of the Normal distribution and its friends. The gamma distribution therefore figures among our core probability distributions.

We will see more about this in Chapter 13, but for now we'll make do with a quick example of calculations.

Example 10.3.3 You have studied *Oh Mega!*, a manufacturer of electronic component, and in particular their production of resistors. Like all such components, their resistors are not infinitely precise, but their resistance values follow a normal distribution $\phi_{(\mu,\sigma)}(x)$. The goal of your study is to estimate σ. You have heard that the most expedient way of doing this is to study $\tau = 1/\sigma^2$ rather than σ itself. You are going to

1) determine the probability that $\tau < 2$;
2) find a value of τ_0 making it 90% probable that the precision τ is at least as large as τ_0;
3) translate the resultats for τ into resultats for σ.

After some measurements of *Oh Mega!*'s line of 100 Ω resistors, you have the following probability distribution for the precision:

$$\tau \sim \gamma_{(31.5,16)}(t).$$

The cumulative probability distribution is then $\Gamma_{(31.5,16)}(t) = X_{63}^2(32t)$. You may then answer your questions.

Answer:

1) $P(\tau < 2) = \underline{0.558\,82}$, which we calculate as follows:

   (a) Mathematica: CDF$\left[\text{GammaDistribution}\left[31.5, \frac{1}{16}\right], 2\right]$;
   (b) CASIO: ChiCD(0, 64, 63);
   (c) TI: χ^2cdf(0, 64, 63);
   (d) HP: chisquare_cdf(63, 64).

2) The τ value τ_0 for which $P(\tau > \tau_0) = 0.9$, is the τ value where $P(\tau < \tau_0) = 0.1$. That is, $\tau_0 = \Gamma_{(31.5,16)}^{-1}(0.1) = X_{63}^{-2}(0.1)/32 = \underline{1.534\,7}$, which we calculate as follows:

   (a) Mathematica: InverseCDF$\left[\text{GammaDistribution}\left[31.5, \frac{1}{16}\right], 0.1\right]$;
   (b) CASIO: InvChiCD(0.9, 63)/32;
   (c) TI: Solver. χ^2cdf(0, 32x, 63) − 0.1 = 0 or Invχ^2(0.1, 63)/32;
   (d) HP: chisquare_icdf(63, 0.1)/32.

3) That $\tau < 2$ means that $\sigma > \sqrt{0.5}$, since $\tau = 1/\sigma^2$ and thereby $\sigma = 1/\sqrt{\tau}$. Then

$$P(\sigma > \sqrt{0.5}) = 0.558\,82.$$

Further, we had that $P(\tau > 1.534\,7) = 90\%$, which is equivalent to

$$P(\sigma < 0.807\,212) = 90\%. \qquad \square$$

10.4 "Student's" t Distribution, $t_{(\mu,\sigma,\nu)}$

---

t distribution: $X \sim t_{(\mu,\sigma,\nu)}(x)$ is a continuous stochastic variable taking values in $\mathbb{R}$; we say that it follows a t distribution with parameters $\mu \in \mathbb{R}$, $\sigma > 0$, and $\nu \in \mathbb{N}$. The parameter ν is traditionally called "degrees of freedom", and is sometimes written df.

pdf: $t_{(\mu,\sigma,\nu)}(x) = k \cdot \left(1 + \dfrac{(x-\mu)^2}{\nu\sigma^2}\right)^{-(\nu+1)/2}$,

where $k = \dfrac{\Gamma\left(\frac{\nu+1}{2}\right)}{\Gamma\left(\frac{\nu}{2}\right)} \cdot \dfrac{1}{\sigma\sqrt{\pi\nu}}$ (for the Γ *function*, see A.2.10).

CDF: $T_{(\mu,\sigma,\nu)}(x)$	iCDF: $T^{-1}_{(\mu,\sigma,\nu)}(p)$
CASIO: tCD$(-10^{99}, (x-\mu)/\sigma, \nu)$ **HP:** student_cdf$(\nu, (x-\mu)/\sigma)$ **TI:** tcdf$(-10^{99}, (x-\mu)/\sigma, \nu)$	$\mu - \sigma * \text{InvTCD}(p, \nu)$ $\mu + \sigma * \text{student\_icdf}(\nu, p)$ $\mu + \sigma * \text{invT}(p, \nu)$

Wolfram: StudentTDistribution$[\mu, \sigma, \nu]$

If your calculation tools do not support t with three parameters, go via the standard version $t_\nu = t_{(0,1,\nu)}$, using the relation $T_{(\mu,\sigma,\nu)}(x) = T_\nu\left(\dfrac{x-\mu}{\sigma}\right)$.

Expected value: $\mu_X = \begin{cases} \mu & \nu > 1 \\ - & \nu \le 1 \end{cases}$ **Variance:** $\sigma_X^2 = \begin{cases} \sigma^2\dfrac{\nu}{\nu-2} & \nu > 2 \\ \infty & \nu \le 2 \end{cases}$

Normal approximation: $\phi_{(\mu,\sigma)}$ is a good approximation when $\nu > 30$

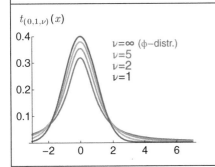

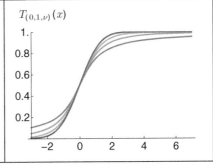

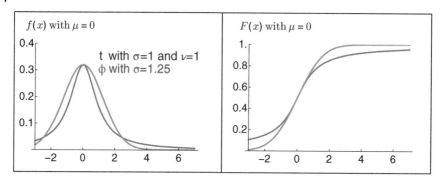

10.4.1 Applications: Almost Everything

The t distribution arises in both Bayesian and frequentist inference when we have measured n values x_k from a Normal distributed population with unknown standard deviation σ, and are estimating the location measure μ, or the next measurement, x_{n+1}. The applications of the t distribution are thus the same as for the Normal distribution.

10.4.2 Calculating t Distributions

In ye olden days, everything was calculated by hand, or looked up in tables. For the Normal distribution, you'd look up $\Phi(x) = p$ in suitable tables of Φ. But $T_\nu(x) = p$ would require one table per degree of freedom ν, making that task a whole lot more arduous. But when working with the t distribution, it is often a workaround to calculate from p to x instead of calculating from x to p. These days, however, both calculations are done with ease in a fraction of a second.

Example 10.4.1 Our estimate of the length (in m) of the next plank from the sawmill, is $X \sim t_{(4.95, 0.1, 15)}$. What is the probability that the next plank is shorter than 5 m?

Answer: This is $P(X \leq 5) = T_{(4.95,\, 0.1,\, 15)}(5) = \underline{0.687\,835}$.

- Mathematica: CDF[StudentTDistribution[4.95, 0.1, 15],5];
- CASIO: $\text{tCD}\left(-10^{99}, \frac{5-4.95}{0.1}, 15\right)$;
- HP: $\text{student\_cdf}\left(15, \frac{5-4.95}{0.1}\right)$;
- TI: $\text{tcdf}\left(-10^{99}, \frac{5-4.95}{0.1}, 15\right)$. $\qquad\qquad\square$

Example 10.4.2 You are out fishing, and are following an estimate of the weight in kilograms of the next fish: $Y \sim t_{(2.4, 0.7, 9)}$. What is the probability that the next fish weighs in excess of 2 kg?

Answer: This is $P(Y \geq 2) = 1 - T_{(2.4,\, 1.2,\, 9)}(2) = \underline{0.709\,152}$. $\qquad\qquad\square$

Example 10.4.3   You are observing traffic noise, and your estimate of the noise from the next truck (in decibels), is $Z \sim t_{(93,12,21)}$. What is the probability that this noise is in the interval from 90 to 100 dB?

Answer: This is $P(90 < Z \le 100) = T_{(93,12,21)}(100) - T_{(93,12,21)}(90) = \underline{0.314\,551}.$

□

In the section on the Normal distribution, we looked at the sum and difference between Normal distributed variables. We need similar operations for the t distribution. We have the following rule for pairs of t distributed variables.

Rule 10.4.4   *If $X \sim t_{(\mu_X,\sigma_X,\nu_X)}$ and $Y \sim t_{(\mu_Y,\sigma_Y,\nu_Y)}$ are independent, and $Z = X \pm Y$, then Z itself does not follow precisely a t distribution, but a very good approximation to the distribution of the difference is* Satterthwaite's approximation:

$$Z \sim f_Z(z) \approx t_{(\mu_Z,\sigma_Z,\nu_Z)}(z),$$

where

- $\mu_Z = \mu_X \pm \mu_Y$
- $\sigma_Z = \sqrt{\sigma_X^2 + \sigma_Y^2}$
- $\nu_Z = \left\lfloor \dfrac{\left(\dfrac{\sigma_X^2}{\nu_X+1} + \dfrac{\sigma_Y^2}{\nu_Y+1}\right)^2}{\left[\dfrac{\left(\dfrac{\sigma_X^2}{\nu_X+1}\right)^2}{\nu_X} + \dfrac{\left(\dfrac{\sigma_Y^2}{\nu_Y+1}\right)^2}{\nu_Y}\right]} \right\rfloor.$

where $\lfloor x \rfloor$ is the integer part of x. If $Y \sim \phi$, then $\nu_Z = \nu_X$.

There are variants of the formula for degrees of freedom that give similar values, and we sometimes speak of the *Welch–Satterthwaite* formula and test.

Example 10.4.5   If $X \sim t_{(-8,3,11)}$ and $Y \sim t_{(2,4,9)}$, then $Z = X - Y \sim t_{(\mu_Z,\sigma_Z,\nu_Z)}$, where $\mu_Z = (-8) - 2 = -10$, $\sigma_Z = \sqrt{3^2 + 4^2} = 5$, and

$$\nu_Z = \left\lfloor \dfrac{\left(\dfrac{3^2}{11+1} + \dfrac{4^2}{9+1}\right)^2}{\left[\dfrac{\left(\dfrac{3^2}{11+1}\right)^2}{11} + \dfrac{\left(\dfrac{4^2}{9+1}\right)^2}{9}\right]} \right\rfloor = \lfloor 16.456\,5 \rfloor = 16.$$

□

10.4.3 $t_{v,\alpha}$, the inverse of $T_{(0,1,v)}(x)$

In statistics, calculating inverse cumulative probabilities is as important as calculating probabilities. This goes for t distributions as well. We are therefore interested in the inverse cumulative t distribution $T_{(\mu,\sigma,v)}$. See the beginning of this section (Section 10.4) for how to calculate it and its inverse.

There is a useful standard variant of the CDF, the cumulative t distribution, $T_v = T_{(0,1,v)}$. But its inverse, the iCDF, has an even more useful standard variant, $t_{v,p} = T_v^{-1}(p) = T_{(0,1,v)}^{-1}(p)$. It uses the same symbol as the pdf, distribution t function itself, but in this book we distinguish them by the fact that the pdf is written with the three parameters as an index with parentheses, thus: $t_{(\mu,\sigma,v)}(x)$, whereas the standard iCDF is written with the parameter and the argument values v and p both in the index (that is: two indexes as opposed to three), thus: $t_{v,p}$.

As with z_α, the standard inverse for the ϕ distribution, $t_{v,p}$ indicates the value of x for which the area under the *left* tail of $t_{(0,1,v)}(\cdot)$, from $-\infty$ to x, has area p.

The relation between $t_{v,p}$ and the general T^{-1} is

$$T_{(\mu,\sigma,v)}^{-1}(p) = \mu + \sigma \cdot t_{v,p}.$$

Example 10.4.6 Find $t_{2,0.1}$.

- In the table of $t_{v,p}$ values, p is given by the choice of column, and v by the choice of row: to find $t_{2,0.1}$, locate the column for $p = 0.1$, and the row for $v = 2$. They intersect in the number $-1.885\,6$, which means that $t_{2,0.1} = -1.885\,6$.
- Mathematica: InverseCDF[StudentTDistribution[0, 1, 2], 0.1] gives $-1.885\,62$.
- CASIO: $-$InvTCD(0.1, 2) gives $-1.885\,62$ (notice the minus sign!).
- HP: student_icdf(2, 0.1) gives $-1.885\,62$.
- TI: invT(0.1, 2) gives $-1.885\,62$. □

Example 10.4.7 Your estimate of X, the weight of the next capercaillie you shoot in the Long Forest, is $X \sim t_{(3.5,0.7,11)}$. Find a fixed weight x such that there is a probability of exactly 5% that the next capercaillie you shoot weighs less than that.

Answer: We show our standard ways of calculating:

- $T_{(\mu,\sigma,v)}^{-1}(0.05) = \mu + \sigma \cdot t_{v,0.05} = 3.5 - 0.7 \cdot t_{11,0.05} = \underline{2.242\,88}$
- Mathematica: InverseCDF[StudentTDistribution[3.5, 0.7, 11], 0.05];
- CASIO: $3.5 - 0.7 \cdot$InvTCD(0.05, 11);
- HP: $3.5 + 0.7 \cdot$student_icdf(11, 0.05);
- TI: $3.5 + 0.7 \cdot$invT(0.05, 11). □

Rule 10.4.8 $t_{v,1-p} = -t_{v,p}$

Example 10.4.9 $X \sim t_{(6,2.1,3)}$. Find x such that $P(X \le x) = 0.9$.

Answer: $\mu + \sigma \cdot t_{v,p} = 6 + 2.1 \cdot t_{3,0.9}$
$$= 6 - 2.1 \cdot t_{3,0.1} = 6 - 2.1 \cdot (-1.637\,74) = \underline{9.439\,26}. \qquad \square$$

10.5 Beta Distribution, $\beta_{(a,b)}$

β **distribution:** $X \sim t_{(\mu,\sigma,v)}(x)$ is a continuous stochastic variable taking values in $[0, 1]$; we say that it follows a β distribution with parameter $a, b > 0$.

pdf: $\beta_{(a,b)}(x) = \dfrac{1}{B(a,b)} \cdot x^{a-1}(1-x)^{b-1}$,

where $B(a, b)$ is the Euler B *function* (A.2.14). The notation for the CDF is thus $I_{(a,b)}(x)$. To calculate CDF and iCDF below, we use the F distribution (A.3.1).

CDF: $I_{(a,b)}(x)$	iCDF: $I_{(a,b)}^{-1}(p)$
CASIO: $\mathrm{FCD}\left(0, \dfrac{bx}{a(1-x)}, 2a, 2b\right)$	$1/\left(1 + \dfrac{b}{a \cdot \mathrm{InvFCD}(1-p,2a,2b)}\right)$
HP: $\mathrm{fisher\_cdf}\left(2a, 2b, \dfrac{bx}{a(1-x)}\right)$	$1/\left(1 + \dfrac{b}{a \cdot \mathrm{fisher\_icdf}(2a,2b,p)}\right)$
TI: $\mathrm{Fcdf}\left(0, \dfrac{bx}{a(1-x)}, 2a, 2b\right)$	**CX:** $1/\left(1 + \dfrac{1}{a \cdot \mathrm{InvF}(p,2a,2b)}\right)$
	83+: Solver: $\mathrm{Fcdf}\left(0, \dfrac{bx}{a(1-x)}, 2a, 2b\right) -$
	$p = 0 \gg x = 0.5$

Wolfram: BetaDistribution[a,b]

Expected value: $\mu_X = \dfrac{a}{a+b}$ **Variance:** $\sigma_X^2 = \dfrac{ab}{(a+b)^2(a+b+1)}$

Normal approximation: $\phi_{(\mu_X,\sigma_X)}$ is a good approximation when $a, b > 10$.

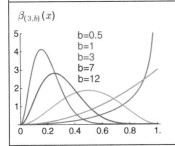

$\beta_{(3,b)}(x)$

b=0.5
b=1
b=3
b=7
b=12

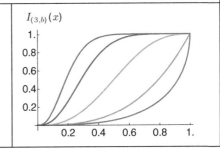

$I_{(3,b)}(x)$

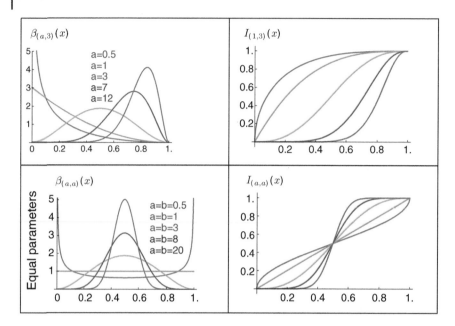

10.5.1 Application: Proportions

The beta distributions are primarily used in the study of *proportions*, including the type of proportion we call probabilities. It is the preferred distribution for estimating the parameter p for Bernoulli processes. Since the study of proportions is central, this makes the beta distribution a similarly central probability distribution, especially in Bayesian analysis. When we employ the β distributions to study a proportion π, this means that we do not know its value for sure, and that instead, the β probability distribution expresses our knowledge of it. π is then itself a stochastic variable, $\pi \sim \beta_{(a,b)}$. The proportion π can be for instance any of the following.

- The probability that an unknown biased coin lands *heads* up.
- The proportion of patients having a certain disease cured by a given medicine.
- The probability that Osman Sow of the MK Dons scores his next goal with the left foot.
- The percentage of voters who will vote Libertarian in the next US presidential election.
- What proportion of the day will be cloudy.
- Pepsi Cola's market share in Russia next year.
- The probability that Manuel Neuer of FC Bayern Munich saves the next shot at his goal.

When we say about a proportion X that it is a stochastic variable $X \sim \beta_{(a,b)}$, we say something about how probable it is that the proportion is within a certain interval. We will look more closely at this in Section 13.2 about statistical inference, where we have rules for Bayesian updating of probabilities for proportions through β distributed prior probabilities.

But let us return to the present: how do we calculate β distributions?

10.5.2 Calculating β Distributions

The first quantities that we want to calculate for $X \sim \beta_{(a,b)}$, are the probabilities $P(X \in (A, B))$, the expected value μ_X, and the variance σ^2. We calculate the latter two from the parameters by means of the formulas on the overview sheet at the beginning of Section 10.5, whereas the probability in principle involves calculating an integral:

$$P(X \in (A, B)) = k_{a,b} \cdot \int_A^B x^{a-1}(1-x)^{b-1}\, dx, \quad \text{where } k_{a,b} = \frac{\Gamma(a+b)}{\Gamma(a)\Gamma(b)}.$$

The resulting integral will be a polynomial, and probably one of a high degree resulting from calculating $(1-x)$ to some power. This is the kind of calculation we leave to our calculation tools.

Example 10.5.1 Bard is looking at π, the proportion of patients who have their fractured tailbone healed within a week of taking Wunderzil^{TM}.

1) After a controlled study, Bard concludes that $\pi \sim \beta_{(17,5)}$.
   (a) What does Bard's result say is the expected proportion of healed patients, $E[\pi]$?
   (b) How sharp is this estimate?
   (c) What is the probability that π is larger than 75%?
2) Bard conducts a second study, and updates his estimate to $\pi \sim \beta_{(40,11)}$.
   (a) What is now the expected proportion of healed patients, $E[\pi]$?
   (b) How precise is the new estimate?
   (c) What is the new probability that π is larger than 75%?

Answer:

1) After the first study:
   (a) $\mu_\pi = \frac{17}{17+5} = 0.772\,727$;
   (b) we recall from Section 7.6.1 that the precision is $\tau = 1/\sigma^2$, so for a $\beta_{(a,b)}$ distributed stochastic variable π we have

$$\tau_\pi = \frac{(a+b)^2(a+b+1)}{ab}$$
$$= \frac{(17+5)^2(17+5+1)}{17 \cdot 5} = \underline{130.965};$$

(c) the probability that π is larger than 75% is then

$$P(\pi \geq 0.75) = 1 - I_{(17,5)}(0.75) = \int_{0.75}^{1} \beta_{(17,5)}(p)dp = \underline{0.632\,58}.$$

2) After the second study:
   (a) $\mu_\pi = \dfrac{40}{40+11} = 0.784\,314;$
   (b) this time,

$$\tau_\pi = \frac{(40+11)^2(40+11+1)}{40 \cdot 11} = \underline{307.391};$$

(c) the probability that π is larger than 75% is now updated to

$$P(\pi \geq 0.75) = 1 - I_{(40,11)}(0.75) = \int_{0.75}^{1} \beta_{(40,11)}(p)\,dp = \underline{0.737\,798}.$$

The second estimate, which Bard got after collecting more data, was (not surprisingly) sharper than the first estimate. □

Example 10.5.2 Your estimate of X, the proportion of students who prefer Pepsi Max to other light sodas, is $X \sim \beta_{(102,53)}(x)$. Find the probability that the proportion who prefer Pepsi Max is less than 70%.

Answer: We are finding $P(X < 0.7)$.

- Mathematica: CDF[BetaDistribution[102, 53],0.7]=0.865 549.

Most desktop calculators do not have the β distribution built in, so we must rewrite to make it fit what's available. It turns out that calculations of the β distribution may be rewritten into calculations concerning the F distribution (Section A.3.1). The recipe is given in the overview at the beginning of Section 10.5.

- CASIO: FCD(0, $\frac{53 \cdot 0.7}{102(1-0.7)}$, $2 \cdot 102, 2 \cdot 53) = 0.865\,549$.
- HP: fisher_cdf($2 \cdot 102, 2 \cdot 53, \frac{53 \cdot 0.7}{102(1-0.7)}$) = 0.865 549.
- TI: Fcdf(0, $\frac{53 \cdot 0.7}{102(1-0.7)}$, $2 \cdot 102, 2 \cdot 53) = 0.865\,549$. □

Example 10.5.3 Your estimate of the proportion of left-foot goals by Arjen Robben (FC Bayern Munich) is $Y \sim \beta_{(15,24)}(x)$. Find the probability that the true proportion lies between 20 and 40%.

Answer: We are calculating $P(0.2 < X < 0.4)$.

- Mathematica: Probability[$0.2 < x < 0.4, x \approx$ BetaDistribution[15,24] gives us 0.582 778.
- HP: fisher_cdf($30, 48, \frac{24 \cdot 0.4}{15(1-0.4)}$) − fisher_cdf($30, 48, \frac{24 \cdot 0.2}{15(1-0.2)}$).

On the CASIO and TI calculators, the cumulative probability distributions have *two x* value entries, *lower* and *upper*, so what we get out is really the probability that the stochastic variable takes a value between *lower* and *upper*. We make use of that feature:

- CASIO: FCD$(\frac{24\cdot0.2}{15(1-0.2)}, \frac{24\cdot0.4}{15(1-0.4)}, 2 \cdot 15, 2 \cdot 24)$;
- TI: Fcdf$(\frac{24\cdot0.2}{15(1-0.2)}, \frac{24\cdot0.4}{15(1-0.4)}, 2 \cdot 15, 2 \cdot 24)$. $\square$

How do we calculate the probability that 5 of Arjen Robben's next 15 goals are scored with his left foot? It is tempting to make a best estimate p of the proportion of his left-foot scorings, and to calculate the answer using the binomial distribution. This solution even sounds correct – and, because it is so convincing, we need to look at it in detail to see why it is wrong. The short of it, in summary, is that the uncertainty of the estimate p cannot be neglected, since it has an effect. We have actually seen this once before, in Note 6.7.2 and the associated problems. We will see this effect even more spectacularly in the applied example (Example 13.2.2) in Chapter 13.

For the correct solution, we look at the general question: when the β distribution describes a game's probability π, what is the probability of a sequence of positive ($\top$) and negative ($\perp$) outcomes? Or for a concrete example (Example 10.5.6): if we have a coin with probability π of H, and $\pi \sim \beta_{(18,22)}$, what is then the probability of *HHTHTHTH*?

Rule 10.5.4 *If $X_1, \ldots, X_n$ are trials in a Bernoulli process with parameter π, and $\pi \sim \beta_{(a,b)}$, then the probability of a given sequence of $k\top$ and $l\perp$ $(k + l = n)$ is given by*

$$P = \frac{k_{a,b}}{k_{a+k,b+l}} = \frac{\left(\frac{\Gamma(a+b)}{\Gamma(a)\Gamma(b)} \right)}{\left(\frac{\Gamma(a+b+k+l)}{\Gamma(a+k)\Gamma(b+l)} \right)} = \frac{\binom{a+b}{a} \cdot \frac{ab}{a+b}}{\binom{a+b+k+l}{a+k} \cdot \frac{(a+k)(b+l)}{a+b+k+l}}.$$

where the last equality is for integers only. For half-integer values, use the fact that

$$\Gamma(n + \frac{1}{2}) = \frac{(2n)!}{n!4^n} \sqrt{\pi}.$$

The probability of a combination with $k\top$ and $l\perp$ equals the probability of such a sequence times the number of such sequences, $\binom{k+l}{k}$.[1]

---

[1]The probability of the combination has its own name, the *beta binomial distribution* (see Appendix A.3.3).

Proof: For each value of p, the probability of a sequence with $k\mathsf{T}$ and $l\perp$ equals $p^k(1-p)^l$. Similar to what we did in 6.4.1, we take the weighted mean of the sequence probabilities, with weight per p equal to $f(p) = \beta_{(a,b)}(p)$ for p. Then

$$
\begin{aligned}
P(k\mathsf{T}, l\perp) &= \int_0^1 p^k(1-p)^l \beta_{(a,b)}(p)\,dp = \int_0^1 k_{a,b}p^{a+k-1}(1-p)^{b+l-1}\,dp \\
&= \frac{k_{a,b}}{k_{a+k,b+l}} \int_0^1 k_{a+k,b+l}\,p^{a+k-1}(1-p)^{b+l-1}\,dp \\
&= \frac{k_{a,b}}{k_{a+k,b+l}} \int_0^1 \beta_{(a+k,b+l)}(p)\,dp = \frac{k_{a,b}}{k_{a+k,b+l}}.
\end{aligned}
$$
□

Example 10.5.5 $\pi \sim \beta_{(12.5,23.5)}$. Find the probability that the next three observations are $\mathsf{T}\perp\mathsf{T}$.

Answer: We employ Rule 10.5.4, but since a and b are half integers, we cannot use the simplified formula for integers. We put $k = 2$ and $l = 1$ into the formula, and get

$$
\begin{aligned}
P(\mathsf{T}\perp\mathsf{T}) &= \frac{\left(\dfrac{\Gamma(12.5+23.5)}{\Gamma(12.5)\Gamma(23.5)}\right)}{\left(\dfrac{\Gamma(12.5+23.5+2+1)}{\Gamma(12.5+2)\Gamma(23.5+1)}\right)} = \frac{\left(\dfrac{\Gamma(36)}{\Gamma(12.5)\Gamma(23.5)}\right)}{\left(\dfrac{\Gamma(39)}{\Gamma(14.5)\Gamma(24.5)}\right)} \\
&= \frac{\left(\dfrac{35!}{\frac{24!}{12!4^{12}}\sqrt{\pi}\cdot\frac{46!}{23!4^{23}}\sqrt{\pi}}\right)}{\left(\dfrac{38!}{\frac{28!}{14!4^{14}}\sqrt{\pi}\cdot\frac{48!}{24!4^{24}}\sqrt{\pi}}\right)} \approx \underline{0.078\,347\,3}.
\end{aligned}
$$
□

Example 10.5.6 We have a coin whose probability π of H is distributed $\pi \sim \beta_{(18,22)}$.

- What is the probability of *HHTHTHTH*? (sequence).
- What is the probability of $5H$ and $3T$? (combination).

Answer: Here, $a = 18$, $b = 22$, $k = 5$, and $l = 3$, so we may use the simplified formula for integers:

$$
P(HHTHTHTH) = \frac{\left(\dbinom{18+22}{18}\right)\cdot\dfrac{18\cdot22}{18+22}}{\left(\dbinom{18+22+5+3}{18+5}\right)\cdot\dfrac{(18+5)(22+3)}{18+22+5+3}} = \frac{1\,254}{414\,305} \approx 0.003\,026\,76;
$$

$$
P(5H,\,3T) = \binom{5+3}{5}P(HHTHTHTH) = \frac{702\,24}{414\,305} \approx 0.169\,498.
$$
□

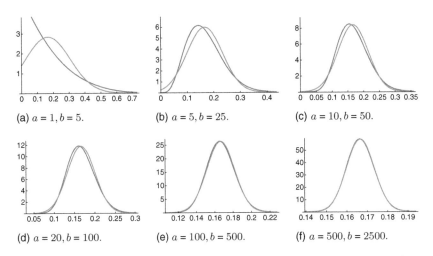

(a) $a = 1, b = 5$. (b) $a = 5, b = 25$. (c) $a = 10, b = 50$.

(d) $a = 20, b = 100$. (e) $a = 100, b = 500$. (f) $a = 500, b = 2500$.

Figure 10.11 Normal approximation and comparison to increasing parameter values for β.

10.5.3 Normal Approximation

To calculate $\beta(a, b)$ for *very* large values of a and b, we must resort to approximations even when we use a calculating tool. For this, we use Rule 10.1.13, *Normal approximation*, from Section 10.1.5. Since $\beta_{(a,b)}$ has expected value $\frac{a}{a+b}$ and standard deviation $\sqrt{\frac{ab}{(a+b)^2(a+b+1)}}$, the correct Normal approximation for $\beta_{(a,b)}$ is $\phi\left(\frac{a}{a+b}, \sqrt{\frac{ab}{(a+b)^2(a+b+1)}}\right)$. In Figure 10.11, we have let β be blue, and have colored N purple. Note the scaling along the two axes; how the curves get narrower and taller as the parameter values for a and b increase.

Example 10.5.7 $X \sim \beta_{(2\,183,\,7\,927)}$. What are

1) μ_X;
2) σ_X;
3) $P(X \in (0.21, 0.22))$?

Answer: $a = 2\,183$ and $b = 7\,927$, so

1) $\mu = \dfrac{2\,183}{2\,183+7\,927} = \underline{0.215\,925}$;

2) $\sigma = \sqrt{\dfrac{2\,183 \times 7\,927}{(2\,183+7\,927)^2(2\,183+7\,927+1)}} = \underline{0.004\,091\,98}$;

3) $P(X \in (0.21, 0.22)) = \Phi_{(0.215\,925,\,0.004\,091\,98)}(0.22)$
$$- \Phi_{(0.215\,925,\,0.004\,091\,98)}(0.21) = \underline{0.766\,528}.$$

There is a 76% probability that $X \in (0.21, 0.22)$. $\square$

10.5.4   Inverse Cumulative β Distribution

In statistics, calculating inverse cumulative probabilities is as important as calculating the probabilities themselves. This applies to the β distributions as well. We will therefore look at the inverse of the cumulative β distribution, $I_{(a,b)}^{-1}$, and show an example of calculating it on our favourite tools, commonly via the inverse F distribution,

$$I_{(a,b)}^{-1}(p) = \frac{1}{1 + \dfrac{b}{a \cdot F_{(2a,2b)}^{-1}(p)}}.$$

Our tools find $I_{(a,b)}^{-1}(p)$ thus:

- Mathematica: InverseCDF[BetaDistribution[a,b],p];
- CASIO: $1/(1 + \dfrac{b}{a \cdot \text{InvFCD}(1-p,\, 2a,\, 2b)})$;
- HP: $1/(1 + \dfrac{b}{a \cdot \text{fisher\_icdf}(2a,\, 2b,\, p)})$
- TI CX: $1/(1 + \dfrac{b}{a \cdot \text{InvF}(p,\, 2a,\, 2b)})$
- TI 83+: Solver: Fcdf$(0, \dfrac{bx}{a(1-x)},\, 2a,\, 2b) - p = 0 \gg x = 0.5$

If direct calculation of $I_{(a,b)}^{-1}(p)$ is inaccessible, and even calculation via rewriting into the F distribution is unavailable, the best remaining route is the Normal approximation, $\Phi_{(\mu_X,\sigma_X)}^{-1}(p)$.

Example 10.5.8   Bard has estimated the proportion ψ of Sith among the Jedi. According to his findings, it is $\psi \sim \beta_{(10,40)}(p)$. He wants to find a number $p_0 \in (0,1)$ so that there is an exact probability of 90% that the proportion of Sith is less than or equal to p_0.

Answer: Bard needs to find a p_0 such that $P(\psi \le p_0) = I_{(10,40)}(p_0) = 0.9$. In other words, he needs to calculate $p_0 = I_{(10,40)}^{-1}(0.9)$. He does that in one of the following ways:

- Mathematica: InverseCDF[BetaDistribution[10,40],0.9] gives $p_0 = \underline{0.274\,411}$;
- CASIO: $1/(1 + \dfrac{40}{10 \cdot \text{InvFCD}(1-0.9,\, 2 \times 10,\, 2 \times 40)})$
- HP: $1/(1 + \dfrac{40}{10 \cdot \text{fisher\_icdf}(2 \times 10,\, 2 \times, 0.9)})$;
- TI: Solver: Fcdf$(0, \dfrac{40x}{10(1-x)},\, 20,\, 80) - 0.9 = 0$ or $1/(1 + \dfrac{40}{10 \cdot \text{InvF}(0.9,20,80)})$;
- Normal approximation: $\mu_\psi = 0.2$, $\sigma_\psi = 0.056\,011\,2$:

$$I_{(10,40)}^{-1}(0.9) \approx \Phi_{(0.2,\, 0.0560112)}^{-1}(0.9) = \underline{0.271\,781}.$$

Hence, it is less than 90% probable that a smaller proportion of the Jedi than 27% are Sith. □

10.6 Weibull Distribution, $weib_{(\lambda,k)}$*

Weibull distribution $T \sim weib_{(k,\lambda)}(x)$ is a continuous stochastic variable taking values in $[0, \infty)$. We say that it follows the Weibull distribution with parameters $\lambda, k > 0$.

pdf: $weib_{(k,\lambda)}(t) = \dfrac{k}{\lambda} \times \left(\dfrac{t}{\lambda}\right)^{k-1} e^{-(t/\lambda)^k}$

CDF:	iCDF:
$\text{WEIB}_{(k,\lambda)}(t) = 1 - e^{-(t/\lambda)^k}$	$\text{WEIB}_{(k,\lambda)}^{-1}(p) = \lambda(-\ln(1-p))^{1/k}$

Wolfram: WeibullDistribution[k, λ]

Other distributions:
Exponential: $weib_{(1,1/\lambda)}(t) = \exp_\lambda(t)$
Rayleigh: $\text{WEIB}_{\left(2, \sqrt{2}\sigma\right)}(t) = \text{RAYL}_\sigma(t) = 1 - e^{-t^2/2\sigma^2}$

Expected value:	**Variance:**
$\mu_T = \dfrac{\lambda}{k} \cdot \Gamma\left(\dfrac{1}{k}\right)$	$\sigma_T^2 = \lambda^2 \left(\Gamma\left(1 + \dfrac{2}{k}\right) - \left(\Gamma\left(1 + \dfrac{1}{k}\right)\right)^2 \right)$

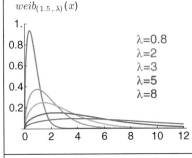

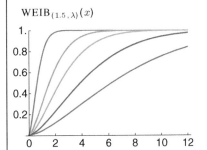

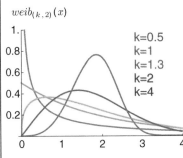

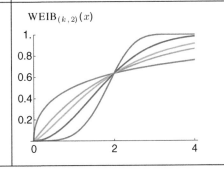

10.6.1 Application: Waiting and Decay Times

The Weibull distribution is, like the Erlang and exponential distributions, used in the study of waiting times. The exponential distribution is a special case of

Weibull with the first parameter $k = 1$. The waiting times typically described by the Weibull distribution are the lifetimes of things, components, or living beings. Since what the Weibull is waiting for is a death, it is said to be a study of *decay times*. As opposed to the variants of the Erlang distribution, the Weibull is not memoryless, but rather knows how much time has passed since the inception. And as opposed to the Erlang distribution, the decay rate is not constant, but is rather proportional to some power of the time that has passed since the inception, t^{k-1}. The length of a human life is Weibull distributed, and $k - 1 > 0$, since the rate of death is higher when you are 89 than when you are 5.

The death and decay rate, that is, the probability density describing how probable it is that a component that has survived up until time t will die at precisely time t, either rises or falls with time or stays constant.

- If the death rate sinks with time, this indicates a high "infant mortality"; for industrial components, this is a sign that many of the components had initial production errors, and therefore failed early. Or with living beings, it indicates complications around birth and early years typical of societies with hard conditions for children. This corresponds to $k < 1$.
- If the death rate sinks with time, this indicates that the deaths are due to wear. In human terms, this is more akin to modern societies, and corresponds to $k > 1$.
- If the death rate is constant, this indicates mortality due to random events. In human terms, this would be like a society where most deaths were by accident, war, or other randomly selected killings that kill irrespective of age – much like, presumably, bronze age cultures. This corresponds to $k = 1$, which is the exponential distribution.

The Weibull distribution's first parameter λ decides the *scale* of the distribution, or in more colloquial terms: how long-lived the population is. The parameter k determines the *shape* of the distribution: if it is heavy on early, late, or medium term deaths. The upper illustration on the overview page for this section (10.6) shows how the shape changes $weib_{(k,\lambda)}$ from a rapidly descending curve that is "thick" around 0 for $k < 1$, via the exponential curve when $k = 1$, and how it progresses towards having a distinct hump as k grows beyond 1. The lower illustration in the overview shows how λ stretches or compresses the curve of $weib_{(k,\lambda)}(x)$ without altering its fundamental shape.

The Weibull distribution with $k = 2$ is the *Rayleigh distribution*, and so, like the Rayleigh distribution, the Weibull distribution is used in the study of weather and related phenomena.

10.6.2 Calculating the Weibull Distribution

Calculating the Weibull distribution is straightforward from the formulas, for pdf, CDF, and iCDF alike. This is one of the few distributions where the Normal approximation is a bad idea, since calculating the Weibull distribution itself is simpler – and of course more exact!

Example 10.6.1 T follows a Weibull distribution with parameters $\lambda = 5.6$ and $k = 41$. Use the Normal approximation to find $P(T \leq 5.6)$, and then check against the exact value.

Answer:
1) The Weibull distribution with parameters $\lambda = 5.6$ and $k = 41$ has expected value $\mu = 5.52$ and standard deviation $\sigma = 0.17$, so its Normal approximation is $\phi_{(5.52,0.17)}$.
2) Rule 10.1.13 then says that

$$P(T \leq 5.6) = \Phi_{(5.52, 0.17)}(5.6) = 0.68.$$

(By comparison, the exact value is 0.63.) □

Example 10.6.2 *Dynamic Acoustics'* top model headphones have a lifetime that follows a Weibull distribution with parameters $\lambda = 15.4$ and $k = 1.7$. They have a 5 year warranty.

1) What is the principal cause of failure in these headphones? Defect, randomness, or wear?
2) What is the probability that the headphones will start failing before the warranty expires?
3) What is the probability that the headphones last for at least 20 years?

Answer:
1) Since $k = 1.7 > 1$, the principal cause of failure is wear.
2) Their lifetime is a stochastic variable T with distribution $weib_{(1.7,15.4)}$. Then

$$P(T \leq 5) = \text{WEIB}_{(1.7,15.4)}(5) = 1 - e^{-\left(\frac{5}{15.4}\right)^{1.7}} \approx 0.14 = 14\%.$$

3) We invoke the rule of complementary probabilities, and then

$$P(T > 20) = 1 - \text{WEIB}_{(1.7,15.4)}(20) = e^{-\left(\frac{20}{15.4}\right)^{1.7}} \approx 0.21 = 21\%. \quad □$$

10.7 Exercises

10.7.1 Normal Distribution, $\phi_{(\mu,\sigma)}$

1 Get to know your calculation tools and tables, to familiarize yourself with computing $\phi_{(\mu,\sigma)}(x)$, $\Phi_{(\mu,\sigma)}(x)$, and z_α. Write down your findings in a note-tobok for probability distributions.

2 Cumulative Normal distribution $\Phi_{(\mu,\sigma)}$ and probability
  (a) $X \sim \phi_{(0,1)}$; what is $P(X \leq 1.43)$?
  (b) $X \sim \phi_{(0,1)}$; what is $P(X > 1.43)$?
  (c) $X \sim \phi_{(0,1)}$; what is $P(X \leq -2.38)$?
  (d) $X \sim \phi_{(0,1)}$; what is $P(X > -2.38)$?
  (e) $X \sim \phi_{(3,\,1.9)}$; what is $P(X \leq 4.35)$?
  (f) $X \sim \phi_{(3,\,1.9)}$; what is $P(X > 4.35)$?
  (g) $X \sim \phi_{(-7,\,4)}$; what is $P(X \leq 0.2)$?
  (h) $X \sim \phi_{(0,1)}$; what is $P(X \in (-2.38,\ 1.43))$?
  (i) $X \sim \phi_{(10,5)}$; what is $P(X \in (8,\ 13))$?

3 Inverse cumulative Normal distribution z
  (a) Find $z_{0.05}$.
  (b) Find $z_{0.95}$.
  (c) Let $X \sim \phi_{(2,1)}$. Find a such that $P(X \leq a) = 0.05$.
  (d) Let $X \sim \phi_{(2,1)}$. Find a such that $P(X \leq a) = 0.95$.

4 The Normal approximation
  (a) A discrete stochastic variable X has expected value $\mu_X = 3$ and $\sigma_X = 1.2$. Use the Normal approximation to find $P(X \leq 4)$.
  (b) A continuous stochastic variable X has expected value $\mu_X = 3$ and $\sigma_X = 1.2$. Use the Normal approximation to find $P(X \leq 4)$.
  (c) A discrete stochastic variable X has expected value $\mu_X = 5.1$ and $\sigma_X = 2.2$. Use the Normal approximation to find $P(X \in \{6,7\})$.
  (d) A continuous stochastic variable X has expected value $\mu_X = 5.1$ and $\sigma_X = 2.2$. Use the Normal approximation to find $P(X \in (5,7))$.
  (e) A discrete stochastic variable X has expected value $\mu_X = 12.1$ and $\sigma_X = 4.7$. Use the Normal approximation to find

$$P = P(X \in \{12, 13, 14, 15, 16, 17, 18, 19\}).$$

5 Sums of Normally distributed stochastic variables
  (a) $X_1 \sim \phi_{(0,1)}$, $X_2 \sim \phi_{(-1,3)}$, $X_3 \sim \phi_{(2,4)}$, $X_4 \sim \phi_{(2,2)}$. These four stochastic variables are independent, and $X = X_1 + X_2 + X_3 + X_4$. What is the distribution of X?
  (b) $X_k \sim \phi_{(2,4)}$, and $X = \sum_{k=1}^{10} X_k$. What is the distribution of X?
  (c) $X_k \sim \phi_{(7,4)}$, $k = 1, 2, 3, 4$. Let X be the average of the X_k. What is the distribution of X?

10.7.2 Binormal Distribution, $\phi_{(\mu,\Sigma)}$

X and Y independent

6 What is the probability that $X < 7$ and $Y < 0$, when $Z = (X, Y)$ is binormally distributed with

$$\mu = \begin{bmatrix} 5 \\ -3 \end{bmatrix} \quad \text{and} \quad \sigma = \begin{bmatrix} 100 & 0 \\ 0 & 81 \end{bmatrix}?$$

X and Y dependent

7 Let X be the stock price of SnowPeak Ltd, and let Y be the stock price of FjordWater Ltd. The prices of the two stocks are correlated, so $Z = (X, Y)$ is binormally distributed with parameters

$$\mu = \begin{bmatrix} 20 \\ 12 \end{bmatrix} \quad \text{and} \quad \sigma = \begin{bmatrix} 49 & -28 \\ -28 & 64 \end{bmatrix}.$$

(a) Find the probability distribution for the price of each stock separately – that is: the marginal probability distributions $f_x(x)$ and $f_y(y)$.

(b) Find the probability distribution of the stock price of each of the stocks as a function of the price of the other – that is: the conditional probability distributions $f_{x|y}(x)$ and $f_{y|x}(y)$.

(c) Find the distribution of Z as a rotation of an independent distribution.

10.7.3 Gamma Distribution, $\gamma_{(k,\lambda)}$, With Family

8 $T \sim exp_5$. What are μ_T, σ_T, and $P(T \leq 4)$?

9 $T \sim exp_{2.3}$. What are μ_T, σ_T, and $P(T \leq 3.2)$?

10 $T \sim exp_{3.1}$. What are μ_T, σ_T, and $P(T > 1.2)$?

11 $T \sim exp_{4.4}$. What are μ_T, σ_T, and $P(T \in \langle 0.15, 0.28])$?

12 $T \sim exp_\lambda$, and $E[T] = 2.9$. What are λ, μ_T, and σ_T?

13 $T \sim exp_\lambda$, and $Var(T) = 2.25$. What are λ, μ_T, and σ_T?

14 $T \sim erl_{(2,5)}$. What are μ_T, σ_T, and $P(T \leq 1)$?

15 $T \sim erl_{(3, 2.3)}$. What are μ_T, σ_T, and $P(T \leq 1.7)$?

16 $T \sim erl_{(5,3.1)}$. What are μ_T, σ_T, and $P(T > 2.3)$?

17 $T \sim erl_{(2,4.4)}$. What are μ_T, σ_T, and $P(T \in \langle 0.2, 2])$?

18 $T \sim erl_{(2,\lambda)}$, and $E[T] = 2.9$. What are λ, μ_T, and σ_T?

19 $T \sim erl_{(3,\lambda)}$, and $Var(T) = 2.25$. What are λ, μ_T, and σ_T?

20 You and Morty Matrix are campaigning on the high street for the coming election, handing out leaflets for the "Mathematics Party, because $2 + 2 = 4$". You meet 1 sympathizer every 23 minutes. This waiting time is exponentially distributed.

(a) What is the probability that the next sympathizer arrives within half an hour?

(b) What is the probability that 2 sympathizers arrive within half an hour?

21 You are on probation with Statisticus Ltd for 4 weeks, as sales engineer. The waiting time between sales is exponentially distributed, but with parameter λ being a characteristic of the salesman. The requirement for being permanently hired is 2 sales within the probation period.

(a) What is the probability of getting the first sale within 2 weeks, as a function of λ, when time is measured in weeks?

(b) What must your λ be if you are going to have at least a 60% probability of permanent employment?

22 You are measuring a radioactive material, and the number of minutes until the next click on the Geiger counter is exponentially distributed with parameter $\lambda = 0.25$.

(a) What is the expected waiting time for 10 clicks?

(b) What is the probability of at least 3 clicks within 2 minutes?

(c) What is the probability of precisely 3 clicks within 2 minutes?

23 At the bookstore where you are working, the number of minutes until the next sale of the book *Home Engineering* is exponentially distributed with parameter $\lambda = \frac{1}{15}$.

(a) What is expected time until the next sale of *Home Engineering*?

(b) What is the probability that the next sale happens in between 5 and 20 minutes?

24 This Christmas, you have a seasonal job in a bookstore. T, the number of minutes until the next customer asks about the book *Home Engineering* is exponentially distributed with parameter $\lambda = \frac{1}{15}$.

(a) Write down the probability distribution of the waiting time until 3 customers have asked about the book *Home Engineering*.

(b) What is the probability that you must wait more than 30 minutes before 3 customers have asked about the book *Home Engineering*? Calculate in both of the following ways.

- Calculate directly on the distribution itself.
- Calculate by using the Normal approximation.

(c) What is the probability that precisely m customers have asked about the book during a 30 minute time period?

(d) What is the expected number of customers who have asked about the book during the period?

25 You have studied *Oh Mega*'s production of capacitors. Just as with the resistors in Example 10.3.3, their values follow a Normal distribution

$\phi_{(\mu,\sigma)}(x)$, and just as in the case of the resistors, you study τ, and get that $\tau \sim \gamma_{(16,4)}(t)$.

(a) Write down the cumulative probability distribution of τ by means of a χ^2 distribution.

(b) Find the probability that $\tau < 6$.

(c) Find a value τ_0 such that $P(\tau < \tau_0) = 90\%$.

(d) Translate the results for τ into results for σ.

26 You have studied the salmon in the river Loppa, and are more interested in the weight *variations* than in the mean weight. The salmon weight follows a Normal distribution $\phi_{(\mu,\sigma)}(x)$, so you study the variance by means of τ. Your investigations have concluded that $\tau \sim \gamma_{(19.5, 44)}(t)$.

(a) Write down the cumulative probability distribution of τ by means of a χ^2 distribution.

(b) Find the probability that $\tau > 0.64$.

(c) Find a value τ_0 such that $P(\tau > \tau_0) = 95\%$.

(d) Translate the results for τ into results for σ.

27 You have studied the brightness of a certain type of star, and are interested in the *variation*. The brightness follows a Normal distribution $\phi_{(\mu,\sigma)}(x)$, so you study the variance by means of τ. Your investigations have concluded that $\tau \sim \gamma_{(101.5, 15)}(t)$.

(a) Write down the cumulative probability distribution of τ by means of a χ^2 distribution.

(b) Find the probability that $\sigma > 2$.

(c) Find a value σ_0 such that $P(\sigma < \sigma_0) = 98\%$.

10.7.4 Student's t Distribution, $t_{(\mu,\sigma,\nu)}$

28 Find $t_{4, 0.1}$.

29 $X \sim t_{(0,1,8)}$. Find an x such that $P(X \leq x) = 0.005$.

30 $X \sim t_{(0,1,6)}$. Find an x such that $P(X \leq x) = 0.9$.

31 $X \sim t_{(7,3,11)}$. Find an x such that $P(X \leq x) = 0.95$.

32 $X \sim t_{(-3.14, 7.2, 23)}$. Find an x such that $P(X \leq x) = 0.999$.

33 $X \sim t_{(4.4, 3.1,7)}$ and $Y \sim t_{(-0.1, 2.2,18)}$. Find the probability distribution of $Z = X - Y$.

10.7.5 Beta Distribution, $\beta_{(a,b)}$

34 $X \sim \beta_{(1,1)}$. Find μ_X, σ_X, $P(X \leq 0.4)$, and graph the probability distribution.

35 $X \sim \beta_{(2,2)}$. Find μ_X, σ_X, $P(X > 0.6)$, and graph the probability distribution.

36 $X \sim \beta_{(2,3)}$. Find μ_X, σ_X, $P(X \in \langle 0.3, 0.6])$, and graph the probability distribution.

37 $X \sim \beta_{(3,2)}$. Find μ_X, σ_X, $P(X \in \langle 0.4, 0.65])$, and graph the probability distribution.

38 $X \sim \beta_{(53,22)}$. Find $P(X \le 0.7)$, and p such that $P(X \le p) = 80\%$.

39 $X \sim \beta_{(108,72)}$. Find $P(X \le 0.5)$, and p such that $P(X \le p) = 95\%$.

40 $X \sim \beta_{(17,42)}$. Find $P(X \le 0.2)$ and p such that $P(X \le p) = 10\%$.

41 $X \sim \beta_{(43,19)}$. Find $P(X \ge 0.8)$, and p such that $P(X \ge p) = 90\%$.

42 $X \sim \beta_{(128,81)}$. Find $P(X \ge 0.6)$, and p such that $P(X \ge p) = 99\%$.

43 $X \sim \beta_{(491,396)}$. Find μ_X and σ_X, and find $P(X \in [0.45, 0.50))$. (You may calculate this both exactly and with the Normal approximation.)

44 In the following problems, the probability distributions for the parameter p of a Bernoulli process are given. Find the probability of event H.
  (a) $p \sim \beta_{(12,17)}(p)$. $H = \top$.
  (b) $p \sim \beta_{(12,17)}(p)$. $H = \bot$.
  (c) $p \sim \beta_{(12,17)}(p)$. $H = \bot\top$.
  (d) $p \sim \beta_{(12,17)}(p)$. $H = \top\bot$.
  (e) $p \sim \beta_{(12,17)}(p)$. $H = \top\top\top\bot\bot\top\bot\bot$.
  (f) $p \sim \beta_{(12,17)}(p)$. $H = \bot\bot\bot\bot\top\top\top\top$.
  (g) $p \sim \beta_{(12,17)}(p)$. $H = 4\bot$ and $4\top$ (combination).
  (h) $p \sim \beta_{(52,12)}(p)$. $H = $ a given *sequence* with $3\top$ and $1\bot$.
  (i) $p \sim \beta_{(52,12)}(p)$. $H = $ *the combination* $3\top$ and $1\bot$.
  (j) $p \sim \beta_{(52,12)}(p)$. $H = $ *the combination* $4\top$ and $0\bot$.
  (k) $p \sim \beta_{(52,12)}(p)$. $H = $ *the combination* $2\top$ and $2\bot$.
  (l) $p \sim \beta_{(52,12)}(p)$. $H = $ *the combination* $1\top$ and $3\bot$
  (m) $p \sim \beta_{(52,12)}(p)$. $H = $ *the combination* $0\top$ and $4\bot$.
  (n) $p \sim \beta_{(52,12)}(p)$. $H = $ at least $3\top$ in 4 trials.
  (o) $p \sim \beta_{(52,12)}(p)$. $H = $ less than $3\top$ in 4 trials.

45 Let X be the proportion of 40 W light bulbs that break when dropped onto a carpet from a height of 1 m. You have tried it out, and your probability distribution for X is now $X \sim \beta_{(23,48)}$.

(a) What is the *expected* proportion of light bulbs μ_X that break when dropped onto a carpet from a height of 1 m?

(b) What is the probability that the next light bulb you drop onto a carpet from a height of 1 m, will break?

(c) What is the probability that X is within 5% of this value?

46 As above, but you have made more trials, and now $X \sim \beta_{(50,103)}$.

47 It is election time, and you are fact checking candidate April Weatherstone's factual claims. Let Y be the proportion of errors in Weatherstone's factual claims. Your estimate of Y, after studying two months of election campaigns, is $\beta_{(17,64)}$. What is the probability that 2 or fewer of Weatherstone's next 10 factual claims are erroneous?

10.7.6 Weibull Distribution, $weib_{(k,\lambda)}$

48 $T \sim weib_{(5,4)}$. What are μ_T, σ_T^2, and $P(T \le 4)$?

49 $T \sim weib_{(0.2,2)}$. What are μ_T, σ_T, and $P(T \ge 1)$?

50 $T \sim weib_{(2,5)}$. What are μ_T, σ_T, and $P(T \in (1,3))$?

51 $T \sim weib_{(0.5,3)}$. What are μ_T, σ_T, and $P(T \in (2,4))$?

52 When Jack goes into the Canadian wilderness, he looks most forward to seeing reindeer. Let T be the time (in hours) that it takes before he sees a reindeer. $T \sim weib_{(2,2)}$.

(a) What is expected waiting time μ_T for Jack to see a reindeer?

(b) What is the probability that Jack sees a reindeer before time μ_T?

53 It is given that T, the time (in seconds) it takes Santa's engineer elves to make a remote controlled car, is Weibull distributed with parameters $\lambda = 2.5$ and $k = 1$.

(a) What is the expected time for the engineering elves to make a car?

(b) What is the standard deviation for the time for the engineering elves to make a car?

(c) What is the probability that they take between 0.5 and 1.5 seconds?

54 The longevity of Solan's motorized vehicles (in years), is $T \sim f(x) = weib_{(1,0.5)}(x)$.

(a) What are μ_T and σ_T^2?

(b) What is the probability that a given motorized vehicle lasts for more than one year?

(c) Write down $f(x)$ as simplified as possible. You will then see that the lifetime distribution of the vehicles is a special case of the Weibull distribution, which is also known under another name. Which probability distribution is this? And what are the parameter(s)?

10.7.7 Continuous Uniform Distribution

The continuous uniform distribution has been omitted as a section so that the student should have at hand a tractable probability distribution to build up and study its properties. The first assignment is the key.

55 X follows a continuous uniform probability distribution over the interval $I = (a, b)$ if $P(X = x) = 1/(b - a)$ whenever $x \in I$, and 0 otherwise. Use the rules from sections 7.3, 7.5 and 7.6 to solve the problems below:

(a) Find μ_X.

(b) Find σ_X^2.

(c) Graph the probability distribution $f(x)$ and the cumulative probability distribution $F(x)$ of the continuous uniform probability distribution.

56 X is (continuously) uniformly distributed over an interval $[a, b]$, and $M \subset [a, b]$ is a disjoint union of intervals whose widths sum up to w. What is $P(X \in M)$?

Part II

Inference

11

Introduction

CONTENTS

11.1 Mindful of the Observations

In the old days, Bayesian statistics was often called *inverse* statistics. That is a rather accurate way of putting it, if by "ordinary" (non-inverse) statistics we mean working *from* a model *to* predictions of observations, as shown in Figure 11.1. The model is then summed up in probability distributions for the observations, as for instance a Normal distribution ϕ, or distributions $bern_p$ for Bernoulli processes, and $erl_{(k,\lambda)}$ and $pois_\lambda$ for Poisson processes.

The "forward" direction in this paradigm is conclusion from model to observation, whereas *inverse* statistics concludes from observation to model, and then after that to the *next* observation. We have borrowed abbot Sōzen from the Norwegian zen temple Bugaku in order to illustrate how we perform Bayesian statistical inference.

We had our first glimpse of statistical inference in Chapter 6. Our core example was the Gamesmaster's dice, Example 6.3.5, which we expanded in Example 6.7. Gamesmaster picked a random die from a bag containing one each of the dice $D_4, D_6, D_8, D_{10}, D_{12}, D_{20}$ that are all painted red on four of the faces and white on the remainder. He then tossed the die behind a screen, and the players were told only if the die had landed red or white, but not which die he had. From this, they were to guess which die Gamesmaster had picked, and to give probabilities for the reds and whites of the next observations.

We then looked at probability distributions, and got a taste of the extended concept of "population": when we have a finite population, an observation is

The Bayesian Way: Introductory Statistics for Economists and Engineers, First Edition.
Svein Olav Nyberg.

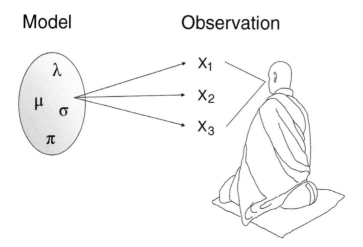

Figure 11.1 The zen monk Sōzen sees both the model and the observations.

equivalent to a random *sampling* from that population. But often, our observations arise from an ongoing process; there is no fixed set from which to sample. It is often still noticeable, however, that the observations follow a statistical distribution $f(x)$, as if they were sampled from an infinitely large population whose measurement distributions are described by $f(x)$. So, in the extended concept of "population", we will say that the essential feature of the model and the population is not the precise elements, but the *distribution $f(x)$* of the values. We say that we *sample from a probability distribution $f(x)$*.

11.1.1 Models and Walls

We let abbot Sōzen illustrate, while Bard and Frederick and their friends make comments.

"So what is the difference between Frederick's frequentism and Bard's Bayesianism here?" Sam asks.

"So far," Frederick replies, "nothing. Or almost nothing. The two of us have different interpretations of the meaning of the basic concept of probability, and in the Gamesmaster's dice example, we frequentists insist that, when the die has been picked, it has been picked. From there on, there is nothing random about it: it was either picked, or it wasn't picked. As you will see if you remember Example 5.1.6, we will say that after the die is picked, the value of the probability $P(D_8)$ is either 0 or 100%, regardless of our state of knowledge about it. But to the Bayesians, probability is something entirely different, so the dice example works nicely within their paradigm.

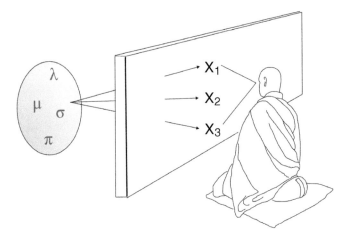

Figure 11.2 Zen monk Sōzen does not see the model, only the observations.

"But the dice example is still a useful one, for it illustrates both our similarities and our differences," Frederick continues, "and what we both do. Each die D_k corresponds to a model, a probability distribution for assigning probabilities to our red–white observations. But we do not know which model is the correct one. Is it D_8? Is it D_{20}? We do not know, for we do not see the probability distribution directly. As illustrated in Figure 11.2, we *only see the observations*. We see red and we see white, but we do not see the die. We do not see how many faces are white, and how many are red."

"Statistical investigations are all essentially like that," Bard says, and takes over. "The underlying reality is not directly accessible to us, but is as if hidden behind a screen, just like the dice. We do not see the probability distribution from which the observations are sampled, but we see the observations. Likewise, we do not *see* the probability p that a coin will give *heads*, but we see the outcomes of individual coin flips. Neither do we see the waiting time parameter λ for a Poisson process, but we may measure individual actual waiting times."

"There are times when we *in principle* could have discerned the underlying model," Frederick interjects. "We consider the weights of salmon in the river Loppa to be Normally distributed $\phi_{(\mu,\sigma)}$, and if we had managed the feat of emptying the river of every single fish, we *would* have the precise values of μ and σ. But note that this would be the weights of the salmon there and then, and not of one minute later, and it would work only if we captured and weighed *all* the salmon at once. So though theoretically thinkable, such exhaustive knowledge will *in practice* be impossible, and the population parameters will be as if hidden behind a screen."

"But what do you do if you can't know?" Sam asks, "do you just give up?"

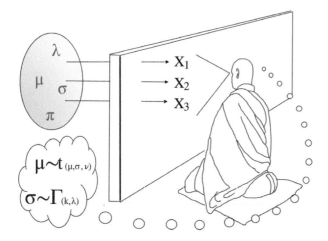

Figure 11.3 Zen monk Sōzen estimates the model parameters based on the observations.

"No," Frederick replies, "we estimate the parameters of the distribution, like Sōzen in his figure (Figure 11.3). But this is where Bard and I part ways. For Bard considers the parameters to be stochastic variables, and sets up probability distributions for them according to his observations. For my own part, I consider only the observations to be stochastic variables, whereas the parameters are fixed but unknown magnitudes whose probability of having such and such values, or of being within such and such intervals, is 0 or 100%. So Sōzen's illustration (Figure 11.3) illustrates Bard's viewpoint only, not mine."

11.1.2 Randomness

"To get good estimates of the parameters," Bard says, "we have to make our observations in the right way. We need representative observations. If we wish to know what the British think about the question of Scottish independence, we will not be doing an overly good job of it if all we do is poll the first 20 Scotsmen exiting a football match between England and Scotland. But what *does* work well, is *random sampling* from the population, since randomness has no preference. Both Frederick's techniques and mine are based on such non-preferential sampling."

"But we still need to stay alert," Frederick takes over, "for what is *random* is not always easy to discern. Picking random persons outside the football match is random enough, but it is the wrong 'random.' We need to make sure our randomness is *unbiased*, not preferring one part of the population to another."

"In all fairness, we should mention that *randomness* is a hard enough problem that it belongs to philosophy as well as to statistics," Bard interjects, "for what does it really mean for anything to be random? As we saw in the discussion on randomness in Section 5.14, a die may be 'random' in an everyday sense. But upon closer examination, when we looked at the physics of the situation, it was anything but random. Maybe randomness is simply another way of saying *I don't know* – an expression of our ignorance or partial ignorance, about the details of the system. Or maybe the proponents of the theory of *propensity* are right: that randomness is a fundamental property of some systems."

"And we tend not to know these underlying mechanisms," Frederick replies, "but what we know, is how we choose our observations. We should therefore examine what sources of bias there may be in this particular system, and then we should strive to eradicate these sources of bias from our samples, to the best of our ability. For instance: if we are looking at salmon weights in Loppa, we should examine whether certain parts of the river have smaller or bigger fish than the rest, or if different sizes of fish are more easily caught at different times of the day. We then remove the possibilities of such bias by not fishing at only one spot, or at only one time of day."

"And *then* we use the weight of the fish caught without such bias to estimate the weight of all the salmon of Loppa?" Sam asks.

"Precisely!" Bard and Frederick intone in unison.

11.1.3   Next Observation

"After that," Bard continues, "we estimate the next observation – the weight of the next salmon. This is what Sōzen does in Figure 11.4. Or maybe we don't. *That* depends what the goals of our investigations are.

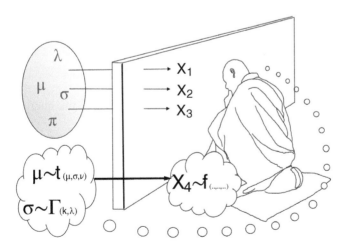

Figure 11.4 Zen monk Sōzen uses his estimate of the model to estimate the next observation.

"Sometimes, the question is whether a person has cancer or not, and we use diagnostic observations to determine that. In such instances, we are solely interested in the question of whether or not they have cancer, not in predicting the result of diagnostic observation.

"The example of Gamesmaster's dice is at the opposite end: here, our real interest is only in whether the next toss will yield a red or a white, whereas estimating the number of faces is interesting only as a means to that end: when I have a probability distribution for that parameter – the number of faces – I may also find a probability distribution for the next observation."

"I notice that you share the same goal, but evaluate the *methods* differently," Sam comments, "but could I ask you both to clarify one thing for me. I don't quite understand what Frederick means by *estimate* if it's not a probability distribution, and I don't understand how Bard is able to transmogrify observations into probability distributions. Would you care to explain?"

"That's why we're here," Bard replies. "The key to answering both questions lies in understanding our views on probability. *My* philosophy allows me a *prior* probability expressing my degree of knowledge and ignorance prior to the investigations. If I know nothing, I choose to work with a prior expressing my ignorance. But regardless of how much or how little I know, I have a prior probability. Having a prior probability allows me to apply Bayes' theorem to the observations to find a *posterior* probability, just as we did in the Gamesmaster's dice example. That is the key to *my* side. Without a *prior* probability, it is not possible to obtain a *posterior* probability to estimate the model."

"That's me!" Frederick says and waves his hands. "I'm the prior-less guy, so that you can see the difference between us. For us frequentists, degree of knowledge or conviction, ignorance, or whatever, is irrelevant. For us, probability is not about degrees of knowledge, hence the kind of *prior probability* Bard talks about makes no sense within our framework. Frequentist estimates are properties of the data alone."

11.1.4 The Data Alone

"But aren't the estimates properties of the model as well?" Sam queries Frederick.

"Actually, no," Frederick replies. "I know it might sound strange, and though I consider my view to be far more correct than Bard's, I have in time realized that quite a few, even professors in other disciplines, never get a good grasp of what my methods are actually all about. We have had our meetings of concern in the statistics association, ..."

"... and the solution is to teach a method that people understand: *Bayes!*" Bard interjects.

"... and the solution is *not agreed upon*," Frederick retorts, "but my father taught me something that might be of use to understand our two worlds. He was in control of one of the major casinos on the island I come from, and he told me that the trick to running a successful casino is not to win every game, but to win *in the long run*. This applies even though you as the controller of the casino often hold the same kind of privileged position that the Gamesmaster has in the dice example."

Frederick stops to see if Sam is paying attention, and continues: "The gambler may win or lose, because gambling is a hobby to him. Or at least ought to be. But if you are running the casino, this is your livelihood, so by the end of the day, in summary, you must be a net winner. You may lose individual games along the way, and this is even in your interest, for that is what attracts gamblers to your casino. But your net total, when the day is done, should be one where your winnings outrank your losses. Are you with me? "

Sam nods, "but what does this have to do with statistical estimates?"

"Everything!" Frederick replies, " at least in the frequentist world. For we use the data, and the data alone, to estimate the model. These estimates are, as I told you, properties of the data, and which data we get from our investigation is random. So whether our estimates capture or do not capture the parameter is ..." says Frederick, and adds a dramatic pause, "random! You see: like the casino, we do not mind missing the target in individual cases. But our techniques are designed to capture the parameters as best as possible *in the long run*. Bard's techniques are more like those of a gambler, trying to win the individual gambles."

Sam's eyes show a glimmer of understanding, so Frederick adds: "But I will readily admit that our techniques do not bear the same semblance of unity that Bard's do."

"That's right," Bard points out, "for whereas I find all that I need through the (*posterior*) probability distribution of the model, Frederick does not have any *core* engine driving all of his methods from a single framework. My colleagues have often called frequentist techniques *ad hoc inventions* ..."

"And some of *mine* call your priors *superstition*," Frederick interjects.

"But despite the differences between our two schools, the two of us have agreed to present statistics together," Bard continues, "for the benefit of you students. We think that regardless of which of our two school you are attracted to, you will understand it better when it is contrasted to the other one. For the difference lies in different philosophies, and they again are best understood in each others' contrast. This is also the root source of our differences. That, and of course disagreements as to the practicality of our two approaches in handling the uncertainty that is after all the subject matter of statistics. And we know and respect each others' techniques for what they are."

"Amen!" Frederick agrees. "But as you surely understand, then, I would like to explain *my* techniques on a case basis, if that is all right with you, Sam. I've even got my own chapter in Bard's book, Chapter 16!"

Sam nods, but at the same time, Frederick and Bard remember that they are competitors in an area that is even closer to the hearts of both of them. For they are both in love with their common friend Mina.

Bard and Frederick are of course unable to separate statistics from their infatuations, and start discussing how large a proportion of her hugs Mina will be granting each of them. The statistical model in question is then her tendency of handing out hugs, parameter **p** for the probability, whereas the observations are the hugs themselves – or rather: to whom the hugs are given. The *future* observation is then who gets the next hug. Bard, who is the Bayesian, has a prior belief that the two of them have an equally large chance of getting the next hug, whereas the frequentist Frederick says he will wait and see until he has some observations to go by.

Then Mina gives the first hug to Frederick.

Frederick observes that he has received 100% of the hugs, and that thus the best estimate is that Mina will hug him only; he therefore predicts that Mina's next hug will be his as well. In Section 16.1, we will learn about the conceptual underpinnings of Frederick's estimate: *unbiased* point estimates. We will also be looking at a more nuanced and less infatuated form of Frederick's reasoning around proportions in Section 16.3.1.

Bard on his side concludes that Mina probably will be giving more hugs to Frederick than to himself, but not all of them. Bard weighs his initial model against the data gathered, and his estimate is that Mina will give $\frac{1}{3}$ of her future hugs to him, and $\frac{2}{3}$ to Frederick. This is a *Bayesian* point estimate. We will learn more about Bayesian estimates of proportions in Section 13.2.

After a week has passed, Bard has received 17 hugs, and Frederick 28. In *Frederick's* model, Bard's future share is $\frac{17}{17+28} \approx 37.8\%$ of the hugs, whereas Frederick himself gets the rest. I *Bard's* model, he will be receiving $\frac{17+1}{17+1+28+1} \approx 38.3\%$ of the hugs. So we see that even though the two friends develop their models in different ways, they converge to some kind of agreement as they gather more data. The end of the story? The end of this story is that Mina dates Sam. She is her own woman, and not the subject of possessive calculations. Or as Bard and Frederick head-shakingly agree, "in statistics and romance, nothing is certain!"

11.2 Technically …

In statistical inference, there are a few technical phrases that are well worth noting, and we have compiled the following list of key terms.

Parameter: We implicitly assume that the underlying population may indeed be well described by means of a probability distribution, and that this distribution again belongs to a certain class and may be specified by means of a few parameters $p_1, p_2, \ldots, p_k$. If, for instance, the population is the salmon weights in Loppa, and we say that the salmon weight follows the probability distribution $\phi_{(3,0.7)}$, then $\mu = 3$ and $\sigma = 0.7$ are the parameters in our investigation. In our Loppa examples, we knew the values of μ and σ, but in statistical inference, the values of these parameters are usually hidden, and known only through *estimates*.

Observation: Before our observations, our future observations are *stochastic variables* $X_1, X_2, \ldots$ following the probability distribution(s) of the population. When we have observed X_k, have a concrete value x_k, we call this concrete value x_k a *realized value* for X_k. The Loppa salmon weights are $X_k \sim \phi_{(3,0.7)}$ prior to weighing. After weighing, their weight are realized values. For instance, the unknown X_5 has materialized as the concrete $x_5 = 4.1$.

Statistic: We rarely need the individual observations when we perform inference; what we need are mathematical summaries of the data; such a summary number is called a *statistic*. Fundamentally, any number that is a function of the observations is a statistic, but the interesting and relevant ones tend to be the ones we used for summing up our data in Chapter 2: mean ($\bar{X} = \frac{1}{n}(X_1 + \cdots + X_n)$), variance, standard deviation, median, and percentile. Here too, we differentiate between the statistic Ψ as a stochastic variable prior to observation, and its realized value ψ afterwards.

A collection of statistics is *sufficient* if they contain enough information for our inferences. For instance, for the Loppa salmon, the mean and the variance are together *sufficient statistics* for inference on the parameters μ and σ. The realized value of $\bar{X}$ is in this instance $\bar{x}$.

Estimator: A statistic $\hat{\Theta}$ is an *estimator* if it is an estimate (a guess) at the value of the parameter θ. For instance, for the Loppa salmon, $\bar{X}$ is an estimator for μ. Since using "the data alone" for estimation is primarily a frequentist notion, estimators are a topic in Section 16.1.

Posterior: Bayesians typically code and extract all information about a parameter through its *posterior* probability distribution. This is uniquely Bayesian. A Bayesian may for instance say, after 15 observations, that μ, the mean salmon weight in Loppa, follows a Student's t distribution, $\mu \sim t_{(2.9, 0.1, 14)}$, whereas he after 200 observations may come up with the more precise estimate $\mu \sim t_{(2.93, 0.04, 199)}$. This is the topic of Chapters 12 and 13.

Prior: Another key Bayesian concept. Whereas the *posterior* codes the total information available after the new observations, the *prior* codes the information prior to these observations. The observations themselves are coded into the *likelihood.*

11.3 Reflections

1 Bayesian/frequentist

(a) Who makes their estimates of the model parameters from the data alone?

(b) Who speaks of $P(\text{observation} \mid \text{model})$?

(c) Who speaks of $P(\text{model} \mid \text{observation})$?

(d) Who presupposes *randomness* for their methods?

(e) Who does all their inference through a probability distribution?

(f) Who speaks of *unbiased* estimates?

2 What are the two main purposes of statistical inference mentioned by Bard and Frederick? What is the difference between these two purposes, and how are the purposes related?

3 Why is *randomness* important?

4 May you observe a population, a model, or the model's parameters directly?

5 Your company has acquired the Chuck Wood's lumber mill. Along with the mill itself, they also got the mill's inventory. Your job is to estimate the humidity of the lumber by measuring 100 units. Discuss in groups which factors may bias the selection and sampling of units.

6 Discuss strengths and weaknesses in Bard's and Frederick's estimates of the proportion of hugs Mina will give to each of them. May one of the ways of analysing fit better in one context, and the other better in another context? If so: which kind of analysis fits which kind of context best?

12

Bayes' Theorem for Distributions

Geophysicist Hannah is impressed by Bayes' theorem, but asks Bard "Isn't it somewhat limited? I overheard you telling Sam that you used it to determine probability distributions, but I think I must have been mistaken. For Bayes' theorem, that's what you do in those little tables with priors and likelihoods and posteriors, when you're given a finite list of alternatives. But probability distributions are often about an infinite number of alternatives ... and sometimes even whole intervals, so that the alternatives are not denumerable."

In response, Bard asks Hannah for examples of probability distributions from her own field of expertise, geophysics. Hannah considers the question for a little while, before she answers: "Right now, I am studying the amount of uranium in a certain area, and from previous studies of this kind, I have found that the amount in grams, in a 10 kg sample, follows the distribution xe^{-x}."

"OK, that will work as an example, actually as a useful *prior* probability distribution," Bard answers, and asks: "Are you able to use Geiger counters on these samples, and get clicks such that the frequency of clicks tells you something about how much uranium is in the sample?"

Hannah nods.

"OK," Bard continues, "you probably have some known probability distribution describing when the next click will come in, given x grams of uranium in your sample. What is it?"

The Bayesian Way: Introductory Statistics for Economists and Engineers, First Edition.
Svein Olav Nyberg.
© 2019 John Wiley & Sons, Inc. Published 2019 by John Wiley & Sons, Inc.

"I really can't recall," Hannah replies, "but can't we for now just say it's ... maybe xe^{-tx} for *two* clicks?"

"That will work," Bard replies. "Then all we need is a measurement."

"Let us say the second click occurs at $t = 7.0$ minutes," Hannah replies; "but how would you use Bayes' theorem to find a new probability distribution for the amount of uranium in the sample then?"

"Good question," Bard replies, "and we will get to the answer. That is, *you* will get to the answer, and use Bayes' theorem to find a new probability distribution for your uranium sample. But before you do so, we will look at some simpler examples. Your uranium example has a continuous prior and a continuous likelihood. It is simpler for understanding to start with something that is essentially Bayes' theorem as you already know it, with a discrete prior and a discrete likelihood."

12.1 Discrete Prior

We start out simply, with an example having four enumerated alternatives, $A_1, \ldots, A_4$. We will continue as hinted, at the end of Chapter 6, and write the row values as function values, so that we are simply multiplying functions.

Example 12.1.1 For each alternative A_k,

- let the *prior* probability be $f_{\text{pre}}(k) = P(A_k) = \frac{1}{30}k^2$;
- let the likelihood be $g(k) = g_B(k) = P(B|A_k) = \frac{1}{k}$.

What is then the *posterior* probability $f_{\text{post}}(k) = P_{\text{post}}(A_k) = P(A_k|B)$?

k	Prior $f_{\text{pre}}(k)$	Likelihood $g(k)$	Joint $P \times L = f_{\text{pre}}(k) \cdot g(k)$	Posterior $f_{\text{post}}(k)$
$I = \{1, 2, 3, 4\}$	$k^2/30$	$1/k$	$k/30$	$k/10$
			$S = 1/3$	

Do you see how S is then the *sum*? This is because that one row of k's is really 4 rows packed into one, so that $S = \sum_{k=1}^{4} k/30 = 1/30 + 2/30 + 3/30 + 4/30 = 1/3$.

We may also graph our functions, as in Figure 12.1. □

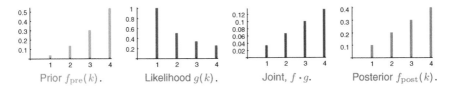

Figure 12.1 Bayes' theorem shown with function graphs.

You probably see how this saves you a fair bit of work, compared to writing each row on its own,[1] but it also opens up the possibility of infinitely many "rows", or alternative values for k.

Example 12.1.2 Let the *prior* probability distribution be $f_0(k) = \frac{1}{2}k2^{-k}$ for all integers k, and let the likelihood be $k4^{-k}$. May we then find a posterior probability distribution for all $k \in \mathbb{N}$? The answer is yes:

Values of k	Prior	Likelihood	Joint	Posterior
$k \in \mathbb{N}$	$\frac{1}{2}k2^{-k}$	$k4^{-k}$	$\frac{1}{2}k^2 8^{-k}$	$\frac{343}{72}k^2 8^{-k}$
			$S = \dfrac{36}{343}$	

where we calculated $S = \sum_{k=1}^{\infty} \frac{1}{2}k^2 8^{-k} = \frac{36}{343}$ on a calculator. □

The general rule for updating a probability distribution is as follows.

Rule 12.1.3 *Bayes' theorem, discrete prior: Let $X \sim f_{\text{pre}}(x)$. We update this probability distribution by means of an observation $Y = y$, thus: for each x, let $g(x) = h_x(y)$ be the probability (density) function for Y, conditional on that $X = x$. Then the updated probability distribution for X, $f_{\text{post}}(x)$ is found via this table:*

---

[1] Try it out: write out this table with four rows, to see for yourself.

Values	Prior	Likelihood	Joint	Posterior
$x \in I$	$f_{\text{pre}}(x)$	$g(x)$	$f_{\text{pre}}(x) \cdot g(x)$	$f_{\text{post}}(x) = \dfrac{f_{\text{pre}}(x) \cdot g(x)}{S}$

$$S = \sum_{z \in I} f_{\text{pre}}(z) \cdot g(z)$$

Working with the above functional version of Bayes' theorem, you calculate the sum by means of some calculation tool, so once that is set up, you should not worry if the formulas *look* complex; it's now a job for the computer.

Example 12.1.4 Once a year, the betting agency Mad Oaks sets up a new round of what they call an "octo lotto": it consists of eight balls that are either red or white. At the beginning of the year, each ball has had its color determined by the toss of a fair die, so that the number of red balls is distributed $bin_{(8, 0.5)}$. Gamblers are not told the colors of these eight balls. Further games are then based on sampling from these balls, *with* replacement, and displaying only one ball at a time.

It's January, and the first three results are in: red–red–red. Estimate the number of red balls in this year's lot of eight.

Your *prior* estimate on the number of reds is $f_{\text{pre}}(x) = bin_{(8,0.5)}(x) = \binom{8}{x}0.5^8$ for $x = 0, \dots, 8$. Further, if there are x reds among the 8, the probability of getting a red ball is $p = \frac{x}{8}$, so the likelihood function is $g(x) = \binom{3}{3}p^3(1 - p)^{3-3} = p^3 = (\frac{x}{8})^3 = 0.5^9 x^3$.

x values	Prior	Likelihood	Joint	Posterior
$0, \dots, 8$	$\binom{8}{x}0.5^8$	$0.5^9 x^3$	$0.5^{17}\binom{8}{x}x^3$	$\dfrac{1}{22\,528}\binom{8}{x}x^3$
			$S = 0.171\,875$	

where $S = \sum_{x=0}^{8} 0.5^{17} \cdot \binom{8}{x}x^3 = 0.171\,875$.

We display this graphically in Figure 12.2, where the colored graphs in 12.2b and in 12.2c correspond to the likelihoods and joint probabilities of the actual

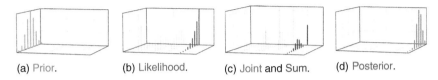

(a) Prior.   (b) Likelihood.   (c) Joint and Sum.   (d) Posterior.

Figure 12.2 Bayes' theorem for Mad Oaks' "octo lotto".

observation, whereas the grayed-out graphs in the same correspond to the likelihoods and joint probabilities of other potential observations.   ☐

12.1.1 ... When Likelihood is a Density

"We employ the same rule when the likelihood is a probability density as when it is a discrete point probability," Bard tells Hannah. "We could do the maths and justify it via a limit argument with increasingly small partitions, but given our thrust, I hope we can write this on the *we know it can be done but don't care about the details* account."

"Absolutely!" Hannah agrees, "I am application oriented, and it suffices for me to get the general gist of it to know that it is indeed so. If a mathematical proof helps understanding, then fine, but just going through the motions because some abstract person expects it is of no use to me."

Example 12.1.5 We have prior distribution function $X \sim f_{\text{pre}}(x) = \frac{1}{55}(10 - x)$ for $x = 0, 1, \ldots, 9$, and the likelihood is the probability density $g(x) = x^2$. We find the updated probability distribution $f_{\text{post}}(x)$ like this:

k	Prior	Likelihood	Joint	Posterior
$\{0, \ldots, 9\}$	$\frac{1}{55}(10 - x)$	x^2	$\frac{1}{55}(10 - x)x^2$	$\frac{1}{825}(10 - x)x^2$
			$S = 15$	

where $S = \sum_{x=0}^{9} \frac{1}{55}(10 - x)x^2 = 15$.   ☐

Bayes' theorem with the likelihood being a probability density applies just as much if the x-dimension is categorical data, for instance if x is one of {Salmon, Seachar, Trout}, as in the following continuation of Example 8.1.8.

Example 12.1.6 In the Loppen watercourses, sports fishers catch three different species of fish. You yourself have caught a fish there, but know so little about fish, you say, that you "couldn't distinguish a cod from a shark". You therefore descide to resort to a stronger suit of yours: statistics. Catches in Loppen are distributed as follows.

- Salmon: 63.7%, with weight distribution $\sim g_L(y) = \phi_{(4.2,1.2)}(y)$.
- Seachar: 22.4%, with weight distribution $\sim g_S(y) = \phi_{(1.7,0.5)}(y)$.
- Trout: 13.9%, with weight distribution $\sim g_O(y) = \phi_{(0.95,0.1)}(y)$.

Your fish weighs 2.5 kg, so the probability densities for the different fish are as follows.

- Salmon: $g_L(2.5) = 0.121\,9$.
- Seachar: $g_S(2.5) = 0.221\,8$.
- Trout: $g_O(2.5) = 0.000\,0$.
- Joint: $g(3) = 0.637 g_L(2.5) + 0.224 g_S(2.5) + 0.139 g_O(2.5) = 0.127\,3$.

The table then becomes:

k values	Prior	Likelihood	Joint	Posterior
Salmon	0.637	0.121 9	0.077 64	0.610
Seachar	0.224	0.221 8	0.049 70	0.390
Trout	0.139	0.000 0	0.000	0.000
			$S = 0.127\,3$	

There is thus a 61% probability that your fish is a salmon, and a 39% probability that it is a seachar. The probability that it is a trout is negligible.

We display this graphically in Figure 12.3, where the colored graphs in 12.3b, 12.3c, and 12.3d correspond to the likelihoods, joint probabilities, and sums of the actual observation, whereas the grayed-out graphs in the same correspond to the likelihoods, joint probabilities, and sums of other potential observations. □

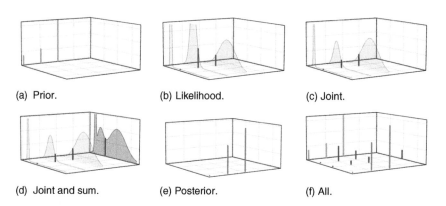

(a) Prior.   (b) Likelihood.   (c) Joint.

(d) Joint and sum.   (e) Posterior.   (f) All.

Figure 12.3 Bayes' theorem for the fish in the Loppen watercourses.

12.2   Continuous Prior

"You handled the two cases of the discrete prior skillfully," Bard complements Hannah.

"Thank you," Hannah replies, "with good help. But the way you say *discrete prior* makes me think there is a case of *continuous prior* as well."

"I'm not very good at keeping secrets," Bard admits, "or playing poker. I am simply too easy to read."

"I assume the transition from discrete to continuous prior works by these limit arguments you mentioned," Hannah continues, or almost asks, "just as with the likelihood. Is that right?"

"Indeed it is," Bard replies, "and the rule for continuous prior is almost like the one for discrete prior, but with one key difference. Can you guess what it is?"

"That the *S* is not a discrete sum, but an integral," Hannah responds almost before Bard has asked his question. "I am a geo-*physicist*, after all!"

"And right you are," Bard confirms.

Rule 12.2.1 *Bayes' theorem, continuous prior: let $X \sim f_{\mathrm{pre}}(x)$. We update this probability distribution by means of an observation $Y = y$, thus: for every x, let $g(x) = h_x(y)$ be the probability (density) of Y, given that $X = x$. Then the updated probability distribution $f_{\mathrm{post}}(x)$ for X is given by the following table:*

Values	Prior	Likelihood	Joint	Posterior
$x \in I$	$f_{\mathrm{pre}}(x)$	$g(x)$	$f_{\mathrm{pre}}(x) \cdot g(x)$	$f_{\mathrm{post}}(x) = \dfrac{f_{\mathrm{pre}}(x)\cdot g(x)}{S}$

$$\int_I f_{\mathrm{pre}}(z) \cdot g(z)\, \mathrm{d}z.$$

Example 12.2.2   The switch on an old diesel generator starts the generator only sometimes. At Motorkopf Ltd, you have a policy to the effect that units whose failure rate exceeds 20%, must be replaced. You decide to perform a statistical test of this switch, and include former experiences with this switch in the form of a *prior* probability distribution for π, the switch's failure rate. It is $f_{\mathrm{pre}}(x) = \frac{1}{e-1}e^x$ (on $[0, 1]$). Then you perform five trials, and the switch works four out of the five times. The success rate is x, which is hence the probability of success in a single trial. The probability of four successes out of five trials then equals $g(x) = \binom{5}{4}x^4(1-x)$. This is the likelihood. We may now find the *posterior* probability distribution:

x-values	Prior	Likelihood	Joint	Posterior
$[0, 1]$	$\dfrac{1}{e-1}e^x$	$5x^4(1-x)$	$\dfrac{5}{e-1}e^x x^4(1-x)$	$\dfrac{1}{53e-144}e^x x^4(1-x)$
			$S = \dfrac{265e-720}{e-1}$	

where

$$S = \int_0^1 f(x)g(x) = \int_0^1 \frac{1}{e-1}e^x x^4(1-x)\,dx = \frac{265e-720}{e-1}.$$

The *posterior* distribution is thus $f_{\text{post}}(x) \approx 14.5e^x x^4(1-x)$.

We display this graphically in Figure 12.4, where the colored graphs in 12.4b and 12.4c correspond to the likelihoods and joint probabilities of the actual observation, whereas the grayed-out graphs in the same correspond to the likelihoods and joint probabilities of other potential observations.

The purpose of using Bayes' theorem is rarely just to find the *posterior* probability distribution. We want to make use of this new probability distribution for some purpose, for instance calculating the probability that the value of the stochastic variable is within a certain interval.

In this case, you will be looking at whether π, the rate at which the switch works, is in excess of 80%. Your investigations have yielded a statistical answer, and you find the probability that the rate exceeds 80% like this:

$$P(\pi > 0.8) = \int_{0.8}^1 f_{\text{post}}(x)\,dx = \int_{0.8}^1 14.5e^x x^4(1-x)\,dx = 0.40.$$

We continue this example in Section 12.3. □

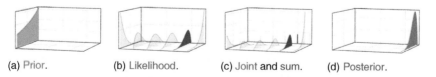

(a) Prior.　　(b) Likelihood.　　(c) Joint and sum.　　(d) Posterior.

Figure 12.4 Bayes' theorem for the switch on the diesel generator.

12.2.1 ... When Likelihood is a Density

"Now we have only one category left," Bard continues, "the continuous prior when the likelihood is a probability density. Your problem is of this kind. What is your *prior*, Hannah?"

"It must be the probability of the amount of uranium in the sample ... No, wait! ... the probability *distribution*, before my measurements with the Geiger counter. And then the *posterior* probability distribution for the amount of uranium – which is what I am looking for – *after* my Geiger measurements. The prior is $f_{\text{pre}}(x) = xe^{-x}$."

"Good," Bard replies, "and the likelihood?"

"... is that which has a different probability for the different x-values... probability *distribution* for the different x-values," Hannah replies, "and this is of course the number of clicks on the Geiger counter. The likelihood is then $g(x) = xe^{-tx}$."

"And you believed this would be difficult?" Bard laughs. "You might as well calculate the posterior while you are at it!"

Hannah fills in Table 12.1:

Table 12.1 Hannah's table for the uranium sample

Values	Prior	Likelihood	Joint	Posterior
$x \in [0, \infty)$	xe^{-x}	xe^{-1x}	x^2e^{-2x}	$4x^2e^{-2x}$
			$S = \frac{1}{4}$	

"I integrated the function in the 'Joint' column from 0 to ∞, since that was the range of possible values," Hannah comments. "The next click can't be sooner than 0 – like 10 minutes ago! And then the integral became $S = \int_0^\infty x^2 e^{-2x} = \frac{1}{4}$."

"Correct!" Bard exclaims, "now everything is in place!"

We display this graphically in Figure 12.5, where the colored parts of the graphs in 12.5b and 12.5c correspond to the likelihoods and joint probabilities

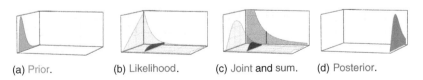

(a) Prior. (b) Likelihood. (c) Joint and sum. (d) Posterior.

Figure 12.5 Bayes' theorem for Hannah's uranium, illustrated.

of the actual observation, whereas the grayed-out parts of the graphs in the same correspond to the likelihoods and joint probabilities of other potential observations.

12.3 Next Observation

In the first Bayes chapter, we had a rule to determine the probability of the next observation (Rule 6.4.1). This rule applies to the function version of Bayes' theorem as well, but with the modification that we find a probability distribution, discrete or continuous.

We have seen, through the kinds and examples we have examined, that in Bayes' theorem, x has taken the place of A_k, and $f_{\mathrm{pre}}(x)$ has replaced $P(A_k)$. But what was $g(x)$ again? Well, let us look at the Gamesmaster examples in the first Bayes chapter. There, A_k was that "die D_k had been picked", whereas B was a totally different event whose probabilities varied depending on which die had been picked.

In the language of probability distributions and stochastic variables, this means that $g(x)$ measures the probability of an observation y you have made (for instance that a die toss gave $y = 7$), *given that* $X = x$. But when we are finding the probability that the next toss of the die ends up thus and thus, then y is not fixed, so we should really write $g_y(x)$. If the observation is $Y \sim h(y)$, then $g_y(x) = h_{|X=x}(y)$, the conditional probability (*density*) of Y, conditioned on the fact that $X = x$. Still, for our purposes it has until now been better to keep the notation x centered, and rather leave the observational variable y implicit.

But in this round, it is precisely the probability (*density*) that $Y = y$ we are after, and then y must be made explicit! We again sum over $g_y(x)$ for the different values of x, with the weights for each x given by $f(x)$. We sum it up as follows.

> **Rule 12.3.1** *Given the (posterior) probability distribution $X \sim f(x)$, we find the probability (density) that the next observation is $Y = y$ by looking at the Sum box, S in our tables, when $f(x)$ is filled into the place of the prior, and $g(x) = g_y(x)$ is the likelihood for an observation y. This is Rule 6.2.1, function version.*
>
> $$f_{|X=x}(y) = \begin{cases} \sum_x f(x) \cdot g_y(x) & \text{(X is discrete)} \\[2ex] \int_{-\infty}^{\infty} f(x) \cdot g_y(x)\, \mathrm{d}x & \text{(X is continuous)}. \end{cases}$$

Example 12.3.2 (Continuation of Example 12.2.2) You will never get a precise value for π, the success rate for the switch. But you may still calculate the probability that it works next time you try it. It is really simple: set up the Bayes' theorem table, minus the last column, and then the answer according to Rule 12.3.1 may be found in the Sum box at the bottom of your table. If this is what you want to find, you do *not* need the *posterior* column.

We write the *posterior* from Example 12.2.2 into the *prior* place of the table, and the probability that the switch works once, given that $\pi = x$, is $g(x) = x$:

x-values	Prior	Likelihood	Joint
$[0, 1]$	$\dfrac{1}{53e-144}e^x x^4(1-x)$	x	$\dfrac{1}{53e-144}e^x x^5(1-x)$
			$S \approx 0.74$

where $S = \int_0^1 \frac{e^x x^5(1-x)}{53e-144}\, dx = \frac{840-309e}{53e-144} \approx 0.74$. Since this is less than 80%, the switch must be replaced. $\qquad\square$

"Interesting," Hannah says, and asks, "may I say something about the probability of when the next click happens on my Geiger counter as well? Not just the probability or probability density of a click at some specified time like for instance $t = 1$, but for general t, as a function?"

"Try," Bard replies, "but look at the diagrams we made in conjunction with our updates first. There, we have marked the conditional probabilities of our *actual* observation, the *likelihood*, in color, whereas the conditional probabilities of the other, potential, observations are depicted in light gray. The sum is then a single number, as also marked."

"OK," Hannah replies, "so if I look at *all* the potential observations rather than just the actual one, I get different values for each possible observation, or more accurately a probability distribution? Like this?" And she redoes the colors, making the diagrams in Figure 12.6.

"Just like that!" Bard replies, "and now the calculations themselves are a mere formality. Calculate as you have illustrated."

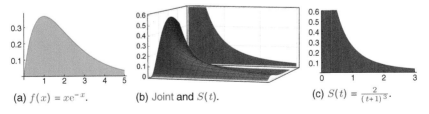

(a) $f(x) = xe^{-x}$. (b) Joint and $S(t)$. (c) $S(t) = \frac{2}{(t+1)^3}$.

Figure 12.6 Hannah's first illustration for the calculation of $S(t)$ for the uranium sample.

Table 12.2 Hannah's table for the uranium sample

x values	Prior $P(x)$	Likelihood $L(t,x)$	$P(x) \times L(t,x)$
$x \in [0,\infty)$	xe^{-x}	xe^{-tx}	$x^2e^{-(t+1)x}$
			$S(t) = \frac{2}{(t+1)^3}$

"Then I will leave t as t instead of fixing its value to be one in the likelihood," I think, "but instead just use $g(x) = xe^{-tx}$ as-is. See Table 12.2."

"Not bad," Bard comments, "but this is the probability distribution for the possible values of the *first* observation. What is the probability distribution of the *next* observation?"

"Of course!" Hannah replies, "I have to use the last probability distribution expressing my current knowledge, that is the *posterior*, and not my initial probability distribution, the *prior*. May I have a second try?"

Bard nods, and Hannah illustrates (Figure 12.7), and calculates (Table 12.3).

"And you got it just right!" Bard comments. "Maybe you should consider becoming a statistician?"

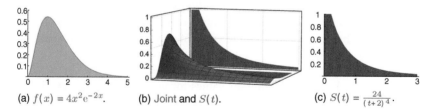

(a) $f(x) = 4x^2e^{-2x}$.   (b) Joint and $S(t)$.   (c) $S(t) = \frac{24}{(t+2)^4}$.

Figure 12.7 Hannah's second illustration for the uranium sample.

Table 12.3 Hannah's second table for the uranium sample

x values	Prior $P(x)$	Likelihood $L(t,x)$	$P(x) \times L(t,x)$
$x \in [0,\infty)$	$4x^2e^{-2x}$	xe^{-tx}	$4x^3e^{-(t+2)x}$
			$S(t) = \frac{24}{(t+2)^4}$

12.4 Repeat Updates

As in the tabular version of Bayes' theorem, we make further updates by setting the new *prior* equal to the old *posterior* distribution. And as in the tabular version of Bayes' theorem, all we need to do is to extend the table for calculating the next observation with a column for the *posterior* distribution. We continue Example 12.3.2

Example 12.4.1   The switch on the old diesel generator:

x	Prior	Likelihood	Joint	Posterior
$[0,1]$	$\frac{e^x x^4 (1-x)}{53e-144}$	x	$\frac{1}{53e-144} e^x x^5 (1-x)$	$\frac{e^x(1-x)x^5}{840-309e}$
			$S = \frac{840-309e}{53e-144}$	

□

12.5 Choice of Prior

How should you select your prior? The prior may be any probability distribution, and ideally, it reflects your entire knowledge and conviction about the parameter you are estimating. For practical purposes, we limit ourselves to probability distributions that lend themselves to reasonable calculations, and we may actually do this, being well assured that the precise shape of the prior plays a diminishing role as our data set grows.

Sensitivity to prior is your first concern in choosing a prior. The *posterior* is a balance between the prior and the likelihood: when the data are few, the prior dominates. When the data set grows, the likelihood dominates, and we speak of the "dominance of the data". This means that your choice of prior matters the most when your data are few. In such instances, a well chosen prior is worth its weight in gold, whereas a poorly chosen prior will give you a posterior according to the GIGO ("Garbage In, Garbage Out") principle. For big data sets, the prior must be very informative if it is to make much of a difference, and such priors usually arise only as the posterior probability distribution from previous investigations.

Let us now look at how we ought to choose our prior.

A **neutral prior** is the best choice if you are ignorant about the parameter, or if your knowledge is very limited. It is also the right choice if you are evaluating a case where you are required to be neutral and not include your own previous knowledge or opinion, and therefore must "let the data speak for themselves". A neutral prior is also a wise choice when your data set is small, and you need to be careful not to "contaminate" the result by previous impressions. This is the starting point of what we call *Objective Bayes*.

When you know something about the parameter you are estimating, it pays to use an informative prior coding that knowledge. The reason is that if you instead use a neutral prior, your analysis will not reflect the entirety of your knowledge, but only the data you last gathered.

Remember that the *posterior* distribution only reflects reality to the extent of the knowledge you code into it. Imagine that you are the captain of a sunken submarine, and that your instruments give certain depth readings, whereas your intuition as a captain tells you something like "we were cruising at a depth of 250 meters before the accident happened, and then we sank for half a minute before we came to rest, which brings us to roughly 300 meters' depth, plus or minus 20 meters". Would you discard your intuition as a captain and solely trust the instruments? Remember that your analysis might determine where the search and rescue will be conducted. What will give you and your crew the greatest chance of being rescued? Trusting your intuition as a captain, or relying solely on what the instruments told you?

A good rule of thumb for choosing a prior is to use "Occam's razor" and not make the prior more complicated than it need be. If you wish to limit your search to a given interval $[A, B]$, use the simplest possible function that is positive inside the interval, but close to zero outside. In the next chapter, we will look at some particularly well suited priors that turn Bayesian updating into very simple calculations, all the while making ample room for coding your previous knowledge into the *prior* probability distributions.

In some cases, the best way to express your prior is graphically, as follows.

(1) Make a function graph $\tilde{f}$ expressing your convictions about the probabilities. The more determinate your convictions are, the more precise your function graph needs to be if it is to reflect your convictions.
(2) Let $I = \int_{-\infty}^{\infty} \tilde{f}(x)\,dx$, and let $f(x) = I^{-1} \cdot \tilde{f}(x)$.
(3) Calculate what the probability distribution f says about $P(X \in \langle a, b])$ for important intervals. If you think the values sound wrong, return to step one, and adjust your prior graph.

Example 12.5.1 Svein is giving a prior estimate of the mean weight in his class. He thinks most of his fellow classmates are around eighty-something kilograms, so he is convinced the mean is eighty-something, "probably near the middle of the eighties", but is otherwise not at all sure. In Figure 12.8, he draws a rough graph he thinks might work: a "sour" parabola intersecting the horizontal axis at 80 and 90 kg. □

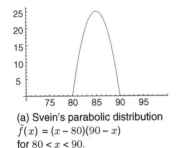

(a) Svein's parabolic distribution
$\hat{f}(x) = (x - 80)(90 - x)$
for $80 < x < 90$.

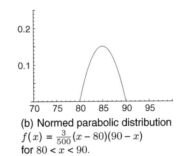

(b) Normed parabolic distribution
$f(x) = \frac{3}{500}(x - 80)(90 - x)$
for $80 < x < 90$.

Figure 12.8 Making a prior distribution function.

12.5.1 Improper Priors

When you are ignorant and know absolutely nothing about the matter at hand, a fully neutral prior giving equal weight to all possible values is a tempting ideal. But is this even possible? For who hasn't in their time desired a probability distribution that would be uniform over the integers, or over all the real numbers? The immediate answer is that there is no such distribution, since a uniform distribution by definition is $f(x) = k$ for all the values x for which it is defined. The total probability is then the sum or integral of f over all the possible values, and thus ends up as infinity. But such a prior still remains an ideal when we need to allow all values and show preference to none.

The good news is that under the right circumstances we *may* employ such a function as a prior. But what are the right circumstances? Let us first look at the following motivating example.

Example 12.5.2 You are a subordinate engineer on a team who has jumped to the Pegasus galaxy, and you have found an abandoned *dart*, a *wraith* space vessel, and you want to know how long the vessel has been abandoned for. Rodney McKay, the team genius, declares that we as rational beings must acknowledge that we are unable to make any pronouncement on the issue until we have made empirical investigations. Ronen Dex finds a fully charged battery next to the dart, and McKay informs us that the probability that the battry should be fully charged after having lain there for t months is $h(t) = e^{-2t}$. Your job is to find a probability distribution for how long the dart has been abandoned.

You recall some statistics you learned back on Earth. You see that X, the time the dart has been abandoned, is a stochastic variable with sample space $U = \mathbb{R}^+ = [0, \infty)$. You take McKay seriously, and decide that the prior $f(t)$ must assign equal probability to all values of t, so that $f(t) = k$ for all $t \in [0, \infty)$. But you run into a problem: the total probability must be 1, which requires that $\int_0^\infty f(t)\, dt = \int_0^\infty k\, dt = 1$. But for any positive k, the integral becomes infinite!

You are therefore left with two options. Your first option is to recall that the wraith dart *can't* have been abandoned for just any length of time; there is some upper limit. A million or a billion years? No, the wraith civilisation is not that old. So the dart can't have been abandoned for longer than that. You don't know precisely how old the wraith civilisation is, but as the resourceful person you are, you simply write "n years", with n to be filled in later. Your prior is

$$f_n(t) = \begin{cases} 0 & t < 0 \\ 1/n & t \in [0, n) \\ 0 & t \geq n. \end{cases}$$

To save some calculations, you cheat and rescale the prior to $f_n = 1$ on $[0, n)$, instead of the unscaled $1/n$. It makes no difference to the *posterior*, since when the prior or likelihood are scaled by a factor of c, that factor disappears in the posterior, since you divide by the total, S, which has been scaled by that factor c as well.

You use McKay's likelihood, $h(t) = e^{-2t}$. Your table is then

t	Prior	Likelihood	Joint	Posterior
$t \in [0, n)$	1	e^{-2t}	e^{-2t}	$\dfrac{2e^{-2t}}{1 - e^{-2n}}$
			$\dfrac{1 - e^{-2n}}{2}$	

X's posterior distribution is thus $\hat{f}_n = 2e^{-2t}/1 - e^{-2n}$. The expression is somewhat cumbersome, and you notice that it is very near $2e^{-2t}$, and that indeed $\lim_{n\to\infty} \hat{f}_n = \lim_{n\to\infty} 2e^{-2t}/(1 - e^{-2n}) = 2e^{-2t}$. You decide to see what happens if you allow yourself to choose $n = \infty$ for your prior, and then use

$$f_{\text{pre}}(t) = \begin{cases} 0 & t < 0 \\ 1 & t \geq 0. \end{cases}$$

The table then becomes:

t	Prior	Likelihood	Joint	Posterior
$t \geq 0$	1	e^{-2t}	e^{-2t}	$2e^{-2t}$
			$\dfrac{1}{2}$	

You have discovered a shortcut! By letting the prior be a function that strictly speaking is no probability distribution, you made your calculations and results quicker and neater than what an exact calculation would have given you, but yet with a result that is very close. Delighted, you decide to tell McKay of your discovery. McKay glances at your calculations, and exclaims "Oh yes, improper priors. Very useful! I have used them myself since I first attended a statistics course at the university at age nine." □

Definition 12.5.3 *A function or operator* $f(x)$ *is an* improper prior *in a calculation of a* posterior *probability distribution if it is non-negative, and the use of it as a prior yields that* $S = \int_I f_{\text{pre}} \times g < \infty$ *(continuous)* $S = \sum_I f_{\text{pre}} \times g < \infty$ *(discrete), making the* posterior *probability distribution found into a probability distribution.*

Bayes' theorem may then be used in precisely the same way as when we employ a proper prior.

12.6 Exercises

1 Santa's workshop makes 10 different types of sack for Santa and his elves, types $A_1, A_2, \ldots, A_{10}$. The number of A_x type sacks made are $17 \times x$. In other words, there exist 17 A_1 sacks, 34 A_2 sacks, etc. The proportion of soft gifts depends on the type of sack they is in. Type A_1 contains 1% soft gifts, A_2 contains 4%, and in general, sacks of type A_x contain x^2% soft gifts. Each sack contains one million gifts. Your elf gets one such sack, picked at random.
 – What is the probability that your elf got a sack of type A_x?
 – The elf pulls out two random gifts for you. Both are soft. Use this information to update the probabilities of which type of sack your elf picked.
 – What is the probability that your elf has a bag of type A_8?

2 You have 100 A_k, numbered from 1 to 100, and $f_{\text{pre}}(k) = k^2/338\,350$. You make observation B_1, and see that $g(k) = P(B_1|A_k) = 1/k$. What are the posterior probabilities $P_{\text{post}}(A_k)$?

3 You participate in a chocolate lottery, where every participant gets a bag. Your Christmas elf has filled the bags like this: he puts a dark chocolate into the bag. Then he tosses a coin. If the coin lands heads, he adds a milk chocolate, and tosses the coin again, and repeats the process. The process stops at the first tails. Let A_1 be that there are zero milk chocolates in the bag, A_2 that there is one, etc. Now, your elf is coming to you. You know how he filled the bag, but the bag is enchanted (of course), so the size does not reveal what's inside.

(a) Show that the *prior* probabilities are: $P(A_1) = 1/2$, whereas $P(A_2) = 1/4$, $P(A_3) = 1/8, \ldots, P(A_k) = 1/2^k$, which means that the prior probability distribution function is $f(k) = 1/2^k$ for all positive integers k.

(b) The elf allows you to shout "Abra cadabra, chocolate come to me!" to your bag, and then a random chocolate from the bag will appear. When you try it, you get a dark chocolate. This is your observation B. Show that the probability of getting a dark chocolate from bag k is $P(B|A_k) = \text{positive/total} = 1/k$. This is your likelihood.

(c) Find the updated (*posterior*) probability of the alternatives A_k. (Hint: $\sum_{j=1}^{\infty} 2^{-j} \times j^{-1} = \ln(2)$.)

4 You have found a magic lamp on the beach. The genie in the lamp is a mathematician, and she has decided to fill a bag with rocks in the following way: she tossed a fair coin repeatedly until she got tails. She then counted the total number of tosses, X, and filled the bag with 2^X rocks. The first rock is an ordinary, gray pebble, but the rest are nuggets of gold.

(a) Let $f(n) = P(X = n)$. Find an expression for the function $f(n)$.

(b) Let $g(n)$ be the probability that a random sampling from the bag yields a gold nugget, given that the genie tossed the coin n times. Find an expression for $g(n)$.

(c) You now sample a rock from the bag, at random. It's a gold nugget. Find the updated probability distribution that the genie had flipped the coin n times.

(d) The genie puts a gray pebble in the bag as replacement for the gold nugget, and asks you to have a go one more time. What is the probability that the next rock is a gold nugget as well?

5 Bernoulli trials
   - The alternatives are indexed by an x running from 1 through 15.
   - The prior probability in choosing among the 15 alternatives is uniform.
   - For alternative x, we have $P(\top) = x/15$ whereas $P(\bot) = 1 - P(\top)$.
   - You perform two trials, "with replacement"; the outcome is $\top\bot$.

(a) Write down the function $f(x)$ describing the prior probabilities of getting alternative x.

(b) Find the likelihood function $g(x)$.

(c) Fill in the table to find the posterior probability that you got alternative x.

(d) What is the probability that the next trial yields $\top$?

6 Sacks with handles; the contents are white (W) and black (B) balls.
   - You have seven sacks with an index x running from 1 through 7.
   - Sack x has x handles, and in a random pick, each *handle* has equal probability of being picked.

- All the sacks are filled with white (W) and black (B) balls, a total of 50 in each sack. In sack x, there are x^2 white balls.
- You pick a random ball from your sack, and get black. You put the ball back in the sack.

(a) Write down the function $f(x)$ expressing the prior probabilities you picked sack x when you pulled a random handle.[2]

(b) Find the likelihood function $g(x)$.

(c) Fill in the table to find the posterior probability that you picked sack x when you pulled a random handle.

(d) What is the probability that the observation of your next trial, "2 samplings with replacement" yields W–W?

7 Nuts come in two different chiralities (threadings); left-handed (L) and right-handed (R). Oleson's hardware store sells packs containing both kinds in one pack.

- Oleson sells 5 types of nut pack. The types are enumerated by an x running from 1 through 5.
- All packs contain 100 nuts. A pack of type x has $10x$ left-handed (L) nuts; the remainder are right-handed (R).
- Oleson has just not sold any of the packs, so he knows he has $30 - 5x$ packs of type x in his store.
- The packs are unfortunately unmarked, so now the entire inventory of them is now on sale. You pick a pack at random, and then pick a nut to study it. It is a left-handed nut (L). You put it back before picking a new random nut; it, too, is left-handed (L).

(a) Write down the function $f(x)$ expressing the prior probabilities[3] you picked a pack of type x.

(b) Find the likelihood function $g(x)$.

(c) Fill in the table to find the posterior probability that you picked a pack of type x.

(d) You put the nut back. What is the probability that the next random nut you pick is left-handed (L)?

8 You are updating a prior probability

$$f_{\text{pre}}(x) = \begin{cases} 0.5 + 0.25x & x \in \langle -2, 0] \\ 0.5 - 0.25x & x \in \langle 0, 2] \\ 0 & \text{otherwise} \end{cases}$$

---

[2]Hint: what is the total number of handles? How many handles does alternative x have? What, then, is $f(x)$?

[3]Hint: what is the total number of nut packs? $\sum_{x=1}^{5} (30 - 5x) = 150 - 5 \sum_{x=1}^{5} x = \cdots$.

and have a likelihood

$$g(x) = \begin{cases} 14 & x \le -1 \\ 2 & x \in (-1, 0] \\ 3 & x > 0. \end{cases}$$

Find the *posterior* probability distribution $f_{\text{post}}(x)$.

9 Make a die or any other physical object that may serve as your "random generator". Divide the possible outcomes into two roughly equal sets. If, for instance, you have made a 6-sided die, you may divide it into A: "low numbers" (1, 2, 3) and B: "high numbers" (4, 5, 6). Your task is to estimate the π, the probability of getting A.

(a) Make a guess at the value of π. *Guess* is the key here, but feel free to make a few preliminary trials. Let p be this value, and let n be the number of tosses you would have to perform to be as certain as you are. Let your prior distribution be $\pi \sim \beta_{(a,b)}(x)$, where $a = n \times p$ and $b = n \times (1 - p)$. If you need easy calculations, round off a and b to the nearest integer.

(b) Use the cumulative beta distribution (Section 10.5) to find the probability that $P(\pi > 0.5)$.

(c) Use the inverse cumulative beta distribution (Subsection 10.5.4) to find an interval $[A_1, B_1]$ such that $P(\pi < A_1) = 10\%$ and $P(\pi > B_1) = 10\%$.

(d) Toss the die a self-chosen number of times, and get k A and l B. What is the probability of this outcome if $\pi = x$? This is your likelihood function $g(x)$.

(e) Find the *posterior* probability distribution for π.

(f) Find the probability that $P(\pi > 0.5)$.

(g) Find an interval $[A_2, B_2]$ such that $P(\pi < A_2) = 10\%$ and $P(\pi > B_2) = 10\%$.

(h) Repeat the last three assignments, and repeat again. Make a guess at how many trials you must make for the interval $[A_n, B_n]$ to be no wider than 0.1.

(i) And repeat.

13

Bayes' Theorem with Hyperparameters

This chapter and the previous one are about the same thing: employing Bayes' theorem to perform what a student once called "a magical transformation of observations into probability distributions". But as opposed to the previous chapter, where we were left to do heavy sums and integrals, here we will be working with techniques where the heavy mathematical lifting has been done in advance, so that all we have to do is to compute a few simple sums and put them into the right formulas. It is the technique of *hyperparameters*.

To explain this magic that transforms observations into probability distributions, and just what a *hyperparameter* really is, we will let ourselves be led out into the woods by a boy we'll call Rocky. To be more precise, we will be looking for Rocky, and in that process we will be discovering our first Bayesian *updating rule for Gaussian processes*.

13.1   Bayes' Theorem for Gaussian Processes

The Normal distribution is frequently also called the Gaussian distribution after its discoverer, Carl Friedrich Gauss. A *Gaussian process* is a sequence of independent observations $X_1, X_2, \ldots$ sampled from a common Normal distribution $\phi_{(\mu,\sigma)}$.

But as opposed to what we did in Section 10.1, we are not going to use the parameters μ and σ to discern the properties of the observations $X_1, X_2, \ldots$ (Figure 11.1). We are going in the opposite direction, and will use the observations

The Bayesian Way: Introductory Statistics for Economists and Engineers, First Edition.
Svein Olav Nyberg.

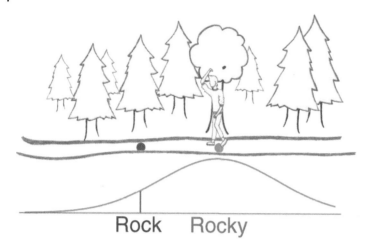

Figure 13.1 The position of the rock follows a Normal distribution $\phi_{(\mu,\sigma)}$.

$X_1, X_2, \dots$ to discern the properties of the unknown parameters μ and σ, as illustrated in Figure 11.3. We are going to find probability distributions for the values of the parameters μ and σ, and it is the parameters of these new probability distributions that are what we call *hyperparameters*.

Now we will look for Rocky. Rocky likes to toss red rocks on the path when he walks in the woods. Precisely where the rock lands, we do not know, but it is always Normally distributed with center μ at Rocky's own position, and standard deviation σ. So when his position is at μ, the position x of his rocks are $x \sim \phi_{(\mu,\sigma)}$, as illustrated in Figure 13.1.

Then Rocky hides in the woods, right off the path from where he stood when he threw the rock. When we arrive, we can't see Rocky, but we can see his red pebble. If we had no reason to suspect any part of the path was a more likely launching base for Rocky's toss than any other part, then "the rock is more than five meters to the left of where Rocky stood" and "Rocky stood more than five meters to the right of where the rock has now landed" have the same probability. This goes for any other distance as well, not just five meters. Indeed,

$$\phi_{(\mu,\sigma)}(x) = \phi_{(x,\sigma)}(\mu).$$

We have illustrated this symmetry in Figure 13.2.

We have now performed the magic trick: transformed the observation x into a probability distribution for μ. The parameters of μ's probability distribution are x and σ, which are then the *hyperparameters*.

Now, this was under the neutral assumption that Rocky could have been anywhere on the path, with equal probability. But what if we had more information? What if something influenced where Rocky stood when he tossed his rocks?

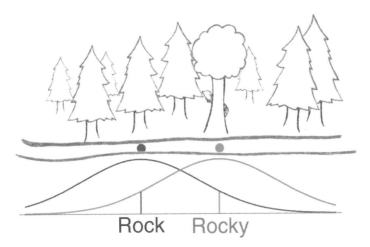

Figure 13.2 *P*(rock is more than five meters to the left of Rocky) = *P*(Rocky is more than five meters to the right of rock).

Pebbles, Rocky's little sister, likes to play with her big brother when the family is out for a hike in the woods. And Rocky, being a good big brother, includes her in his play. This time, they have agreed that Pebbles gets to throw the rock first, and that Rocky then tosses the rock from where it landed. We arrive later, and Pebbles tells us where she stood when she tossed the rock. We know the standard deviation for her throws as well, σ_{Pebbles}, and as with Rocky, where they land is centered on where they were thrown from. As illustrated in Figure 13.3,

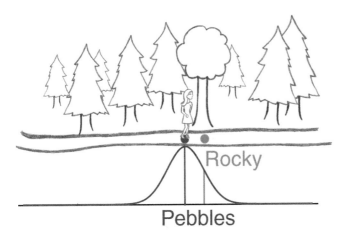

Figure 13.3 Pebbles sets the hyperparameters for the prior probability.

this gives us a probability distribution for Rocky's launch base, μ, even before we have found the pebble:

$$\mu \sim \phi_{(\mu_{\text{Pebbles}},\, \sigma_{\text{Pebbles}})}.$$

Since we now have a probability distribution for Rocky's position μ given in advance of finding the rock, we will call μ_{Pebbles} and σ_{Pebbles} the *prior* hyperparameters for μ.

The numbers from the observations of where Rocky's rock landed makes up the *likelihood*, and when we combine prior and likelihood, we get the *posterior* probability distribution for Rocky's position μ. We may use the techniques from the previous chapter to calculate this posterior probability distribution by multiplying *prior* with *likelihood* and dividing by the total.

However, we have thus far kept the calculations out in the open as sums and integrals, but do we really need to? The calculations we need to perform are quite advanced, especially from the stand-point of bachelor's degree studies in engineering, economics, or medicine, for example, but they end in almost identical results every time. Identical except for the alterations of a few key numbers. So these operations are really also quite unnecessary to perform more than once. And that one time is for the theoretician when he sets up the framework for the formulas.

We have by now understood how data become probability distributions, and we have understood what *hyperparameters* are. So let us leave the advanced calculations to the theorist, just as we leave car maintenance to the mechanic, and instead focus on how to operate the gas pedal, the brakes, and the steering wheel.

This chapter consists of Bayesian driving lessons.

First, we will look at three driving lessons for Gaussian processes, with slightly different rules according to whether σ and μ are known in advance or need to be managed. A bit like the difference in handling a car with manual or with automatic gear shift. After that, we will learn how to drive Poisson processes and Bernoulli processes. Common to all the driving lessons is a notation for keeping tidy track of the hyperparameters:

$P \vDash p_1$ (we say something about the first parameter)

$ p_2$ (we say something about the second parameter)

$ \vdots$

where the socket-plug-like "$P \vDash$" means that "In the model with probability distribution P, the hyperparameters, or intermediate values we use to calculate the hyperparameters, are thus: ".

To update the probabilities and hyperparameters, we make use of the *statistics* we mentioned in Section 11.2, and recall that a collection of statistics is *sufficient* for our updates if we need no further information about our observations in order to perform the update.

We then conclude with new hyperparameters and new probability distributions for the parameters, $p_i \sim f(x)$, where $f(x)$ is p_i's (new) probability distribution. After that, we may also indicate the probability distribution of the next observation, named for instance X_+, or – if we are looking at the next m observations – X_{+m}.

13.1.1 Unknown μ But Known σ

You are estimating the parameter μ for a Normal distribution $\phi_{(\mu,\sigma)}$ where the standard deviation σ has known value s_0.

Choice of prior for μ: Let your best estimate on the parameter μ be m_0. Then indicate how certain you are of your estimate m_0 as follows: let κ_0 be *how many measurements would be needed* for you to be as certain as you are. If you have somehow found your prior for μ expressed as a Normal distribution $\phi_{(m_{\text{pre}}, s_{\text{pre}})}$, let $m_0 = m_{\text{pre}}$ and $\kappa_0 = s_0^2 / s_{\text{pre}}^2$.

Don't panic if you don't know which values to use. *Not knowing* is simply nothing other than saying that you are as certain as if you had zero observations. You code this in numbers as $\kappa_0 = 0$ and $m_0 = 0$. These are the hyperparameter values of *Jeffreys' reference prior*,[1] which is an objective, neutral prior for the hyperparameters.

In both cases, let $\Sigma_0 = m_0 \kappa_0$. Your *prior* hyperparameters are then given by

$$P_0 \models \kappa_0 \quad \text{(the strength of you prior, in observation equivalents)}$$

$$\Sigma_0 \quad \text{(prior's counterpart to the "sum of observations").}$$

Updating: After n observations $x_1, \ldots, x_n$, the sufficient statistics are n and $\Sigma_x = \sum_{k=1}^n x_k$. You *posterior* hyperparameters are then

$$P_1 \models \kappa_1 = \kappa_0 + n$$
$$\Sigma_1 = \Sigma_0 + \Sigma_x.$$

This process may be repeated as many times as you want, on to $P_2, P_3, \ldots$ as far as you may wish to go, where the *prior* hyperparameters for the new update equal the *posterior* hyperparameters from the previous update. When we wish to obtain the *posterior* probability distributions for μ and the (posterior) predictive probability distribution for the next observation X_+, let $m_1 = \Sigma_1 / \kappa_1$.

---

[1] Jeffreys' prior for μ here is an *improper* prior (see Section 12.5.1) giving equal probability to all $x \in \mathbb{R}$. It is often written $f(x) = \phi_{(0,\infty)}(x) \equiv 1$, a Normal distribution with precision $\tau = 0$.

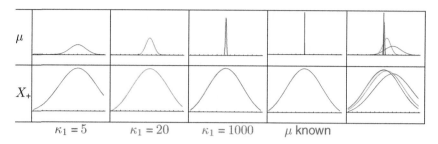

Figure 13.4 The probability distributions for different numbers of observations.

Then

$$\mu \sim \phi_{\left(m_1, s_0 \sqrt{1/\kappa_1}\right)}(x)$$

$$X_+ \sim \phi_{\left(m_1, s_0 \sqrt{1+1/\kappa_1}\right)}(x),$$

which also gives us $m_1 = E[\mu] = \frac{\Sigma_1}{\kappa_1}$. See Figure 13.4 for an illustration of how these probability distributions change as you get more observations.

We consider σ to be a known quantity when we know σ with sufficient precision for our purposes. If you do not know σ with the desired precision, you should go to Section 13.1.3, where we treat both μ and σ as unknown quantities.

Typical cases will be processes where the uncertainty resides in the act or apparatus of measurement itself, rather than in what is measured, such as in Example 13.1.1. In such cases, we may get to know the uncertainty σ with great precision, all the while the value μ that we want to estimate arises new and unknown for every new object we want to measure. Other cases are when the uncertainty describes a standard deviation in a production with similar equipment, where over time we get to know the uncertainty σ of the production equipment better and better, all the while the concrete values of the produced units are new and unknown for every case, as in Example 13.1.2.

Example 13.1.1 *Eggeseth's laser speed gauges* measure the speed of an object by sending out one or more pulses of laser light. The gauge then measures the Doppler shift in the reflection of each pulse, and calculates the speed from that shift. The uncertainty for a single pulse measurement is stated as 5 km/h. The traffic police are trying out Eggeseth's smallest model, which makes its measurements with three independent pulses. The built-in calculator finds the average assesment from the three pulse measurements, which it then shows on the display.

On a straight stretch on the M8 between Paisley and Glasgow, they spot their first speeder driving at 137 km/h. Considering the measurement uncertainty: what is the probability that the actual speed was in excess of 135 km/h?

Answer: Here, μ is the actual speed, whereas σ is the uncertainty in speed measurement from each pulse, which means $\sigma = 5$.

The traffic police are bound to judge by the measurements alone, so they set $\kappa_0 = 0$ and therefore also $m_0 = 0$. Then, $\Sigma_0 = \kappa_0 m_0 = 0$. Then the prior hyperparameters are

$$P_0 \models \kappa_0 = 0$$
$$\Sigma_0 = 0.$$

For the updating, they need the sum of the pulse measurements. Since the average of the 3 measurements is 137, Definition 2.3.10 tells us that $\Sigma_x = 3 \times 137 = 411$. The updated hyperparameters are then

$$P_1 \models \kappa_1 = 0 + 3 = 3$$
$$\Sigma_1 = 0 + 411 = 411.$$

That gives the following probability distribution for the speed:

$$\mu \sim \phi_{(411/3,\, 5\sqrt{1/3})}(x) = \phi_{(137,\, 2.886\,75)}(x).$$

The probability that the actual seed was in excess of 135 km/h is then

$$P(\mu > 135) = 1 - \Phi_{(137,\, 2.886\,75)}(135) = 0.755\,789 \approx \underline{76\%}. \qquad \square$$

"Does it matter how many sensors he has," Sam queries. "The lot of them are, after all, equally inaccurate."

"It actually does," Bard replies, "for they even each other out in the sum, such that the precision of the average is greater than each individual precision. If the traffic police had used a 20 pulse gauge when they got 137 km/h, they would have had $P(X > 135) \approx 96\%$, and if they had bought Eggeseth's latest model with 500 pulses, the probability that the speed was *less* than 135 km/h would have been roughly 2×10^{-19}. See Figure 13.5."

Example 13.1.2 You produce audiophile miniature amplifiers. You have designed several series with different output effects, and therefore know well from experience that the standard deviation of your production is $\sigma = 0.7\,W$.

1) At the beginning of the production, you measure the first five amplifiers you produced. The sum of their output effects is $\Sigma_x = 11.729\,W$. Use a neutral prior, and find the probability distribution for the value of μ, the mean output power (in watts) of the production. Find the probability distribution of

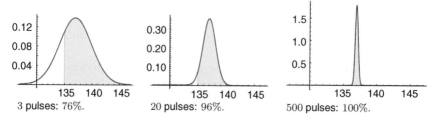

3 pulses: 76%.   20 pulses: 96%.   500 pulses: 100%.

Figure 13.5 The probabilities of $X > 135$ km/h for laser gauges with different numbers of pulses.

the effect the next amplifier produced, and finish off by calculating the probabilities $P(\mu < 2)$ and $P(X_+ < 2)$.

2) Later on in your production, you measure seven new units, and the output power measurements are $x_1 = 2.360, x_2 = 2.359, x_3 = 2.335, x_4 = 2.345, x_5 = 2.354, x_6 = 2.355$, and $x_7 = 2.357$ W. Use this to update the probability distribution of μ, and find the new probability distribution for the output effect of the next amplifier. Then, calculate the probabilities $P(\mu < 2)$ and $P(X_+ < 2)$.

Answer: The unknown mean output power is μ, and the known standard deviation is $\sigma = 0.7$.

(1) You are using a neutral prior, so then

$$P_0 \models \kappa_0 = 0$$

$$\Sigma_0 = 0.$$

The updated values are then

$$P_1 \models \kappa_1 = \kappa_0 + n = 0 + 5 = 5$$

$$\Sigma_1 = \Sigma_0 + \Sigma_x = 0 + 11.729 = 11.729.$$

This gives the following probability distributions for μ and X_+:

$$\mu \sim \phi_{(11.729/5, 0.7\sqrt{1/5})}(x) = \phi_{(2.345\,8, 0.313\,05)}(x)$$

$$X_+ \sim \phi_{(11.729/5, 0.7\sqrt{1+1/5})}(x) = \phi_{(2.345\,8, 0.766\,812)}(x).$$

This in turn gives us that

$$P(\mu < 2) = \Phi_{(2.345\,8, 0.313\,05)}(2) = 0.134\,663 \approx 13.5\%$$

$$P(X_+ < 2) = \Phi_{(2.345\,8, 0.766\,812)}(2) = 0.326\,01 \approx 32.6\%.$$

(2) You employ Rule 2.3.9, and find from your measurements that[2] $\Sigma'_x = 16.465$ and $n' = 7$. Since P_1 describes what you knew before the measurements, its hyperparameters are your priors:

$$P_1 \vDash \kappa_1 = 5$$
$$\Sigma_1 = 11.729.$$

The updated values are then

$$P_2 \vDash \kappa_2 = \kappa_1 + n' = 5 + 7 = 12$$
$$\Sigma_2 = \Sigma_1 + \Sigma'_x = 11.729 + 16.465 = 28.194.$$

The new probability distributions for μ and X_+ are then

$$\mu \sim \phi_{(28.194/12,\, 0.7\sqrt{1/12})}(x) = \phi_{(2.3495,\, 0.202\,073)}(x)$$
$$X_+ \sim \phi_{(28.194/12,\, 0.7\sqrt{1+1/12})}(x) = \phi_{(2.3495,\, 0.728\,583)}(x),$$

which means that

$$P(\mu < 2) = \Phi_{(2.3495,\, 0.202\,073)}(2) = 0.041\,853 \approx 4.2\%$$
$$P(X_+ < 2) = \Phi_{(2.3495,\, 0.728\,583)}(2) = 0.315\,721 \approx 31.6\%. \qquad \square$$

13.1.2 Known μ But Unknown σ

You are estimating the parameter σ for a Normal distribution $\phi_{(\mu,\sigma)}$, when the mean μ has known value m_0.

Prior for σ: Let your best estimate of the parameter σ be s_0. Then indicate how certain you are of your estimate s_0 as follows: let n_0 be *how many measurements would be needed* for you to be as certain as you are. Don't panic if you don't know which values to use. *Not knowing* is simply nothing other than saying that you are as certain as if you had zero observations. This gives you the *reference priors*, in other words objective, neutral priors for the hyperparameters. To code this in numbers, you let $v_0 = n_0 - 1$ and $SS_0 = s_0^2 \times \max(0, v_0)$.
Then, your *prior* hyperparameters are given by

$$P_0 \vDash v_0$$
$$SS_0.$$

Updating: After n observations $x_1, \ldots, x_n$, the sufficient statistics are n, Σ_x, and SS_x. You get another set of sufficient statistics if you use SB_x instead of SS_x. We

---

[2] Use the prime " $\prime$ " on Σ'_x to mark a new round of measurements and distinguish its sum Σ'_x from that of the first round, Σ_x.

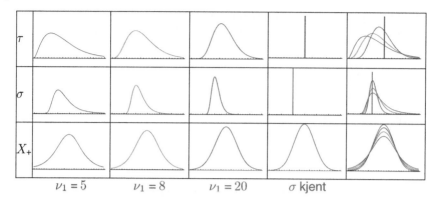

$\nu_1 = 5 \qquad \nu_1 = 8 \qquad \nu_1 = 20 \qquad \sigma \text{ kjent}$

Figure 13.6 The probability distributions for different numbers of observations.

recall that $\Sigma_x = \sum_{k=1}^{n} x_k$, $\bar{x} = \Sigma_x / n$, and $SS_x = \sum_{k=1}^{n} (x_k - \bar{x})^2$, whereas the new statistic is $SB_x = \sum_{k=1}^{n} (x_k - m_0)^2 = SS_x + n \times (\bar{x} - m_0)^2$.

Your *posterior* hyperparameters are then

$$P_1 \vDash \nu_1 = \nu_0 + n$$
$$SS_1 = SS_0 + SB_x.$$

This process may be repeated as many times as you want, on to $P_2, P_3, \ldots$ far as you may wish to go, where the *prior* hyperparameters for the new update equal the *posterior* hyperparameters from the previous update. We might then wish to obtain

(1) the (marginal posterior) probability distribution for the precision $\tau = \sigma^{-2}$;
(2) the (posterior) predictive probability distribution for the next observation X_+.

Let $s_1 = \sqrt{SS_1 / \nu_1}$. Then the probability distributions of τ and X_+ are given as

$$\tau \sim \gamma_{(\nu_1/2,\, SS_1/2)}$$
$$X_+ \sim t_{(m_0, s_1, \nu_1)},$$

which also gives us $\tau_1 = E[\tau] = \nu_1 / SS_1$. Notice that $s_1 = \sqrt{SS_1 / \nu_1} = 1/\sqrt{\tau_1}$. See Table 13.6 for an illustration of how these probability distributions change as you get more observations.

It is a bit unusual to know μ but not σ, but it may still occur, such as for instance when

- we may find Σ_x (the sum of the observations) and n in a very simple manner. If we have caught $n = 2000$ fish in a net, finding Σ_x is easy since it is the total weight of the catch. With such a large number of individual values, Σ_x/n is a sufficiently precise estimate of μ that we may consider $\mu = \Sigma_x/n$ to be a known value. To find the estimate of σ, however, we need to weigh the fish one by one;

- we have pairwise, independent measurements of differences. If $X_1, X_2 \sim \phi_{(\mu_X,\sigma_X)}(x)$, Rule 10.1.10 says that $X_1 - X_2 = W \sim \phi_{(0, \sqrt{2}\sigma_X)}(w)$, which means that $\mu_W = 0$ is known, and we may use the rule above to find a probability distribution for $\sigma_W = \sqrt{2}\sigma_X$, and that by dividing by $\sqrt{2}$ we may find a probability distribution for σ_X.

Example 13.1.3 In your small fishing boat, you have been working off the coast of Helgeland in Norway, and are now coming in to the small fishing village Sandessjøen to sell your cod. A marine biologist wants to know the standard deviation of cod in the area. You don't know, but your total catch weighs in at 6541 kg, and as the filleting factory counts the fish, they find that you had 1055 cod. This gives you confidence to state that the mean weight of Helgeland cod is known, and is $m_0 = 6541/1055 = 6.2\,\text{kg}$.

To find the standard deviation requires weighing cod individually, but this is time consuming, and the cod have already been filleted anyway. But you tell the marine biologist that you had taken home eight randomly picked cod for your spouse. You weigh them together, and the weights are, respectively, 5.45, 4.1, 6.6, 4.9, 7.1, 6.9, 5.95, and 6.95 kg. What is the probability that the standard deviation σ for Helgeland cod exceeds 2 kg?

Answer: The marine biologist asks you to use a neutral prior, that is $n_0 = 0$, so your *prior* hyperparameters are given by

$$P_0 \vDash v_0 = -1$$
$$SS_0 = 0.$$

We recall that the known value of μ is $m_0 = 6.2\,\text{kg}$. For the eight cod you brought home, you find $\Sigma_x = 47.95$ and thus $\bar{x} = 5.993\,75$. Using Rule 2.4.1, you find $SS_x = 8.407\,19$. Your *posterior* hyperparameters are then

$$P_1 \vDash v_1 = v_0 + n = -1 + 8 = 7$$
$$SS_1 = SS_0 + SS_x + n(\bar{x} - m_0)^2$$
$$= 0 + 8.407\,19 + 8(5.993\,75 - 6.2)^2 = 8.747\,5.$$

The probability distribution of $\tau = \sigma^{-2}$ is then

$$\tau \sim \gamma_{(72, 8.747\,5/2)}(t) = \gamma_{(3.5, 4.373\,75)}(t)$$

This means the probability that $\sigma > 2$ (see Figure 13.7) is

$$P(\sigma > 2) = P(\sigma^{-2} < 0.25) = \Gamma_{(3.5, 4.373\,75)}(0.25) = 0.051\,222\,8 \approx 5\%. \quad \square$$

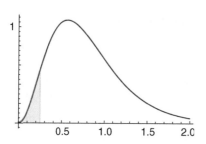

Figure 13.7 $P(\sigma > 2)$ when $\tau \sim \gamma_{(3.5,\,4.373\,75)}(t)$.

13.1.3   Unknown μ and σ

Choice of prior hyperparameters: You are estimating the parameters μ and σ for a Normal distribution $\phi_{(\mu,\sigma)}$. You find the values κ_0, m_0, and Σ_0 in the same way as in Section 13.1.1, and the values v_0 and SS_0 in the same way as in Section 13.1.2. For our calculations, we will also make use of $C_0 = SS_0 + \kappa_0 \times m_0^2$.

Then, your *prior* hyperparameters are given by

$$P_0 \models \kappa_0$$
$$\Sigma_0$$
$$v_0$$
$$SS_0 \text{ or } C_0.$$

Updating: After n observations $x_1, \dots, x_n$, your sufficient statistics are n and $\Sigma_x = \sum_{k=1}^{n} x_k$, plus a choice of either $\Sigma_{x^2} = \sum_{k=1}^{n} x_k^2$, or $SS_x = \Sigma_{x^2} - \Sigma_x^2/n$.

Your *posterior* hyperparameters are then given by

$$\left. \begin{aligned} P_1 \models \kappa_1 &= \kappa_0 + n \\ \Sigma_1 &= \Sigma_0 + \Sigma_x \end{aligned} \right\} \qquad m_1 = \Sigma_1/\kappa_1$$

$$\left. \begin{aligned} v_1 &= v_0 + n \\ C_1 &= C_0 + \Sigma_{x^2} \\ SS_1 &= C_1 - \Sigma_1^2/\kappa_1 \\ &= SS_0 + SS_x + (n\kappa_0/\kappa_1) \times (S_x/n - m_0)^2 \end{aligned} \right\} \qquad s_1^2 = SS_1/v_1.$$

This process may be repeated as many times as you want, on to $P_2, P_3, \dots$ as far as you may wish to go, where the *prior* hyperparameters for the new update equal the *posterior* hyperparameters from the previous update. We may wish to obtain probability distributions, and we have available

(1) the (marginal posterior) probability distribution for the precision $\tau = \sigma^{-2}$;
(2) the (marginal posterior) probability distribution for the mean μ;
(3) the (posterior) predictive probability distribution for next observation X_+.

The probability distributions are

$$\tau \sim \gamma_{(v_1/2,\, SS_1/2)}$$
$$\mu \sim t_{(m_1,\, s_1 \times \sqrt{1/\kappa_1},\, v_1)}$$
$$X_+ \sim t_{(m_1,\, s_1 \times \sqrt{1+1/\kappa_1},\, v_1)}.$$

Example 13.1.4 Tara, Penelope, and Beatrice want to estimate the distribution of income among their fellow students. Their first guess is that the mean is £16 000, but they are no more certain of this than if they had asked three students. They know nothing about the standard deviation. They proceed to ask 50 students about their annual incomes, and get a mean of $\bar{x} = 15\,734.80$ and the sample standard deviation is $s_x = 8\,655.30$. What are the probability distributions for the mean μ and the precision $\tau = \sigma^{-2}$ for the distribution of income among their fellow students? Also, what is the probability distribution for the income of the next student, were they to ask?

Answer: We begin with the *prior*:

$$P_0 \vDash \kappa_0 = 3$$
$$\Sigma_0 = 3 \times 16\,000 = 48\,000$$
$$v_0 = -1$$
$$SS_0 = 0$$
$$C_0 = 0 + 3 \times 16\,000^2 = 768\,000\,000.$$

The students have already calculated $\bar{x}$ and s_x, so we employ Definitions 2.3.10 and 2.4.3, and Rule 2.4.5 in reverse, to find $\Sigma_x = 786\,740$, $SS_x = 3\,670\,796\,686.41$, and $\Sigma_{x^2} = 16\,049\,993\,238.41$. Then the updated hyperparameters are

$$P_1 \vDash \kappa_1 = \kappa_0 + n = 3 + 50 = 53$$
$$\Sigma_1 = \Sigma_0 + \Sigma_x = 48\,000 + 786\,740 = 834\,740$$
$$m_1 = \frac{\Sigma_1}{\kappa_1} = \frac{834\,740}{53}$$
$$v_1 = v_0 + n = -1 + 50 = 49$$
$$C_1 = C_0 + \Sigma_{x^2} = 768\,000\,000 + 16\,049\,993\,238.41$$
$$= 16\,817\,993\,238.41$$
$$SS_1 = C_1 - \frac{\Sigma_1^2}{\kappa_1} = 16\,817\,993\,238.41 - \frac{834\,740^2}{53}$$
$$\approx 3\,670\,995\,736.52$$
$$s_1^2 = \frac{SS_1}{v_1} = \frac{3\,670\,995\,736.52}{49}.$$

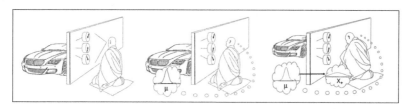

Figure 13.8 Sōzen performs his measurements in two rounds, and with him, we will perform our updates in two rounds as well.

We get that

$$\tau \sim \gamma_{(49/2,\, 3\,670\,995\,736.52/2)}(t) = \gamma_{(24.5,\, 1\,835\,497\,868.26)}(t)$$

$$\mu \sim t_{(834\,740/53,\, \sqrt{3\,670\,995\,736.52/49}\, \times\, \sqrt{1/53},\, 49)}(x) = t_{(15\,749.80,\, 1\,188.93,\, 49)}(x)$$

$$X_+ \sim t_{(834\,740/53,\, \sqrt{3\,670\,995\,736.52/49}\, \times\, \sqrt{1+1/53},\, 49)} = t_{(15\,749.80,\, 8\,736.81,\, 49)}(x).$$

$\square$

Example 13.1.5 In Figure 13.8, we see abbot Sōzen being concerned with mindfulness in traffic. He wants to gauge the speed of the cars driving past in the 50 km/h zone outside Bugaku temple. He decides to use an Eggeseth laser speed gauge with 500 pulses, so that the measurement uncertainty becomes a negligible factor. He wants to see how the speeds are distributed. As part of his investigations, he wants to know the probability that the mean speed is 10% or more above the speed limit, and also the probability that any random car would speed at more than 20% above the speed limit.

(1) During his first round, he measures 10 cars passing by with speeds (in kilometers per hour) of 59.1, 60.1, 60.1, 57.1, 59.9, 58.1, 47.6, 56.1, 56.0, and 57.8, which gives us the following statistics to work with: $n = 10$, $\Sigma_x = 571.9$, and $\Sigma_{x^2} = 32\,830.7$.

(2) In his next round, he measures 40 cars. We mark the new statistics with a prime to distinguish them from those of the first round. They are $n' = 40$, $\Sigma'_x = 2254.2$, and $\Sigma'_{x^2} = 127\,436.06$.

We are now going to calculate the probability distributions for μ, σ, and X_+, and then the probabilities $P(\mu > 60)$ and $P(X_+ > 60)$. For our first update, we use a neutral prior.

Solution: Neutral prior means

$$P_0 \models \kappa_0 = 0$$
$$\Sigma_0 = 0$$
$$\nu_0 = -1$$
$$C_0 = 0.$$

When we make our update with Sōzen's first set of measurements, we get

$$P_1 \vDash \kappa_1 = 0 + 10 = 10$$
$$\Sigma_1 = 0 + 571.9 = 571.9$$
$$m_1 = \frac{571.9}{10} = 57.19$$
$$v_1 = -1 + 10 = 9$$
$$C_1 = 0 + 32\,830.7 = 32\,830.7$$
$$SS_1 = 32\,830.7 - 10 \times 57.19^2 = 123.7.$$

The first-round posterior probability distributions are then

$$\tau \sim \gamma_{(9/2,\,123.7/2)}(t) = \gamma_{(4.5,\,61.85)}(t)$$
$$\mu \sim t_{(571.9/10,\,\sqrt{123.7/9}\times(1/\sqrt{10}),\,9)}(x) = t_{(57.19,\,1.172,\,9)}(x)$$
$$X_+ \sim t_{(571.9/10,\,\sqrt{123.7/9}\times\sqrt{1+1/10},\,9)} = t_{(57.19,\,3.888,\,9)}(x)$$

giving us probabilities

$$P(\mu > 60) = 1 - T_{(57.19,\,1.172,\,9)}(60) \approx 2.0\%$$
$$P(X_+ > 60) = 1 - T_{(57.19,\,3.888,\,9)}(60) \approx 24.4\%.$$

We then perform our second update with Sōzen's second set of measurements, using the posterior hyperparameters of the first round as our priors:

$$P_2 \vDash \kappa_2 = 10 + 40 = 50$$
$$\Sigma_2 = 571.9 + 2\,254.2 = 2\,826.1$$
$$m_2 = 56.522$$
$$v_2 = 9 + 40 = 49$$
$$C_2 = 32\,830.7 + 127\,436.06 = 160\,266.76$$
$$SS_2 = 160\,266.76 - 50 \times 56.522^2 = 529.9.$$

The second-round posterior probability distributions are then

$$\tau \sim \gamma_{(49/2,\,529.9/2)}(t) = \gamma_{(24.5,\,264.95)}(t)$$
$$\mu \sim t_{(2\,826.1/50,\,\sqrt{529.9/49}\times 1/\sqrt{50},\,49)}(x) = t_{(56.522,\,0.465,\,49)}(x)$$
$$X_+ \sim t_{(2\,826.1/50,\,\sqrt{529.9/49}\times\sqrt{1+1/50},\,49)} = t_{(56.522,\,3.321,\,49)}(x)$$

giving us probabilities

$$P(\mu > 60) = 1 - T_{(56.522,\,0.465,\,49)}(60) \approx 0.0\%$$
$$P(X_+ > 60) = 1 - T_{(56.522,\,3.321,\,49)}(60) \approx 15.0\%.$$

We see that μ, the cars' mean speed, becomes very precise, whereas our estimate of the speed of the next car remains spread out. This corresponds to our intuition of the matter, for it would indeed be strange if the passing cars started driving at speeds increasingly close to the mean speed just because Sōzen had started gauging their speeds. So while the precision of our μ estimate keeps growing, the probability distribution of our estimates of how fast the next car passes by Bugaku, X_+, instead converges to the underlying but unknown probability distribution of the actual seeds, $\phi_{(\mu,\sigma)}$. So the limiting standard deviation for our estimate of μ is zero, whereas for X_+ the limit is the underlying σ. □

13.1.4 Summary

We esimate the parameters μ and σ for a Normal distribution $\phi_{(\mu,\sigma)}$ by means of n observations $x_1, \ldots, x_n$ sampled from this Normal distribution. The sufficient statistics are n, Σ_x, and SS_x, where $\Sigma_x = \sum_{k=1}^n x_k$, and $SS_x = \sum_{k=1}^n (x_k - \bar{x})^2$. For known $\mu = m_0$, we may replace the statistic SS_x by $SB_x = \sum_{k=1}^n (x_k - m_0)^2 = SS_x + n \times (\bar{x} - m_0)^2$. For unknown μ, we may use the statistic $\Sigma_{x^2} = \sum_{k=1}^n x_k^2$ instead of SS_x.

- Let m_0 be your best estimate of μ, and let κ_0 be *how many measurements would be needed* for you to be as certain as you are. Let $\Sigma_0 = \kappa_0 m_0$.
- Let s_0 be your best estimate of σ, and let n_0 be *how many measurements would be needed* for you to be as certain as you are. Let $v_0 = n_0 - 1$, and $SS_0 = \max(0, v_0) \times s_0^2$. If you are using the statistic Σ_{x^2}, let $C_0 = SS_0 + \kappa_0 \times m_0^2$.
- If you are ignorant of either μ or σ, indicate your certainty by letting the observation equivalents (respectively κ_0 and v_0) be zero. This corresponds to an objective and neutral prior in the absence of observation.
- If you are certain of the values of μ or σ, choose the corresponding rectangle in Table 13.1.

Notice that there really are *three* degrees of knowledge for each parameter: fully known, fully unknown, and *partially unknown*. However, since the updating rules are the same for *fully unknown* and *partially unknown*, the two are merged into the category *unknown*.

13.2 Bayes' Theorem for Bernoulli Processes

We recall from Section 9.2.2 that a Bernoulli process is a sequence of independent Bernoulli distributed stochastic variables with common parameter p.

If the parameter p of a Bernoulli process is unknown, we may estimate it by means of Bayes' theorem. For individual observations, p is the parameter of a Bernoulli distribution $bern_p$, and for n observations, p is the second parameter of a binomial distribution $bin_{(n,p)}$. The hyperparameter version of Bayes' theorem gives an estimate of the parameter p, and of the next observations, according to the following rules.

Table 13.1 Summary of the Gaussian updating rules

	$\sigma = s_0$ (known)	σ unknown
$\mu = m_0$ (known)	No updating. Posterior values: $\tau = 1/s_0^2$ $\mu = m_0$ $X_+ \sim \phi_{(m_0, s_0)}(x)$	*Posterior* hyperparameters: $\left.\begin{array}{l} v_1 = v_0 + n \\ SS_1 = SS_0 + SB_x \end{array}\right\} s_1^2 = \dfrac{SS_1}{v_1}$ Posterior values: $\tau \sim \gamma_{(v_1/2, B_1/2)}(t)$ $\mu = m_0$ $X_+ \sim t_{(m_0, s_1, v_1)}(x)$
μ unknown	*Posterior* hyperparameters: $\left.\begin{array}{l} \kappa_1 = \kappa_0 + n \\ \Sigma_1 = \Sigma_0 + \Sigma_x \end{array}\right\} m_1 = \dfrac{\Sigma_1}{\kappa_1}$ Posterior values: $\tau = 1/s_0^2$ $\mu \sim \phi_{(m_1, s_0 \sqrt{1/\kappa_1})}(x)$ $X_+ \sim \phi_{(m_1, s_0 \sqrt{1+1/\kappa_1})}(x)$	*Posterior* hyperparameters: $\left.\begin{array}{l} \kappa_1 = \kappa_0 + n \\ \Sigma_1 = \Sigma_0 + \Sigma_x \end{array}\right\} m_1 = \dfrac{\Sigma_1}{\kappa_1}$ $\left.\begin{array}{l} v_1 = v_0 + n \\ C_1 = C_0 + \Sigma_{x^2} \\ SS_1 = C_1 - \kappa_1 m_1^2 \end{array}\right\} s_1^2 = \dfrac{SS_1}{v_1}$ Alternatively: $SS_1 = SS_0 + SS_x + n \times \dfrac{\kappa_0}{\kappa_1}(x - m_0)^2$ Posterior values: $\tau \sim \gamma_{(v_1/2, B_1/2)}(t)$ $\mu \sim t_{(m_1, s_1 \times \sqrt{1/\kappa_1}, v_1)}(x)$ $X_+ \sim t_{(m_1, s_1 \times \sqrt{1+1/\kappa_1}, v_1)}(x)$

Choice of prior: You are estimating the parameter p for a Bernoulli distribution $bern_p$. You then find two parameters: a_0 is the weight of *success*, $\top$, whereas b_0 is the weight of *failure*, $\bot$. You choose either a *neutral* prior expressing ignorance, or an *informative* prior expressing your knowledge.

- The *neutral* prior has parameters $a_0 = b_0 = u \in [0, 1]$. There are three common choices of value for u.
  - $u = 0$: "Total ignorance" (*Novick & Hall, Haldane, Jaynes*). Use this if you do not even know whether both $\top$ and $\bot$ are possible outcomes.
  - $u = 0.5$: "Ordinary ignorance" (*Jeffreys*). Use this if you know that both $\top$ and $\bot$ are possible outcomes, but otherwise know nothing about p.
  - $u = 1$: "Informed ignorance" (*Laplace, Bayes*). Use this if you initially, before observations, consider all values of p to be equally probable.

If, in a concrete case, you find yourself unable to choose between two of the alternative values for u above: choose the lowest one!

- An *Informative* prior may be chosen either from *posterior* hyperparameters from a previous update, or by coding your informal knowledge as numbers, thus: make a guess p_0 of p, and let κ_0 be how many observations you would need to be as certain as you are. Then, $a_0 = \kappa_0 p_0$ and $b_0 = \kappa_0(1 - p_0)$. Notice that you should end up with $a_0, b_0 \geq 1$, since otherwise you would have a value p_0 too extreme to be justified by as few as κ_0 observations.

Your *prior* hyperparameters are then given by

$$P_0 \vDash a_0$$
$$b_0.$$

Updating: After n observations $x_1, \ldots, x_n$, you have found k = number of T, and l = number of $\perp$. Then, k and l are sufficient statistics. Your *posterior* hyperparameters are then

$$P_1 \vDash a_1 = a_0 + k$$
$$b_1 = b_0 + l.$$

This process may be repeated as many times as you want, on to $P_2, P_3, \ldots$ as far as you may wish to go, where the *prior* hyperparameters for the new update equal the *posterior* hyperparameters from the previous update. We may then read off the following posterior and predictive probability distributions.

1) The (posterior) probability distribution of the parameter p.
2) K_{+m} – the number of successes (T) during the next m observations.
3) L_{+s} – the number of failures ($\perp$) before s new successes (T).

The probability distributions are then (the probability distributions βb and βnb are described in Appendix A.3):

$$p \sim \beta_{(a_1, b_1)}$$
$$K_{+m} \sim \beta b_{(a_1, b_1, m)}$$
$$L_{+s} \sim \beta nb_{(a_1, b_1, s)},$$

which also gives us $p_1 = E[p] = \frac{a_1}{a_1 + b_1}$.

A special case is if a_1 or b_1 is zero. In that case p follows a discrete probability distribution with either $P(\text{T}) = 0$ or $(\text{T}) = 1$, with the opposite probability for $\perp$. For all other cases, see Table 13.9 for an illustration of how these probability distributions change as you make more observations.

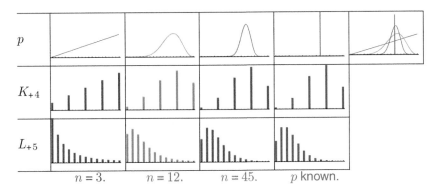

Figure 13.9 The probability distributions for different numbers of observations.

Example 13.2.1 "SeedSmart Ltd" has planted 18 new rose bushes for Sam. They did so last year as well, and then 6 of the bushes yielded red roses, whereas 12 yielded white. "How many will there be of each kind?" Sam wonders.

"Let's calculate," Bard replies. "First, we must set the *prior* distribution parameters for the proportion of red roses. What do you know about roses, Sam? And what did you know before you started growing them in your own garden?

"I know they come in both a red and a white variety – well, at least the ones from SeedSmart," Sam replies, "and they seem to have white and red only – and already last year I knew that SeedSmart's system of keeping track of which seeds are which variety is no better than random, so any bush from them could be either red or white.

"To me," Bard comments, "it sounds like your knowledge is best expressed through a neutral prior with $a_0 = b_0 = u = 1$. Agreed, Sam?"

Sam nods, and Bard starts calculating:

Prior $P_0 \models a_0 = 1$
$$b_0 = 1.$$

Updating: $P_1 \models: a_1 = a_0 + 6 = 1 + 6 = 7$
$$b_1 = b_0 + 12 = 1 + 12 = 13.$$

So the proportion of rose bushes that are red is $p \sim \beta_{(7,13)}$. The best estimate of p is then $E[p] = \frac{7}{7+13} = 0.35$.

"And now you probably wonder how many red rose bushes you will be growing for your Rosie?" Bard asks.

Sam nods; "she would have preferred three red and two white. What is the probability of that? Do we use a binomial distribution with $p = 0.35$ and $n = 5$?"

Bard reminds Sam that he has solved problems with an uncertain p before (Chapter 6: problems (12)–(14)), and says: "Just use the formula for *predictive distributions* when p is uncertain. Let us do the calculations together. And ... I left my pocket calculator with all the pre-programmed functions at home today, so we'll have to make do with the ugly formula."

Bard writes:

$$P(K_{+5} = 3) = \beta b_{(7,13,5)}(3) = \binom{5}{3} \times \frac{\binom{7+13}{7} \times \frac{7 \times 13}{7+13}}{\binom{7+13+5}{7+3} \times \frac{(7+3)(13+5-3)}{7+13+5}}$$

"0.179 842," Sam bursts out before Bard has had time to calculate anything at all. "I've got Wolfram Alpha on my phone, so I only needed to type

PDF[BetaBinomialDistribution[7, 13, 5, 3]

and the answer popped out together with a fair bit of explanation and other links!" □

We shall look at a more demanding example to illustrate how much of a difference it may make if we commit the error of calculating with a binomial distribution with fixed rate parameter value $p_1 = E[p]$ instead of taking into account the uncertainty in the rate parameter p.

Example 13.2.2 Troll Motors have tested their assembly line to test production quality. They want fewer than 1% of their cars to have a critical number of production errors, and, during the testing of 500 cars, 4 had the critical amount of such errors. That is, $4/500 = 0.08$, which is not too bad. But what is the probability that fewer than 1% of the cars produced will end up with the critical number of errors if Troll produces 20 000 cars?

We must first choose our prior, and we here invoke the rule that *if in doubt, choose the lowest u value*. Then,

Prior $P_0 \vDash a_0 = 0$
$\qquad b_0 = 0.$

Updated, $P_1 \vDash a_1 = a_0 + 4 = 4$
$\qquad b_1 = b_0 + 496 = 496.$

Let X be the number of cars with a critical number of errors. We know that 1% of a production of 20 000 cars is 200. What, then, is $P(X \le 200)$?

Wrong solution: A cumulative binomial distribution with parameter $p = 4/500 = 0.08$ gives

$$P(X \leq 200) = BIN_{(20\,000,\,0.08)}(200) = 0.999\,049 \approx 99.9\%.$$

That is, a mere 0.1% probability that the number of cars with critically many errors exceeds 1% of the production. Pretty good, isn't it? There should be no doubt that the assembly line is ready for full production! It is a pity, then, that this was the wrong way to calculate this probability, since we forgot to include the effect of the uncertainty in the value of p!

Correct solution: We include the uncertainty in p by calculating using a beta binomial distribution:

$$P(X \leq 200) = P(K_{+200\,00} \leq 200) = BB_{(4,\,496,\,20\,000)}(200) = 0.734\,991 \approx 73.5\%.$$

There is in other words a 26.5% probability that more than 1% of the 20 000 cars will have critically many errors, and there is a significant difference between 26.5 and 0.1%.

N.B. Notice that CDF[BetaBinomialDistribution[4, 496, 20000], 200]

exceeds the capacity of the free version of Wolfram Alpha, so you need to rewrite to

Sum PDF[BetaBinomialDistribution[4, 496, 20000], x] x from 0 to 200. □

The general lesson to take home from this example is that, when the number of future occurrences to be predicted is far larger than the number in the preliminary test, and the p value lies somewhere in the extremes (near zero or one), then the error of approximating the correct solution, which uses the βb distribution, by a simple *bin* distribution may become spectacularly large. In the case above, the error would have been very costly, and one may wonder if such a decision maker would get to keep his job.

For a given $\varepsilon > 0$ we may check if

$$\left| \frac{k^2}{a} + \frac{l^2}{b} - \frac{(k+l)^2}{a+b} \right| < 2\varepsilon.$$

A reasonable rule-of-thumb limit for when to approximate the βb and βnb distributions by the *bin* and *nb* distributions is $\varepsilon = 0.05$. When that rule is satisfied,

- $bin_{(k+l,\,a/(a+b))}(k)$ is a good approximation to $\beta b_{(a,b,k+l)}(k)$
- $nb_{(k,\,a/(a+b))}(l)$ is a good approximation to $\beta nb_{(a,b,k)}(l)$.

13.3 Bayes' Theorem for Poisson Processes

We recall from Sections 9.6.1 and 10.3.2 that a Poisson process is a waiting and counting process with rate λ, where the waiting time for the next occurrence is exponentially distributed exp_λ, whereas the number of occurrences during the next (time) unit is Poisson distributed $pois_\lambda$.

If the parameter λ of a Poisson process is unknown, we may estimate it by means of Bayes' theorem. We then count the number of occurrences n in the space of t (time) units, and estimate λ according to the following rules.

Prior for λ: Let λ_0 be your best guess at the value of the rate parameter λ. Then indicate how certain you are of your estimate λ_0 by stating how many occurrences κ_0 you would need to be as certain as you are, or for how many (time) units τ_0 you would have needed to observe to be as certain as you are. These values are related through the formula $\lambda_0 = \frac{\kappa_0}{\tau_0}$. If you are completely uncertain, or for other reasons want a neutral reference prior,[3] you simply refrain from making a guess at λ_0, and set $\kappa_0 = \tau_0 = 0$.

Then your *prior* hyperparameters are given by

$$P_0 \vDash \kappa_0$$
$$\tau_0.$$

Updating: You have registered n occurrences during t units. For an estimate of the rate λ for a Poisson process, n and t are sufficient statistics. Your *posterior* hyperparameters are then

$$P_1 \vDash \kappa_1 = \kappa_0 + n$$
$$\tau_1 = \tau_0 + t.$$

This process may be repeated as many times as you want, on to $P_2, P_3 \ldots$ as far as you may wish to go, where the *prior* hyperparameters for the new update equal the *posterior* hyperparameters from the previous update. When we wish to read the posterior probability distribution for λ and the predictive probabilities for the next observations, we find that

$$\lambda \sim \gamma_{(\kappa_1, \tau_1)}(l).$$

As for the predictions, we have *two* different predictions we may make:

- $N_{+\theta}$, the number of occurrences in the next θ units;

$$N_{+\theta} \sim nb_{(\kappa_1, \frac{\tau_1}{\tau_1+\theta})}(\eta),$$

---

[3] Another common reference prior corresponds to $\kappa_0 = \frac{1}{2}$ and $\tau_0 = 0$, but we will stick with the simplest one.

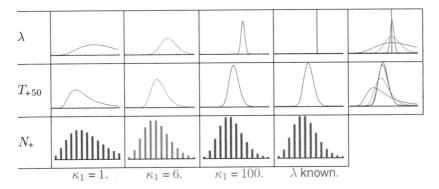

Figure 13.10 The probability distributions for different values of the hyperparameters.

where nb is the negative binomial distribution (9.5), using the generalized binomial (Section A.2.12) when the numbers are non-integer.

- T_{+k}, the waiting time for the next k occurrences:

$$T_{+k} \sim g\gamma_{(k, \kappa_1, \tau_1)}(t),$$

where $g\gamma$ is the gamma–gamma distribution (Section A.3.2).

See Table 13.10 for an illustration of how these probability distributions change as you make more observations.

Example 13.3.1 We look a hundred years back in time, to Norway, to study the number of offspring for women who had passed child bearing age (we use 50 as a cut-off). The number of offspring in societies with high childhood mortality are generally considered to be Poisson distributed. We are estimating λ for this Poisson distribution. For our prior, we use the numbers from a similar study from Sweden at the same time, where they counted a total of 26 offspring, distributed between 5 women. We use for our prior

$$P_0 \vDash \kappa_0 = 26$$
$$\tau_0 = 5.$$

In our Norwegian study, we look at 10 women, and note that they had a total of 49 offspring. Our updated probabilities are then

$$P_1 \vDash \kappa_1 = 26 + 49 = 75$$
$$\tau_1 = 5 + 10 = 15.$$

We then get a *posterior* distribution for the parameter λ:

$$\lambda \sim \gamma_{(75, 15)}(l).$$

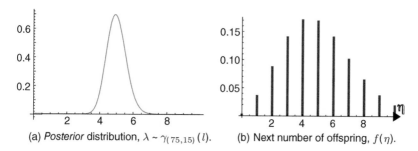

(a) *Posterior* distribution, $\lambda \sim \gamma_{(75,15)}(l)$.

(b) Next number of offspring, $f(\eta)$.

Figure 13.11 Distributions used for studying the number of offspring.

In this case, we have no waiting time T, since our units are not time, but the individual women. We would, however, like to estimate the number of offspring the next woman has. That estimate is

$$N_+ \sim f(\eta) = nb_{(75, \frac{15}{16})}(\eta).$$

From these distributions we may read all that we need to know, such as for instance that the expected number of offspring is $75(1 - \frac{15}{16})/\frac{15}{16} = 5$. See Figure 13.11 for illustration. □

Example 13.3.2 You work for the venture capital investment company Reodora, and you are responsible for their portfolio of technologically innovative companies. In a model inspired by Warren Buffett, these companies will pay Reodora \$10 million every time their free funds are in excess of \$20 million – unless they can make good use of them for internal investment. Your experience with these payments is that they follow a Poisson process. Your job right now is to estimate that process's λ, and then to estimate the waiting time for the next two payments.

The portfolio is almost four years old, but since the money flow for the first year tends to go out rather than in, you limit your investigations to the past three years. You find that Reodora have received 39 payments during the period. Since this is roughly one payment per month, you decide to use months as your time units. You use a neutral prior:

$$P_0 \vDash \kappa_0 = 0$$
$$\tau_0 = 0.$$

Your updated probabilities are then

$$P_1 \vDash \kappa_1 = 0 + 39 = 39$$
$$\tau_1 = 0 + 36 = 36 \text{ (since 3 years = 36 months).}$$

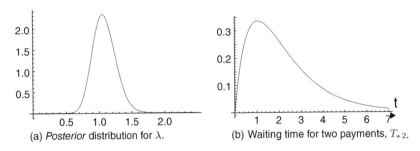

(a) *Posterior* distribution for λ. (b) Waiting time for two payments, T_{+2}.

Figure 13.12 Distributions used for studying Reodora's portfolio payments.

We then get the following *posterior* distribution for the parameter λ:

$$\lambda \sim \gamma_{(39,36)}(l).$$

The waiting time for the next two payments is then distributed

$$T_{+2} \sim g\gamma_{(2,39,36)}(t) = \frac{65t}{54\left(\frac{t}{36}+1\right)^{41}}.$$

See Figure 13.12 for illustration. □

13.4 Exercises

Bayes' Theorem for Gaussian Processes

1 You are given a prior or prior hyperparameters for a Gaussian process, and observational data, and values z and s. Find (if relevant) the posterior (predictive) probability distribution of μ, τ, and X_+. Further, calculate (if relevant) the probabilities $P(\mu < z)$, $P(\sigma < s)$, and $P(X_+ < z)$. To build understanding, it pays off to draw at least a few of the function graphs.

Unknown μ, known σ.
(a) Prior: $\kappa_0 = 0$, $\Sigma_0 = 0$. $s_0 = 2.5$. Statistics: $n = 5$, $\Sigma_x = 23.43$. $z = 6$.
(b) Prior: $\kappa_0 = 0$, $\Sigma_0 = 0$. $s_0 = 6$. We have 7 observations with $\bar{x} = 45.414\,3$. $z = 40$.
(c) Prior: $\kappa_0 = 0$, $\Sigma_0 = 0$. $s_0 = 15.9$. Data: $\{319.3, 369.4, 327.3\}$. $z = 350$.
(d) Prior: $\kappa_0 = 0$, $\Sigma_0 = 0$. $s_0 = 200$. Statistics: $n = 12$, $\Sigma_x = 35\,723.7$. $z = 3\,125$.
(e) Prior: $\phi_{(5,1)}$. $s_0 = 2$. We have 8 observations with $\bar{x} = 5.45$. $z = 6$.
(f) Prior: $\phi_{(510,15)}$. $s_0 = 30$. Statistics: $n = 7$, $\Sigma_x = 3568.2$. $z = 500$.
(g) Prior: $\phi_{(0.19,0.03)}$. $s_0 = 0.03$. Data: $\{0.174, 0.144, 0.255, 0.237\}$. $z = 0.15$.

Known μ, unknown σ.
(h) Prior: $\nu_0 = -1$, $SS_0 = 0$. $m_0 = 23$. Statistics: $n = 5$, $\Sigma_x = 119.03$, $SS_x = 3.433\,77$. $z = 25$ and $s = 1.25$.

(i) Prior: $v_0 = -1$, $SS_0 = 0$. $m_0 = 50$. We have 8 observations with average $\bar{x} = 49.871\,6$ and sample standard deviation $s_x = 3.424\,29$. $z = 45$ and $s = 10$.

(j) Prior: $v_0 = 2$, $SS_0 = 0.16$. $m_0 = 1.2$. Data: $\{1.18, 1.14, 1.22, 1.25, 1.15, 1.18, 1.22, 1.15, 1.24, 1.26\}$. $z = 1$ and $s = 0.1$.

Unknown μ, unknown σ.

(k) Prior: $\kappa_0 = 0$, $\Sigma_0 = 0$, $v_0 = -1$, $SS_0 = 0$. Statistics: $n = 8$, $\Sigma_x = 318\,32$, $\Sigma_{x^2} = 126\,742\,346$. $z = 4\,000$ and $s = 100$.

(l) Prior: $\kappa_0 = 0$, $\Sigma_0 = 0$, $v_0 = -1$, $SS_0 = 0$. We have 7 observations with average $\bar{x} = 11.528\,6$ and sample standard deviation $s_x = 24.234\,2$. $z = 0$ and $s = 15$.

(m) Prior: $\kappa_0 = 1$, $\Sigma_0 = 0.007$, $v_0 = 1$, $SS_0 = 1.042$. Data: $\{0.232, -1.587, -0.986\}$. $z = 0$ and $s = 0.5$.

(n) Prior: $\kappa_0 = 0$, $\Sigma_0 = 0$, $v_0 = -1$, $SS_0 = 0$. Statistics: $n = 13$, $\Sigma_x = 295.839$, $SS_x = 174.964$. $z = 20$ and $s = 5$.

(o) Prior: $\kappa_0 = 2$, $\Sigma_0 = 40$, $v_0 = 2$, $SS_0 = 3$. We have 8 observations with average $\bar{x} = 19.671\,9$ and sample standard deviation $s_x = 1.044\,14$. $z = 17.9$ and $s = 1.179$.

(p) Prior: $\kappa_0 = 0$, $\Sigma_0 = 0$, $v_0 = -1$, $SS_0 = 0$. Statistics: $n = 8$, $\Sigma_x = 737.1$, $SS_x = 6\,604.9$. $z = 50$ and $s = 25$.

(q) Prior: $\kappa_0 = 6$, $\Sigma_0 = 0.000\,019\,8$, $v_0 = 5$, $SS_0 = 1.5 \cdot 10^{-7}$. We have 30 observations with average $\bar{x} = 3.306\,35 \cdot 10^{-6}$ and sample standard deviation $s_x = 2.997\,56 \cdot 10^{-8}$. $z = -0.000\,04$ and $s = 0.000\,04$.

2 (*Theoretical*) Prove the formula for the probability distribution of X_+ in Section 13.1.1. (Hint: Use that μ and $(X_+ - \mu) \sim \phi_{(0,\sigma)}$ are independent.)

3 **Capacitors:** You have measured the capacitance of FR Electronics's smallest capacitors. From a sample of 25 measurements, you got an average of $\bar{c} = 49.19\,\mu F$, and a sample standard deviation of $s_c = 2.15\,\mu F$. Assume the capacitance of this kind of capacitor follows a Normal distribution $\phi_{(\mu,\sigma)}$ with unknown values for μ and σ. Use a neutral prior, and find the probability distributions of μ and τ, and find the probability that a randomly selected capacitor of this kind has a capacitance of more than $50\,\mu F$.

4 (*Challenge*) The tensile strength of cables of the same type and thickness typically follows some Normal distribution $\phi_{(\mu,\sigma)}$, where μ and σ depend only on type and thickness. A colleague of yours has pulled apart wires of a certain type and thickness to find their tensile strength. The only thing you know is that your colleague made use of a neutral prior, and that the *posterior* probability distribution for μ (in kilonewtons) was $\mu \sim t_{(943,11,6)}$. What is the probability that the next cable deforms at a smaller load than 900 kN?

Bayes' Theorem for Bernoulli Processes

5 *Posterior:* You are given the prior hyperparameters of a binomial process, observation data, and a value p. Find the posterior distribution for π and its Normal approximation. Further, calculate the probability $P(\pi \le p)$ both by exact calculation on β, and by using the Normal approximation.

   (a) Prior hyperparameters: $a_0 = 2$, $b_0 = 2$. Observations: 17 positive, 29 negative. $p = 0.4$.

   (b) Prior: $\beta_{(1,7)}$. Observed: $k = 4$ positive and $l = 89$ negative. $p = 0.07$.

   (c) Prior hyperparameters: $a_0 = 0$, $b_0 = 0$. Observed: $k = 42$ positive and $l = 13$ negative. $p = 0.7$.

   (d) Prior: $\beta_{(0.5, 0.5)}$. Observed: $k = 434$ positive and $l = 177$ negative. $p = 0.7$.

6 *Predictive:* [requires a good calculation tool] You are given the posterior for the Bernoulli parameter π, and numbers m, s, k, and l. Find the predictive distributions for K_{+m} and L_{+s}, and calculate the probabilities $P(K_{+m} \le k)$ and $P(L_{+s} < l)$.

   (a) Posterior: $a_1 = 19$, $b_1 = 31$. $m = 5$, $s = 5$, $k = 3$, $l = 7$.

   (b) Posterior: $a_1 = 5$, $b_1 = 96$. $m = 20$, $s = 4$, $k = 3$, $l = 12$.

   (c) Posterior: $a_1 = 42$, $b_1 = 13$. $m = 7$, $s = 14$, $k = 4$, $l = 5$.

   (d) Posterior: $a_1 = 434.5$, $b_1 = 177.5$. $m = 32$, $s = 20$, $k = 23$, $l = 8$.

7 The Bayern Munich player Arjen Robben scores most of his goals with the left foot. The statistics of the goals he has scored by foot are as follows:

Season	left	right
2013/14	9	1
2014/15	12	2
2015/16	3	0

Let π be the proportion of Robben's foot scorings that he does with his left. Find the probability distribution of π when...

   (a) you use *Jeffreys' prior* for proportions, and update the probability distribution for π with the observations from 2013/14;

   (b) you go a new round, and update with the observations from 2014/15;

   (c) you further update with the observations from 2015/16;

   (d) What is the probability that π is in excess of 75%?

   (e) Find the probability distribution of the number of left-foot goals in Robben's next 3 foot scorings.

   (f) Find the probability that 2 out of the next 3 foot scorings are done with his left.

8 Bard has given you a biased coin, and you wonder what the probability π of heads is. He replies that he doesn't quite know, but that his friend Sam,

who gave it to him, once estimated the probability of heads to be $\frac{3}{7}$, and that Sam was as certain of that as if he had flipped the coin 21 times.

(a) What are your *prior* hyperparameters a_0 and b_0?

(b) You get 23 heads and 18 tails. What are the *posterior* hyperparameters?

(c) You decide to flip the coin a few more times, to get an even sharper estimate on π. What are your *prior* hyperparameters this time?[4]

(d) You now get 458 heads and 366 tails. What are your *posterior* hyperparameters now?

9 You are looking at the proportion π of consumers who prefer MegaCola to its competitors. You use Jeffreys' prior hyperparameters, $a_0 = b_0 = 0.5$.

(a) You arrange blind tastings of MegaCola and and its competitors, and then ask the participants to indicate which one they preferred. After having tested 50, the feedback is that 41 of them prefer MegaCola. What are your *posterior* hyperparameters, and what is the probability distribution of π?

(b) What is the probability that the proportion who prefer MegaCola exceeds 75%?

10 You are estimating the proportion of Macintoshes among the laptops of a rather large company. Your prior hyperparameters are $a_0 = 7$ and $b_0 = 3$. You ask 10 laptop-using colleagues; 8 of them are on a Macintosh. What are now your *posterior* hyperparameters for the proportion of MacBooks?

11 You are looking into the quality of the diamonds of the diamond mines in a new area. You have a special interest in "Fancy diamonds"[5] of quality IF and VVS, and you are estimating π, the proportion of diamonds from the new mines that fit one of these descriptions. After having evaluated 172 diamonds, you have found 19 that are either IF or VVS, while the rest are of lower quality grades.

(a) You are unsure which of the 3 neutral priors to choose. Find the *posterior* distributions you find for π, for each of them.

(b) Calculate, for each of the 3 priors, the probability that $\pi > 0.1$.

12 In the two-player board game Go, black and white take turns putting a stone on the board, with black having the first move. Sondre Glimsdal, the *Og* Go club chairman, wonders what percentage of the games is won by each color. He believes it is fairly even, so his prior for the proportion of games won by white is $\beta_{(7,7)}$.

---

[4]Yes, the answer is obvious; mathematics and statistics problems don't *have to* be hard!
[5]Naturally colored diamonds.

(a) Sondre looks at 20 randomly chosen games, and sees that 13 were won by white, and 7 were won by black. What is Sondre's *posterior* probability distribution for the proportion of games won by white?

(b) Sondre looks at another 20 random games, and finds that 11 of these were won by white, and 9 were won by black. What is Sondre's *new* posterior probability distribution for the proportion of games won by white?

(c) What is Sondre's probability that white wins at least half of the games?

Bayes' Theorem for Poisson Processes

13 *Posterior:* You are given *prior* hyperparameters for a Poisson process, and observational data. Find the posterior distribution for the rate parameter λ.

(a) Prior: $\kappa_0 = 0$, $\tau_0 = 0$. Observed: $n = 7$ occurrences during $t = 5$ units.

(b) Prior: $\kappa_0 = 5$, $\tau_0 = 10$. Observed: $n = 3$ occurrences during $t = 7$ units.

(c) Prior: $\lambda \sim \gamma_{(12,3)}$. Observed: $n = 17$ occurrences during $t = 5$ units. $k = 2$.

(d) Prior: $\kappa_0 = 3$, $\tau_0 = 73$. Observed: $n = 6$ occurrences during $t = 119$ units.

14 *Predictive:* [requires a good calculating tool] You are given *posterior* distributions or hyperparameters for the rate λ of a Poisson process, and values k, l, m. Calculate the predictive distribution for N_+ and T_{+k}, and calculate $P(T_{+k} \leq l)$ and $P(N_{+1} \leq m)$.

(a) Posterior hyperparameters: $\kappa_1 = 7$, $\tau_1 = 5$. $k = 2$, $l = 3$, $m = 4$.

(b) Posterior: $\lambda \sim \gamma_{(8,17)}$. $k = 1$, $l = 2$, $m = 1$.

(c) Posterior hyperparameters: $\kappa_1 = 29$, $\tau_1 = 8$. $k = 3$, $l = 1$, $m = 5$.

(d) Posterior: $\lambda \sim \gamma_{(9,192)}$. $k = 2$, $l = 30$, $m = 0$.

15 ($t =$ units $=$ number of hunters) The daily catch for grouse hunters in an area is considered to be Poisson distributed with rate λ. One day, you talked to 23 hunters from a certain part of the Lowlands, and their total catch was 111 grouse. Use a neutral prior, and find the *posterior* probability distribution for λ.

16 ($t =$ time $=$ number of weeks) The number of plumbing gaskets that need changing every week in an apartment complex is assumed to follow a Poisson process with rate λ. You are estimating this need for an apartment complex with 70 flats, and have looked into the documentation for the last semester (26 weeks), and find 53 gasket changes registered. Use a neutral prior, and calculate the *posterior* probability distribution for λ.

17 ($t =$ distance $=$ number of kilometers) The number of cracks in the tarmac per kilometer of road is assumed to follow a Poisson process with rate λ.

Your job is to find this rate for a lesser highway, and you have found 13 cracks in 10 km. Use neutral prior, and calculate the *posterior* probability distribution of λ.

18 ($t =$ area $=$ number of square meters) How many four-leaf clovers are there per square meter in a field of leaf clovers? Assume that the occurrence follows a Poisson process with rate λ. You look at three independent leaf clover fields. The first field is $t_1 = 1.9\,\text{m}^2$ in area, and has $n_1 = 0$ four-leaf clovers. The second field had an area of $t_2 = 0.7\,\text{m}^2$, and has $n_2 = 3$ four-leaf clovers, whereas the third spot of area $t_3 = 1.2\,\text{m}^2$ had $n_3 = 1$ four-leaf clovers. Use a neutral prior, and find the *posterior* probability distribution of λ.

19 ($t =$ volume $=$ number of cubic meters) The number of bacterial colonies per cubic centimeter in a certain polluted lake is assumed to be Poisson distributed with parameter λ. You sample 1 deciliter, and find 157 bacterial colonies. Use a neutral prior, and calculate the *posterior* probability distribution of λ.

14

Bayesian Hypothesis Testing

"Sam?" Bard asks, "do you want to take part in a free bet? Free for you, that is, not me. I'll be tossing this D_8 die into a box – *there we go!* – where none of us can see it, and then you will bet on one of the alternatives: was it a 5, or was it a non-5?"

"Well then, I bet on the non-5," Sam replies, "since it has the greater probability of the two. Logical!"

"I forgot to tell you something," Bard continues. "If you bet on 5 and were right, I'll give you £200. If you bet on non-5 and were right, I will pay you £1. A wrong guess gives you nothing, of course."

"Hold on," Sam says hurriedly, "that is a quite different bet. I am changing my bet to 5. Logical!"

"Good," Bard replies, "Let's have a look, and ... ah, it was a 4."

"I don't regret my choice," Sam responds, "and I would bet likewise if we did it again." He looks suggestively at Bard before he continues: "For it was logical to bet as I did. $\frac{1}{8}$ times 200 is 25, whereas $\frac{7}{8}$ times 1 is less than a pound. So in the long run my betting strategy will pay off."

"Good again," Bard announces, "you have just made a decision by means of a utility function!"

14.1 The Utility Function *u*

"A decision is a choice between different alternatives $\theta_1, \theta_2, \ldots$," Bard explains. "An alternative θ has a *utility function* $u_\theta(x)$ describing the consequences of

The Bayesian Way: Introductory Statistics for Economists and Engineers, First Edition.
Svein Olav Nyberg.
© 2019 John Wiley & Sons, Inc. Published 2019 by John Wiley & Sons, Inc.

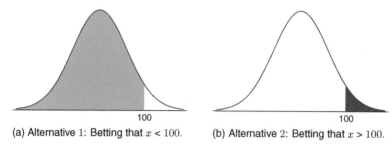

(a) Alternative 1: Betting that $x < 100$. (b) Alternative 2: Betting that $x > 100$.

Figure 14.1 Two alternatives, illustrated together with a probability distribution.

choosing θ for the different values of a variable x. But x is an unknown quantity, like the outcome when I tossed the D_8, so the rational choice therefore depends on the probability distribution over the x values. Look at the probability distribution in Figure 14.1, Sam: Would you bet that x is larger or smaller than 100?"

"I see that $x < 100$ is a lot more probable," Sam replies, "but you're not fooling me this time! Before I give my final answer, I would like to know the payoff for each of the two alternatives!"

"I like your way of thinking," Bard smiles, "for you have realized that you need to take *both* probability *and* utility into account. The yardstick of how much an alternative is worth is *expected utility*."

> **Definition 14.1.1** *For a continuous stochastic variable $X \sim f(x)$ taking values in (A, B), and a utility function $u_\theta(x)$, the expected utility is*
>
> $$U_\theta = E[u_\theta(X)] = \int_A^B u_\theta(x) \cdot f(x)\, \mathrm{d}x.$$

"And for a discrete X, we naturally use the sum instead of the integral," Bard comments, "in any case we will stick with two kinds of utility: *stepwise* functions and *linear* functions. These kinds of function lend themselves easily to calculation and theoretical reasoning."

Stepwise functions

A stepwise function divides $\mathbb{R}$ into n intervals, and is constant within each interval. (See Figure 14.2.)

$$u_\theta(x) = \begin{cases} v_1 & x \in I_1 \\ v_2 & x \in I_2 \\ \vdots & \vdots \\ v_n & x \in I_n. \end{cases}$$

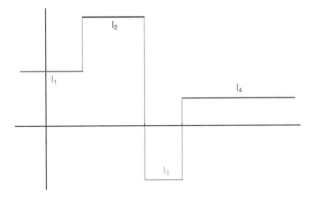

Figure 14.2 A stepwise function.

The expected utility of a stepwise function u, given a stochastic variable X, is

$$U_\theta = \sum_{k=1}^{n} v_k \times P(X \in I_k) = v_1 \times P(X \in I_1)$$

$$+ v_2 \times P(X \in I_2) + \cdots + v_n \times P(X \in I_n).$$

Linear functions

A linear function

$$u_\theta(x) = a + bx$$

is a utility function for X when $|\mu_X| < \infty$. (See Figure 14.3.) The expected utility of a linear function u, given a stochastic variable X, is

$$U_\theta = a + b\mu_X.$$

Example 14.1.2 Let $X \sim \phi_{(2,5)}$, and let u_1 and u_2 be two utility functions,

$$u_1(x) = \begin{cases} 3 & x < -1 \\ 2 & -1 < x < 2 \\ 5 & 2 < x < 6 \\ -1 & x > 6 \end{cases} \quad \text{and } u_2(x) = -1 + 3x, \quad \text{and find } U_1 \text{ and } U_2.$$

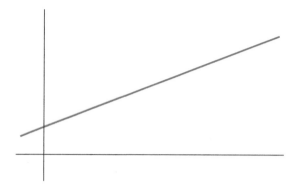

Figure 14.3 A linear function.

Answer: The expected utility values of X under respectively u_1 and u_2 are then

$$U_1 = 3P(X < -1) + 2P(-1 < X < 2) + 5P(2 < X < 6) - 1P(X > 6)$$
$$= 3 \times 0.274253 + 2 \times 0.225747 + 5 \times 0.288145 - 1 \times 0.211855 = \underline{2.503\,12}.$$
$$U_2 = -1 + 3\mu_X = -1 + 3 \times 2 = \underline{5}.$$

$\square$

Example 14.1.3 Let $Y \sim \gamma_{(4,1)}$, and let u_1 and u_2 be two utility functions,

$$u_1(y) = \begin{cases} 2 & y < 5 \\ 8 & y > 5 \end{cases} \quad \text{and } u_2(y) = 27 - 5y, \quad \text{and find } U_1 \text{and } U_2.$$

Answer: The expected utilities of Y, under respectively u_1 and u_2, are then

$$U_1 = 2P(Y < 5) + 8P(Y > 5) = 2 \cdot 0.734974 + 8 \cdot 0.265026 = \underline{3.59016}$$

$$U_2 = 27 - 5\mu_Y = 27 - 5 \cdot \frac{4}{1} = \underline{\underline{7}}$$

$\square$

Example 14.1.4 Let $Z \sim \beta_{(7,13)}$, and let u_1 and u_2 be two utility functions,

$$u_1(z) = \begin{cases} -1 & z < 0.2 \\ 4 & 0.2 < z < 0.5 \\ 10 & z > 0.5 \end{cases} \quad \text{and } u_2(z) = 2.3 + 0.8z, \quad \text{and find } U_1 \text{ and } U_2.$$

Answer: The expected utility values for Z under respectively u_1 and u_2 are then

$$U_1 = -1P(Z < 0.2) + 4P(0.2 < Z < 0.5) + 10P(Z > 0.5)$$
$$= -1 \cdot 0.067\,600\,1 + 4 \cdot 0.848\,866 + 10 \cdot 0.083\,534\,2 = \underline{4.163\,21}.$$

$$U_2 = 2.3 + 0.8\mu_Z = 2.3 + 0.8 \cdot \frac{7}{7 + 13} = \underline{\underline{2.58}}.$$

$\square$

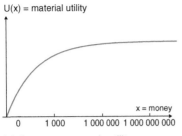

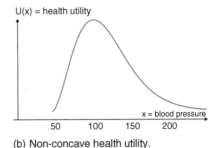

(a) Concave economic utility. (b) Non-concave health utility.

Figure 14.4 Two other utility models.

For practical purposes, it often makes sense to tie the concept of *utility* directly to something measurable. In economics, a simple model is the simple equality utility = pay-off. A more common, but unfortunately more complicated, model is letting the utility be a concave function of the pay-off, as illustrated in Figure 14.4.

Different choices and contexts result in different utility profiles. The utility of a lower exchange rate between the UK £ and the US $ is different for a woman from the USA planning a trip to London and a British man planning a vacation in Boston. We will now look at two examples with stepwise and linear utility functions where utility = pay-off, and we will then compare the values of the two choices.

Example 14.1.5 Frederick is buying shares in ACME Dynamics Ltd. His investigations tell him that X, the *now value* in one year from now, follows a Normal distribution $X \sim \phi_{(47.2, 4.7)}$. What is the utility function $u(x)$ of buying k shares if the current price is £40.0, and what is the expected pay-off U of that purchase?

Answer: The pay-off is the sales value minus the cost. The sales value is £x per share, whereas the cost is £40.0. For k shares, the utility function is then

$$u_1(x) = k(x - 40.0).$$

This is a linear function, so since $\mu_X = 47.2$, the expected utility is

$$U_1 = k(\mu_X - 40.0) = k(47.2 - 40.0) = \underline{7.2k}. \qquad \square$$

Example 14.1.6 Frederick considers options[1] as an alternative to buying the shares themselves. The options in this instance are our bets, and cost £10 a

---

[1] An option is an agreement about a future transaction, based on the future value of a share, commodity, or other item. The simplest options are agreements that give the holder of the option the right but not the duty to buy (or to sell) a given number of shares of a certain company at a previously agreed price at a certain set time T, regardless of the actual price at time T.

piece. In one year from now, they yield £0 if the share price X is less than or equal to £47.5, but £25 if $X > 47.5$. What is the utility function $u_2(x)$ of buying k such options, and what is the expected pay-off of buying these k options when $X \sim \phi_{(47.2, 4.7)}$?

Answer: Pay-off equals yield minus expense. The separating value is a share price of $x = £47.5$, so

$$u_2(x) = \begin{cases} k(0 - 10) = -10k & x \le 47.5 \\ k(25 - 10) = 15k & x > 47.5. \end{cases}$$

We see that

$$P(X \le 47.5) = \Phi_{(47.2,4.7)}(47.5) = 0.525\,447$$

$$P(X > 47.5) = 1 - \Phi_{(47.2,4.7)}(47.5) = 0.474\,553.$$

This gives the expected utility

$$U_2 = -0.525\,447 \times 10k + 0.474\,553 \times 5k = \underline{1.863\,82k}.$$

□

Example 14.1.7 Frederick ponders whether he should buy shares or options, given that $X \sim \phi_{(47.2, 4.7)}$. He has £200 to spend, so he can either buy $\frac{200}{40} = 5$ shares for them, or $\frac{200}{10} = 20$ options. The expected pay-off for buying shares is then $U_1 = 7.2 \times 5 = £36$, whereas the pay-off of buying options is $U_2 = 1.863\,82 \times 20 = £37.276\,4$. This means $U_2 > U_1$, so in this case, the rational choice is for Frederick to spend his £200 buying options rather than shares.

□

14.2 Comparing to a Fixed Value

A common comparison is between two stepwise utility functions u_0 and u_1 that are split once, at the same x value. Often, u_0 then corresponds to a choice *not to act*, and we call this the *null alternative*. The study of the difference between the two utility functions boils down to studying their difference, $u(x) = u_1(x) - u_0(x)$. That is, the difference in utility between acting on alternative 1 on the one hand, and on the other hand of "letting everything stay the same" and accordingly not acting to change anything (alternative 0). This is called a one sided *hypothesis test*.

The act of choice in a one sided hypothesis test boils down to a very simple question: will you act on the value of a certain stochastic variable Θ being *larger* or *smaller* than a given reference value θ_0? The two beliefs in themselves have neither cost nor value, so the utility depends what *actions* accompany the two choices.

Example 14.2.1 As an illustration, let us look at Mr. Königsegg, who is accused of speeding at 140 km/h on a small road in Norway. The consequences of a *guilty* and a *not guilty* verdict are very different, and they depend on whether we act on the belief that his speed X was in excess of this reference speed $x_0 = 140$ km/h, or not.

A golden rule of thumb for a fair court system is that "it is better that 10 guilty men go free than 1 innocent man be wrongly convicted". This may be justified from considerations of utility for society. In this instance, alternative 0 is a *not guilty* verdict for Königsegg, whereas alternative 1 is to convict him of speeding at past 140 km/h. The utility difference between the two alternatives is $u(x) = u_1(x) - u_0(x)$. A precise assessment of this utility is difficult, but we will nevertheless have to choose some numbers to illustrate the method. Let x be the speed.

- If Königsegg is guilty, that is, $x > 140$ km/h, declaring him guilty rather than innocent is worth 1 unit of utility.
- If Königsegg is not guilty, that is, $x \leq 140$ km/h, declaring him innocent rather than guilty is worth 10 units of utility.

This means that

$$u(x) = \begin{cases} -10 & x \leq 140 \\ 1 & x > 140. \end{cases}$$

Since $U_1 = E[u_1(X)]$ and $U_0 = E[u_0(X)]$, then $U = E[u(X)] = U_1 - U_0$. We will then choose alternative 1 over alternative 0 only if $U_1 > U_0$, which is equivalent to $U > 0$. Let X be Königsegg's speed, and $p = P(X \leq 140)$. Then

$$U = -10p + 1(1-p) = 1 - 11p.$$

This means that $U > 0$ when $p < \frac{1}{11} \approx 0.09$. We should in other words not find Königsegg guilty of speeding at more than 140 km/h unless the probability that his speed was *less than* 140 km/h, is less than 9%. At this limit, we ensure that the limit goes at 10 guilty going free for every innocent man found guilty. □

This kind of evaluation is known under another name, *hypothesis testing*, where we boil down the procedure to its bare bones, as follows: you have a null value θ_0, and are going to determine whether A: $\Theta < \theta_0$, or B: $\Theta > \theta_0$.

- If A is the case, it is w_A utility units better to choose alternative A than to choose alternative B.
- If B is the case, it is w_B utility units better to choose alternative B than to choose alternative A.

The *null alternative* is the alternative with the largest w value. In hypothesis *testing*, we call this alternative the null *hypothesis*, and name it H_0. The other alternative, called the *alternative* hypothesis, is written H_1. We then find U thus: let w_0 be the largest of the two values, and w_1 the smallest, and let $p = P(H_0)$. Then

$$U = w_1 \cdot P(H_1) - w_0 \cdot P(H_0) = w_1(1-p) - w_0 p = w_1 - (w_0 + w_1)p.$$

We choose H_1 over H_0 only if there is positive expected utility in doing so, in other words if $U > 0$. That is the case only when $p < \frac{w_1}{w_0 + w_1}$.

Note 14.2.2 *The significance of a hypothesis test is $\alpha = \frac{w_1}{w_0 + w_1}$, where w_0 and w_1 are as defined in the section above.*

We formally define a hypothesis test in the following way.

Definition 14.2.3 *A one sided hypothesis test of a stochastic variable Θ with significance α is a test of whether $\Theta < \theta_0$ or $\Theta > \theta_0$. One alternative is specified as the* null *hypothesis H_0, whereas the other is specified as the* alternative *hypothesis H_1. We then say that we test hypothesis H_1, and the conclusion is that we*

- *choose H_1 if $P(H_0) < \alpha$. ("We reject the null hypothesis H_0.")*
- *choose H_0 if $P(H_0) \geq \alpha$. ("We do not reject the null hypothesis H_0.")*

A hypothesis test is left sided *if H_1 is specified as $\Theta < \theta_0$, and* right sided *if H_1 is specified as $\Theta > \theta_0$.*
The significance α is either stated directly, or we calculate it ourselves as

$$\alpha = \frac{w_1}{w_0 + w_1},$$

where w_0 is the utility of choosing H_0 over H_1 if H_0 actually is the case, and w_1 is the utility of choosing H_1 over H_0 if H_1 actually is the case.

It is worth noting the following.

- Θ is usually a parameter, and its probability distribution is usually a *posterior* probability distribution after some measurements, like the ones we find in Chapters 12 and 13.
- The significance value α often follows a customary rule of thumb within the area of application, rather than being explicitly calculated from precise utility considerations of the problem at hand. Typical customary values for α are 0.1, 0.05, and 0.01.

- H_1 is often an alternative to the ruling consensus, and if assumed true would require a change in current practice. Conversely, H_0 often equals the consensus. Hence the name *the conservative hypothesis*.
- H_1 is often the alternative we *desire* to be true, but that requires costly action.

Example 14.2.4 (Purely numerical example) $\theta \sim t_{(12,3,5)}$. Test the alternative hypothesis $\theta > 7$ with significance value $\alpha = 0.075$.

Answer: H_1 is that $\theta > 7$, so then the null hypothesis is H_0: $\theta \leq \theta_0 = 7$. The calculation then gives us that $P(H_0) = P(\theta \leq 7) = T_{(12,3,5)}(7) = 0.078 > \alpha$. So we choose H_0. Or in other words: we *don't reject H_0* in favor of H_1. $\square$

Example 14.2.5 Hypothesis test and action: The Norwegian pharmaceutical Lyvjaberg AS are testing out their blood pressure reducing medicine Odrøre. How much blood pressure is lowered after one month's use of the medicine will follow a Normal distribution $\phi_{(\mu,\sigma)}$, and they want to estimate μ. More specifically, they wish to see if they beat the best of their competition, whose medicine results in a reduction of blood pressure of 31 mmHg after one month's use.

The alternative hypothesis H_1 is that $\mu > h_0 = 31$, in other words that Odrøre on average is better than its competitors' products. H_0 is then that $\mu \leq 31$, which is to say that Odrøre is not better on average, but worse than (or at best: equally good as) the others.

From previous market experience, Lyvjaberg thinks that the cost w_0 of starting production if Odrøre turns out not to be better than the competition after all, when it is tried out on a grand scale, is roughly 20 times larger than the loss w_1 of not starting production if Odrøre happened to actually be the best product. The test significance is then $\alpha = \frac{1}{20+1} \approx 0.05$. The choice of rounding off to 0.05 is motivated both by this being a round number close enough to $\frac{w_1}{w_0+w_1}$, and that 0.05 is acknowledged as a decent and easily recognizable significance value.

Lyvjaberg has reviewed the change of blood pressure of 25 volunteers after a month of trying out their medicine. For objectivity's sake, they have chosen a neutral prior, and their *posterior* probability distribution for μ is then $t_{(39, 3.9, 24)}$. The hypothesis test then consists of the following calculation:

$$P(H_0) = P(\mu \leq 31) = T(39, 3.9, 24)(31) \approx 0.026 < 0.05 = \alpha.$$

Given this result, Lyvjaberg *rejects the null hypothesis H_0*. Or in our words: they choose alternative H_1, and act on that. That is, they are now sufficiently convinced that Odrøre is better than the competing products, and they start production. $\square$

14.2.1 Hypothesis Testing for Gaussian Processes

For a Gaussian process, we perform inference on the parameters μ and σ, and on the next observation X_+. In practice, this means calculating on γ, ϕ, and t distributions.

Example 14.2.6 You have been asked officially to certify the plums of a fruit farmer from Hardanger. More specifically, you will assess whether his *Mallard* plums may be sold under the protected label *Hardanger plums*. The size requirement is that the diameter be at least 40 mm. You assume that the weight follows a Normal distribution $\phi_{(\mu,\sigma)}$. The key variable to assess here is neither of the parameters μ and σ, but rather the size of the next plum, X_+. However, since you are at it, you decide to test the parameters μ and σ as well. You perform the hypothesis test with significance $\alpha = 0.05$, and ask yourself three questions. The first two for comments and tips to the farmer, and the last one for the certification itself.

1) Is the mean diameter μ more than 45 mm? ($H_0^{\mu}: \mu \leq 45$.)
2) Is the standard deviation σ less than 5 mm? ($H_0^{\sigma}: \sigma \geq 5$.)
3) Does a randomly sampled plum have a diameter in excess of 40 mm? ($H_0^+: X_+ \leq 40$.)

Since this is an official certification, you should use a neutral prior, meaning $\kappa_0 = \Sigma_0 = n_0 = SS_0 = 0$. You measure the diameters of 30 randomly picked plums, and derive the statistics $n = 30$, $\Sigma_x = 1445.22$, and $SS_x = 401.179$. From the statistics, you get the distributions $\tau \sim \gamma_{(14.5, 200.59)}(t)$ and $\mu \sim t_{(48.174, 0.679062, 29)}(x)$, whereas $X_+ \sim t_{(48.174, 3.78086, 29)}(x)$. Then

1) $P(H_0^{\mu}) = T_{(48.174, 0.679062, 29)}(45) = 0.000031 < \alpha$;
2) $P(H_0^{\sigma}) = P(\sigma \geq 5) = P(\tau \leq 0.04) = \Gamma_{(14.5, 200.59)}(0.04) = 0.0250 < \alpha$;
3) $P(H_0^+) = T_{(48.174, 3.78086, 29)}(40) = 0.0195 < \alpha$.

You therefore reject all three null hypotheses, and choose H_1. You tell the farmer that his plums will overall be within the required values, but that he may expect one in fifty plums to be a bit too far on the small side. □

14.2.2 Hypothesis Testing for Bernoulli Processes

With Bernoulli processes, we test the parameter p. Using the methods of Section 13.2, the *posterior* probabilities we obtain for p become β distributions. So in this section, we do calculations on β distributions.

Example 14.2.7 Arboreal scientist Bryn believes that maple leaves have a tendency to land underside up rather than sunnyside up, but decides that this is worthy of an empirical investigation. Let π be the proportion of maple leaves landing with the underside up. Bryn's null hypothesis H_0 is that $\pi \leq 0.5$, while the alternative hypothesis is that $H_1: \pi > 0.5$. Bryn starts with a flat prior, $\beta_{(1,1)}$, Bayes' prior for *informed ignorance* for Bernoulli processes. Bryn decides to test with significance $\alpha = 0.06$. He looks at 50 leaves on the ground, and find that 31 lie underside up, whereas the remaining 19 lie sunnyside up. Bryn's *posterior* distribution is then $\pi \sim \beta_{(32,20)}$. That means

$$P(H_0) = I_{(32,20)}(0.5) = \int_0^{0.5} \beta_{(32,20)}(x)\, dx = 0.046 < 0.06.$$

Since $P(H_0) < \alpha$, Bryn *rejects* the null hypothesis in favor of the alternative hypothesis, and may from now on confidently state that maple leaves land more frequently underside up than sunnyside up. □

14.2.3 Hypothesis Testing for Poisson Processes

For a Poisson process, we test the parameter λ. Using the methods of Section 13.3, the *posterior* probabilities we obtain for λ become γ distributions. So in this section, we do calculations on γ distributions.

Example 14.2.8 The consultancy Aleatica are analysing a call center for Ratatosk Marketing Ltd. They work from the standard assumption that the incoming calls are a Poisson process, and set out to find the rate parameter λ. Specifically, they want to test if λ, the rate of incoming calls per minute, is more than 0.2. In the language of statistics, their alternative hypothesis H_1 is that $\lambda > 0.2$. If they reject H_0 and accept H_1, they will recommend restructuring. Restructuring is costly, so they choose a significance value of $\alpha = 0.1$ for the test. Their custom is to use a neutral prior ($\kappa_0 = \tau_0 = 0$), and the observational time span is two hours. They note 28 incoming calls during that time, which means their statistics are $n = 28$ and $t = 120$ (minutes). Then

$$P_1 \vDash \kappa_1 = \kappa_0 + n = 0 + 28 = 28$$
$$\tau_1 = \tau_0 + t = 0 + 120 = 120,$$

so $\lambda \sim \gamma_{(28, 120)}$. This means

$$P(H_0) = P(\lambda \leq 0.2) = \Gamma_{(28, 120)}(0.2) = 0.232\,258 > 0.1 = \alpha.$$

As a result of this, Aleatica does *not* reject H_0, and therefore do not recommend that Ratatosk restructure their business. □

14.3 Pairwise Comparison

In the previous section, we looked at comparison of a stochastic variable against a fixed magnitude. But we are more often interested in comparing two stochastic variables against each other – usually comparing two parameters θ_1 and θ_2. In the previous section, we grouped this study by process. Here, we will group by probability distribution.

14.3.1 Normal and *t* Distributed Variables

We have at hand two unknown Normal distributions $\phi_{(\mu_x, \sigma_x)}$ and $\phi_{(\mu_y, \sigma_y)}$, and are going to decide which is the largest: μ_x, or μ_y. How we calculate this depends on the degree and type of dependence between the two magnitudes. The two simplest cases to handle are as follows.

- The values come pairwise, $\{(x_i, y_i)\}$. We may then often assume that the differences $z_i = x_i - y_i$ follow a Normal distribution, $\phi_{(\mu_z, \sigma_z)}$, where $\mu_z = \mu_x - \mu_y$. The case then reduces to hypothesis testing the single variable μ_z, using $\{z_i\}$ as data, so we return to Section 14.2.1, and notice that testing $\mu_x < \mu_y$ reduces to testing if $\mu_z < 0$. Similarly for testing $\mu_x > \mu_y$.

An example of such a study would be if y were husband height and x were wife height, and you wanted to assess the question of whether husbands can be said to be taller than their wives.

- The *posterior* probability distributions for μ_x and μ_y are independent. This typically comes about when the sampled values and the priors are independent, and is thus a reasonable assumption when the inferences for μ_x and μ_y have been performed independently. Whereas for the pairwise collected data we reduced the two variables to one *a priori*, we will for the independent variables perform that reduction *a posteriori*. In other words, we will take the *posterior* probability distributions of μ_x and μ_y and use the formulas of Rules 10.1.10 and 10.4.4 to find a new stochastic variable $\theta = \mu_x - \mu_y$. To test whether $\mu_x < \mu_y$, we then simply test whether $\theta > 0$. Similarly for testing $\mu_x > \mu_y$.

Example 14.3.1 A Finnish forester wants to test where the pines grow best, and has made a test planting of 17 trees at "Plot A", and 23 trees at "Plot B". One year later, he measures how much taller the trees have grown, and brings his data to a statistician in Helsinki. The forester says he has the most faith in plot B, but that it will be costly to move his main production there, and away from A. The statistician considers the costs, and says he will test the alternative hypothesis $\mu_B > \mu_A$ with significance $\alpha = 0.1$. The statistician calculates, and using neutral priors, he gets that $\mu_A \sim t_{(48.1, 4.3, 16)}$, whereas $\mu_B \sim t_{(51.7, 4.1, 22)}$. He then uses Rule 10.4.4 to derive the probability distribution of $\theta = \mu_A - \mu_B$:

1) $48.1 - 51.7 = -3.6$;
2) $\sqrt{4.1^2 + 4.3^2} = 5.941\,38$;

3) $\left\lfloor \dfrac{\left(\dfrac{4.3^2}{16+1} + \dfrac{4.1^2}{22+1} \right)^2}{\left(\dfrac{\left(\dfrac{4.3^2}{16+1} \right)^2}{16} + \dfrac{\left(\dfrac{4.1^2}{22+1} \right)^2}{22} \right)} \right\rceil = 33,$

so $\theta \sim t_{(-3.6, 5.941\,38, 33)}$. The alternative hypothesis H_1 is then that $\theta < 0$, whereas the conservative hypothesis is $H_0: \theta \geq 0$. We calculate

$$P(H_0) = P(\theta \geq 0) = 1 - T_{(-3.6, 5.941\,38, 33)}(0) = 0.274\,357 > 0.1 = \alpha.$$

The statistician therefore tells the forester that he should continue using A as his main plot, rather than plot B. □

14.3.2 γ Distributed Variables

This section concerns parameters τ for Gaussian processes, and parameters λ for Poisson processes. Their *posterior* probability distributions are γ distributions. If A and B are γ distributed, then $A - B$ does not have any nice distribution, so we can't solve the problem by comparing two γ distributed by subtraction, the way we did in Section 14.3.1.

But another fundamental arithmetic operation saves the day: division. Rule A.3.1 lets us study $Q = A/B$, and through that finding the probability that $A < c \times B$. Indeed, if $A \sim \gamma_{(k_A, l_A)}(t)$ and $B \sim \gamma_{(k_B, l_B)}(t)$, then

$$P(A < c \times B) = P\left(\frac{A}{B} < c \right) = F_{(2k_A, 2k_B)} \left(c \times \frac{k_B l_A}{k_A l_B} \right).$$

On our favorite calculating tools, this is

- Mathematica: $\text{CDF}\left[\text{FRatioDistribution}[2k_A, 2k_B], c \times \frac{k_B l_A}{k_A l_B} \right]$
- CASIO: $\text{FCD}\left(0, c \times \frac{k_B l_A}{k_A l_B}, 2k_A, 2k_B \right)$
- TI: $\text{Fcdf}\left(0, c \times \frac{k_B l_A}{k_A l_B}, 2k_A, 2k_B \right)$
- HP: $\text{fisher\_cdf}\left(2k_A, 2k_B, c \times \frac{k_B l_A}{k_A l_B} \right).$

Example 14.3.2 The consultancy Aleatica are helping a call center; they are going to determine if Group B has an incoming call rate so much higher than that of Group A that the call center should reorganize. The alternative hypothesis (implying reorganization) is then $H_1: \lambda_B > \lambda_A$, and cost considerations give them a significance value of $\alpha = 0.2$.

- Group A gets 48 incoming calls during a two-hour span, and, following the calculations in Example 14.2.8, we get the posterior distribution $\lambda_A \sim \gamma_{(48,120)}$.
- Group B gets 78 incoming calls during a three-hour span, which results in the posterior distribution $\lambda_B \sim \gamma_{(78,180)}$.

We then calculate

$$P(H_0) = P(\lambda_A \geq \lambda_B) = P(\lambda_B \leq \lambda_A)$$

$$= F_{(2 \cdot 78, 2 \cdot 48)} \left(\frac{48 \cdot 180}{78 \cdot 120} \cdot 1 \right)$$

$$= F_{(156,96)} \left(\frac{12}{13} \right)$$

$$= 0.325\,872 > 0.2 = \alpha.$$

Aleatica does therefore *not reject* H_0, and therefore does not recommend restructuring.

You perform the calculation in our example tools as follows:

- Mathematica: CDF $\left[\text{FRatioDistribution}[156, 96], \frac{12}{13} \right]$
- CASIO: FCD $\left(0, \frac{12}{13}, 156, 96 \right)$
- TI: Fcdf $\left(0, \frac{12}{13}, 156, 96 \right)$
- HP: fisher_cdf $\left(156, 96, \frac{12}{13} \right)$. □

This test may not only tell us the probability of A being larger than B, but will also find the probability that A is larger than a certain multiple of B, as in the following example.

Example 14.3.3 As in Example 14.3.2, Aleatica is helping a call center evaluate whether they should reorganize. However, at this call center, Group B is 1.5 times larger than Group A. Aleatica here recommends reorganizing only if B's incoming call rate λ_B is at least twice as large as A's rate λ_A. Their alternative hypothesis H_1 is therefore $\lambda_B > 2\lambda_A$. Calculating costs leads them to conclude a significance of $\alpha = 0.05$. the *posterior* distributions of the rate parameters are

- $\lambda_A \sim \gamma_{(97,90)}$
- $\lambda_B \sim \gamma_{(241,90)}$.

Then

$$P(H_0) = P(\lambda_B \leq 2\lambda_A)$$

$$= F_{(2 \cdot k_B, 2 \cdot k_A)} \left(2 \cdot \frac{k_A \cdot l_B}{k_B \cdot l_A} \right)$$

$$= F_{(2 \cdot 241, 2 \cdot 97)} \left(2 \cdot \frac{97 \cdot 90}{241 \cdot 90} \right)$$

$$= 0.032\,414\,5 < 0.05 = \alpha.$$

This means that Aleatica recommends that the call center should act on H_1 being true, and therefore reorganize. □

14.3.3 β Distributed Variables

This section concerns parameters p and π for Bernoulli processes. Their *posterior* probability distributions are β distributions. For the Gaussian variables in Section 14.3.1, we could compare two variables by subtraction, calculating probabilities for the difference. For the γ distributed variables in Section 14.3.2, we could compare two variables by division, calculating probabilities for the ratio. For β distributed variables, we have no such solution, but we nevertheless *do* have a means – a formula – for comparing the probability of one β distributed variable being greater than the other. The formula for comparing β distributions is stated in Rule A.3.3.

Example 14.3.4   Zen abbot Sōzen and protestant priest Skippervold both enjoy quirky European football statistics. Sōzen says fotball is a matter of heart, and roots for the Edinburgh team *Heart of Midlothian*. Skippervold, on the other hand, thinks a team named after the mother of Jesus is a better choice, and roots for the team *Motherwell* from the city of the same name. They know they will never agree which team is *best*, so they compete by means of idiosyncratic statistical comparisons. The usual routine is that one of them makes a statement about some statistic he thinks favors his own team, and then they investigate the alternative hypothesis that it indeed *does* favor the proponents team, with significance $\alpha = 0.2$.

Today, it's Sōzen's turn. He claims that Hearts's U20 youth team have a higher percentage of left-foot shots at the goal than does Motherwell's U20 team. Sōzen in addition claims that left-foot scorings indicate something positive about the culture and tolerance in a football team.

For their empirical investigation, the two friends have chosen to watch the game between two youth teams in the FA Youth Cup finals in 2016. In this game, there were a total of 41 shots at the goals, of which 19 were by Hearts, and 22 by Motherwell. Hearts's players shot 12 times at Motherwell's goal with

Figure 14.5 Sözen observes, calculates *posterior* distributions, and concludes.

their right foot, and 7 times with their left, whereas Motherwell's players shot 14 shots at Hearts's goal with their right, and 8 with their left.

We were going to look at the long-term proportion of left-foot goal shots for the two teams. Let this proportion be p_H for the Hearts youth, and let it be p_M for Motherwell. In the inference, use Novick and Hall's neutral prior for both. That gives us the *posterior* probability distributions for these two parameters

$$p_H = \beta_{(7,12)}$$
$$p_M = \beta_{(8,14)}.$$

The alternative hypothesis H_1 is that $p_H > p_M$, which means $H_0: p_H \leq p_M$. Then, what remains is to find $P(H_0)$. We will do an exact calculation for both, and do them with a Normal approximation using the formulas of Section A.3.4. Exact first:

$$P(H_0) = P(p_H \leq p_M) = p(p_M > p_H)$$
$$= \sum_{k=0}^{8-1} \frac{\Gamma(7+k)\Gamma(12+14)\Gamma(k+14+1)\Gamma(7+12)}{(14+k)\Gamma(7)\Gamma(12)\Gamma(14)\Gamma(k+1)\Gamma(7+12+k+14)}$$
$$= 0.488\,756 > 0.2 = \alpha.$$

We see with Sözen in Figure 14.5 that we can't reject H_0. This means Sözen has no support for his claim that the Hearts youth use their left foot more than the kids from Motherwell when shooting a goal.

We may also calculate via the Normal approximation: $\delta = p_H - p_M$. Then,

$$P(H_0) = P(p_H < p_M) = P(\delta < 0)$$
$$= \Phi\left(\frac{7}{7+12} - \frac{8}{8+14}, \sqrt{\frac{7 \cdot 12}{(7+12)^2(7+12+1)} + \frac{8 \cdot 14}{(8+14)^2(8+14+1)}}\right)(0)$$
$$= 0.487\,043 > 0.2 = \alpha.$$

In this case, the Normal approximation was sufficiently close to the exact calculation that the two yielded the same result. We don't reject H_0. $\qquad \square$

14.4 Exercises

Utility Functions

1 In the problems below, you are given the probability distribution of a stochastic variable X and a utility function $u(x)$. Find the expected utility U.

(a) $X \sim \beta_{(17,9)}$ and $u(x) = 3x + 2$;

(b) $X \sim \gamma_{(7,21)}$ and $u(x) = \begin{cases} 9 & x < 0.3 \\ -4 & x > 0.3; \end{cases}$

(c) $X \sim \phi_{(5.3,1.9)}$ and $u(x) = \begin{cases} -1 & x < 3 \\ 1.5 & 3 < x < 6 \\ 4 & x > 6; \end{cases}$

(d) $X \sim t_{(10,5,2)}$ and $u(x) = \begin{cases} -1 & x < 15 \\ 7 & x > 15; \end{cases}$

(e) $X \sim \phi_{(41.3,9.1)}$ and $u(x) = -2x + 90$;

(f) $X \sim t_{(-2.73,1.21,8)}$ and $u(x) = -5x + 8$.

2 You are going to decide whether $A: \Theta < \theta_0$ or $B: \Theta > \theta_0$. The gain in utility of choosing A instead of B is

$$u(x) = \begin{cases} w_A & x < \theta_0 \\ -w_B & x > \theta_0. \end{cases}$$

In the first three subproblems below, you are given θ_0, w_A, and w_B. Formulate the decision problem as a hypothesis test by indicating significance level α and stating the alternative hypothesis H_1.

(a) $\theta_0 = 7$, $w_A = 9$, $w_B = 1$;

(b) $\theta_0 = -3$, $w_A = 25$, $w_B = 175$;

(c) $\theta_0 = 100$, $w_A = 1$, $w_B = 100$.

(d) A pharmaceutical company has developed a medicine that changes certain blood values. They have two choices: to produce, or not to produce. If the mean change in that blood value is more than 17, they earn 1 unit more by producing than by not producing. If the change is below 17, they save 19 units by *not* producing. Convert this into statements about A, B, and θ_0, and proceed as in the other subproblems.

Hypothesis Test for Gaussian Processes

3 You are given the *posterior* distribution $\theta \sim f(x)$, the significance α, and alternative hypothesis H_1. Test, and decide between the competing hypotheses.

(a) $\theta \sim \phi_{(7,2)}$, $\alpha = 0.05$, and $H_1: \theta > 3$.

(b) $\theta \sim \phi_{(9,2)}$, $\alpha = 0.05$, and $H_1: \theta > 7$.

(c) $\theta \sim \phi_{(8,3)}$, $\alpha = 0.006$, and $H_1: \theta < 16$.
(d) $\theta \sim t_{(7,2,3)}$, $\alpha = 0.05$, and $H_1: \theta > 3$.
(e) $\theta \sim t_{(9,2,5)}$, $\alpha = 0.05$, and $H_1: \theta > 4$.
(f) $\theta \sim t_{(24,3,1)}$, $\alpha = 0.075$, and $H_1: \theta < 16$.

4 You have a job controlling how well pubs fill pint servings. More precisely, you sample to evaluate if the mean servings μ are at least 1.0 pint. For your job, you use a neutral prior. At one particular pub one evening, you have sampled 10 pints, and measured: {0.98, 0.98, 0.96, 1.02, 0.98, 1.0, 1.02, 0.96, 0.96, 0.98}. Using these data, determine whether $H_1: \mu < 1.0$, with significance $\alpha = 0.1$.

5 **Capacitors:** Measuring 25 of FR Electronics's smallest capacitors, you got $\bar{c} = 49.19 \, \mu F$ and sample standard deviation $s_c = 2.15 \, \mu F$. Assume the capacitances follow a Normal distribution $\phi_{(\mu,\sigma)}$, and determine, with significance $\alpha = 0.02$ and neutral prior, the alternative hypothesis that the mean μ is less than $50 \, \mu F$.

Hypothesis Test for Bernoulli Processes

6 You are given a (posterior) distribution for $\pi \sim \beta_{(a,b)}$, a significance α, and H_1. Test the following competing hypotheses, to decide between them, both by direct calculation and by Normal approximation.
(a) Posterior: $\pi \sim \beta_{(35,24)}$, $\alpha = 0.1$, and $H_1: \pi > 0.5$.
(b) Posterior: $\pi \sim \beta_{(78,21)}$, $\alpha = 0.05$, and $H_1: \pi < 0.85$.
(c) Prior: $\pi \sim \beta_{(1,1)}$. Observations: 22 positive outcomes and 51 negative. $\alpha = 0.02$, and $H_1: \pi > 0.2$.

7 You are estimating π, the proportion who prefer MegaCola to its competitors, and your *posterior* hyperparameters for π are $a_1 = 41.5$ and $b_1 = 9.5$. If the proportion who prefer MegaCola is more than 75%, MegaCola will launch a costly campaign. A consideration of the utilities on both sides results in your having to decide whether $\pi > 0.75$ with significance $\alpha = 0.1$. What will be your recommendation to MegaCola?

8 Your are estimating π, the proportion of "Fancy diamonds" of quality IF and VVS, in a diamond mining project where they are considering buying new and expensive mining equipment if this proportion exceeds 0.1. Owing to the high costs, they will be determining whether $H_1: \pi > 0.1$ with significance $\alpha = 0.1$. For your report, you use Jeffreys' prior ($u = 0.5$). After examining 172 rocks, you have found 19 that are either IF or VVS, whereas the rest are of lower grade. What will be your recommendation for the mining project?

Hypothesis Test for Poisson Processes

9 You are given a probability distribution for $\tau \sim \gamma_{(k,\lambda)}$, a significance α, and H_1. Determine the hypothesis test, both by direct calculation and by using the Normal approximation.

(a) Posterior: $\tau \sim \gamma_{(6,3)}$, $\alpha = 0.04$, and H_1: $\tau < 4$.

(b) Posterior: $\tau \sim \gamma_{(10,29.7)}$, $\alpha = 0.05$, and H_1: $\tau > 0.2$.

(c) Posterior: $\tau \sim \gamma_{(17.5,53.4)}$, $\alpha = 0.1$, and H_1: $\tau > 0.5$.

(d) Posterior: $\tau \sim \gamma_{(20/3, \sqrt{32})}$, $\alpha = 0.001$, and H_1: $\tau < 3.2$.

10 ($t = $ time) The number of plumbing gaskets that need changing every week in an apartment complex is assumed to follow a Poisson process with rate λ. You are estimating this need for an apartment complex with 70 flats, and have looked into the documentation for the last semester (26 weeks), and find registered 53 gasket changes. For your hypothesis test, use a neutral prior, and let hypothesis H_1 for the rate be that $\lambda > 1.5$. Decide between H_0 and H_1 with significance $\alpha = 0.06$.

11 ($t = $ volume in cubic centimeters) The number of bacterial colonies per cubic centimeter in a certain polluted lake is assumed to be Poisson distributed with parameter λ. You sample 1 deciliter, and find 157 bacterial colonies. Use neutral prior, and test the alternative hypothesis H_1: $\lambda < 1.75$ with significance $\alpha = 0.04$.

Pairwise Comparison for Gaussian Processes

12 The brothers Odd and Kjell Aukrust lie home in bed with whooping cough, and as Kjell rattles off a particularly long-lasting cough, Odd exclaims: *That one lasted for rather a long time, but not as long as mine do!* Kjell disagrees, so they decide to measure coughing times (in seconds):

- Odd: 22, 20, 21, 20, 21, 21, 19, 21
- Kjell: 15, 12, 32, 12, 11, 13, 14.

The budding statistician Odd assumes the lengths of the coughing bouts follow Normal distributions, respectively $\phi_{(\mu_K,\sigma_K)}$ (Kjell) and $\phi_{(\mu_O,\sigma_O)}$ (Odd). He would like to establish *with significance* $\alpha = 0.2$ that his own bouts of coughing last longer than Kjell's, in other words he wants to test H_1: $\mu_O > \mu_K$. But Odd is ill today, so he leaves it up to us to do the work. We use neutral priors.

13 Nicholas believes that the tomcat Baggins purrs for longer than the female cat Perry, but Caroline, who is a student keenly interested in statistics, asks him to back up his claim by hypothesis testing it with significance $\alpha = 0.1$.

Nicholas then times how long each cat purrs after one single stroke, and gets the following purring durations (in seconds):

- Baggins: 59, 71, 102, 64, 56, 83
- Perry: 48, 43, 51, 48, 54.

Assume that the purring durations follow Normal distributions, with respective mean values μ_B for Baggins and μ_P for Perry. Use neutral priors and assume σ unknown. Which side has the burden of proof here? To answer this, establish which hypothesis is H_0 and which is H_1, and decide between the two competing hypotheses.

Pairwise Comparison for Poisson Processes

14 Your are comparing two γ distributed variables $\Theta \sim \gamma_{(k,l)}$ and $\Psi \sim \gamma_{(m,n)}$ to decide between hypotheses H_1 (as specified below) and H_0, with significance α.

(a) $\Theta \sim \gamma_{(7,70)}$, and $\Psi \sim \gamma_{(4,80)}$. Significance $\alpha = 0.05$, and $H_1: \Theta > \Psi$.

(b) $\Theta \sim \gamma_{(9,20)}$, and $\Psi \sim \gamma_{(11,20)}$. Significance $\alpha = 0.1$, and $H_1: \Theta < \Psi$.

(c) Let the parameters in the Problem 14.b be 10 times as large. This corresponds to 10 times as many observations of each kind as in the previous problem. Do you think there should be a difference? Think about it for a while, and then calculate to see whether it matters or not!

15 ($t =$ units $=$ hunters) You have been told that 23 grouse hunters from a certain part of the Lowlands had a total catch of 111 grouse in a single day, and concluded that the catch rate there was $\lambda_{\text{Lowlands}} \sim \gamma_{(111,23)}$. You have now spoken to 8 hunters from the Highlands, and their catch on the same day was 65 grouse. Use a neutral prior and find the *posterior* probability distribution for $\lambda_{\text{Highlands}}$. Then decide, with significance $\alpha = 0.1$, whether $H_1: \lambda_{\text{Highlands}} > \lambda_{\text{Lowlands}}$.

Pairwise Comparison for Bernoulli Processes

16 You are comparing two β distributed variables $\psi \sim \beta_{(k,l)}$ and $\pi \sim \beta_{(m,n)}$ to determine the hypothesis H_1 (which is either $\psi > \pi$, or $\psi < \pi$) with significance α. Do this *both* by exact calculation using Rule A.3.3, *and* by Normal approximation. (The exact calculation requires good calculation tools.)

(a) $\psi \sim \beta_{(2,5)}$ and $\pi \sim \beta_{(4,3)}$, $\alpha = 0.1$, $H_1: \psi < \pi$.

(b) $\psi \sim \beta_{(23,17)}$ and $\pi \sim \beta_{(17,23)}$, $\alpha = 0.1$, $H_1: \psi > \pi$.

(c) $\psi \sim \beta_{(20,20)}$ and $\pi \sim \beta_{(17,23)}$, $\alpha = 0.05$, $H_1: \psi > \pi$.

(d) Let the parameters of the Problem 16.c be 10 times as large. This corresponds to ten times as many positive observations and ten times as many negative observations. Do you think there should be a difference? Think about it for a while, and then calculate to see whether it matters or not!

17 Your company has for a long time used Imperial Deliveries for freight. Lately, however, a promising new competitor has surfaced: Centurium Falcon Freight. You decide to test the rate of delivery errors to compare the two. Let π_{ID} be the proportion of erroneous deliveries at Imperial Deliveries, and let π_{CFF} be the rate for Centurium Falcon Freight.

Your internal routines are built around delivery by Imperial Deliveries, making a change of freight company cost a bit. However, at the same time there is a fair bit to save if the new alternative is an improvement. A cost analysis using utility functions indicates that you should test the hypothesis $H_1: \pi_{CFF} < \pi_{ID}$ with significance $\alpha = 0.15$.

You run 200 trial deliveries with both companies. Imperial Deliveries make four delivery errors, whereas Centurium Falcon Freight make one error. Investigate, with prior $\beta_{(1,1)}$, and decide between the competing hypotheses.

15

Estimates

> **CONTENTS**
>

15.1 Introduction

We are going to estimate an unknown parameter θ from a model from some observations. A *point* estimate of a parameter θ is a single value that is thought to be the best representative of our knowledge about θ.

The *interval* estimate for a parameter is the tolerant relative of the point estimate, in that it gives a certain slack and indicates an entire interval of possible values. The interval estimates are stated in terms of their sizes: if the size is in terms of probability, we speak of a *P% credible interval*, and if the size is in terms of the interval width, we speak of an *HPD interval of width l*. (HPD means *Highest Posterior Density*.)

We have interval estimates for the next observation X_+ as well, and these are the *(posterior) predictive* intervals.

15.2 Point Estimates

A point estimate for a parameter θ is a summary of our knowledge about the parameter in a single number $\hat{\theta}$, which is our best guess at the value of θ. In the

The Bayesian Way: Introductory Statistics for Economists and Engineers, First Edition.
Svein Olav Nyberg.
© 2019 John Wiley & Sons, Inc. Published 2019 by John Wiley & Sons, Inc.

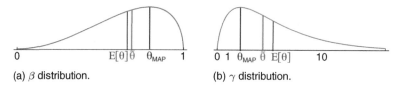

(a) β distribution.　　　　(b) γ distribution.

Figure 15.1 Unequal point estimates.

Bayesian school, we find the point estimates from the probability distribution for the parameter, $f(x)$. The three most important point estimates are

- MAP, the Maximum A Posteriori estimate θ_{MAP}, is simply the max value of the probability distribution (also known as its *mode*);
- The median $\tilde{\theta}$;
- The expected value $\mu_\theta = E[\theta]$.

These three are "best guesses", each in its own way: θ_{MAP} is the maximum value of the probability distribution, $\tilde{\theta}$ minimizes $l(t) = E[|\theta - t|]$, whereas $\mu_\theta = E[\theta]$ minimizes $k(t) = E[(\theta - t)^2]$.

For symmetrical distributions, like the Normal and t distributions, all three point estimates coincide, whereas for the β and γ distributions, we usually have either $E[\theta] \geq \tilde{\theta} \geq \theta_{MAP}$, or $\theta_{MAP} \geq \tilde{\theta} \geq E[\theta]$, as illustrated in Figure 15.1.

Rule 15.2.1 *The three point estimates for ϕ, t, β, and γ distributions are:*

Distribution	MAP	Median	Expected value
$\phi_{(\mu,\sigma)}$	μ	μ	μ
$t_{(\mu,\sigma,\nu)}$	μ	μ	μ
$\beta_{(a,b)}$	$\frac{a-1}{a+b-2}$	$I^{-1}_{(a,b)}(0.5)$	$\frac{a}{a+b}$
$\gamma_{(k,\lambda)}$	$\frac{k-1}{\lambda}$	$\Gamma^{-1}_{(k,\lambda)}(0.5)$	$\frac{k}{\lambda}$

15.3 Interval Estimates

Definition 15.3.1 *A P% interval estimate for a stochastic variable Θ is an interval (a, b) where $P(a < \Theta < b) = P\%$.*

- *A P% interval estimate for a parameter θ is called a* credible interval. *We will denote this by an upper index θ, that is, I^θ.*
- *A P% interval estimate for the next observation, X_+, is called a* predictive interval. *We will denote this by an upper index $+$, that is, I^+.*

The point of interval estimates is to indicate a range of values for a variable Θ so that the reader may feel confident that there is a $P\%$ probability that Θ is within this range. Which range, and thereby which interval limits serve that purpose best, depends on the application. One-sided limits and intervals are best when only one of the directions is critical to hedge against. For instance: you are more interested in knowing that there is a 99% probability that your Christmas present from grandma in Australia will arrive before the December 22nd than in knowing that there is a 99% probability that it will arrive after the December 15th. Or the other way round: you are told you may get a reward, and might be more interested in knowing that there is a 90% probability that your reward will be in excess of a thousand dollars than that there is a 90% probability that your reward will be *less* than two thousand dollars.

But the one-sided intervals are not the most frequently used ones. The most used are the "symmetric" or "two-sided with equal tails". A *symmetric interval* $I_{2\alpha} = (a, b)$ for Θ is characterized by the fact that $P(\Theta < a) = P(\Theta > b) = \alpha$, which means that the probability of each "tail" is α. These interval estimates limit the range of possible values in both directions, which makes sense for most applications: it is not sufficient to know that a hammer, an electronic component, or a shoe are larger than some given magnitude. Neither is it sufficient that they are smaller than such a magnitude. Like Goldilocks, we want our numbers limited in both directions, and find a range where they are *just right*.

Let F be a cumulative probability density. Then the three most common interval estimates, with their lower indexes, are as follows.

1) Left-sided $(1 - \alpha)100\%$ interval: $I_{\alpha,l} = \left(F^{-1}(0),\ F^{-1}(1 - \alpha)\right)$.
2) Symmetric $(1 - 2\alpha)100\%$ interval with equal tails: $I_{2\alpha} = \left(F^{-1}(\alpha),\ F^{-1}(1 - \alpha)\right)$.
3) Right-sided $(1 - \alpha)100\%$ interval: $I_{\alpha,r} = \left(F^{-1}(\alpha),\ F^{-1}(1)\right)$.

Notice how the limits of $I_{2\alpha}$ coincide with the limits of $I_{\alpha,l}$ and $I_{\alpha,r}$. If we find $I_{2\alpha}$, we have the one-sided intervals for free. We illustrate $L_{0.1}$, $I_{0.2}$, and $R_{0.1}$ for our four most used probability distributions in Figure 15.2.

The one-sided intervals are specified either by only an upper limit, or by only a lower limit. The two-sided intervals are specified by both limits, implicitly giving us the two one-sided intervals. So after the first example, we will stick with two-sided intervals only.

15.3.1 HPD Intervals

We have a fourth kind of interval estimate that we call HPD (*Highest Posterior Density*) intervals. This is also a two-sided interval. For a given probability $P\%$ the HPD interval H is the narrowest possible interval where $P(\Theta \in H) = P\%$.

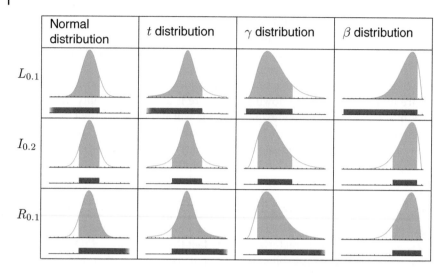

Figure 15.2 One- and two-sided intervals given by probability *P*% compared.

This also means that among intervals of width *l*, *H* is the interval with the highest probability of containing Θ. Note, as illustrated in Figure 15.3, that the probability *density* is the same at both endpoints of an HPD interval.

For symmetric distributions like ϕ and *t*, the HPD intervals are identical to the symmetric intervals. For other distributions, like the γ and β distributions, they differ, and there, the HPD intervals are most easily found from the width *l*, rather than from the probability *P*%. We therefore find it expedient to write the HPD intervals for Θ of width *l* as H_l^{Θ}.

If H^{τ} is an HPD interval for τ, then the interval of the corresponding values for σ is *not* an HPD interval for σ. This is as opposed to the $(1 - 2\alpha)100\%$ symmetric interval for τ, where the interval of the corresponding values of σ is a $(1 - 2\alpha)100\%$ symmetric interval as well.

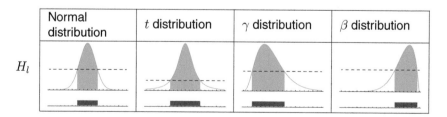

Figure 15.3 HPD intervals.

15.3.2 Estimating Sample Size

When you are testing the quality of cars in commercial production, the cost of a single measurement may very well be an entire car, as in a crash test. At the same time, achieving accurate results from the tests carry very high utility for both the manufacturer and the customer. The same applies to medical research. In medical trials, the test subjects face a potential hazard to their health – and in the worst case, maybe life. At the same time, accurate test results save lives. In both of these cases, we need to balance the cost of testing against the utility of the results.

We may explore the full blooded version of the balance between the utility of the result and the cost of the testing if we have a utility function weighing the utility of the width of the interval against the cost of further observation. This approach belongs to advanced studies, but such advanced statistics studies should be undertaken if you want to plan very expensive test regimes.

In this book, we will keep to the basics, and employ formulas that start with either a desired interval width l, or a fixed probability $1 - 2\alpha$ that Θ is in the interval. Where possible, we will find the number of observations n required for the interval to have probability $(1 - 2\alpha)100\%$ or width l (or narrower). Where this is not obtainable, our formulas will give a statistically best guess of the n required for the interval to have (roughly) width l, or probability $1 - 2\alpha$ of containing Θ.

15.4 Estimates for the φ Distribution

Problem: We have a Gaussian process with observables $X_k \sim \phi_{(\mu,\sigma)}$. The parameter μ itself is unknown, whereas $\sigma = s_0$ is known. Using the methods of Section 13.1.1, we find the *posterior* distribution of μ and the predictive distribution of X_+. Both follow Normal distributions. We then find estimates for μ and X_+.

Example 15.4.1 (Continuation of Example 13.1.1) The traffic police had measured speeds on a straight stretch on the M8 between Paisley and Glasgow, where they spotted their first possible speeder. Their probability distribution of his speed μ was

$$\mu \sim \phi_{(137, 2.886\,75)}(x).$$

We will find both right- and left-sided 95% credible intervals for the actual speed μ, as well as a 90% symmetric credible interval for μ. We will also find the three kinds of point estimate for μ.

Solution: To find the credible intervals, we first need to find α. For the one-sided intervals, that means solving the equation $(1 - \alpha)100\% = 95\%$, and for the symmetric ones, solving $(1 - 2\alpha)100\% = 90\%$. In both cases, $\alpha = 0.05$. The intervals are then

$$I^\mu_{0.05,l} = \left(\Phi^{-1}_{(137,\,2.886\,75)}(0),\ \Phi^{-1}_{(137,\,2.886\,75)}(0.95) \right) = \underline{(-\infty,\ 141.748)}$$

$$I^\mu_{0.10} = \left(\Phi^{-1}_{(137,\,2.886\,75)}(0.05),\ \Phi^{-1}_{(137,\,2.886\,75)}(0.95) \right) = \underline{(132.252,\ 141.748)}$$

$$I^\mu_{0.05,r} = \left(\Phi^{-1}_{(137,\,2.886\,75)}(0.05),\ \Phi^{-1}_{(137,\,2.886\,75)}(1) \right) = \underline{(132.252,\ \infty)}.$$

The point estimates are straightforward:

$$\mu_{MAP} = \tilde{\mu} = E[\mu] = \underline{137}.$$

15.4.1 Estimating Sample Size

Rule 10.1.6 provides a useful quick formula for interval estimates: if $\Theta \sim \phi_{(m,s)}$, then the $(1 - 2\alpha)100\%$ symmetric interval estimate is

$$I^\Theta_{2\alpha} = m \pm z_\alpha \times s.$$

For the posterior and predictive distributions, we may use the hyperparameters in Section 13.1.1 directly. The formulas then become

$$[t]I^\mu_{2\alpha} = m_1 \pm z_\alpha s_0 \sqrt{\frac{1}{\kappa_1}}$$

$$I^+_{2\alpha} = m_1 \pm z_\alpha s_0 \sqrt{1 + \frac{1}{\kappa_1}}.$$

We recall that $\kappa_1 = \kappa_0 + n$, which means that as n, the number of observations, increases, the width of the credible interval $I^\mu_{2\alpha}$ shrinks to 0, while the width of the predictive interval $I^+_{2\alpha}$ converges to a stable width of $2z_\alpha s_0$. We use these facts to establish the required sample sizes for a given interval width.

> **Rule 15.4.2** *(Required sample size with fixed α) If $\sigma = s_0$ is known, the $(1 - 2\alpha)100\%$ credible interval for μ will be narrower than l if the number of observations is at least*
>
> $$n = \frac{4z_\alpha^2}{l^2} \times s_0^2 - \kappa_0.$$

Notice that this also means that, when we have n or more observations, $P(\mu \in H^\mu_l) \geq 1 - 2\alpha$. We will look at an example of an interval estimate for μ and X_+.

Example 15.4.3 (Continuation of Example 13.1.2) You have tested the output effect of your audiophile miniature amplifiers in two rounds, and you consider σ to be known, with value $s_0 = 0.7$ mW.

- Find 95% interval estimates for μ and X_+ from both rounds.
- How wide will the predictive interval for X_+ be when n approaches infinity?
- For which number of observations n will the width of $I^\mu_{0.1}$ be less than 0.1 mW?
- For which number of observations n will $P(\mu \in H^\mu_{0.1})$ be larger than 0.95?

Answer: After the first round, we had $\mu \sim \phi_{(2.345\,8,\,0.313\,05)}$ and $X_+ \sim \phi_{(2.345\,8,\,0.766\,812)}$. We see that $95\% = (1 - 2 \cdot 0.025) \times 100\%$, and therefore use that $z_{0.025} = -1.959\,96$, which gives the intervals

$$I^\mu_{0.05} = 2.345\,8 \pm 1.959\,96 \times 0.313\,05 = \underline{(1.732\,23,\ 2.959\,37)}$$

$$I^+_{0.05} = 2.345\,8 \pm 1.959\,96 \times 0.766\,812 = \underline{(0.842\,879,\ 3.848\,72)},$$

where the respective widths of the intervals are 1.227 13 and 3.005 84.

After the second round, we had $\mu \sim \phi_{(2.349\,5,\,0.202\,073)}$ and $X_+ \sim \phi_{(2.349\,5,\,0.728\,583)}$. Then

$$I^\mu_{0.05} = 2.349\,5 \pm 1.959\,96 \times 0.202\,073 = \underline{(1.953\,45,\ 2.745\,55)}$$

$$I^+_{0.05} = 2.349\,5 \pm 1.959\,96 \times 0.728\,583 = \underline{(0.921\,503,\ 3.777\,5)},$$

where the respective widths of the intervals are 0.792 11 and 2.855 99.

The width of the predictive interval converges to $2 \times 1.959\,96 \times 0.7 = 2.743\,94$ as $n \to \infty$, whereas the width of the credible interval for μ converges to 0, as it does for any such credible interval. Since $\kappa_0 = 0$, the width of $I^\mu_{0.05}$ will be less than 0.1 when

$$n = \frac{4 \times 1.959\,96^2}{0.1^2} \times 0.7^2 - 0 = 752.923.$$

Since we may only make whole observations, we round up, so $n = 753$ is the smallest number of observations for which the width of $I^\mu_{0.05}$ is less than 0.1, and also $P(\mu \in H^\mu_{0.1}) > 0.95$.

15.5 Estimates for the *t* Distribution

Problem: We have a Gaussian process with observations $X_k \sim \phi_{(\mu,\sigma)}$. The parameters μ and σ are both unknown. Using the method in Section 13.1.3, we find the *posterior* distribution of μ and the predictive distribution of X_+. Both are *t* distributions. In addition, we find the *posterior* distribution of $\tau = \sigma^{-2}$, which is a γ distribution. Here, we will find estimates for μ and X_+.

From Section C.2, we may deduce a useful formula for symmetric interval estimates when $\Theta \sim t_{(m,s,v)}$. Then

$$I^{\Theta}_{2\alpha} = m \pm t_{v,\alpha} \times s.$$

If we return to the hyperparameters in Section 13.1.3, the formulas become

$$I^{\mu}_{2\alpha} = m_1 \pm t_{v_1,\alpha} s_1 \sqrt{\frac{1}{\kappa_1}}$$

$$I^{+}_{2\alpha} = m_1 \pm t_{v_1,\alpha} s_1 \sqrt{1 + \frac{1}{\kappa_1}},$$

where $m_1 = \Sigma_1/\kappa_1$ and $s_1^2 = SS_1/v_1$. We have no corresponding expression for the credible interval of σ.

Example 15.5.1 (Continuation of Example 13.1.5) Abbot Sōzen has measured the speed of the cars driving past in the 50 zone outside his temple. He did this in two rounds, and we will help him by calculating the 80% interval estimates for the parameter μ and for the next observation X_+.

First round: The *posterior* probability distributions were $\tau \sim \gamma_{(4.5, 61.8545)}$ and $\mu \sim t_{(57.19, 1.17241, 9)}$, whereas the predictive distribution was $X_+ \sim t_{(57.19, 3.88844, 9)}$. Since $80\% = (1 - 2 \times 0.1) \times 100\%$, and $t_{9,0.1} = 1.38303$, we have

$$I^{\mu}_{0.2} = 57.19 \pm 1.38303 \times 1.17241 = \underline{(55.5685, 58.8115)}$$

$$I^{+}_{0.2} = 57.19 \pm 1.38303 \times 3.88844 = \underline{(51.8122, 62.5678)}.$$

Second round: The *posterior* probability distributions were $\tau \sim \gamma_{(24.5, 264.953)}$ and $\mu \sim t_{(56.522, 0.465068, 49)}$, whereas the predictive distribution was $X_+ \sim t_{(56.522, 3.32125, 49)}$. Since $t_{49,0.1} = 1.29907$, we have

$$I^{\mu}_{0.2} = 56.522 \pm 1.29907 \times 0.465068 = \underline{(55.9178, 57.1262)}$$

$$I^{+}_{0.2} = 56.522 \pm 1.29907 \times 3.32125 = \underline{(52.2075, 60.8365)}.$$

We see that just as in the case of the known σ, the width of I^{μ}, the credible interval for the cars' mean speed, gets narrower as the number of observations increases, whereas the predictive interval I^+ for the speed of the next car stays roughly as wide (and will indeed converge on some fixed positive value).

15.5.1 Estimating Sample Size

For t distributions, there is no formula corresponding to Rule 15.4.2 giving a credible interval that has a guaranteed maximum width for a sample size of n

observations. Since σ is unknown, the best we are able to achieve is an estimate that itself is subject to probabilities.

This kind of estimate requires that we have *some* information about σ. The simplest relation requires prior hyperparameters κ_0, SS_0, and $v_0 > 0$ (see 13.1.2). We then have the following rule.

Rule 15.5.2 *The expected probability that $\mu \in H_l^\mu$ is equal to or greater than $1 - 2\alpha$ when*

$$n \geq \frac{4t_{2v_0,\alpha}^2}{l^2} \times \frac{SS_0}{v_0} - \kappa_0.$$

This is also the probability that $I_{2\alpha}^\mu$ is narrower than l.

Example 15.5.3 (Continuation of 15.5.1 and 13.1.5) How many new measurements must Sōzen make of the cars driving past his temple if he wants a credible interval of width $l = 0.5$ and at the same time an expected probability of 95% for it to contain μ, the cars' actual mean speed?

Answer: Here, the *prior* hyperparameters equal the *posterior* hyperparameters from the previous round. We have $\kappa_2 = 50$, $SS_2 = 529.906$, and $v_2 = 49 > 0$. Using the fact that $t_{98,0.025} = -1.984\,47$, we get that n must be at least

$$n \geq \frac{4t_{2v_2,\alpha}^2}{l^2} \cdot \frac{SS_2}{v_2} - \kappa_2 = \frac{4t_{98,0.025}^2}{0.5^2} \cdot \frac{529.906}{49} - 50 = 631.413,$$

which means Sōzen must observe the speed of at least 632 more cars to get the desired values for the interval.

15.6 Estimates for γ Distributions

Problem 1: We've got a Gaussian process with observations $X_k \sim \phi_{(\mu,\sigma)}$. The parameters μ and σ are unknown. Using the method of Section 13.1.3, we find the *posterior* distribution of $\tau = 1/\sigma^2$, which is a γ distribution. We are going to find estimates for τ and σ.

Problem 2: We've got a Poisson process with unknown parameter λ. Using the method of Section 13.3, we find the *posterior* distribution of λ, which is a γ distribution. We are going to find estimates for λ.

The credible interval with equal tails follows the general rule, with $F^{-1}(p) = \Gamma_{(k,\lambda)}^{-1}(p)$. From 10.3, we have the rule that $\Gamma_{(k,\lambda)}^{-1}(p) = \frac{1}{2\lambda} \times \mathbb{X}_{2k}^{-2}(p)$, which facilitates the use of a calculator for handling the γ distribution. These two rules combined imply that

$$I_{2\alpha}^{\lambda} = \left(\Gamma_{(k,\lambda)}^{-1}(\alpha), \Gamma_{(k,\lambda)}^{-1}(1-\alpha) \right)$$

$$= \left(\frac{1}{2\lambda} \mathbb{X}_{2k}^{-2}(\alpha), \frac{1}{2\lambda} \mathbb{X}_{2k}^{-2}(1-\alpha) \right).$$

The width of $I_{2\alpha}^{\Theta}$ is then $\frac{1}{2\lambda} \left(\mathbb{X}_{2k}^{-2}(1-\alpha) - \mathbb{X}_{2k}^{-2}(\alpha) \right)$, and hence a multiple of $E[\Theta] = \frac{k}{\lambda}$. We will make use of this fact when we get to the estimation of sample sizes.

For a distribution $\Theta \sim \gamma_{(k,\lambda)}(x)$ and a given length l, the HPD interval $H_l^{\Theta} = (a, a+l)$ is given by

$$a = \frac{l}{e^{\lambda l/(k-1)} - 1}.$$

Example 15.6.1 (For problem 1: continuation 2 of Example 13.1.5) Abbot Sōzen has measured the speed of the cars driving past in the 50 zone outside his temple. He did this in two rounds, and we will now help him by calculating the estimates for the *second* round: both the interval estimates $I_{0.1}^{\tau}$, $I_{0.1}^{\sigma}$, and $H_{0.07}^{\tau}$ and the point estimates τ_{MAP}, $\tilde{\tau}$, and $E[\tau]$.

The *posterior* probability distribution from the second round is $\tau \sim \gamma_{(24.5, 264.953)}$.

Point estimates:

$$\tau_{MAP} = \frac{24.5 - 1}{264.953} = \underline{0.088\,695}$$

$$\tilde{\tau} = \Gamma_{(24.5, 264.953)}^{-1}(0.5) = 0.091\,214\,2$$

$$E[\tau] = \frac{24.5}{264.953} = \underline{0.092\,469\,2}.$$

Interval estimates: the interval limits for $I_{0.1}^{\tau}$ then become $\Gamma_{(24.5, 264.953)}^{-1}(0.05) = 0.064\,030\,8$ and $\Gamma_{(24.5, 264.953)}^{-1}(0.95) = 0.125\,189$. Since $\tau = 1/\sigma^2$, and thereby $\sigma = 1/\sqrt{\tau}$, the interval limits for $I_{0.1}^{\sigma}$ are then $1/\sqrt{0.064\,030\,8} = 3.951\,9$ and $1/\sqrt{0.125\,189} = 2.826\,29$, which means that

$$I_{0.1}^{\tau} = \underline{(0.064\,030\,8,\ 0.125\,189)}$$

$$I_{0.1}^{\sigma} = \underline{(2.826\,29,\ 3.951\,9)},$$

whose widths are, respectively, 0.061 158 7 and 1.125 61. If we were to insist on a width of $l = 0.07$, and find the HPD interval $H_{0.07}^{\tau}$, we would get the left endpoint

$$a = \frac{0.07}{e^{264.953 \times 0.07/(24.5-1)} - 1} = 0.058\,2517.$$

The right endpoint is then $0.058\,2517 + 0.07 = 0.128\,2517$, which means $H_{0.07}^{\tau} = (0.058\,252, 1\,282\,517)$.

Example 15.6.2 (For problem 2) Archbishop Itabashi thinks our friend abbot Sōzen needs some extra mindfulness training. At the same time, he worries that his cat Dai-ichi is becoming radioactive, so Sōzen is told to follow Dai-ichi with a Geiger counter for an hour. How radioactive Dai-ichi is, remains to be established, but the cat is obviously very active, so Sōzen has to pay attention to getting his Geiger clicks. After 60 minutes, Sōzen has registered 28 Geiger clicks from the cat.

The number of Geiger clicks over a time period follows a Poisson distribution with parameter λ. The archbishop thinks that we, the readers, also need some mindfulness training, so he puts us to work to find interval estimates for λ, starting with neutral hyperparameters. He specifies that he first wants $I_{0.08}^{\lambda}$. He then wants us to find l, the width of $I_{0.08}^{\lambda}$, and then to find the most probable interval of width l; that is, H_l^{λ}.

Answer: $\kappa_0 = \tau_0 = 0$, so

$$P_1 \vDash \kappa_1 = 0 + 28 = 28$$
$$\tau_1 = 0 + 60 = 60.$$

Then

$$\lambda \sim \gamma_{(28,60)}(l).$$

The credible interval with equal tails is then

$$I_{2 \times 0.04}^{\lambda} = \left(\Gamma_{(28,60)}^{-1}(0.04), \Gamma_{(28,60)}^{-1}(0.96) \right) = (0.324\,318, 0.631\,909),$$

whose width is $l = 0.307\,591$. The HPD interval $H_{0.307\,591}^{\lambda} = (a, a + l)$ is then given by

$$a = \frac{0.307\,591}{e^{60 \times 0.307\,591/(28-1)} - 1} = 0.313\,59.$$

Then $H_{0.37}^{\lambda} = (0.313\,59, 0.621\,181)$, and $P(\lambda \in H_{0.307\,591}^{\lambda}) = 0.922\,391$.

15.6.1 Estimating Sample Size

For the Poisson process parameter λ, the most accessible formula for sample size is a formula that looks at the *relative* size; that is, it looks at the ratio of the interval width l to $E[\lambda]$. For these estimates, it is expedient to use the fact that the Normal approximation to $\gamma_{(\kappa_1, \tau_1)}$ is $\phi_{(\mu, \sigma)}$, with $\mu = \frac{\kappa_1}{\tau_1}$ and $\sigma^2 = \frac{\kappa_1}{\tau_1^2}$.

Then

$$I_{2\alpha}^{\Theta} \approx \frac{\kappa_1}{\tau_1} \pm z_\alpha \cdot \frac{\sqrt{\kappa_1}}{\tau_1}$$

giving an interval width of $l = 2z_\alpha \cdot \frac{\sqrt{\kappa_1}}{\tau_1}$.

Rule 15.6.3 *Define the relative interval width of $I_{2\alpha}^{\Theta}$ to be $r = \frac{l}{E[\lambda]}$. Since $E[\lambda] = \frac{\kappa_1}{\tau_1}$, we have $r = \frac{2z_\alpha}{\sqrt{\kappa_1}}$. This means that in order for $r < R$ for some fixed magnitude R, we need*

$$\kappa_1 \geq \frac{4z_\alpha^2}{R^2},$$

which means that the number of new observations must be at least

$$n \geq \frac{4z_\alpha^2}{R^2} - \kappa_0.$$

Example 15.6.4 (Continuation of Example 15.6.2) How many more clicks must Sōzen get on his Geiger counter if the relative interval width should be $r < 0.1$ for $I_{0.08}^{\lambda}$?

Answer: Sōzen has already gathered $\kappa_0 = 28$ clicks. Since $z_{0.04} = -1.75069$,

$$n \geq \frac{4 \cdot 1.75069^2}{0.1^2} - 28 = 1197.97.$$

This means that Sōzen needs at least 1198 more clicks.

15.7 Estimates for β Distributions

Problem: We have a Bernoulli process with unknown parameter p, where we have made n observations with outcomes $\top$ or $\bot$.

In Section 13.2, we found the *posterior* probability distribution $p \sim \beta_{(a,b)}$. The symmetric $(1 - 2\alpha)100\%$ credible interval for p is then

$$I^p_{2\alpha} = \left(I^{-1}_{(a,b)}(\alpha),\ I^{-1}_{(a,b)}(1 - \alpha) \right).$$

For larger parameter values, typically $a, b > 10$, we may use the Normal approximation $\phi_{(\mu,\sigma)}$ where $\mu = \frac{a}{a+b}$ and $\sigma^2 = \frac{ab}{(a+b)^2(a+b+1)}$. This gives the approximate $(1 - 2\alpha)100\%$ credible interval

$$I^p_{2\alpha} \approx \left(\Phi^{-1}_{(\mu,\sigma)}(\alpha),\ \Phi^{-1}_{(\mu,\sigma)}(1 - \alpha) \right).$$

The HPD interval $H^p_l = (k, k + l)$ for $p \sim \beta_{(a,b)}$ is given by solving for k in the following equation (where k is a real solution in the unit interval $(0, 1)$):

$$(k + l)^{a-1}(1 - k - l)^{b-1} - k^{a-1}(1 - k)^{b-1} = 0.$$

Except for very low values of a and b, this equation must be solved numerically, requiring a decent tool. A good starting point for a numerical search is $k_0 = E[p] = \frac{a}{a+b}$. In Mathematica and Wolfram Alpha, the command is

$$\text{FindRoot}\left[(k + l)^{a-1}(1 - k - l)^{b-1} - k^{a-1}(1 - k)^{b-1}, \{k, k_0\} \right].$$

Example 15.7.1   Hot rod: Sam has looked at 36 old Volvo Amazon cars, and has recorded that 22 of them have been treated with hot rod, bodywork repairs using liquid lead, while 14 had not received such treatment. The only thing Sam knew about hot rodding in advance was that there *was* such a thing, so he chose for his prior hyperparameters $a_0 = 1$ and $b_0 = 1$. We are going to estimate π, the total proportion of hot rodded Volvo Amazon cars.

We will first find the three point estimates. Then, we will calculate $I^\pi_{0.05}$, using both exact calculation and Normal approximation. Finally, we will find the HPD interval $H^\pi_{0.3}$.

Answer: The *posterior* distribution for the proportion of hot rodded old Volvo Amazon cars is $\pi \sim \beta_{(1+22,1+14)} = \beta_{(23,15)}$.

   The point estimates are then

$$\pi_{MAP} = \frac{23 - 1}{23 + 15 - 2} = \underline{0.611\,111}$$

$$\tilde{\pi} = I^{-1}_{(23,15)}(0.5) = \underline{0.607\,129}$$

$$E[\pi] = \frac{23}{23 + 15} = \underline{0.605\,263}.$$

Exact calculation of the β distribution (see Section 10.5) gives

$$I^p_{2 \cdot 0.025} = \left(I^{-1}_{(23,15)}(0.025), I^{-1}_{(23,15)}(0.975) \right) = \underline{(0.447\,568, 0.752\,458)},$$

an interval whose width is $l = 0.304\,89$.

For the *Normal approximation*, $\mu = 0.605\,263$ and $\sigma = 0.078\,269\,7$, which gives

$$I^p_{2 \cdot 0.025} \approx \left(\Phi^{-1}_{(\mu,\sigma)}(0.025), \Phi^{-1}_{(\mu,\sigma)}(0.975) \right) = \underline{(0.451\,857, 0.758\,669)},$$

an interval whose width is $l = 0.306\,812$.

For the *HPD interval*, we solve

$$(k + 0.3)^{23-1}(1 - k - 0.3)^{15-1} = k^{23-1}(1 - k)^{15-1}$$

and get the left limit of the interval $k = 0.453\,97$. The right end of the interval is then $k + l = 0.453\,97 + 0.3 = 0.753\,97$, so

$$H^\pi_{0.3} = \underline{(0.453\,97, 0.753\,97)}.$$

For this interval, $P(\pi \in H^\pi_{0.3}) = 0.946\,308$.

15.7.1 Estimating Sample Size

For the β distributions, we choose a "worst case" sample size formula guaranteeing that a credible interval of width l has a probability of at least $1 - 2\alpha$ of containing π.

> **Rule 15.7.2** *If the parameter π has prior probability distribution $\beta_{(a,b)}$, $a, b > 1$, then n or more new observations, where*
>
> $$n = \frac{z^2_\alpha}{l^2} - a - b,$$
>
> *will ensure that $P(\pi \in H^\pi_l) \geq 1 - 2\alpha$, and that the width of $I^\pi_{2\alpha}$ will be l or less.*

Example 15.7.3 (Continuation of Example 15.7.1) How many new observations must Sam make to ensure that the width of a 95% credible interval for π is at most 0.1?

Answer: Sam's old *posterior* will be his new *prior*, so for the formula, $a = 23$ and $b = 15$. From Example 15.7.1, we have $\alpha = 0.025$ and $l = 0.1$, so the number of new observations Sam needs to make is at least

$$n = \frac{z_{0.025}^2}{0.1^2} - 23 - 15 = 346.146,$$

that is, 347 new observations.

15.8 Exercises

Gaussian Processes With Known σ

1 From distribution to interval.
   (a) $\mu \sim \phi_{(14, 3)}$. Find $I_{0.025,l}^{\mu}$ and $I_{0.025,r}^{\mu}$ and $I_{0.05}^{\mu}$.
   (b) $\mu \sim \phi_{(-4.3, 7.2)}$. Find $I_{0.005,l}^{\mu}$ and $I_{0.005,r}^{\mu}$ and $I_{0.01}^{\mu}$.
   (c) $\mu \sim \phi_{(48, \sqrt{19})}$. Find $I_{0.1}^{\mu}$.
   (d) $X_+ \sim \phi_{(0.018, 0.000\,134)}$. Find $I_{0.005}^{+}$.
   (e) $\mu \sim \phi_{(4.3, -7.2)}$. Find $I_{0.03}^{\mu}$.

2 From data + prior to interval. Find $I_{2\alpha}^{\mu}$ and $I_{2\alpha}^{+}$.
   (a) Data: $\{0, 1, 2, 3, 4, 5, 6, 7, 8, 9, 10\}$, neutral prior, known $\sigma = 2.3$; $2\alpha = 0.05$.
   (b) Prior: $\kappa_0 = 7$, $\Sigma_0 = 1\,253$, $s_0 = 15$. $2\alpha = 0.1$. We have 23 observations with $\bar{x} = 185.814$.
   (c) Prior: $\kappa_0 = 7$, $\Sigma_0 = 35.7$, $s_0 = 0.1$. $2\alpha = 0.02$. Statistics: $n = 5$, $\Sigma_x = 25.615$.

3 Sample size:
   (a) Given known $\sigma = 5$, and *prior* hyperparameter $\kappa_0 = 0$, how many observations n do you have to make to ensure $I_{0.01}^{\mu}$ is narrower than 0.5?
   (b) With known $\sigma = 0.42$, and *prior* hyperparameter $\kappa_0 = 8$ how many observations n do you have to make to ensure $I_{0.07}^{\mu}$ is narrower than 0.123?

4 You have tried weighing your dog, knowing well that it is unable to stand still on the scales. You have done this four times, and based on the wobbling of the weight dial, you assume the weighings correspond to $\sigma = 0.4$ kg; the mean weight was $\bar{y} = 17.5$ kg. Using a neutral prior, indicate a 90% credible interval for your dog's weight.

5 Your son is doing athletics, and his performance varies from day to day. He wants to compete, and has asked you to help him by assessing his high jumps. His coach is also a gymnastics and mathematics teacher, and tells

you that jump heights of each athlete follows a Normal distribution $\phi_{(\mu,\sigma)}$, and that in his club, $\sigma = 3$ cm. You measure your son's high jumps 10 times, and the total is 15.7 meters. Using a neutral prior, find 90% interval estimates for the mean jump height, μ, and for your son's next jump, X_+.

Gaussian Processes With Unknown σ

6 From distribution to interval.
  (a) $\mu \sim t_{(68.1,11.9,17)}$. Find $I^\mu_{0.001}$.
  (b) $\mu \sim t_{(68.1,11.9,4)}$. Find $I^\mu_{0.001}$ and $I^\mu_{0.1}$.
  (c) $X_+ \sim t_{(5,1.2,7)}$. Find $I^+_{0.02}$.

7 From data + prior to interval. Find $I^\mu_{2\alpha}$ and $I^+_{2\alpha}$. In addition, find $I^\tau_{2\alpha}$ and $I^\sigma_{2\alpha}$.
  (a) Data: $\{0.896, 0.279, 0.865, 0.955, 0.936, -0.046\}$; use a neutral prior. $2\alpha = 0.05$.
  (b) Prior: $\kappa_0 = 6$, $m_0 = 150$, $\nu_0 = 5$, $s_0 = 10$. $2\alpha = 0.1$. Statistics: $n = 18$, $\Sigma_x = 2\,644.9$, $SS_x = 3\,978.54$.
  (c) Prior: $\kappa_0 = 4$, $\Sigma_0 = 80$, $\nu_0 = 3$, $SS_0 = 27$. $2\alpha = 0.05$. You have 38 observations with average $\bar{x} = 19.823\,7$ and sample standard deviation $s_x = 3.551\,43$.

8 Sample size:
  (a) Using *prior* hyperparameters $\kappa_0 = 7$, $\nu_0 = 6$, and $SS_0 = 17$, how many observations n do you need to make in order for $P(\mu \in H^\mu_{1.3}) \geq 0.9$?
  (b) Using *prior* hyperparameters $\kappa_0 = 5$, $\nu_0 = 4$, and $SS_0 = 50$, how many observations n do you need to make in order for $P(\mu \in H^\mu_2) \geq 0.86$?

9 A standard European football goal is 732 cm wide and 244 cm high, so if the goalkeeper is standing in the middle of the goal, he needs to be able to throw himself far enough to the sides that his hands are 366 cm away from the middle, if he wants to cover the entire goal. We have data from the small Norwegian football club Jerv's junior goalkeeper in 1986, Christian Finne. The average reach for 8 throws is $\bar{x} = 378$ cm, and their sample standard deviation is $s_x = 14$ cm. Find a 96% credible interval for Finne's reach when using a neutral prior.

10 You are out diving with five friends, and stop to admire a school of goldfish. At that point, you are all at the same depth. Your ACME depth gauges respectively 33.1, 28.3, 29.0, 29.7, 33.2, and 30.9 meters. Find a 90% credible interval for your actual depth. Use a neutral prior.

11 You are following the band Pünk Flöyd, and just attended a concert with a rather permeating smell of sweet smoke. The police had conducted random checks of 10 audience members, and had in total impounded 54.4

grams of hashish. The individual weights were as follows: 5.8, 6.0, 1.8, 3.4, 6.8, 4.7, 7.8, 6.1, 6.0, and 6.0 grams. Use a neutral prior, and find a 90% credible interval for the mean possession in the audience. In addition, find a 90% predictive interval estimate for the amount of hashish possessed by a random member of the audience (X_+), and credible intervals for τ and σ.

12 In the Norwegian box office hit film *Il Tempo Gigante*, it was said that on Reodor Felgen's first run of his racing car Il Tempo Gigante, the seismograph in Bergen registered it as an earthquake in Flåklypa of magnitude 7.8 on the Richter scale. For extra observations, the seismograph in Reykjavik registered it as 7.6, and the one in Helsinki had it at 7.9. Use a neutral prior, and find a 95% credible interval for the magnitude of the earthquake in Flåklypa when Felgen drove off in his Il Tempo Gigante.

Poisson Process

13 From distribution to interval.

(a) $\lambda \sim \gamma_{(4,17)}$. Find $I^\lambda_{0.05}$.

(b) $\lambda \sim \gamma_{(7,128)}$. Find $I^\lambda_{0.001}$.

(c) $\lambda \sim \gamma_{(2,8)}$. Find $I^\lambda_{0.1}$.

14 From data + prior to interval. Find $I^\mu_{2\alpha}$ and $I^+_{2\alpha}$.

(a) Prior: $\kappa_0 = 3$, $\tau_0 = 5$. Observed: $n = 13$ occurrences during $t = 20$ units. $2\alpha = 0.02$.

(b) Prior: $\kappa_0 = 4$, $\tau_0 = 8$. Observed: $n = 48$ occurrences during $t = 100$ units. $2\alpha = 0.1$.

(c) Prior: $\kappa_0 = 0$, $\tau_0 = 0$. Observed: $n = 7$ occurrences during $t = 128$ units. $2\alpha = 0.001$.

15 Sample size: Your prior hyperparameter is $\kappa_0 = 5$. How many observations do you need to make for the relative interval width r for a 80% credible interval to be less than 0.2?

16 ($t =$ units) The daily catch for grouse hunters in an area is considered to be Poisson distributed with rate λ. One day, you talked to 23 hunters from a certain part of the Lowlands, and their total catch was 111 grouse. Use a neutral prior, and find an 85% credible interval for λ.

17 ($t =$ time) The number of plumbing gaskets that need changing every week in an apartment complex is assumed to follow a Poisson process with rate λ. You are estimating this need for an apartment complex with 70 flats, and have looked into the documentation for the last semester (26 weeks), and find registered 53 gasket changes. Use a neutral prior, and calculate an 80% credible interval for λ.

18 ($t = $ length) The number of cracks in the tarmac per kilometer of road is assumed to follow a Poisson process with rate λ. Your job is to find this rate for a lesser highway, and you have found 13 cracks in 10 km. Use a neutral prior, and calculate a 92% credible interval for λ.

19 ($t = $ area $= $ number of square meters) How many four-leaf clovers are there per square meter in a field of leaf clovers? Assume that the occurrence follows a Poisson process with rate λ. You look at three independent leaf clover fields. The first field is $t_1 = 1.9\,\mathrm{m}^2$ in area, and has $n_1 = 0$ four-leaf clovers. The second field has an area of $t_2 = 0.7\,\mathrm{m}^2$, and has $n_2 = 3$ four-leaf clovers, whereas the third spot of area $t_3 = 1.2\,\mathrm{m}^2$ has $n_3 = 1$ four-leaf clovers. Use a neutral prior, and find a 90% credible interval for λ.

20 ($t = $ volume $= $ number of cubic centimeters) The number of bacterial colonies per cubic centimeter in a certain polluted lake is assumed to be Poisson distributed with parameter λ. You sample 1 deciliter, and find 157 bacterial colonies. Use a neutral prior, and calculate a 95% credible interval for λ.

Bernoulli Process

21 From distribution to interval.
   (a) $\pi \sim \beta_{(43,96)}$. Find $I_{0.1}^\pi$, both by exact calculation of β, and by using Normal approximation.
   (b) $\pi \sim \beta_{(7,128)}$. Find $I_{0.001}^\pi$, both by exact calculation of β, and by using Normal approximation.
   (c) $\pi \sim \beta_{(2,2)}$. Find $I_{0.05}^\pi$.
   (d) $\pi \sim \beta_{(1,1)}$. Find $I_{0.12}^\pi$ (hint: sketch the graph).

22 From data + prior to interval. Find $I_{2\alpha}^\pi$.
   (a) $2\alpha = 0.07$. Prior: $a_0 = 3$, $b_0 = 5$.
      Observed: $k = 13$ positives and $l = 20$ negatives.
   (b) $2\alpha = 0.1$. Prior: $a_0 = 0$, $b_0 = 0$ (Novick and Hall).
      Observed: $k = 4$ positives and $l = 9$ negatives.
   (c) $2\alpha = 0.05$. Prior: $a_0 = 0.5$, $b_0 = 0.5$ (Jeffreys).
      Observed: $k = 44$ positives and $l = 19$ negatives.
   (d) $2\alpha = 0.02$. Prior: $a_0 = 1$, $b_0 = 1$ (Laplace).
      Observed: $k = 79$ positives and $l = 198$ negatives.

23 Sample size: π has prior hyperparameters $a_0 = 12$ and $b_0 = 31$. How many new observations do you need to make to ensure that $I_{0.2}^\pi$ is narrower than 0.05?

24 You are studying Bard's biased coin. Your *prior* hyperparameters for π, the probability of heads, is $a_0 = 9$ and $b_0 = 12$. Find the 90% symmetric credible interval for π after each update.

(a) You flip 23 heads and 18 tails.

(b) You flip again, and get 458 heads and 366 tails.

(c) You finally flip 5 571 heads and 4 429 tails.

(d) How many times do you need to flip the coin *in total* for the width of $I_{0.1}^{\pi}$ to be less than 0.01?

25 Sondre Glimsdal wants to estimate π, the proportion of *Go*[1] games won by white. His prior is $\pi \sim \beta_{(7,7)}$. Sondre updates his estimate by observing new games. Find the 80% symmetric credible interval for π after each update.

(a) 1st observation: white wins 13 and black wins 7.

(b) 2nd observation: white wins 11 and black wins 9.

(c) How many *new* games does he need to observe to ensure that the width of $I_{0.2}^{\pi}$ is less than 0.05?

(d) 3rd observation: white wins 348 and black wins 255.

---

[1] Also known in Korean as *baduk* and in Chinese as *Wei Qi*.

16

Frequentist Inference⋆

CONTENTS

"This is my chapter," Frederick exclaims proudly. "Here, I will teach you my way of thinking and calculating, so that you Bayesian students may speak and cooperate freely with my frequentist students."

"The first expression I want to teach you," he continues, "is that of being *unbiased*. My most popular techniques for point and interval estimates are built around this concept."

16.1 Unbiasedness and Point Estimates

"So what does it mean to be unbiased?" Frederick asks. "It is to make the sample represent the population. For us frequentists, the concrete sample is a random trial. One out of many possible samples from the population. We want the aggregate of samples to yield the right values – in the long run."

More precisely, we recall from Section 11.2 that the population has parameter θ. From the samples, which are our observations, we may calculate an *estimator* $\hat{\Theta} = h(X_1, \ldots, X_n)$, a stochastic variable which is a function of the X values.

Definition 16.1.1 *Let* $\hat{\Theta}$ *be an estimator for the parameter* θ. *Then,* $\hat{\Theta}$ *is unbiased if* $E[\hat{\Theta}] = \theta$.

The Bayesian Way: Introductory Statistics for Economists and Engineers, First Edition.
Svein Olav Nyberg.

When we have the concrete observations $x_1, \ldots, x_n$, we get a concrete value $\widehat{\theta}$ that we call a *point estimate* of θ. An estimator that is not unbiased is, not surprisingly, called *biased*.

"Note that not all our estimators are unbiased," Frederick continues, "that is, not all the estimators we frequentists make use of. The Bayesian have a great many biased estimators. But back to us: the reason for this is that the property of being unbiased often gets lost when you put the numbers through a function: $E[f(\Theta)] \neq f(E[\Theta])$."

Sam looks at Frederick with a puzzled look, so Frederick explains: "Let me take an example of a function: $f(x) = \sqrt{x}$. The sample variance $Var(X)$ is an unbiased estimator of the population variance. However, its square root, the sample standard deviation $\sigma_X = \sqrt{Var(X)}$, is *not* an unbiased estimator of the population standard deviation. We *could* make an unbiased estimator of the standard deviation. But then the square of *it* would not be an unbiased estimator of the population variance."

Rule 16.1.2   *The three most important unbiased point estimates are*

- $\widehat{\mu}$, *estimate of μ, the mean of a population, is* $\widehat{\mu} = \bar{x}$
- $\widehat{\sigma^2}$, *estimate of σ^2, the variance of a population, is* $\widehat{\sigma^2} = s_x^2$
- $\widehat{p}$, *estimate of p, the proportion of* T *in a population, is* $\widehat{p} = \frac{k}{n}$, *where k is the number of* T *in n observations.*

"So you actually started this course," Frederick smiles at Sam, "by calculating unbiased point estimates!"

16.1.1 $\bar{x}$ is an Unbiased Estimator of μ

For a single observation X, we see that $E[X] = \mu$, the mean of the population. So for a single observation, the observation itself is an unbiased estimator of μ. For larger samples,

$$E[\bar{X}] = E\left[\frac{X_1 + \cdots + X_n}{n}\right] = \frac{\mu + \cdots + \mu}{n} = \mu$$

(according to Rule 8.3.3), so the average $\bar{X}$ is an unbiased estimator of μ as well.

16.1.2 s_x^2 is an Unbiased Estimator of σ^2

The population variance is σ^2. We are going to show that the sample variance $s_x^2 = \frac{1}{n-1} SS_x$ is an unbiased estimator of σ^2. To do this in a simple and tidy manner, we need to make some observations.

Notice that $X_k - \overline{X} = (X_k - \mu) - (\overline{X} - \mu)$, which means that

$$SS_x = \sum_{k=1}^{n} \left((X_k - \mu) - (\overline{X} - \mu) \right)^2$$

$$= \sum_{k=1}^{n} (X_k - \mu)^2 + \sum_{k=1}^{n} (\overline{X} - \mu)^2 + 2 \sum_{k=1}^{n} (\overline{X} - \mu)(X_k - \mu)$$

$$= \sum_{k=1}^{n} (X_k - \mu)^2 - n (\overline{X} - \mu)^2.$$

Show the details of the last equality for yourself. The key to this exercise is to calculate that $\sum_{k=1}^{n} (\overline{X} - \mu)(X_k - \mu) = n(\overline{X} - \mu)^2$. Given this equality, we get $E[SS_x] = (n - 1)\sigma^2$.

This means that $E[s_x^2] = E\left[\dfrac{SS_x}{n-1}\right] = \sigma^2$, and hence the sample variance is unbiased.

"This was the theoretical way. Many of my students, however, prefer to build their understanding through seeing a live example rather than a theoretical deduction," Frederick points out, "so I will now show you the unbiasedness of the sample variance through a simple example. Simple, but with a bit of calculation; you never get away from that."

Example 16.1.3 We have a population consisting of 3 values: $P = \{x_1 = 0, x_2 = 0, x_3 = 6\}$. We will first calculate the (population) mean and variance. After that, we are going to calculate the sample means and variances for all the possible samples of 2 elements from this population. Note that there are a total of 9 possible samples, since we may choose the same element repeatedly. The possible 2-element samples (with replacement), are $u_1 = (x_1, x_1), u_2 = (x_1, x_2), u_3 = (x_1, x_3), u_4 = (x_2, x_1), u_5 = (x_2, x_2), u_6 = (x_2, x_3), u_7 = (x_3, x_1), u_8 = (x_3, x_2), u_9 = (x_3, x_3)$.

<u>Population measures:</u> $\mu = \dfrac{0+0+6}{3} = 2$ and $\sigma_x^2 = \dfrac{0^2+0^2+6^2}{3} - 2^2 = 8$.

<u>Sample measures:</u> We calculate the following for each of the 9 samples.

- The averages for these samples are $\bar{x}_1 = \dfrac{0+0}{2} = 0$, and (calculated in the same way), $\bar{x}_2 = 0, \bar{x}_3 = \dfrac{0+6}{2} = 3, \bar{x}_4 = 0, \bar{x}_5 = 0, \bar{x}_6 = 3, \bar{x}_7 = 3, \bar{x}_8 = 3, \bar{x}_9 = 6$.
- The *population* variances for these samples are $\sigma_1^2 = \dfrac{0^2+0^2}{2} - 0^2 = 0$, and (calculated in the same way), $\sigma_2^2 = 0, \sigma_3^2 = \dfrac{0^2+6^2}{2} - 3^2 = 9, \sigma_4^2 = 0, \sigma_5^2 = 0, \sigma_6^2 = 9, \sigma_7^2 = 9, \sigma_8^2 = 9, \sigma_9^2 = 0$.
- The *sample* variances for these samples are $s_1^2 = 0, s_2^2 = 0, s_3^2 = 18, s_4^2 = 0, s_5^2 = 0, s_6^2 = 18, s_7^2 = 18, s_8^2 = 18, s_9^2 = 0$.

The average of the sample averages is $\frac{0+0+3+0+0+3+3+3+6}{9} = 2$, which equals the population mean, μ. So we see here that the sample average $\hat{\mu} = \bar{x}_i$ is an unbiased estimator of μ.

The average of the population variances of these samples is $\frac{0+0+9+0+0+9+9+9+0}{9} = 4$, which is smaller than the population variance of P, since $\sigma_x^2 = 8$. So we see here that taking the population variance of the sample data is *not* an unbiased estimator of the population variance of the original population P.

The average of the sample variances is $\frac{0+0+18+0+0+18+18+18+0}{9} = 8$, which equals the population variance σ_x^2 for the population P. So we see here that the sample variance of the samples is an unbiased estimator of the population variance of the original population. □

16.1.3 MLE★

"Even though unbiasedness gives the most important point estimates," Frederick says, "we have others. The second most important estimate is the *maximal likelihood estimate*, or *MLE* for short."

"The MLE is quite similar to our MAP," Bard interjects, "the *maximum a posteriori estimate*. The difference between the MLE and the MAP is that the MAP includes a prior."

Frederick continues: "The Bayesians find their MAP from a *posterior* distribution, whereas we frequentists find the MLE from what you have come to know as the likelihood function. In both cases, the estimate is simply the argument value that gives the maximum value, which again puts these estimates in the same category as the *mode*."

"MAP and MLE are often good estimates," Bard says, "and they are a bit more robust under functions like the square root. But for very skewed distributions, or for *multimodal* distributions, that is, distributions with multiple maxima, these estimates may turn out very different from the other point estimates, as we see in Figure 16.1."

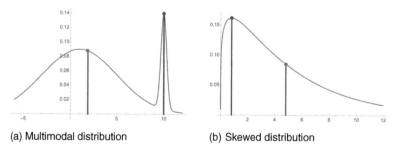

(a) Multimodal distribution   (b) Skewed distribution

Figure 16.1 Mode in red, mean in green.

16.2 Interval Estimates

"Now it's time for my interval estimates." Frederick smiles, "We frequentists call our interval estimates *confidence intervals*. They often coincide with the Bayesians' credible intervals. But despite this, the intervals don't mean the same."

- A given $(1 - \alpha) \times 100\%$ credible interval I_α^θ for a parameter θ is an interval of Bayesian probability $P(\theta \in I_\alpha^\theta) = 1 - \alpha$.
- It is, on the other hand, meaningless to speak of any "frequentist probability" that a given *confidence interval* $\widehat{I_\alpha^\theta}$ contains the parameter θ. Probabilities of this kind follow Example 5.1.6, and are either 0 or 100%, but we do not know which.
- However, what we *can* speak of is the frequentist probability that *the method will result in an interval that contains* θ: using the method, each new data set will result in a new interval $\widehat{I_\alpha^\theta}$. Even if we are not allowed to say anything about each individual interval, we may still speak of *the proportion of intervals that contain* θ, and firmly say that this proportion will approximate $1 - \alpha$ as the number of such data sets and confidence intervals approaches infinity. We have illustrated this process of creating 80% confidence intervals in Figure 16.2.

"We are going to stick with two-sided interval estimates in this lecture," Frederick explains, "and to one parameter only, for simplicity's sake."

16.2.1 Interval Estimates for Gaussian Processes

"For us, the variance σ^2 is either fully known, or fully unknown," Frederick begins, "with no middle ground of a *partly known* σ. This goes for all population parameters."

"We are going to look at some of the most common interval estimates," Frederick continues, "and the first thing you Bayesians will notice, is that we do not ascribe any probability distribution to the parameters. When we have n observations, we find the sufficient statistics n, Σ_x, and SS_x, and calculate $v = n - 1, \bar{x} = \Sigma_x/n$, and $s_x^2 = SS_x/(n - 1)$. For Gaussian processes, our $(1 - \alpha)100\%$ confidence intervals are given in Table 16.1."

"Even though this does not depend on the school, frequentist or Bayesian, that is," Frederick says, "it is more common in my camp to find estimates for σ^2, whereas the Bayesians tend to prefer estimating τ. It really doesn't matter, since $\tau = 1/\sigma^2$ so that you can find the one from the other, but it's useful to know.

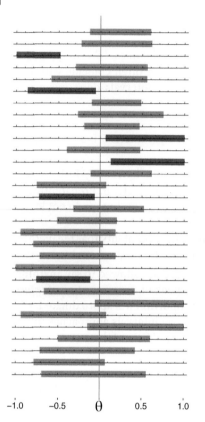

Figure 16.2 80% confidence interval for μ, constructed from data sampled from a Normal distribution $\phi_{(0,1)}$. The intervals capture the true value ($\mu = 0$) roughly 80% of the time.

"But in my chapter, you will get the rules that work best for you," he continues. "In my school, we like to express the confidence interval $\widehat{I_\alpha^{\sigma^2}}$ thus:

$$\frac{SS_x}{\chi_{\nu,1-\alpha/2}^2} < \sigma^2 < \frac{SS_x}{\chi_{\nu,\alpha/2}^2}.$$

Table 16.1 Gaussian confidence intervals

	$\sigma = \sigma_0$ (known)	σ unknown
μ	$\widehat{I_\alpha^\mu} = \bar{x} \pm z_{\alpha/2} \cdot \sigma_0 \cdot \sqrt{\frac{1}{n}}$	$\widehat{I_\alpha^\mu} = \bar{x} \pm t_{\nu,\alpha/2} \cdot s_x \cdot \sqrt{\frac{1}{n}}$
τ	–	$\widehat{I_\alpha^\tau} = \left(\Gamma^{-1}_{(\nu/2,SS_x/2)}(\alpha/2), \Gamma^{-1}_{(\nu/2,SS_x/2)}(1-\alpha/2) \right)$
X_+	$\widehat{I_\alpha^+} = \bar{x} \pm z_{\alpha/2} \cdot \sigma_0 \cdot \sqrt{1+\frac{1}{n}}$	$\widehat{I_\alpha^+} = \bar{x} \pm t_{\nu,\alpha/2} \cdot s_x \cdot \sqrt{1+\frac{1}{n}}$

Note 16.2.1 *Here, χ^2 is a shorthand overload of the χ^2 symbol: $\chi^2_{v,\alpha}$ denotes the inverse cumulative χ^2 distribution with argument α, with v degrees of freedom. The commands for $\chi^2_{v,\alpha}$ are*

- Mathematica: InverseCDF[ChiSquareDistribution[v], α]
- Casio: InvChiCD(α, v)
- TI CX: Invχ^2(α, v)
- TI 83+: Solver: χ^2cdf$(0, x, v) - \alpha = 0 \gg x = 1$
- HP: chisquare_icdf(v, α).

Example 16.2.2 In the time after Example 15.4.1, engineering student Rolf has hacked Eggeseth's smallest laser speed gauge so that it now uses 20 pulses for its measurements, with each individual measurement following a Normal distribution with $\sigma = 5$. Rolf has also made the gauge calculate $P\%$ interval estimates automatically, and he has put a switch on the side of the gauge, to switch between *Bayesian* (with neutral prior) and *frequentist* calculation. Eggeseth sets the switch to *frequentist* right before his favorite speeder Klaus Königsegg whizzes past. Rolf reads the display: "137 km/h", and also gives the 98% (symmetric) confidence interval for Klaus Königsegg's actual speed. What is this interval?

Answer: We use the formula for a 98% confidence interval for μ with known σ. Using $\alpha = 1 - 0.98 = 0.02$, we get the interval

$$137 \pm z_{0.01} \times \frac{5}{\sqrt{20}} \approx \underline{\underline{(134.4, 139.6)}}.$$

Additional question: *Does Rolf's new switch really add anything?*[1] □

Example 16.2.3 Mulligan & Daughters Ltd is a stable company with a 200 year history. Adjusted for inflation, the earnings have been stable and at the same level for the last 50 years, and the average income for the last 30 years in current value, is $\bar{x} = £37.1$ million, while the sample standard deviation is $s_x = £2.73$ million. Find (with all numbers in current value)

1) the 90% confidence interval for the mean income;
2) the 90% prediktivt intervall for next year's income;
3) the 90% confidence interval for the variance in the income.

---

[1] **Answer:** No. For Gaussian processes, the frequentist confidence intervals are identical to the corresponding Bayesian credible intervals (with neutral prior) when σ is fully known or fully unknown.

Answer: $n = 30$, so $v = 29$, which means that $SS_x = 29 \times 2.73^2 = 216.134$.

1) $\widehat{I_{0.1}^\mu} = 37.1 \pm t_{29, 0.05} \times 2.73 \times \sqrt{1/30} \approx \underline{(36.3,\ 37.9)}$;

2) $\widehat{I_{0.1}^+} = 37.1 \pm t_{29, 0.05} \times 2.73 \times \sqrt{1 + 1/30} \approx \underline{(32.4,\ 41.8)}$;

3) we calculate the confidence interval for σ^2 in both ways:

- $\widehat{I_{0.1}^\tau} = \left(\Gamma^{-1}_{(14.5,\ 108.067)}(0.05),\ \Gamma^{-1}_{(14.5,\ 108.067)}(0.95) \right)$

   $= \underline{(0.081\,932\,3,\ 0.196\,901)}$,

   which gives

   $\widehat{I_{0.1}^{\sigma^2}} = \left(\dfrac{1}{0.196\,901},\ \dfrac{1}{0.081\,932\,3} \right) = \underline{(5.078\,7,\ 12.205\,2)}$

- $\widehat{I_{0.9}^{\sigma^2}} = \left(\dfrac{216.134}{\chi^2_{29,\,0.95}},\ \dfrac{216.134}{\chi^2_{29,\,0.05}} \right)$

   $= \left(\dfrac{216.134}{42.557},\ \dfrac{216.134}{17.708\,4} \right) = \underline{(5.078\,7,\ 12.205,\ 2)}$.

Additional question: *How does the frequentist predictive interval differ from the Bayesian predictive interval with neutral priors?*[2]   ☐

16.2.2 Confidence Intervals for Bernoulli Processes

"We have different techniques for calculating confidence intervals for the parameter π for a Bernoulli process," Frederick explains, "some slow but exact, and others fast but a bit imprecise. Statistics is after all the art of being imprecise in a precise way.

"The most common method is the fast one," he continues, "and it is the preferred one when we have more than 30 observations. After n observations, you have k positives, and let $\hat{\pi} = \dfrac{k}{n}$. A $(1 - \alpha)100\%$ confidence interval for the true proportion (the parameter π for a Bernoulli process) is according to this method given by

$$\hat{\pi} \pm z_{\alpha/2} \sqrt{\dfrac{\hat{\pi}(1 - \hat{\pi})}{n}}.$$

---

[2] **Answer:** Numerically, not at all. For Gaussian processes, the frequentist predictive intervals are identical to the corresponding Bayesian intervals (with neutral priors) when σ is fully known or wholly unknown.

Example 16.2.4 The Motherwell striker Louis Moult scored 15 goals in the 2015/16 season of the Scottish Premier League, whereof 10 with his right foot. Find a 90% confidence interval for π, the proportion of goals Moult scores with his right foot in the long run.

Answer: $\hat{\pi} = \frac{10}{15} = \frac{2}{3}$ and $\frac{\alpha}{2} = (1 - 0.9)/2 = 0.05$, so

$$\widehat{I_{0.1}^{\pi}} = \frac{2}{3} \pm z_{0.05} \sqrt{\frac{\frac{2}{3}(1 - \frac{2}{3})}{15}} = \underline{(0.466, \ 0.867)}.$$

$\square$

16.3 Hypothesis Testing

Frederick can't quite hide his pride when he reveals to Sam that the technique of hypothesis testing originated with the frequentist school, "but it is unfortunately also the most misinterpreted one."

Frederick continues: "So let us first see what classical, frequentist, hypothesis testing is about, to make sure you understand hypothesis testing right."

Sam nods: "Sounds great to me; I like understanding.

"I have understood what Bayesian hypothesis testing is," he continues. "I find the *posterior* probability distribution for θ, and then I check the probability that θ is smaller than some reference value θ_0. If that probability is less than a critical value α, I conclude that $\theta > \theta_0$."

Frederick approves: "You understand that well. Our school does not speak of any probability distributions for parameters like θ, but instead regards the conditional probability distributions of the data *given that* $\theta = \theta_0$. For us, the conservative hypothesis H_0 is that $\theta = \theta_0$, rather than that θ is in some given interval in the way of the Bayesians. The alternative hypothesis H_1 comes in three types: (i) $\theta > \theta_0$, (ii) $\theta < \theta_0$, (iii) $\theta \neq \theta_0$."

"OK," Sam says, "I see that one. Any other differences?"

"Yes," Frederick replies, "this is all about making decisions. In our school, we weigh this decision between the possibilities of two types of error. We look at the probability of committing the worse of these two kinds. Just as with confidence intervals, it is all about probability in the aggregate, that is, how large a share of our trials will end in this type of error in the long run, as opposed to the Bayesian probability of making an error in a single trial. The two types of error are type I and type II errors, as shown in Table 16.2.

"As long as you don't actually *know*," Frederick explains, "there is the possibility of error. If you reject H_0, you may make an error. If you *don't reject H_0*, you may also make an error. But one of these errors is in general considered more serious: rejecting H_0 when it is true. This is why the strictness of the hypothesis test is based on how likely it is that we are making this type of error, measured against a significance level α."

Table 16.2 Error types

	H_0 is true	H_0 is false
Not reject H_0	Correct decision	**Type II error**
Reject H_0	**Type I error**	Correct decision

"So only when the probability of H_0 is less than α, you may reject H_0," Sam intones.

"No, the probability of a parameter value is a Bayesian idea," Frederick explains. "We of the frequentist school don't judge by any probability of the parameter value, but rather by the probability of the data given the parameter. Let me explain ... "

"Oh, the other way around," Sam nods. "Please do!"

"Our method starts from the default model, the null hypothesis H_0," Frederick begins, "We want to devise our test so that, if H_0 is true, the probability that our data will make us reject is less than some significance value α. This is tantamount to finding a *rejection region* of data values that are in some way too extreme for the model to be seen as a good explanation. The rejection region will be the most extreme results, and the probability that the data end up in that region will be less than α."

"So in other words," Sam interjects, "H_0 is rejected when the data are too extreme?"

"Yes," Frederick confirms, "and we may determine this in one of two ways: either we establish the region in advance, by finding a first *too extreme* value. Then we reject H_0 if our data are equally or more extreme. The other way is by waiting for the actual data, and finding the probability p of getting data as extreme as our actual data, or worse, given H_0."

"Hmm. I see that one," Sam says, "for then what you call *data as extreme as ours, or worse* will either be a subset of the rejection region, and then the probability p will be at most α, or it will not, and then p will be bigger. But in that case, our data are not in the rejection region."

"I am impressed by your learning, Sam," Fredrick smiles.

"Thank you," Sam smiles in return, "then I understand about the type I errors. But given again that H_0 is true," Sam queries, "what is the acceptable probability of a *type II* error? Is that α as well, or is it $1 - \alpha$?"

"*That* value is totally dependent on the exact value of θ," Frederick replies, "which we don't know. But it does in any case not matter for our decision process. We focus on the type I error only, since that is the serious error."

"Then I think I understand," Sam responds thoughtfully, "but then it's time to learn the ropes – performing the techniques – isn't it? *Knowing is doing*, as they say."

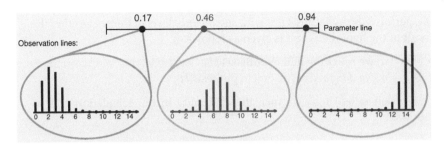

Figure 16.3 Parameter and observation line for a Bernoulli process.

"I like the way you are thinking," Frederick applauds, "so let us look at the techniques. To understand the technique, it helps to draw two parallel lines: one for the possible values of the parameter θ, and one for the values of the observations. I will illustrate this in the hypothesis test of the parameters p and μ. Then, I will do a hypothesis test of σ as well, and you will be equipped with a good frequentist starter kit."

16.3.1 Hypothesis Testing of Proportions

When we look at the parameter π for a Bernoulli process, the *parameter line* is the range of possible values π, the probability of success, may take. In the case of π, that means the interval $[0, 1]$. The other line, the *observation line*, consists of the integers from 0 through n, where n is the number of trials. We illustrate this in Figure 16.3.

"The exact calculation for this hypothesis test of p is as follows," Frederick says. "We will look at the quicker method via approximation afterwards. But the exact method first, for the sake of understanding."

Sam agrees.

Method 16.3.1 In a (right) one-sided frequentist hypothesis test of a proportion π, the *alternative hypothesis* H_1 is that $\pi > \pi_0$, whereas the conservative hypothesis H_0 is that $\pi = \pi_0$. The observations are from n Bernoulli trials, and the number of successes are a stochastic variable X following a binomial distribution with parameters n and the unknown π. From an observation of k successes, we find the *p-value*

$$p = P(X \geq k | \pi = \pi_0) = \sum_{m=k}^{n} \binom{n}{m} \pi_0^m (1 - \pi_0)^{n-m}.$$

We perform the hypothesis test with *significance* level α; this means that the result of the hypothesis test is given as follows.

- If $p < \alpha$, we *reject* the null hypothesis H_0.
- If $p \geq \alpha$, we *don't reject* the null hypothesis H_0.

In a (left-tailed) test, we turn the inequality in the calculation of p as follows:

$$p = P(X \leq k | \pi = \pi_0) = \sum_{m=0}^{k} \binom{n}{m} \pi_0^m (1 - \pi_0)^{n-m}.$$

In a two-sided test, we reject H_0 with significance α iff (iff = "if and only if") we reject H_0 in one of the two corresponding one-sided tests with significance $\frac{\alpha}{2}$.

Example 16.3.2 Sam arranges a Christmas dance party for a large company, and thinks it would be good if there were roughly as many men as women. He decides to investigate, and does this with significance $\alpha = 0.1$, as nothing stricter is required for a Christmas party. Let π be the proportion of men. His alternative hypothesis is then that $\pi \neq \frac{1}{2}$. The conservative hypothesis is that $\pi = \pi_0 = \frac{1}{2}$. So far, $n = 14$ people have signed up for Sam's party, whereof $k = 8$ men. We look at both tails:

$$P(8 \text{ or more men}) = \sum_{m=8}^{14} \binom{14}{m} 0.5^m (1 - 0.5)^{14-m} = 0.3953 > \alpha/2$$

$$P(8 \text{ or fewer men}) = \sum_{m=0}^{8} \binom{14}{m} 0.5^m (1 - 0.5)^{14-m} = 0.7880 > \alpha/2 \qquad \square$$

... and none of the numbers give any reason to reject H_0, which means we *don't reject* the null hypothesis H_0, which says that there is an equal chance that the next guest joining will be a man as that it will be a woman.

"The quicker, approximate, calculation uses the Normal approximation (10.1.13)," Frederick explains, "and since most textbooks drop the continuity correction $\pm\frac{1}{2}$, so will we. But note that if we had calculated Example 16.3.2, the correct Normal approximation would have needed to include the continuity correction $\pm\frac{1}{2}$, giving us the respective numbers 0.394 6 and 0.788 7, which are very close to the exact answer. Note that *without* the continuity correction, we get 0.2965 and 0.703 5, which is quite a lot more off."

"But recall again that statistics is the art of being imprecise in a precise manner," Frederick smiles, "and as a rule of thumb, the quickest method is *good enough* when we have in excess of 30 observations. This goes even without the continuity correction, which after all takes some time to apply, and thus the quick and *good enough* method is to use the uncorrected Normal approximation when you have more than 30 observations."

Method 16.3.3 After n observations with k successes, let $\hat{\pi} = k/n$, and let

$$w = \frac{\hat{\pi} - \pi_0}{\sqrt{\frac{\pi_0(1-\pi_0)}{n}}}.$$

Then z decides the hypothesis test thus:

Alternative hypothesis H_1	Reject H_0 in favor of H_1 if	
	Direct test (alt. 1)	Indirect test (alt. 2)
$\pi < \pi_0$	$\Phi(w) < \alpha$	$w < z_\alpha$
$\pi > \pi_0$	$\Phi(-w) < \alpha$	$-w < z_\alpha$
$\pi \neq \pi_0$	$\Phi(-|w|) < \frac{\alpha}{2}$	$-|w| < z_{\alpha/2}$

Notice that most textbooks use $z_\alpha = -\Phi^{-1}(\alpha)$, which gives positive values for $\alpha < 0.5$, while we in this book use the inverse directly with $z_\alpha = \Phi^{-1}(\alpha)$, which gives negative values for $\alpha < 0.5$.

Example 16.3.4 The Aberdeen striker Adam Rooney scored 38 goals in the last two seasons, whereof 29 with his right foot. Decide with significance $\alpha = 0.1$ whether Rooney scores the majority of his goals with the right foot.

Answer: We are going to look at π, the proportion of goals he scores with his right foot. The null hypothesis is that $\pi = \pi_0 = 0.5$, whereas H_1 is that $\pi > \pi_0$. Here, $\hat{\pi} = \frac{29}{38}$, so

$$w = \frac{\frac{29}{38} - 0.5}{\sqrt{\frac{0.5(1-0.5)}{38}}} \approx 3.244\,43.$$

We calculate this as follows both directly and indirectly, to show both methods.

- Directly: $\Phi(-w) = \Phi(-3.244\,43) = 0.000\,59 < 0.1$, so we *reject* the null hypothesis in favor of the alternative hypothesis that he scores the majority of his goals with his right foot.
- Indirectly: $-w = -3.244\,43 < -1.281\,55 = z_{0.1}$, so we *reject* the null hypothesis in favor of the alternative hypothesis that he scores the majority of his goals with his right foot. $\square$

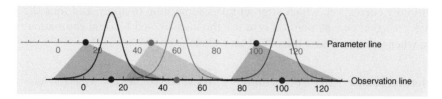

Figure 16.4 Parameter and observation line for a Gaussian process.

16.3.2 Hypothesis Testing for the Mean Value of Gaussian Processes

"When we looked at hypothesis testing of proportions," Frederick says, "we saw that the parameter line was the interval $[0, 1]$, whereas the observation line was the integers $0, \ldots, n$. The two could not be confused. But when we are working with the Gaussian case, which means the Normal or t distribution, both of these lines are the entire set of real numbers, $\mathbb{R}$, and the conditional distribution, given that $\mu = \mu_0$, is a Normal distribution centered at μ_0. We have illustrated this in Figure 16.4.

"In a (right) one-sided hypothesis test for the mean μ in a Normal distributed population with variance σ^2," Frederick explains, "an *alternative hypothesis* H_1 is that $\mu > \mu_0$, whereas the conservative hypothesis H_0 is that $\mu = \mu_0$. After n observations, we calculate the p-value, the probability p that the average of the measurements is at least as large as $\bar{x}$, and reject the null hypothesis if $p < \alpha$.

"The common way to calculate this," Frederick informs us, "is by defining

$$
w = \begin{cases} \dfrac{\bar{x} - \mu_0}{\sigma/\sqrt{n}} & \text{if } \sigma \text{ is known} \\[2ex] \dfrac{\bar{x} - \mu_0}{s_x/\sqrt{n}} & \text{if } \sigma \text{ is unknown} \end{cases}
$$

and then deciding the hypothesis in the same way as for the parameter π for Bernoulli processes.

"Recall again that many textbooks use $z_\alpha = -\Phi^{-1}(\alpha)$ and $t_{\nu,\alpha} = -T_\nu^{-1}(\alpha)$," Frederick points out, "which gives positive values for $\alpha < 0.5$, but that we in this book use $z_\alpha = \Phi^{-1}(\alpha)$ and $t_{\nu,\alpha} = T_\nu^{-1}(\alpha)$, which gives negative values for $\alpha < 0.5$."

Method 16.3.5 For known σ, we use w to decide the hypothesis as follows:

Alternative hypothesis H_1	Reject H_0 in favor of H_1 if	
	Direct test (alt. 1)	Indirect test (alt. 2)
$\mu < \mu_0$	$\Phi(w) < \alpha$	$w < z_\alpha$
$\mu > \mu_0$	$\Phi(-w) < \alpha$	$-w < z_\alpha$
$\mu \neq \mu_0$	$\Phi(-|w|) < \dfrac{\alpha}{2}$	$-|w| < z_{\alpha/2}$

Method 16.3.6 For unknown σ, we use w to decide the hypothesis as follows, with $v = n - 1$:

Alternative hypothesis H_1	Reject H_0 in favor of H_1 if	
	Direct test (alt. 1)	Indirect test (alt. 2)
$\mu < \mu_0$	$T_v(w) < \alpha$	$w < t_{v,\alpha}$
$\mu > \mu_0$	$T_v(-w) < \alpha$	$-w < t_{v,\alpha}$
$\mu \neq \mu_0$	$T_v(-|w|) < \frac{\alpha}{2}$	$-|w| < t_{v,\alpha/2}$

Example 16.3.7 Cruising enthusiast Jamal claims that the dB pressure in Newark's cruising cars is higher than in East Orange's. The cruisers of East Orange have measured dB tested all their cruiser cars, and the mean dB pressure was $\bar{x} = 112.3$ dB. Jamal won't get any support for a comprehensive test unless he can prove with reasonable certainty that he will get a better result. He therefore decides to go for a hypothesis test of his claim. He chooses significance level $\alpha = 0.1$.

He measures the dB pressure in 16 cars, and gets $\bar{x} = 113.4$ dB and $s_x = 6.4$ dB. This means

$$w = \frac{113.4 - 112.3}{6.4/\sqrt{16}} = 0.687\,5.$$

The reference value is the East Orange result of $\mu_0 = 112.3$ dB, and the alternative hypothesis is that μ, the mean dB pressure in Newark, is greater than μ_0, in other words we have $H_1: \mu > \mu_0$. Direct calculation gives $T_{15}(-0.687\,5) = 0.251\,13 > \alpha$. Indirect calculation gives $t_{v,\alpha} = t_{15,0.1} = -1.340\,61 < -0.687\,5$, so $-w < t_{v,\alpha}$ doesn't hold. Jamal sees from both the direct and the indirect calculation that he *can't reject* the null hypothesis with significance $\alpha = 0.1$. This means he can't expect support for a comprehensive test of the dB pressure in Newark's cruiser cars. □

16.3.3 Hypothesis Testing for the Variance of Gaussian Processes

"Consider a Normal distributed population where both μ and σ are unknown," Frederick begins. "When the variance σ^2 is unknown, we are often interested in knowing whether it is above or below some critical 'null' value σ_0^2. To test the variance against the null value after n observations, we calculate SS_x (or $s_x^2 = \frac{SS_x}{n-1}$).

"You determine the hypothesis either directly or indirectly," Frederick says, "and the most common method for us frequentists is the indirect one, with χ^2, as shown in Note 16.2.1, whereas you Bayesians are more used to using Γ and the direct method. So let us here, too, use Γ for the direct method, and χ^2 for the indirect method."

Method 16.3.8 **Hypothesis test for variance:** Use the statistics n and SS_x, and let $v = n - 1$. With reference value σ_0^2 for the null hypothesis $H_0 : \sigma^2 = \sigma_0^2$, and $\tau_0 = 1/\sigma_0^2$, the hypothesis tests are determined as follows:

Alternative hypothesis H_1	Reject H_0 in favor of H_1 if	
	Direct test (alt. 1)	Indirect test (alt. 2)
$\sigma^2 < \sigma_0^2$	$\Gamma_{(v/2, SS_x/2)}(\tau_0) < \alpha$	$\dfrac{SS_x}{\sigma_0^2} < \chi_{v,\alpha}^2$
$\sigma^2 > \sigma_0^2$	$1 - \Gamma_{(v/2, SS_x/2)}(\tau_0) < \alpha$	$\dfrac{SS_x}{\sigma_0^2} > \chi_{v,1-\alpha}^2$
$\sigma^2 \neq \sigma_0^2$	$\Gamma_{(v/2, SS_x/2)}(\tau_0) < \dfrac{\alpha}{2}$ or $1 - \Gamma_{(v/2, SS_x/2)}(\tau_0) < \dfrac{\alpha}{2}$	$\dfrac{SS_x}{\sigma_0^2} < \chi_{v,\alpha/2}^2$ or $\dfrac{SS_x}{\sigma_0^2} > \chi_{v,1-\alpha/2}^2$

Example 16.3.9 A new EU directive is being proposed, concerning the amount of soup in canned soups. The directive is currently being tried out in Ireland, and Wyvern Broth has an inspector visiting to inspect their canned Mulligatawny soups. The directive puts an upper limit on the variance of the contents by insisting that the standard deviation be at most five grams. Does Wyvern Broth's Mulligatawny soup production satisfy the requirements? The alternative hypothesis is that it does. H_1 is then that $\sigma < \sigma_0 = 5$, which we may reformulate as $\sigma^2 < 25$ or $\tau > \tau_0 = 0.04$.

The inspector weighs $n = 10$ cans, and gets $\{x_i\} = \{398.0, 395.0, 399.0, 400.5, 398.5, 398.0, 401.5, 401.0, 402.0, 401.0\}$. This gives the other statistic, $SS_x = 41.725$.

Direct calculation gives $\Gamma_{(9/2, 41.725/2)}\left(\dfrac{1}{25}\right) = \Gamma_{(4.5, 20.8625)}(0.04) = 0.00431187$, which is less than $\alpha = 0.05$, so we reject the null hypothesis.

If we instead choose the indirect test, $\chi_{9,0.05}^2 = 3.32511$, whereas $\dfrac{SS_x}{\sigma_0^2} = \dfrac{41.725}{25} = 1.669$. Since $\dfrac{SS_x}{\sigma_0^2} = 1.669 < 3.32511 = \chi_{v,\alpha}^2$, the indirect method also tells us to reject the null hypothesis.

The inspector therefore gives Wyvern Broth a pass for the variance of the cans. □

16.4 Exercises

Interval Estimates

1 Find the $P\% = (1 - \alpha)100\%$ interval estimates $\widehat{I_\alpha^\mu}$ and $\widehat{I_\alpha^+}$; σ is known.
 (a) $\sigma_0 = 2$. Data: $\{5.9, 5.8, 4.8, 4.7, 1.6, 2.8, 2.6, 5.8, 5.1, 4.1\}$. $P\% = 90\%$.

(b) $\sigma_0 = 12.1$. $\alpha = 0.05$. We have 47 observations with $\bar{x} = 66.008\,5$.
(c) $\sigma_0 = 3.73$. $P\% = 99.9\%$. Statistics: $n = 100\,00$, $S_x = 274\,046$.

2 Find the $P\% = (1 - \alpha)100\%$ interval estimates $\widehat{I_\alpha^\mu}$, $\widehat{I_\alpha^\sigma}$ (use that $\sigma = 1/\sqrt{\tau}$)
   and $\widehat{I_\alpha^+}$; σ is unknown.
   (a) $\alpha = 0.1$. We have 29 observations with average $\bar{x} = 8.206\,9$ and sample
   standard deviation $s_x = 1.688\,18$.
   (b) $P\% = 98\%$. We have 398 observations with average $\bar{x} = 29.688\,6$ and
   sample standard deviation $s_x = 4.724\,27$.
   (c) You have examined mechanical wear on a certain flooring, and want
   to find a 98% confidence interval for the mean wear. You examined 8
   floors, and measured wear depths of respectively 49.3, 65.8, 55.5, 54.4,
   58.7, 61.7, 63.2, 57.8 μm.

3 Find the $(1 - \alpha)100\%$ confidence interval $\widehat{I_\alpha^p}$.
   (a) $k = 17$ positive and $l = 25$ negative. $\alpha = 0.05$.
   (b) You have heard that Coca and Pepsi have an equal share in the Cola
   market at your university, and decide to investigate if this is true. Your
   investigations find 46 Pepsi drinkers and 54 Coca drinkers. Give a 90%
   confidence interval for π, the proportion of Coca drinkers in your entire
   university.

Hypothesis Testing

4 Determine the hypothesis test outcome about the parameter π for a
   Bernoulli process; significance α.
   (a) $H_1: \pi > 0.5$. $\alpha = 0.05$. Observations: $k = 8$ positive and $l = 6$ negative.
   (b) $H_1: \pi < 0.25$. $\alpha = 0.1$. Observations: $k = 2$ positive and $l = 18$ negative.
   (c) $H_1: \pi < 0.1$. $\alpha = 0.05$. Observations: $k = 4$ positive and $l = 96$ negative.
   (d) $H_1: \pi \neq 0.5$. $\alpha = 0.1$. Observations: $k = 40$ positive and $l = 60$ negative.
   (e) You have heard that Coca and Pepsi have an equal share in the Cola
   market at your university, and decide to investigate if this is true. Your
   investigations find 46 Pepsi drinkers and 54 Coca drinkers. Does this
   suffice if you want to say their markets shares are unequal, with significance $\alpha = 0.1$?

5 Determine the hypothesis test outcome about the mean μ for a Gaussian
   process; significance α.
   (a) $H_1: \mu \neq 25$. $\alpha = 0.05$. Statistics: $n = 27$, $S_x = 715.333$. $\sigma_0 = 3.73$
   (known).
   (b) $H_1: \mu > 80$. $\alpha = 0.01$. We have 200 observations with $\bar{x} = 80.674\,1$.
   $\sigma_0 = 5.1$ (known).
   (c) $H_1: \mu \neq 25$. $\alpha = 0.02$. Data: $\{28.1, 42.1, 22.7, 38.8, 28.8, 37.0, 20.4, 37.3\}$.
   σ unknown.

(d) $H_1: \mu < 100$. $\alpha = 0.005$. Statistics: $n = 500$, $S_x = 49\,650$, $SS_x = 53\,803.8$. σ unknown.

6 It is claimed that the mean compression strength for a certain kind of steel beam exceeds 60 000 psi, and you have decided to determine the test outcome of this alternative hypothesis with $\alpha = 0.1$. Your observations are {60 060, 59 580, 60 498, 60 071, 60 593, 60 384, 60 013, 60 491, 60 321, 60 626, 59 897, 61 002, 60 149, 61 058, 60 901}.

7 Determine the hypothesis test outcome concerning the variance σ^2 of a Gaussian process; significance α.
(a) $H_1: \sigma^2 < 100$. $\alpha = 0.05$. Data: {92.0, 86.3, 93.5, 111.4, 69.8, 106.0, 97.3, 78.7, 102.1, 106.6, 72.6, 107.4, 80.8, 87.8, 97.7}.
(b) $H_1: \sigma^2 > 25$. $\alpha = 0.02$. We have 100 observations with sample standard deviation $s_x = 6.412\,36$.

8 We look at the steel beams in (Exercise 6) again. This time, we are testing the variance, and the alternative hypothesis is $H_1: \sigma \neq 666$. Determine the hypothesis test outcome with $\alpha = 0.1$.

17

Linear Regression

CONTENTS

In Section 3.3, we introduced the *linear regression line*, which was the straight line with the smallest total square distance to the observations. We are now going to study the regression line by means of statistical inference.

The basic assumption is simply that there is a relation between the two magnitudes x and y that ideally fits a straight line

$$y(x) = a + bx.$$

This is thought to describe a real (but not directly observable) relation between the variables. The difference between the value given by the line, and the actual observation, may have many reasons: maybe there are other influencing factors, or maybe there is an uncertainty in the measurement process – or maybe it's just some noise or uncertainty inherent in the value itself. But regardless of the underlying reason for the deviations, we may treat them as noises following a common Normal distribution, $\varepsilon \sim \phi_{(0,\sigma)}$. This means that we see the observed value as the value given by the linear relation, plus the noise. We have illustrated this in Figure 17.1, and in the equation

$$y(x) = a + bx + \varepsilon.$$

Linear regression is in other words described by 3 parameters: height a, slope b, and precision $\tau = 1/\sigma^2$. Given these three parameters, the measured value Y, given x, follows a Normal distribution:

$$Y(x) \sim \phi_{(a+bx,\sigma)}.$$

The Bayesian Way: Introductory Statistics for Economists and Engineers, First Edition.
Svein Olav Nyberg.
© 2019 John Wiley & Sons, Inc. Published 2019 by John Wiley & Sons, Inc.

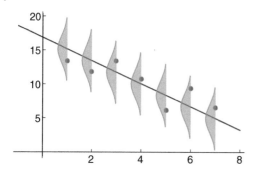

Figure 17.1 Linear relation, plus noise.

17.1 Linear Regression With Hyperparameters

When we perform inference for the regression line, we go the opposite way, as illustrated in Figure 17.2: instead of looking at how y sample values statistically arise from the parameters a, b of the line, and from the noise parameter τ, we start with the observation pairs $\{(x_i, y_i)\}$ – and any prior knowledge we might have – and from that we estimate the precision parameter τ and the line parameters a and b. This gives us our best guess at the regression line

$$\hat{y}(x) = \alpha_0 + \beta x = \alpha_* + \beta(x - \bar{x})$$

itself, but also a probability distribution for the possible values of $y(x)$ given our observations, centered at $\hat{y}(x)$.

The data for linear regression are n observations $(x_1, y_1), \ldots, (x_n, y_n)$. The sufficient statistics are n, Σ_x, SS_x, Σ_y, SS_e, and Σ_{xy}. You find these statistics by calculating simple linear regression (Section 3.3).

The inference requires *prior* probability distributions in addition to the observations. This means prior *hyperparameters* for a, b, and τ. We should take into account, however, that most of us are notoriously poor at estimating the line parameters a and b, and even less reliable when it comes to stating how certain these estimates are. And then the informative prior quickly turns into a disinformative prior instead. For this reason, we will stick with neutral priors for these parameters.

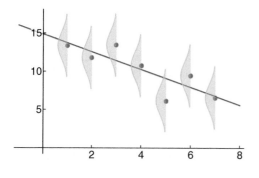

Figure 17.2 Observations, and regression line as a best guess given the uncertainty.

The uncertainty parameter σ, on the other hand, is curiously enough more accessible to us. This is because it often concerns uncertainties in the measurement process itself, but also because we do have a firmer grasp of the variation of the y-values that go with some fixed x-value. We have therefore chosen to retain the option of informative priors for σ.

17.1.1 How-to

Prior for τ or σ: Let s_0 be your best estimate on the uncertainty σ, and let n_0 be how certain you are, expressed in observation equivalents. We will work with $v_0 = n_0 - 2$ and $SS_0 = s_0^2 \times \max(0, v_0)$. If you don't know, simply let $v_0 = -2$ and $SS_0 = 0$. This makes your *prior* hyperparameters for σ (τ) expressed by

$$P_1 \vDash \begin{array}{l} v_0 \\ SS_0. \end{array}$$

The other hyperparameters are implicitly given by the model, since we do not allow informative priors for a or b.

Updating: The prior, together with the statistics from the matrix regression, gives the *posterior* hyperparameters:

$$P_1 \vDash \bar{x} = \Sigma_x/n$$

$$\beta = \begin{bmatrix} \alpha_0 \\ \beta \end{bmatrix} = (X^T X)^{-1} X^T y$$

$$v_1 = v_0 + n$$

$$SS_1 = SS_0 + SS_e$$

or α_* instead of α_0 if you are working with centered x data. Since $\alpha_0 + \beta x = \alpha_* + \beta(x - \bar{x})$, and therefore $\alpha_* = \alpha_0 + \beta\bar{x}$, we may easily switch our notation between the two.

We update in a single round only. If we add new data, the mean $\bar{x}$ moves, making the formulas for updating overly complicated. So you should instead recalculate the relevant statistics for *all* the data, both old and new. Since this is the final step, we may now read the probability distributions for the precision $\tau = \sigma^{-2}$, the slope b, the values of the regression line $y(x)$, and the next observation $Y_+(x)$. We get

$$\tau \sim \gamma\left(\tfrac{v_1}{2}, \tfrac{SS_1}{2}\right) \quad s_1^2 = \frac{SS_1}{v_1}$$

$$b \sim t_{\left(\beta, s_1 \times \sqrt{\frac{1}{SS_x}}, v_1\right)}$$

$$y(x) \sim t_{\left(\alpha_0 + \beta x, s_1 \times \sqrt{\frac{1}{n} + \frac{1}{SS_x}(x-\bar{x})^2}, v_1\right)}$$

$$Y_+(x) \sim t_{\left(\alpha_0 + \beta x_+, s_1 \times \sqrt{1 + \frac{1}{n} + \frac{1}{SS_x}(x-\bar{x})^2}, v_1\right)}.$$

Notice that the neutral prior for σ ($v_0 = -2$ and $SS_0 = 0$), gives $s_1^2 = \frac{SS_e}{n-2} = s_e^2$.

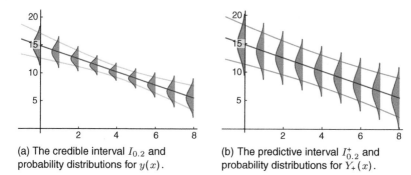

(a) The credible interval $I_{0.2}$ and probability distributions for $y(x)$.

(b) The predictive interval $I_{0.2}^+$ and probability distributions for $Y_+(x)$.

Figure 17.3 Plots of $y(x)$ and $Y_+(x)$ with uncertainty and interval bands.

The $100 \times (1 - 2\theta)\%$ symmetric credible and predictive intervals, as illustrated in Figure 17.3, are given by

$$I_{2\theta}(x) = \alpha_0 + \beta x \pm t_{\nu_1,\theta} \times s_1 \times \sqrt{\frac{1}{n} + \frac{1}{SS_x}(x - \bar{x})^2}$$

$$I_{2\theta}^+(x) = \alpha_0 + \beta x \pm t_{\nu_1,\theta} \times s_1 \times \sqrt{1 + \frac{1}{n} + \frac{1}{SS_x}(x - \bar{x})^2}.$$

If σ is known well enough that we may count it as certain for all practical purposes, let "known well enough" be approximated by "certainty", and set $s_1 = s_0$. With this approximation, $\nu_0 = \infty$, and the t distributions in the formulas above become Normal distributions with the same parameters, since $\lim_{\nu \to \infty} t_{(\mu,\sigma,\nu)} = \phi_{(\mu,\sigma)}$.

The difference between the curves for the credible intervals I_θ for $y(x)$, and the predictive intervals I_θ^+ for $Y_+(x)$, is that the latter are always wider. Whereas the precision of $y(x)$ may be arbitrarily large, $Y_+(x)$ will never be more precise than what the uncertainty/noise $\varepsilon \sim \phi_{(0,s_0)}$ allows. We see a credible and a predictive band of the same percentage in Figure 17.4.

Example 17.1.1 (Continuation of Example 3.3.7) We revisit the height/weight data, and believe we know something about the variability of weights, so we set $s_0 = 4$. We tell ourselves that we are as sure of this as if we had made 5 to 10 observations. It is better to choose the lower number so as not to overshoot, so we set $n_0 = 5$. We want to find

(a) the *posterior* distribution of τ;
(b) the *posterior* distribution of b;
(c) the *posterior* distribution of $y(x)$;
(d) the 95% credible interval for $y(x)$; and
(e) the 90% credible interval for $Y_+(x)$ when $x = 181$.

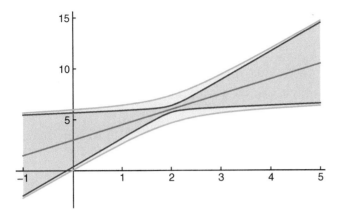

Figure 17.4 Comparison of $I_{0.3}(x)$ (blue, inner) and $I^+_{0.3}(x)$ (green, outer).

The **prior** hyperparameters are then

$$P_0 \vDash \quad v_0 = n_0 - 2 = 5 - 2 = 3$$
$$SS_0 = s_0^2 \times \max(0, v_0) = 4^2 \times 3 = 48.$$

Simple linear regression: We have $n = 15$ and $\Sigma_h = 2\,668.5$, which means $\bar{x} = 177.9$. We write the x data in centered form, $x_i^* = x_i - \bar{x}$. We have calculated the simple linear regression before, in Example 3.3.7, and gotten

$$H_*^T H_* = \begin{bmatrix} 15 & 0 \\ 0 & 1\,271.1 \end{bmatrix} \quad H_*^T v = \begin{bmatrix} 1\,262 \\ 1\,875.7 \end{bmatrix} \quad \beta_* = \begin{bmatrix} 84.133\,3 \\ 1.475\,65 \end{bmatrix}.$$

Updating requires calculating

$$SS_e = v^T v - \beta_*^T H_*^T H_* \beta_* = 7\,127.85.$$

Then, $P_1 \vDash \quad \bar{x} = 177.9$

$$\beta_* = \begin{bmatrix} \alpha_* \\ \beta \end{bmatrix} = \begin{bmatrix} 84.133\,3 \\ 1.475\,65 \end{bmatrix}$$
$$v_1 = v_0 + n = 3 + 15 = 18$$
$$SS_1 = SS_0 + SS_e = 48 + 7\,127.85 = 7\,175.85.$$

We may then answer the questions:

1) $\tau \sim \gamma\left(\frac{18}{2}, \frac{7\,175.9}{2}\right)$

2) $b \sim t_{\left(1.475\,7, 19.966 \times \sqrt{\frac{1}{1271.1}}, 18\right)}$

3) $y(x) \sim t_{\left(84.133\,3 + 1.475\,65(x - 177.9), 19.966\,4 \times \sqrt{\frac{1}{15} + \frac{1}{1271.1}(x - 177.9)^2}, 18\right)}$

4) $Y_+(x) \sim t_{\left(84.133\,3+1.475\,65(x-177.9),\,19.966\,4\times\sqrt{1+\frac{1}{15}+\frac{1}{1271.1}(x-177.9)^2},\,18\right)}$

5) $I_{0.05} = 84.133\,3 + 1.475\,65(x-177.9) \pm 41.947\,8 \times \sqrt{\frac{1}{15}+\frac{1}{1271.1}(x-177.9)^2}$

6) $I_{0.1}^+ = 84.133\,3 + 1.475\,65(x-177.9) \pm 34.622\,9 \times \sqrt{\frac{16}{15}+\frac{1}{1271.1}(x-177.9)^2}$,

and so $I_{0.1}^+(181) = 88.707\,9 \pm 35.884\,9 = \underline{\underline{(52.823, 124.593)}}$.

Example 17.1.2 (Continuation of Examples 3.3.1 and 3.3.10) We were going to restore old platforms. This time around, we are interested in finding

(a) a 90% credible interval for $y(x)$; and
(b) a 90% predictive interval for the cost of the next platform, given x type 1 errors.

Prior: We consider σ to be totally unknown, and so we set $n_0 = SS_0 = 0$. Then,

$$P_0 \vDash \quad v_0 = n_0 - 2 = -2$$
$$SS_0 = 0.$$

Data: The raw data are the pairs $\{(x_i, y_i)\} = \{(20, 121.56), (4, 104.23), (5, 108.08), (19, 119.01), (10, 110.16)\}$, giving us $\bar{x} = 11.6$. We subtract $\bar{x}$ from the x values, and get centered data $\{(x_i^*, y_i)\} = \{(8.4, 121.56), (-7.6, 104.23), (-6.6, 108.08), (7.4, 119.01), (-1.6, 110.16)\}$. We leave the details of the calculations to the reader, and note that, in centered form, we get

$$X_*^T X_* = \begin{bmatrix} 5 & 0 \\ 0 & 229.2 \end{bmatrix} \quad X_*^T y = \begin{bmatrix} 563.04 \\ 220.046 \end{bmatrix} \quad \beta_* = \begin{bmatrix} 112.416 \\ 0.939\,328 \end{bmatrix}.$$

The total squared error, where the error is the distance between the observed and the predicted value, is independent of whether we describe the line in centered form or not. From Example 3.3.10, we have $SS_e = 6.552\,7$. In the lower right corner of $X_*^T X_*$, we find $SS_x = 229.2$.

Updating : $P_1 \vDash \quad \bar{x} = 11.6$

$$\beta_* = \begin{bmatrix} \alpha_* \\ \beta \end{bmatrix} = \begin{bmatrix} 112.416 \\ 0.939\,328 \end{bmatrix}$$

$$v_1 = v_0 + n = -2 + 5 = 3$$

$$SS_1 = SS_0 + SS_e = 0 + 6.552\,7 = 6.552\,7.$$

Using that $s_1 \times t_{3,\,0.05} = 3.478\,07$, we get the intervals

(a) $I_{0.1}(x) = 112.608 + 0.960\,061(x - 58/5) \pm 3.478\,07 \times \sqrt{0.2 + \frac{(x-11.6)^2}{229.2}}$;

(b) $I_{0.1}^+(x) = 112.608 + 0.960\,061(x - 11.6) \pm 3.478\,07 \times \sqrt{1.2 + \frac{(x-11.6)^2}{229.2}}$.

17.2 Frequentist Estimates for Linear Regression

The frequentist intervals for the regression line are identical to the corresponding Bayesian intervals when σ is completely unknown, which means that Example 17.1.2 is an example of frequentist linear regression as well. Exchange $v_1 \to n - 2$, and $s_1 \to s_e = \sqrt{SS_e/(n-2)}$ and you get the frequentist formulas:

$$\widehat{I}_\theta(x) = \alpha_0 + \beta x \pm t_{n-2, \theta/2} \times s_e \times \sqrt{\frac{1}{n} + \frac{1}{SS_x}(x - \bar{x})^2}$$

$$\widehat{J}_\theta(x) = \alpha_0 + \beta x \pm t_{n-2, \theta/2} \times s_e \times \sqrt{1 + \frac{1}{n} + \frac{1}{SS_x}(x - \bar{x})^2}.$$

The theoretical difference lies, as always, in the interpretation of what the intervals *mean*.

17.3 A Logarithmic Example

Related data are not always linearly related, but may be related via some function, for instance a root, a logarithm, or a polynomial. We are not going into the full study of all these regressions, but will look at an example that may whet your appetite for further study. We are going to look at an industrial example where we take the logarithms of *both* variables, and then find the linear relation between these. Working back from that, we then get a relation between the original variables. The calculations for this example will be identical for a frequentist and for a Bayesian using a neutral prior for an unknown σ.

Example 17.3.1 We studied a welded joint in a certain type of machinery to see how many cycles N this component lasts before breaking, given a tension S on the component. The relation between the variables S and N is known to be *logarithmically linear*, and our task is to find this relation, with uncertainty. For Table 17.1, we add columns for $\log S$ and $\log N$ (here, $\log = \log_{10}$).

- The first extra consideration is incomplete data. In Table 17.1, "↓" indicates that the component had not failed after the completed set of five million cycles. This means that we do not know how many cycles it would have taken for it to fail, *except* that this number is in excess of the number of cycles it had gone through. How do we handle such information? There are several ways of handling incomplete data, and which one you choose depends on the purpose of your calculations.
  - Ideally, you use a weighting over the possible values, in other words a probability distribution. This does have the disadvantage of leading us far astray from the neat formulas of an introductory statistics course and text.

Table 17.1 Tension and lifecycles for the welded joint

	S	N	$X = \log S$	$Y = \log N$
1	265	42 000	2.423 25	4.623 25
2	265	70 000	2.423 25	4.845 1
3	265	79 000	2.423 25	4.897 63
4	202	107 000	2.305 35	5.029 38
5	202	188 000	2.305 35	5.274 16
6	202	204 000	2.305 35	5.309 63
7	139	537 000	2.143 01	5.729 97
8	139	597 000	2.143 01	5.775 97
9	108	800 000	2.033 42	5.903 09
10	108	1 077 000	2.033 42	6.032 22
11	108	5 000 000↓	2.033 42	6.698 97↓
12	108	5 200 000↓	2.033 42	6.716 00↓
13	108	5 400 000↓	2.033 42	6.732 39↓
14	74	5 000 000↓	1.869 23	6.698 97↓
15	74	5 200 000↓	1.869 23	6.716 00↓

- You may choose to use the time the component survived as its lifetime. Since the actual lifetime would have been at least as large as this number, you will at least not overshoot the regression line.
- The incomplete data *may* indicate a "region of immortality", where the actual lifetime differs drastically from the relation binding the other data. Such virtual immortality would, if actually explored, pull the regression line far out from the line indicated by the data in the "main" region. If we suspect this is the case, we should remember that most linear regressions are nothing but good approximations valid only inside precisely just a limited region.

  This indicates that it may be a good idea to omit the incomplete data, and this is also what practitioners do. *But* we need to do this with care, so as to avoid a bias in the opposite direction. Look at Figure 17.5. In our case, if we omit the immortals (marked as red dots), we get for $S = 108$ that the average lifetime is roughly 900 000 cycles. But we then get that, if the three components that did not fail had failed just a little sooner, for instance at 4.9 million cycles, $S = 108$ would have implied an average lifetime of over 3 million cycles. This gives us the seeming paradox that *earlier* failure for some components would imply a *longer* average lifetime! This means that if we omit observations, we must also omit the *other* observations with $S = 108$ or lower (blue dots, circled in red).

- The second extra consideration is the logarithmic relation. It may be logarithmic in only one variable, in which case we are looking for the linear regression between X and $\log Y$, or between $\log X$ and Y, or it may be double, as in this case, where we must find the linear regression between $\log X$ and $\log Y$.

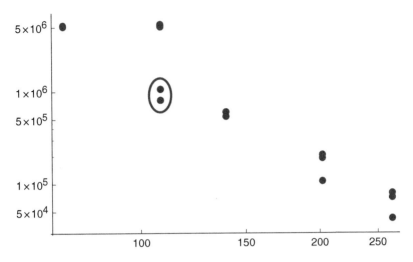

Figure 17.5 The logarithmic data.

After these initial considerations, we proceed with the calculations, using the logarithmic data. We choose to narrow the domain of the linear regression, and keep only the first 8 results. We then calculate the centered x values (using $\bar{x} = 2.30875$), and get the design matrix X_* and the response vector y for the data:

$$
X_* = \begin{bmatrix}
1 & 0.114268 \\
1 & 0.114268 \\
1 & 0.114268 \\
1 & -0.0036263 \\
1 & -0.0036263 \\
1 & -0.0036263 \\
1 & -0.165963 \\
1 & -0.165963
\end{bmatrix}
\qquad
y = \begin{bmatrix}
4.62325 \\
4.8451 \\
4.89763 \\
5.02938 \\
5.27416 \\
5.30963 \\
5.72997 \\
5.77597
\end{bmatrix}.
$$

This means that

$$
X_*^T X_* = \begin{bmatrix} 8 & 0 \\ 0 & 0.0942985 \end{bmatrix}
$$

$$
\left(X_*^T X_* \right)^{-1} = \begin{bmatrix} 0.125 & 0 \\ 0 & 10.6046 \end{bmatrix}
$$

$$
X_*^T y = \begin{bmatrix} 41.4851 \\ -0.324604 \end{bmatrix}.
$$

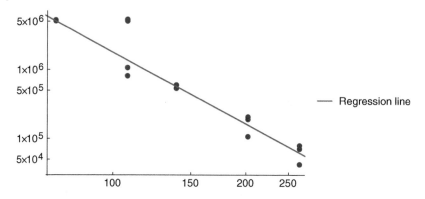

Figure 17.6 The logarithmic data with regression line.

We then get[1]

$$\beta_* = \left(X_*^T X_*\right)^{-1} X_*^T y = \begin{bmatrix} 5.185\,64 \\ -3.442\,3 \end{bmatrix}$$
$$SS_e = y^T y - \beta_*^T X_*^T X_* \beta_* = 0.090\,236\,9.$$

The regression line between x and y, illustrated in Figure 17.6, is then

$$y = 5.185\,64 - 3.442\,3(x - 2.308\,98).$$

We insert $x = \log S$ and $y = \log N$, and get the following relation between S and N:

$$\log N = 5.185\,64 - 3.442\,3(\log S - 2.308\,98)$$
$$N = 10^{\log N} = 10^{5.185\,64 - 3.442\,3(\log S - 2.308\,98)} = 10^{13.133\,8 - 3.442\,3\log S}$$
$$= 10^{13.133\,8} S^{-3.442\,3} = 1.360\,95 \times 10^{13} S^{-3.442\,3}.$$

We now wish to find 95% interval estimates for the linear regression, as illustrated in Figure 17.7. We then use that $s_e^2 = SS_e/(n-2) = 0.090\,236\,9/6 = 0.015\,039\,5$, which means $s_e = 0.122\,636$, and that $t_{6,0.025} = 2.446\,91$, and get

$$I_\theta(x) = \alpha_* + \beta(x - \bar{x}) + 2.446\,9 \times 0.122\,636 \sqrt{\frac{1}{8} + \frac{(x - \bar{x})^2}{SS_x}}$$
$$= 13.133\,8 - 3.442\,3x \pm 0.300\,078 \sqrt{0.125 + \frac{(x - 2.308\,98)^2}{0.094\,298\,5}}$$
$$I_\theta(S) = 1.360\,95 \times 10^{13} S^{-3.442\,3} \times 10^{\pm 0.300\,078 \sqrt{0.125 + (\log S - 2.308\,98)^2/0.094\,298\,5}}.$$

---

[1] These calculations are sensitive to round-off errors, so your decimals might deviate somewhat from our results, which were calculated using Mathematica®.

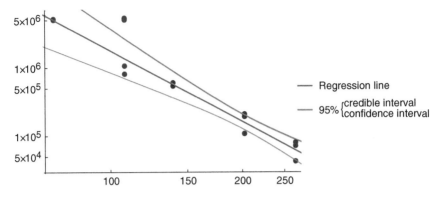

Figure 17.7 The logarithmic data with regression line and 95% interval curves.

The 95% *predictive* interval, as illustrated in Figure 17.8, is then

$$I_\theta^+(x) = 13.133\,8 - 3.442\,3x \pm 0.300\,078\sqrt{1.125 + \frac{(x - 2.308\,98)^2}{0.094\,298\,5}}$$

$$I_\theta^+(S) = 1.360\,95 \times 10^{13}\,S^{-3.442\,3} \times 10^{\pm 0.300\,078\sqrt{1.125 + (\log S - 2.308\,98)^2/0.094\,298\,5}}.$$

Epilogue

We have travelled to the mountains of statistics and back. As befits the theme, our travelling companions sit contemplating the big ship at the Smyril Line's ferry terminal. It is soon ready for departure to Iceland.

"My hard working friends," Bard says as he gets up to board the ferry, "now it's time for us to say goodbye. My work is done. We have reached the shores of the ocean of statistics, and here at the harbour ... our journey together ends.

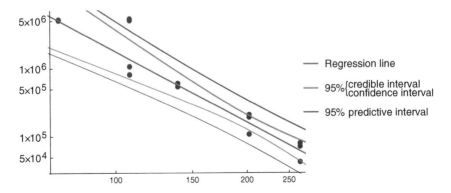

Figure 17.8 The logarithmic data with regression line and 95% predictive curves.

It feels sad to part from you, but it is not a sadness that springs from hurt and harm, for you have grown on your journey and are ready to unfold your own wings and soar."

Bard gathers his suitcases and nods to his statistics colleague, "It's time, Froderick."

"What does he mean by that?" Sam asks Froderick.

"We went on this journey together, Sam, to save the world from frequentism," Froderick replies, "and the world has indeed been saved ... but not I."

"You can't mean what you're saying. You can't leave!" Sam says with eyes growing misty.

Froderick gives Sam his notebook from their journey, before he too boards the boat: "The last few pages are for you, Sam."

17.4 Exercises

1 **Volume training 1:** In the problems below, you are given observational data $\{(x_i, y_i)\}$, and information about σ. Find the ...
 − *posterior* distribution of τ.
 − *posterior* distribution of $y(x)$.
 − posterior *predictive* distribution of $Y_+(x)$.
 − $P\% = (1 − \alpha_1)100\%$ credible interval I_{α_1} for the regressions line $y(x)$.
 − $Q\% = (1 − \alpha_2)100\%$ predictive interval $I_{\alpha_2}^+$ for the next observation $Y_+(x)$.

  (a) Data: $\{(33.74, 260.1), \ (28.71, 226.7), \ (39.9, 300.3), \ (43.29, 321), \ (14.2, 112.3)\}$ Uncertainty: $\sigma_0 = 4$ and $n_0 = 5$. $\alpha_1 = 0.05$, $\alpha_2 = 0.1$.

  (b) Data: $\{(1, 2), (2, 4), (3, 3)\}$. σ unknown. $P\% = Q\% = 90\%$.

  (c) Data: $\{(2, 10), (3, 8), (4, 8), (6, 7)\}$. $\sigma_0 = 0.5$ and $n_0 = 4$. $\alpha_1 = \alpha_2 = 0.05$.

  (d) Data: $\{(0, 0), (1, 2), (2, 7), (3, 5)\}$. σ unknown. $P\% = 90\%$, $Q\% = 95\%$.

  (e) Data: $\{(−2, 7), (1, 5), (8, 6)\}$. σ unknown. $\alpha_1 = \alpha_2 = 0.02$.

  (f) Data: $\{(28, 24), \ (66, 69), \ (44, 48), \ (39, 44), \ (9, 9), \ (1, 15), \ (73, 64), \ (41, 44)\}$. $\sigma_0 = 7$, $n_0 = 10$. $P\% = Q\% = 80\%$.

  (g) Data: $\{(−2.9, 0.8), \ (43.7, 36.9), \ (16.4, 11.7), \ (47.3, 41.9), \ (25.5, 16.9), \ (22.1, 23.1), \ (38.4, 42.4), \ (35.3, 38.8)\}$. $\sigma_0 = 5.5$, $n_0 = 6$. $\alpha_1 = \alpha_2 = 0.02$.

  (h) Data: $\{(135.1, 8.9), (208.7, 16.), (241.4, 16.5), (217.1, 6.6), (215.3, 7.7), (154.8, 2.4), (263.8, 18.6)\}$. σ unknown. $P\% = 95\%$, $Q\% = 90\%$.

2 **Volume training 2:** In the problems below, you are given observational data $\{(x_i, y_i)\}$, and information about σ. Find the ...
 − *posterior* distribution of τ.
 − *posterior* distribution of $y(x)$.
 − posterior *predictive* distribution of $Y_+(x)$.

(a) Data: $\{(77, 3), (94, 19), (87, 8), (86, 21), (70, 1), (75, 8), (81, 1), (80, 16),$ $(91, 11), (98, 22), (82, 4), (86, 21), (101, 15), (75, 2), (83, 18), (87, 7),$ $(78, 15), (93, 12), (77, 8), (80, 12)\}.$ σ unknown.

(b) Data: $\{(23, -80), (2, 2), (22, -79), (20, -67), (10, -37), (-8, 41)\}.$ $\sigma_0 = 3$ and $n_0 = 100\,000.$

(c) Data: $\{(46, 50), (100, 105), (64, 68), (3, 8), (82, 88), (48, 57)\}.$ $\sigma_0 = 2$ and $n_0 = 1000.$

3 **Volume training 3:** In the problems below, you are given observational data $\{(x_i, y_i)\}$, and information about σ. Find the ...

– $P\% = (1 - \alpha_1)100\%$ credible interval I_{α_1} for the regressions line $y(x)$.

– $Q\% = (1 - \alpha_2)100\%$ predictive interval $I_{\alpha_2}^+$ for the next observation $Y_+(x)$.

(a) Data: $\{(3, 22), (7, 16), (11, 8), (4, 21), (12, 13), (17, 3), (7, 11), (6, 14),$ $(13, 10), (2, 20), (15, 9), (13, 11), (14, 14), (15, 4), (8, 8), (12, 5), (16, 3),$ $(6, 9), (15, 0), (2, 19)\}.$ $\sigma_0 = 4$ and $n_0 = 7.$ $\alpha_1 = 0.08,$ $\alpha_2 = 0.01.$

(b) Data: $\{(10, 16), (26, 24), (22, 20), (23, 15), (26, 28)\}.$ $\sigma_0 = 0.5$ and $n_0 = 30.$ $P\% = Q\% = 93\%.$

(c) Data: $\{(10, 16), (26, 24), (22, 20), (23, 15), (26, 28)\}.$ σ unknown. $\alpha_1 = \alpha_2 = 0.07.$

4 The *Norway Cup* is a European football youth cup and training camp that has arranged at Ekebergsletta in Oslo every year since 1972 (except for in 1976, when the organizers, Bækkelagets Sportsklub, arranged the Oslo Handball Cup instead). We are going to look at the trend in the number of participating teams. To make this problem easy to calculate by hand, we have chosen just a few of the years, 10 years apart:

Year	Teams
1975	710
1985	1204
1995	1228
2005	1535
2015	1639

Find the linear regression line for the number of teams as a function of the year. Then find the interval estimate and the predictive interval when you presuppose a neutral prior – both intervals at 90%.

5 In American football, there was a scandal where the New England Patriots had inflated their balls rather poorly ahead of a decisive game against the Indianapolis Colts. Our local football coach informs us that balls behave differently when they are flaccid than when they are hard, and he wants to demonstrate this to us with a ball that has been unused for a few months.

He drops it from a height of 227 cm, and measures its bounce after having inflated it with, respectively, 0, 1, 2, 3, and 4 pumps.

Pumps, x	Rebound height, y (cm)
0	95
1	111
2	115
3	122
4	129

The coach tells us the uncertainty in his measurements was $\sigma_0 = 4$ cm. He was confident that this was an accurate assessment, so we assume $n_0 = \infty$, and use the Normal approximation $t_{(m,s,\infty)} = \phi_{(m,s)}(x)$.

(a) Find the linear regression line $y = \alpha + \beta(x - \bar{x})$.

   In the rest of this exercise, we will estimate the regression line $y(x) = a_* + b(x - \bar{x})$.

(b) Find the *posterior* distribution of b.

(c) Find the *posterior* distribution of a_*.

(d) Find the probability distribution of the next observation of the rebound height, given a total of 20 pumpings.

(e) Find a 90% credible interval for Y_+, given $x_+ = 20$ pumpings.

(f) Why is the linear regression a good model for the rebound height as a function of the number of times the ball has been pumped? Why is the linear regression a *bad* model?

6 The *spectral slope* is an important characteristic of voices and musical instruments. The spectral slope is the slope b of the regression line between $\ln x$ and y, where x is the frequency (measured in hertz), and y is the amplitude (measured in decibels). You have made such measurements from a recording of a singer who sings the vowel E. The measurements are

Measurement	1	2	3	4	5
x (Hz)	360	540	720	900	1080
y (dB)	−3	−13	−22	−25	−28

σ is unknown. Let $z = \ln x$. We are going to look at the regression between z and y.

(a) Find the regression line $y = \alpha_* + \beta(z - \bar{z})$.

(b) Find the *posterior* probability distribution of the *spectral slope b*.

(c) A mathematics savvy vocal afficionado claims that this particular singer has a greater spectral slope than −25 when he sings E. Determine this hypothesis with significance $\alpha = 0.1$.

7 Investigate whether there is a (linear) relation between the weekly salaries of the best UK football strikers and the number of goals they score, with significance $\alpha = 0.2$. We will be using data from 2015.

Player (Team)	Goals, x	Salary, y
Sergio Aguero (Manchester City)	25	£204 000
Harry Kane (Tottenham)	20	£35 000
Raheem Sterling (Liverpool)	7	£30 000
Peter Crouch (Stoke)	7	£43 000

8 You are investigating the solubility of xylose in water inside a pumping system. The water is pressurized, so you get temperatures well in excess of 200°C. The variables are temperature (x) and (maximum) grams of xylose per liter (y). Your measurements are

x (°C)	24	42	62	83	103	124	143	164	185
y (g/L)	497	811	1061	1592	2190	2543	3131	3601	4236

Present your conclusion in the form of a linear regression line, with uncertainty. Indicate a 90% credible interval for solubility at 100°C. Also indicate a 90% predictive interval for a *measurement* of the solubility at 100°C. Explain the difference between the two.

A

Appendix

CONTENTS

A.1 Project

Statistics is a central tool in professional studies, so it makes sense to tie what could be an abstract subject concretely to the profession it is meant to serve. One of the best ways to do this is simply to make concrete use of the statistical tools in a project related to the profession itself. If possible, the project should run through the entire statistics course, from data gathering and summary in the first chapters to inference at the end of the course. We have the following three main types of project.

Empirical

This is the basic form of project, and for most students also the best. The project starts with data collection and basic calculations. You may also present your data in tables and diagrams, all according to Chapters 2 and 3.

The grand analysis of the data comes later, in Part II of the book: Inference. You or the lecturer may sneak a peek to see which *probability distributions* you will need, and keep this in mind when you get to the part about probability distributions. The probability distributions will then take on a more concrete character for you, in that they come alive to you as they relate to your project.

However, the project does not start with gathering data. The project starts when you, maybe with the help of an instructor, identify a professional question that statistics may answer for you. The first thing you should write is therefore a brief *pre project report*. It would be of help to get feedback and possibly course corrections from an instructor.

The pre project report should first of all state *what* your project wants to achieve, and *why*. Do you, for instance, want to know if oak is harder than birch? Or maybe you want to know if flats consume less electricity per square meter than houses do? Or maybe you suspect that there is some approximately linear relation between the hardness and the durability of flooring? Though this is very simple, and you may think this could be left until the end of the project, this is actually the key part of the report, and needs to be done thoroughly before the remainder of the project is undertaken.

The pre project report should also state *how* you are going to obtain your data, and roughly how many measurements you are planning to do. An instructor may adjust this number. As a rule of thumb, never go below 30 measurements. Exceptions are if the measurements are costly or time consuming. For less demanding kinds of observations, you may well aim for a few hundred; this does in particular go for questions of *proportion*.

Finally, state which inferential techniques you will be using. This *is* possible to figure out on your own, but an instructor will be of valuable help here. You may find that what you actually do along the way ends up differing from your pre project, but it is always better to have a plan to guide your aim. Finally, write down which probability distributions you believe you will be making the most use of.

How long should your pre project report be? No more than one or two pages! The pre project may then serve as a template for the final report.

Could or should you do the project in small groups? Absolutely! But for the collective: only if each group member pulls their own weight. And for the individual: only if you are active in all phases of the project, so that it serves as an aid to learning statistics throughout the entire course and not just a small part.

Which kind of data will you be looking at? In principle, anything you wish, but the main types are the following.

- **Magnitudes.** For instance weight, distance, resistance, money, lifetime, blood pressure, and number of new customers. These kinds of data come under *Gaussian processes*, and the key probability distributions are ϕ, t, γ, *bin*, and F.
- **Proportions.** Here, the key measurements are simply success ($\top$) and failure ($\bot$), and the summary looks at the total number of each kind, but nothing more. These kinds of data come under *Bernoulli processes*, and the key probability distributions are β, ϕ, F, *bin*, *nb*, βb, and βnb.
- **"Rare occurrences".** These may be waiting times, like waiting for clicks on a Geiger counter, or the occurrence of some rare toad in a swamp, or the number of offspring in a given species. These kinds of data come under *Poisson processes*, and the key probability distributions are γ, *pois*, $g\gamma$, *nb*, ϕ and F.

- **Pairwise data.** Here, you gather data in pairs, for instance (height, weight) or (cost, effect). These kinds of data come under *linear regression*, and the key probability distributions are t, γ, and ϕ.

The project is rounded off by writing a project report, which should consist of two parts. The first part is a presentation of what and why: *what* you found out, with conclusions and illustrations, and an explanation of *why* you wanted to find out what you did. Think of this part of your project as written for a customer, or possibly for the management of your company. This part is what you aim for when you write the *what-and-why* part of the pre project.

The other part of the project report is the *how*. Think of this part as something not intended for the customer, the management or any other end user, but rather for technical experts who may be hired to evaluate the quality and soundness of your results. This part should contain the raw data, and the calculations, presented to a sufficient degree of detail that another student at your own level would be able to follow what is stated. It may be helpful for groups to meet up in pairs to read each other's reports with this in mind.

Programming

Programming may mean many things, so this should be clarified by an instructor in advance. It may be anything from programming statistical functionality on calculators or other tools that other course participants may also enjoy (we are keenly interested in sharing such programs at http://bayesians.net, so please contact us if you have anything to share), to programming statistical simulations. These simulations may also be related to empirical investigations, and so two groups may cooperate – each with their own project – one simulating and the other empirical.

A programming project should also have a pre project report about the *how* and *why*, and should of course end with a project report.

Theoretical

Theoretical projects at this level do not mean developing new theories, but rather gathering information and evaluating the extent to which a company or institution may need statistical analyses, and to compile an overview of how statistics is already used. You will find that, more often than not, they are in need of such analyses and calculations without being quite aware that this is what they need, or what such an analysis may do for the company.

Here, too, a pre project report should indicate which areas of the company or institution you believe would be most fruitful to study, and how you have planned to evaluate the company's needs for statistical analyses.

This kind of project requires a high degree of self-drivenness and motivation, or a strong pre-existing conviction that the company has statistical needs that are not being met, so we would recommend choosing the theoretical project only if you have this degree of motivation.

A.2 Notation, Formulas, Functions

A.2.1 An index is a unique identification tag for an element. Common indexes are number, date, place, and kind. Indexes may be multiple, like $a_{2,3}$ or a_{23}, or $a_{\text{Grimstad}, 2009.08.27:18:27:31.05}$.

A.2.2 The index set is the set I of elements we wish to include.

A.2.3 Union $\displaystyle\bigcup_{j=m}^{n} A_j = \bigcup_{j\in I=\{m,\dots,n\}} A_j = A_m \cup A_{m+1} \cup \cdots \cup A_n.$

A.2.4 Intersection $\displaystyle\bigcap_{j=m}^{n} A_j = \bigcup_{j\in I=\{m,\dots,n\}} A_j = A_m \cap A_{m+1} \cap \cdots \cap A_n.$

A.2.5 Sum $\displaystyle\sum_{j=m}^{n} a_j = \sum_{j\in I=\{m,\dots,n\}} a_j = a_m + a_{m+1} + \cdots + a_{n-1} + a_n.$

A.2.6 Product $\displaystyle\prod_{j=m}^{n} a_k = \prod_{j\in I=\{m,\dots,n\}} a_j = a_m \cdot a_{m+1} \cdot \cdots \cdot a_{n-1} \cdot a_n.$

If $I = \varnothing$, then $\displaystyle\prod_{k\in\varnothing} a_k = 1.$

A.2.7 Faculty is defined for all $n \in \mathbb{N}_0$, and is: $n! = \displaystyle\prod_{k=1}^{n} k.$

A.2.8 Multinomial: $\displaystyle\binom{n}{k_1, k_2, \dots, k_m} = \frac{n!}{k_1! \cdot k_2! \cdot \cdots \cdot k_m!}.$

A.2.9 Binomial: $\displaystyle\binom{n}{k} = \frac{n!}{k! \cdot (n-k)!}.$ Note that $\displaystyle\binom{n}{k} = \binom{n}{k, n-k}$, and that

$$\binom{n}{k_1, k_2, \dots, k_m} = \binom{n}{k_1} \cdot \binom{n-k_1}{k_2} \cdot \cdots \cdot \binom{n - k_1 - k_2 - \cdots - k_{m-2}}{k_{m-1}}.$$

In calculator notation:

$$\frac{n!}{(n-k)!} = n\, \boxed{\text{nPr}}\, k \qquad \binom{n}{k} = n\, \boxed{\text{nCr}}\, k$$

$$\binom{n}{k_1, k_2, \dots, k_m} = (n\, \boxed{\text{nCr}}\, k_1) \times \cdots \times ((n - k_1 - \cdots - k_{m-2}) \boxed{\text{nCr}}\, k_{m-1}).$$

A.2.10 Gamma function $\Gamma(z)$:

$$
\begin{aligned}
\Gamma(n) &= (n-1)! & \text{when } n \in \mathbb{N} \\
\Gamma(z) &= \int_0^{\infty} t^{z-1} e^{-t}\, dt & \text{when } -z \notin \mathbb{N}_0 \\
\Gamma(z) &\approx \left(\frac{z}{e}\right)^z \sqrt{\frac{2\pi}{z}} & \text{for large } z \\
\Gamma(n + \tfrac{1}{2}) &= \frac{(2n)!}{n! 4^n} \sqrt{\pi} & \text{when } n \in \mathbb{N}
\end{aligned}
$$

A.2.11 The little gamma function $\gamma(z, x)$:

$$\gamma(z, x) = \int_0^x t^{z-1} e^{-t} \, dt \quad \text{when} \ -z \notin \mathbb{N}_0$$
$$\gamma(z, \infty) = \quad \Gamma(z).$$

A.2.12 Generalized binomial: $\text{nCr} = \begin{pmatrix} n \\ r \end{pmatrix} = \dfrac{\Gamma(n+1)}{\Gamma(r+1)\Gamma(n-r+1)}.$

A.2.13 Pochhammer function: $(p)_q = \dfrac{\Gamma(p+q)}{\Gamma(p)}.$

A.2.14 Euler B function $B(a, b)$:

$$B(a, b) = \int_0^1 t^{a-1}(1-t)^{b-1} \, dt = \frac{\Gamma(a)\Gamma(b)}{\Gamma(a+b)}.$$

A.2.15 The incomplete Euler Beta function $B_{(a,b)}(x)$:

$$B_{(a,b)}(x) = \int_0^x t^{a-1}(1-t)^{b-1} \, dt \quad \text{for } x \in [0, 1].$$

A.2.16 β function $\beta_{(a,b)}$:

$$\beta_{(a,b)}(t) = \frac{1}{B_{(a,b)}} \cdot t^{a-1}(1-t)^{b-1} \quad \text{for } t \in [0, 1].$$

A.2.17 Regularized Beta function $I_{(a,b)}(x)$:

$$I_{(a,b)}(x) = \frac{B_{(a,b)}(x)}{B_{(a,b)}} = \int_0^x \beta_{(a,b)}(x).$$

A.2.18 Geometric sum $\sum\limits_{k=A}^{B} r^k$:

$$\sum_{k=A}^{B} r^k = \frac{r^A - r^{B+1}}{1 - r}$$
$$\sum_{k=A}^{\infty} r^k = \frac{r^A}{1 - r} \quad (\text{when } |r| < 1).$$

A.3   Other Probability Distributions

In addition to the probability distributions we looked at in detail in Chapters 9 and 10, we have a few other probability distributions that come in handy for our calculations. They are mainly variants of probability distributions where some parameter itself follows a probability distribution rather than having a fixed value.

We have no need of studying them in detail, but only need to know how to use them and know their relations to other distributions. It is recommended to program these functions into your calculator, or to see if there are any ready programs in http://bayesians.net for your calculator.

A.3.1 Snedecor–Fisher Distribution ("*F* distribution")

The very first distribution we will look at is a distribution that is very useful for calculating other distributions, since it is ubiquitous on advanced pocket calculators, and many other probability distributions may be transformed into it.

F **distribution:** $T \sim f_{(\alpha,\beta)}(t)$ is a continuous stochastic variable taking values in $[0, \infty)$. The probability distribution is

$$f_{(\alpha,\beta)}(t) = \frac{\sqrt{\dfrac{(\alpha t)^\alpha \times \beta^\beta}{(\alpha t + \beta)^{\alpha+\beta}}}}{t \times B_{(\alpha/2, \beta/2)}},$$

where B is the *Euler Beta* function (A.2.14). To express the cumulative probability distribution, we use the *incomplete* Euler B function (A.2.15),

$$F_{(\alpha,\beta)}(t) = \frac{B_{(\alpha/2, \beta/2)}(z)}{B_{(\alpha/2, \beta/2)}}, \quad \text{where } z = \frac{\alpha t}{\alpha t + \beta}.$$

Expected value: $\mu_T = \frac{\beta}{\beta-2}$ **Variance:** $\sigma_T^2 = \frac{2\beta^2(\alpha+\beta-2)}{\alpha(\beta-4)(\beta-2)^2}$

Mathematica: FRatioDistribution[α, β]

CASIO: $F_{(\alpha,\beta)}(t)$ is FCD(0, t, α, β)

TI: $F_{(\alpha,\beta)}(t)$ is Fcdf(0, t, α, β)

HP: $F_{(\alpha,\beta)}(t)$ is fisher_cdf(α, β, t)

Calculators that have the F distribution typically also have f and F^{-1}. A different distribution $h(t)$ may often be calculated through the F distribution, if the cumulative distribution satisfies $H(t) = F_{(\alpha,\beta)}(m(t))$. Then the probability distribution itself is the derivative, $h(t) = m'(t) \cdot f_{(\alpha,\beta)}(m(t))$, whereas $H^{-1}(p) = m^{-1}(F_{(\alpha,\beta)}^{-1}(p))$. In the case where $m(t) = kt$, this reduces to $h(t) = k \cdot f_{(\alpha,\beta)}(kt)$ and $H^{-1}(p) = \frac{1}{k} \cdot F_{(\alpha,\beta)}^{-1}(p)$.

A.3.2 Two Compound Gamma Distributions

The first way to compound two gamma distributions is to divide one stochastic variable by the other. This gives us the following rule.

Rule A.3.1 Pairwise gamma distributed variables: *Let* $T_1 \sim \gamma_{(k,\lambda)}(t)$ *and* $T_2 \sim \gamma_{(\kappa,\tau)}(t)$ *be independent, and let* $m = \frac{\kappa\lambda}{k\tau}$. *Define the derived stochastic variables* $Q = \frac{T_1}{T_2}$ *and* $Q_* = mQ$. *Then*

$$Q_* \sim f_{(2k,2\kappa)}(t) \qquad Q \sim m \cdot f_{(2k,2\kappa)}(mt)$$
$$P(Q_* < t) = F_{(2k,2\kappa)}(t) \qquad P(Q < t) = F_{(2k,2\kappa)}(mt).$$

The other way to compound two gamma distributions is when $T \sim \gamma_{(k,\lambda)}$ and λ itself is not constant, but rather $\lambda \sim \gamma_{(\kappa,\tau)}$. Then, $T \sim g\gamma_{(k,\kappa,\tau)}(t)$. This distribution is called the gamma–gamma distribution. It goes under other names, not least the "Beta Prime distribution", in which case it is written β^{II}.

Gamma–gamma distribution: $T \sim g\gamma_{(k,\kappa,\tau)}(t)$ is a continuous stochastic variable taking values in $[0, \infty)$. The probability distribution is

$$g\gamma_{(k,\kappa,\tau)}(t) = \frac{1}{B_{(k,\kappa)}} \cdot \frac{\tau^\kappa \cdot t^{k-1}}{(\tau + t)^{\kappa+k}},$$

where B is the *Euler Beta* function (A.2.14). To express the cumulative probability distribution, we use the *incomplete* Euler B function (A.2.15),

$$G\Gamma_{(k,\kappa,\tau)}(t) = \frac{B_{(k,\kappa)}\left(\frac{t}{t+\tau}\right)}{B_{(k,\kappa)}}.$$

Expected value: $\mu_T = \frac{k\tau}{\kappa-1}$ **Variance:** $\sigma_T^2 = \frac{k\tau^2(\kappa+k-1)}{(\kappa-2)(\kappa-1)^2}$

Mathematica: BetaPrimeDistribution$[k, \kappa, \tau]$

CASIO: $G\Gamma_{(k,\kappa,\tau)}(t)$ is FCD$(0, \frac{\kappa t}{k\tau}, 2k, 2\kappa)$

TI: $G\Gamma_{(k,\kappa,\tau)}(t)$ is Fcdf$(0, \frac{\kappa t}{k\tau}, 2k, 2\kappa)$

HP: $G\Gamma_{(k,\kappa,\tau)}(t)$ is fisher_cdf$(2k, 2\kappa, \frac{\kappa t}{k\tau})$

A.3.3 (Negative) Binomial Distribution When p is Uncertain: βb and βnb

When the binomial parameter p itself is uncertain, following a probability distribution $p \sim \beta_{(a,b)}$, we must refine our formulas for calculating binomial and negative binomial probability. The new formulas exchange the parameter p with the pair a, b, and we get the formulas below.

Beta-binomial distribution: when $X \sim bin_{(n,p)}$, and $p \sim \beta_{(a,b)}(x)$, we may express the probability distribution of X directly, as $X \sim \beta b_{(a,b,n)}(x)$, where

$$\beta b_{(a,b,n)}(x) = \binom{n}{x} \frac{(a)_x (b)_{n-x}}{(a+b)_n} \quad \text{(See A.2.13 for } (p)_q.)$$

The cumulative probability distribution is

$$\beta B_{(a,b,n)}(x) = \sum_{z=0}^{x} \beta b_{(a,b,n)}(z).$$

Expected value: $\mu_X = \frac{na}{a+b}$ **Variance:** $\sigma_X^2 = \frac{nab(a+b+n)}{(a+b)^2(a+b+1)}$

Mathematica: BetaBinomialDistribution$[a,b,n]$

Beta-negative-binomial distribution: when $X \sim nb_{(k,p)}$ and $p \sim \beta_{(a,b)}(x)$, we may express the probability distribution of X directly, as $X \sim \beta nb_{(a,b,k)}(x)$, where

$$\beta nb_{(a,b,k)}(x) = \binom{x-1}{k-1} \frac{(a)_k (b)_x}{(a+b)_{x+k}} \quad \text{(See A.2.13 for } (p)_q.)$$

The cumulative probability distribution is

$$\text{ßNB}_{(a,b,k)}(x) = \sum_{z=0}^{x} \beta nb_{(a,b,k)}(z).$$

Expected value: $\mu_X = \dfrac{bk}{a-1}$ **Variance:** $\sigma_X^2 = \dfrac{kb(a+b-1)(a+k-1)}{(a-2)(a-1)^2}$

Mathematica: BetaNegativeBinomialDistribution[a,b,k]

The following rule may come in handy if you need to calculate manually, or you are programming a calculator.

Rule A.3.2 *The formulas for the βb and βnb distributions, broken down to Γ functions, faculty, and binomial, look like this:*

$$\beta b_{(a,b,n)}(x) = \binom{n}{x} \cdot \frac{\left(\dfrac{\Gamma(a+b)}{\Gamma(a)\Gamma(b)}\right)}{\left(\dfrac{\Gamma(a+b+n)}{\Gamma(a+x)\Gamma(b+n-x)}\right)}$$

$$= \binom{n}{x} \cdot \frac{\dbinom{a+b}{a} \cdot \dfrac{ab}{a+b}}{\dbinom{a+b+n}{a+x} \cdot \dfrac{(a+x)(b+n-x)}{a+b+n}}$$

$$\beta nb_{(a,b,k)}(x) = \binom{x+k-1}{k-1} \cdot \frac{\left(\dfrac{\Gamma(a+k)}{\Gamma(a)} \dfrac{\Gamma(b+x)}{\Gamma(b)}\right)}{\left(\dfrac{\Gamma(a+b+k+x)}{\Gamma(a+b)}\right)}$$

$$= \binom{x+k-1}{k-1} \cdot \frac{\dbinom{a+b}{a} \cdot \dfrac{a+b+k+x}{a+b}}{\dbinom{a+b+k+x}{a+k} \cdot \dfrac{(a+k)(b+x)}{ab}}.$$

On some calculators, the latter formulations for each of the probability distributions will work for integers only, since the binomial and faculty functions are defined for integers only. However, see A.2.12 for how to handle non-integers as arguments for those functions.

The Γ version works for all calculators that have the Γ function.

In practice, these functions occur after Bayesian updating for proportions, and then the *prior* determines whether the *posterior* will be an integer or not. If the prior hyperparameters are integers, then so are the *posterior* hyperparameters, meaning that a and b in the distributions above are integers. The same goes for half-integers (integer plus $\frac{1}{2}$).

For half-integers, we may use from formula A.2.10 that $\Gamma(n + \frac{1}{2}) = \frac{(2n)!}{n!4^n}\sqrt{\pi}$.

A.3.4 Comparison of β Distributed Variables

Comparison of β distributed stochastic variables requires a bit of computing power, but if you use the formula below, you will get an exact answer rather than an approximation.

Rule A.3.3 *If* $\psi \sim \beta_{(a,b)}$ *and* $\pi \sim \beta_{(\theta,\rho)}$ *are independent stochastic variables, then*

$$P(\psi \leq \pi) = \sum_{k=0}^{\theta-1} \frac{B(a+k,\, b+\rho)}{(\rho+k)\cdot B(k+1,\, \rho)\cdot B(a,b)},$$

where the B function is as described in A.2.14 in Appendix A.2: $B(a,b) = \frac{\Gamma(a)\Gamma(b)}{\Gamma(a+b)}$, so

$$P(\psi \leq \pi) = \sum_{k=0}^{\theta-1} \frac{\Gamma(a+k)\Gamma(b+\rho)\Gamma(k+\rho+1)\Gamma(a+b)}{(\rho+k)\Gamma(a)\Gamma(b)\Gamma(\rho)\Gamma(k+1)\Gamma(a+b+k+\rho)}.$$

We may rewrite this into a formula using faculty or binomials:

$$P(\psi \leq \pi) = \sum_{k=0}^{\theta-1} \frac{\binom{a+b}{a}\cdot\binom{k+\rho}{k}\cdot ab\rho(a+b+k+\rho)}{\binom{a+b+k+\rho}{a+k}\cdot(a+b)(k+\rho)(a+k)(b+\rho)}.$$

If you don't have access to advanced calculating tools, you should settle for a Normal approximation to the β distributions. The Normal approximation works best when all the four parameters a, b, θ, ρ are larger than 10, but if you are cut off from exact calculation, the Normal approximation is the only game in town, so you might as well use it. Just keep in mind that it is rather inaccurate for small parameter values. With the Normal approximation, we get that the difference $\delta = \psi - \pi$ between (the Normal approximations to) two β distributions follows a Normal distribution, thus:

$$\delta \sim \phi\left(\frac{a}{a+b} - \frac{\theta}{\theta+\rho},\, \sqrt{\frac{ab}{(a+b)^2(a+b+1)} + \frac{\theta\rho}{(\theta+\rho)^2(\theta+\rho+1)}}\right).$$

Then

$$P(\psi \leq \pi) = \Phi_{\left(\frac{a}{a+b} - \frac{\theta}{\theta+\rho},\, \sqrt{\frac{ab}{(a+b)^2(a+b+1)} + \frac{\theta\rho}{(\theta+\rho)^2(\theta+\rho+1)}}\right)}(0).$$

A.4 Processes

In this book, we are looking at three stochastic *processes*:

- Bernoulli processes;
- Poisson processes; and
- Gaussian processes.

These are *observational* processes, describing the magnitudes we observe under certain conditions. We will briefly look at their properties, and which probability distributions belong to the processes. What these three processes have in common is that the observations are *independent*.

A.4.1 Bernoulli Processes

A Bernoulli process consists of a sequence of independent Bernoulli *trials*, each of which has the same two outcomes $\top$ and $\bot$, and the same probability $p = P(\top)$. When we speak of a *sequence*, we tend to think of it as a temporal sequence where one trial happens after the other, as illustrated in Figures A.1 and A.2.

However, though the Bernoulli process is best studied when the trials come in a sequence, much of the theory will come into play if the trials are ordered in other ways, for instance on a surface, in a volume, or in *any* other way. As long as you know there are n trials, it may be studied via the Bernoulli process (see Figure A.3).

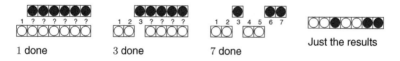

1 done   3 done   7 done   Just the results

Figure A.1 Bernoulli process: choices being made, in sequence.

Figure A.2 Bernoulli process: … and on it goes …

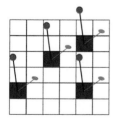

Figure A.3 Bernoulli on a surface.

We will now look at the key probability distributions involved with the Bernoulli process. The first is the Bernoulli distribution itself (9.2), which describes each of the trials, and which we looked at above.

A.4.1.1 Known p

The key to the Bernoulli process is counting, and with a known Bernoulli parameter p, we have the following two counting distributions.

1) $bin_{(n,p)}(k)$: the binomial distribution (Section 9.3) counts k, the number of successes T, given n Bernoulli trials with parameter p (Figure A.10). So $bin_{(n,p)}(k)$ is the probability that n Bernoulli trials with parameter p will yield k T.

2) $nb_{(k,p)}(m)$: the negative binomial distribution (Section 9.5) counts m, the number of failures ⊥, given that the Bernoulli process with parameter p runs until k successes T have been achieved (Figure A.5). So $nb_{(k,p)}(m)$ is the probability that Bernoulli trials with parameter p yield m ○ before they yield its kth ●.

Addition rules

Rule A.4.1 *If $X_1 \sim bin_{(m,p)}(x)$ and $X_2 \sim bin_{(n,p)}(x)$ are independent, and $X = X_1 + X_2$, then*

$$X \sim bin_{(m+n,p)}(x).$$

In particular, if $X \sim bin_{(n,p)}(x)$, then X may be considered the sum of n Bernoulli variables $X_k \sim bin_{(1,p)}(x) = bern_p(x)$, as $X = X_1 + X_2 + \cdots + X_n$.

Figure A.4 Twenty trials yielded five successes (●).

Figure A.5 Waiting for five successes (●): we first got twelve failures (○).

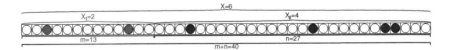

Figure A.6 $X = X_1 + X_2$.

Figure A.7 Number of ○ before the next ● is Y_k.

Rule A.4.2 *If $Y_1 \sim nb_{(k,p)}(y)$ and $Y_2 \sim nb_{(m,p)}(y)$ are independent, and $Y = Y_1 + Y_2$, then*

$$Y \sim nb_{(k+m,p)}(y).$$

In particular, if $Y \sim bin_{(k,p)}(y)$, then Y may be considered the sum of k geometrically distributed variables $Y_i \sim nb_{(1,p)}(x) = geom_p(x)$, as $Y = Y_1 + Y_2 + \cdots + Y_k$. We see this illustrated in Figure A.7.

A.4.1.2 Relations

Rule A.4.3 *For a Bernoulli process with parameter p,*

$$BIN_{(n,p)}(k) = 1 - NB_{(k+1,p)}(n - k - 1),$$

or equivalently,

$$NB_{(k,p)}(n) = 1 - BIN_{(n+k,p)}(k - 1).$$

Proof: The proof of this is simple if you read what the distributions say. Both expressions concern a Poisson Bernoulli with rate p, and what happens up until the nth trial. $BIN_{(n,p)}(k)$ is the probability that there will be k or fewer ⊤ within trial n.

For the other expression, $NB_{(k+1,p)}(n - k - 1)$ is the probability that there will be at most $n - k - 1$ failures before the $(k + 1)$th success. Or in other words that the $(k + 1)$th success will come no later than at trial number … n. Why n? Because the number of trials equals the number of ⊤ plus the number of ⊥, and the number of ⊤ will be precisely $k + 1$ since this *is* the $(k + 1)$th ⊤, and the number of ⊥ was just stated to be at most $n - k - 1$. And $(n - k - 1) + (k + 1) = n$.

The events "k or fewer ⊤ within trial n" and "⊤ number $k + 1$ comes at or before trial n" are complementary, and hence

$$BIN_{(n,p)}(k) + NB_{(k+1,p)}(n - k - 1) = 1,$$

which proves the rule. □

This rule also implies that $pois_{\lambda t}(k) = ERL_{(k,\lambda)}(t) - ERL_{(k+1,\lambda)}(t)$.

A.4.1.3 Unknown p

If p is not known, the most useful distribution for studying p itself is the beta distribution (Section 10.5). When $p \sim \beta_{(a,b)}(t)$, we get two counting distributions that are the counterparts of the two counting distributions for a known p, but with $p \sim \beta_{(a,b)}(t)$. The distributions are as follows.

1) $\beta b_{(a,b,n)}(k)$: the beta binomial distribution is the counterpart of the binomial distribution, where the parameters a and b replace the parameter p and express how p is distributed.
2) $\beta nb_{(a,b,k)}(m)$: the beta binomial distribution is the counterpart of the negative binomial distribution, where the parameters a and b replace the parameter p and express how p is distributed.

See Appendix A.3.3 for details about these two distributions.

A.4.2 Poisson Process

The Poisson process is best understood as a refinement of the Bernoulli process. The primary mode is when the events are ordered along a time line, but for our spatially ordered brain, seeing the process as spread out over a surface gives a good intuition of what is at play. We illustrate this in Figure A.8, where the grid of a Bernoulli process is refined until, at the end, with infinite refinement, we have a Poisson process.

For the purposes of process, however, we will stay with the temporal sequence, and thus the line. The same limiting process takes place there, and so we have an infinitely fine division of the timeline, where positive events occur according to a certain rate λ. In the steps of the limiting process, $\lambda = np$ where n and p are binomial parameters. We illustrate the limit in Figure A.9.

We will now look at the key probability distributions involved with the Poisson process. Note that the aspects of the Poisson process particularly

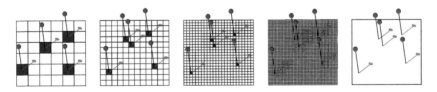

Figure A.8 Poisson on a surface, as a limit of Bernoulli processes.

Figure A.9 Poisson process along a line.

relevant to these distributions are described, together with the distributions themselves, in Sections 9.6.1 and 10.3.2.

A.4.2.1 Known Rate λ

The Poisson process also has discrete occurrences, but, as opposed to the Bernoulli process, the times, spaces, volumes, and intervals containing and separating the occurrences are not integers, but positive real numbers. For a known rate parameter λ, we then have the following two distributions.

1) $pois_{\lambda t}(x)$: the Poisson distribution (Section 9.6) counts the number of occurrences for a Poisson process with rate parameter λ after t time units. More precisely, $pois_{\lambda t}(x)$ is the probability of precisely x units when t time units have passed. See Figure A.10 for an illustration of counting occurences in a Poisson process over 26 time units. Note that this differs from the simpler standard notation for the Poisson distribution, where the time is implicitly set to "1 unit", and the number of occurrences are distributed $pois_{\lambda}(x)$. The Poisson process may also describe searches over volumes V, surfaces A, and so on, where the rate of occurrence per unit remains a constant λ. For the Poisson distribution, the parameter is then λV, or λA, and so on. See also the addition rule A.4.4 below. It is the continuous counterpart of what the binomial distribution is for the Bernoulli process.
2) $erl_{(k,\lambda)}(t)$: the Erlang distribution (Section 10.3) measures t, the time k of occurrences take for a Poisson process with rate parameter λ (Figure A.11). So $erl_{(k,\lambda)}(t)$ is the probability density that the kth occurrence happens at time t. It is the continuous counterpart of what the negative binomial distribution is for the Bernoulli process.

Addition rules

> **Rule A.4.4** *If $Z_1 \sim pois_{\lambda_1}(x)$ and $Z_2 \sim pois_{\lambda_2}(x)$ are independent, and $Z = Z_1 + Z_2$, then*
>
> $$Z \sim pois_{\lambda_1 + \lambda_2}(x).$$

Figure A.10 The number of occurrences during 26 time units is distributed $Z \sim pois_{26\lambda}$. Here, there were $Z = 5$ occurrences.

Figure A.11 The waiting time for four occurrences is distributed $W = erl_{(4,\lambda)}$. Here, it took $W = 13.7$ time units.

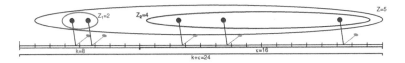

Figure A.12 $Z = Z_1 + Z_2$.

Figure A.13 The Erlang waiting time for five occurrences is the sum of five exponential waiting times for the individual occurrences.

This rule says that, for a Poisson process with parameter λ, the number of occurrences in two distinct intervals (or areas, volumes, and so on) equals the number of occurrences in the first, plus the number of occurrences in the other. We see this illustrated in Figure A.12.

Rule A.4.5 *If $W_1 \sim erl_{(k,\lambda)}(t)$ and $W_2 \sim erl_{(\kappa,\lambda)}(t)$ are independent, and $W = W_1 + W_2$, then*

$$W \sim erl_{(k+\kappa,\lambda)}(t).$$

In particular, if $W \sim erl_{(k,\lambda)}(t)$, then W may be considered the sum of k exponentially distributed variables $W_i \sim erl_{(1,\lambda)}(t) = exp_\lambda(t)$, as $W = W_1 + W_2 + \cdots + W_k$. We see this illustrated in Figure A.13.

A.4.2.2 Relations
An important relationship is that when the waiting time for one occurrence in the Poisson process is $T_j \sim exp_\lambda(t)$, the waiting time for k occurrences is $T = T_1 + \cdots + T_k$, and then $T \sim erl_{(k,\lambda)}(t)$. We see that the Poisson process is the continuous counterpart of the Bernoulli process:

Process: Bernoulli(p) $\leftrightarrow$ Poisson(λ)
Waiting time for *one* occurrence: $geom_p(x) \leftrightarrow exp_\lambda(x)$
Waiting time for *n* occurrences: $nb_{(n,p)}(x) \leftrightarrow erl_{(n,\lambda)}(x)$
Number of occurrences: $bin_{(n,p)}(x) \leftrightarrow pois_{\lambda t}(x)$.

Rule A.4.3 for the Bernoulli process has the following counterpart for the Poisson process, which is actually neater and more elegant in the continuous Poisson case.

Rule A.4.6 *For a Poisson process with parameter λ,*

$$POIS_{\lambda t}(k) = 1 - ERL_{(k+1,\lambda)}(t),$$

or equivalently,

$$ERL_{(k,\lambda)}(t) = 1 - POIS_{\lambda t}(k-1).$$

Proof: The proof of this is simple if you read what the distributions say. Both expressions concern a Poisson process with rate λ, and what happens before time t. $POIS_{\lambda t}(k)$ is the probability that there will be k or fewer occurrences before time t. For the other expression, $ERL_{(k+1,\lambda)}(t)$ is the probability that the time it took for $k+1$ occurrences was at most t, or in other words that, by time t, there were $k+1$ or more occurrences. So the first expression says k or fewer, and the second says $k+1$ or more, which means they are complementary, and so

$$POIS_{\lambda t}(k) + ERL_{(k+1,\lambda)}(t) = 1,$$

which proves the rule. $\qquad\square$

This rule also implies that $pois_{\lambda t}(k) = ERL_{(k,\lambda)}(t) - ERL_{(k+1,\lambda)}(t)$.

A.4.2.3 Unknown Rate λ

If λ is not known, the most useful distribution for studying λ itself is the gamma distribution (Section 10.3). When $\lambda \sim \gamma_{(\kappa,\tau)}(t)$, we get two distributions which are the counterparts of the two distributions for a known λ, but with $\lambda \sim \gamma_{(\kappa,\tau)}(t)$. The distributions are as follows.

1) $nb_{(k,p)}(x)$, the negative binomial distribution, where the parameters $k = \kappa$ and $p = \frac{\tau}{\tau+\theta}$, is the counterpart of the Poisson distribution $pois_{\lambda\theta}(x)$ as the distribution counting how many occurrences x you may find in θ units.
2) $g\gamma_{(k,\kappa,\tau)}(t)$, the gamma–gamma distribution, is the counterpart of the Erlang distribution $erl_{(k,\lambda)}(t)$ measuring how long a time it takes for k occurrences of the Poisson process.

See Section 13.3 for details.

A.4.3 Gaussian Process

The Gaussian distribution, or as we prefer to call it, the *Normal distribution*, may also be seen as a limiting case of the binomial distribution, as we saw in Section 9.3.4. The Gaussian processes are, however, best studied in their own right.

There are, also, many processes that are called *Gaussian*. We shall concentrate on only one type: a sequence of independent stochastic variables that

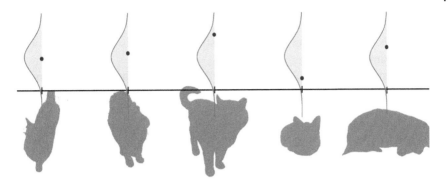

Figure A.14 The model: five weights independently sampled from a Gaussian distribution $\phi_{(\mu,\sigma)}$.

follow identical Gaussian distributions, $\{X_k\}_{k=1}^{\infty}$, $X_k \sim \phi_{(\mu,\sigma)}$. So we have one measurement per positive integer, where each measurement is sampled from the same Normal distribution $\phi_{(\mu,\sigma)}$. This is illustrated in Figure A.14.

A.4.3.1 Known Parameter σ

The two relevant distributions in this case are the Normal distribution itself (Section 10.1), and by extension, the multiNormal distribution with a mean with entries all μ, and diagonal covariance matrix with diagonal entries all equal to σ^2.

A.4.3.2 Unknown Parameter σ

In statistical inference, we do not know the distribution they were sampled from, and so at least one of the parameters is partially or wholly unknown. We only have the data themselves, and maybe some prior information, to work with. This is illustrated in Figure A.15.

Figure A.15 The reality: five cats whose weights follow a Gaussian distribution $\phi_{(\mu,\sigma)}$ with unknown parameters μ and σ.

The two relevant distribution in this case are as follows.

1) $\gamma_{(k,\lambda)}(t)$ is the most useful probability distribution for describing the precision parameter $\tau = 1/\sigma^2$.
2) The t distribution: if X follows a Normal distribution with precision parameter τ, and τ itself is γ distributed, then X actually follows a t distribution.

A.4.3.3 The Parameter μ?

Whether the parameter μ is known or unknown does not influence which distributions are relevant.

B

Solutions to Exercises

B.1 Introduction

1 **Review:** Read the chapter itself until you have found your answers.

2 The symmetrical alternatives are sequences of H and T. The best way to sort them is alphabetically: $HHH, HHT, HTH, HTT, THH, THT, TTH, TTT$. We count a total of eight sequences, of which three (HHT, HTH, and THH) have two heads and one tails, so

$$P(2H \text{ and } 1T) = \frac{3}{8}.$$

The Bayesian Way: Introductory Statistics for Economists and Engineers, First Edition.
Svein Olav Nyberg.
© 2019 John Wiley & Sons, Inc. Published 2019 by John Wiley & Sons, Inc.

3 We make a table like the one we did for D_4 and D_6 above, and get 36 squares. We write in the outcomes of the two D_6 dice. The squares with sum three are $(1, 2)$ and $(2, 1)$, so

$$P(3) = \frac{2}{36} = \frac{1}{18}.$$

The squares with sum seven are far more numerous. They are $(1, 6)$, $(2, 5)$, $(3, 4)$, $(4, 3)$, $(5, 2)$, and $(6, 1)$, so

$$P(7) = \frac{6}{36} = \frac{1}{6}.$$

So if you are betting on the outcome of a toss of two 6-sided dice, it pays to bet on seven rather than on three.

4 Yes. By the frequentist definition, this means that the probability of burning a given pancake is 0.137. We will later see how the Bayesian probability also converges to 0.137 as the number of pancakes increases.

B.2 Data

1 Review: Read the chapter itself until you have found your answers.

2 Measures of location:

(a) i. Mode = 2443 (Drevsjø; central location near Femunden, and has the most members)

  ii. Median = 6020 (Ålesund; is in the list, but with few members, and far away from Femunden)

  iii. Mean = 4859 (Postal code Nedenes, a small place down south far away from Femunden, not in list)

  iv. Mode is best.

(b) i. Mode = 0 NOK

  ii. Median = 359 500 NOK

  iii. Mean = 1.74 NOK

  iv. Median is the most representative.

(c) i. Mode = 0 NOK

  ii. Median = 0 NOK

  iii. Mean = 194 000 NOK

  iv. Mean is the most representative.

3 The position number ... The vowels are *a, e, i, o, u*, and their positions are $1, 5, 9, 15, 21$. $\Sigma_x = 51$, so the mean = 10.2, the median = 9, and since $\Sigma_{x^2} = 773$, then $SS_x = 252.8$, which means the population standard deviation is $\sigma_x = 15.9$.

4 To find the answers in Mathematica/Wolfram Alpha, write Median[list],…, Mean[list], StandardDeviation[list]. The interquartile range is a bit heftier, since there are about 20 different versions of it, and you have to specify which:

Quantile[theList, 3/4, {{1, −1}, {0, 1}}] − Quantile[theList, 1/4, {{1, −1}, {0, 1}}].

For the exercises, we get:

(a) $\tilde{x} = -1$, $Q_3 - Q_1 = 7$, $\bar{x} = 0$, $s_x = \sqrt{13}$

(b) $\tilde{y} = \underline{0.4}$, $Q_3 - Q_1 = 10.55$, $\bar{y} = 3.577\,78$, $s_y = 5.314\,55$

(c) $\tilde{z} = \underline{68}$, $Q_3 - Q_1 = 58.25$, $\bar{z} = 78.5$, $s_z = 36.582\,8$

(d) $\tilde{w} = \underline{0.659\,64}$, $Q_3 - Q_1 = 0.519\,295$, $\bar{w} = 0.684\,33$, $s_w = 0.267\,304$.

5 The numbers are:

(a) $\tilde{y} = 5.8$, $Q_3 - Q_1 = 0.8$, $\bar{y} = 5.675$, $s_y = 0.498\,428$

(b) $\tilde{z} = 7$, $Q_3 - Q_1 = 4$, $\bar{z} = 6$, $s_z = 3.265\,99$

(c) $\tilde{w} = 0.375\,471$, $Q_3 - Q_1 = 0.331\,882$, $\bar{w} = 0.368\,707$, $s_w = 0.196\,231$.

6 You are in charge of …

(a) …

(b) $b_S = 7$, $b_M = 8$, $b_L = 8$, $b_{XL} = 11$, $b_{XXL} = 12$, $b_{3XL} = 12$, and $v_S = 90.5$, $v_M = 98$, $v_L = 106$, $v_{XL} = 115.5$, $v_{XXL} = 127$, $v_{3XL} = 139$

(c) $\tilde{x} = 112.789$, $Q_1 = 101.068$, and $Q_3 = 121.71$

(d) $\bar{x} = 112.803$, and $s_x^2 = 171.877$.

7 We give the values only – no diagrams or tables:

(a) $\tilde{x} = 18$, $Q_3 - Q_1 = 6$, $\bar{x} = 18$, $s_x = 3.481\,55$

(b) $\tilde{x} = 45$, $Q_3 - Q_1 = 30$, $\bar{x} = 45$, $s_x = 22.972\,8$

(c) $\tilde{x} = 0.343\,636$, $Q_3 - Q_1 = 0.486\,920$, $\bar{x} = 0.401\,4$, $s_x = 0.288\,17$.

8 A randomized survey …

(a) …

(b) Intervals: $(0, 13), (13, 19), (19, 35), (35, 51), (51, 65), (65, 81)$ (If you are 12.95 years old, your stated age is 12.) Finding b_k and v_k is straightforward.

(c) $\tilde{x} = 40.793\,1$ and $Q_1 = 29.146\,3$ and $Q_3 = 50.724\,1$

(d) $\bar{x} = 40.357\,6$ and $s_x = 14.324\,8$.

The solution here will depend on your specific jelly worms.

B.3 Multidimensional Data

1 **Review:** Read the chapter itself until you have found your answers.

2 You are given the data set …

(a)

k	x_k	y_k
1	−1	3
2	0	5
3	3	9
4	5	7

(b) Diagram: Plot the four points in an xy coordinate system.

(c) Covariance: $\sigma_{xy} = \frac{17}{4} = 4.25$ and $s_{xy} = \frac{17}{3} \approx 5.666\,67$.

(d) Correlation: $\rho_{xy} = r_{xy} = \frac{17}{\sqrt{455}} \approx 0.796\,972$.

3 You are given the data set …

(a) $D^* = \{(x^*, y)\}_{i \in I} = \{(-2.75,\ 3),\ (-1.75,\ 5),\ (1.25,\ 9), (3.25,\ 7)\}$

(b) $\beta = 0.747\,253$

(c) $y = 6 + 0.747\,253(x - 1.75)$

(d) $y = 4.692\,31 + 0.747\,253x$

(e) Multivariate calculus:

    i. $f(a, b) = (y_{a,b}(x_i) - y_i)^2 = \sum_{j=1}^{4}(a + bx_i - y_i)^2 = (a - b - 3)^2 +$ $(a + 3b - 9)^2 + (a + 5b - 7)^2 + (a - 5)^2 = 4a^2 + 14ab - 48a +$ $35b^2 - 118b + 164;$

    ii. $f_a = \frac{\partial}{\partial a}f = 8a + 14b - 48$ and $f_b = \frac{\partial}{\partial b}f = 14a + 70b - 118$, so $f_a = f_b = 0$ gives the solution $a = \frac{61}{13} \approx 4.692\,31$ and $b = \frac{68}{91} \approx 0.747\,253;$

    iii. $y = 4.692\,31 + 0.747\,253x.$

(f) Yes, they are identical, and *shall* be identical. Always.

(g) $y(10) = 4.692\,31 + 0.747\,253 \times 10 = 12.164\,8.$

(h)

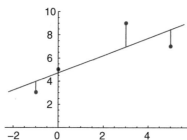

4 For the data sets below, calculate the following.

- The covariances σ_{xy} and s_{xy}.
- The correlation between x and y.
- The linear regression line. Use matrix regression.
- The square of the standard error, s_e^2.
- Illustrate at least one of them with a regression line and data points.

(a) $\sigma_{xy} = 182.928$, $\quad s_{xy} = 203.253$, $\quad \rho_{xy} = 0.351\,942$, $\quad y = 182.72 +$
$1.748\,17(x - 93.6) = 19.091\,7 + 1.748\,17x$ and $s_e^2 = 2\,827.5$

(b) $\sigma_{xy} = 21.007\,2$, $\quad s_{xy} = 26.259$, $\quad \rho_{xy} = 0.880\,048$, $\quad y = 2.136 +$
$0.549\,377(x - 4.688) = -0.439\,478 + 0.549\,377x$, and $s_e^2 = 5.600\,83$

(c) $\sigma_{xy} = 462.359$, $\quad s_{xy} = 528.411$, $\quad \rho_{xy} = 0.966\,852$, $\quad y = 39.625 +$
$0.847\,613(x - 37.625) = 7.733\,58 + 0.847\,613x$, and $s_e^2 = 36.443\,7$.

5 For the data sets below, calculate the following.
- The linear regression surface. Use matrix regression.
- The square of the standard error, s_e^2.
  - **(a)** $y = 38.271\,6 - 3.896\,61x_1 + 15.477\,4x_2$, and $s_e^2 = 1\,819.11$
  - **(b)** $y = 75.815\,5 + 0.167\,863x_1 - 0.124\,208x_2$, and $s_e^2 = 481.789$
  - **(c)** $y = 67.736\,2 + 0.201\,929x_1 + 0.345\,562x_2$, and $s_e^2 = 6.576\,54$.

B.4 Set Theory and Combinatorics

Set Theory

1 **Review:** Read the chapter itself until you have found your answers.

2 $A \setminus (B \cap C) = \{a, b, c, d, e, h, i, j\}$ (equal to A itself).

3 ...
(a) 21.
(b) KM.
(c) $K \cup M$.
(d) $|K \cup M| = |K| + |M| - |KM|$.

4 ...
(a) 7.
(b) $P \cup J$.
(c) PJ.
(d) $|P \cup J| = |P| + |J| - |PJ|$.

5 ...
(a) $\frac{47}{60}$.
(b) $p(D \cup L) = p(D) + p(L) - p(DL)$.

6 ...
(a) $\frac{1}{5}$.
(b) $p(SL) = p(S) + p(L) - p(S \cup L)$.

7 ...

(a) $A^c = \{$yellow, blue$\}$.

(b) $A \cup B\{$red, orange, yellow, green, blue, indigo, violet$\}$, and thus equal to Ω.

(c) $A \cap B = \{$green$\}$.

(d) $A \setminus B = \{$red, orange, indigo, violet$\}$.

(e) $B \setminus A = \{$yellow, blue$\}$.

(f) Yes.

(g) No.

(h) No.

(i) No.

(j) No.

Combinatorics

8 $\binom{28}{2,3,5,7,11} = \underline{1\,052\,427\,228\,652\,800}$.

9 $\binom{231}{4} = \underline{115\,584\,315}$.

10 $\binom{80}{58} = \binom{80}{22} = \underline{27\,088\,786\,024\,742\,634\,400}$.

11 ... $\binom{12}{7} = \underline{792}$.

12 $\binom{7}{3} = \underline{\underline{35}}$.

13 $\binom{60}{13} = \underline{5\,166\,863\,427\,600}$.

14 $6 \cdot 5 = \underline{\underline{30}}$.

15 $\binom{5}{3} = \underline{\underline{10}}$.

16 $\binom{7}{3} = \underline{\underline{35}}$.

17 ...

(a) 90.

(b) ...

(c) ...

18 Cinema: …
   (a) $25 \cdot 24 = \underline{\underline{600}}$.
   (b) $24 + 24 = \underline{\underline{48}}$.
   (c) $\frac{48}{600} = \underline{\underline{0.08}}$.

19 …
   (a) 3.
   (b) 3.
   (c) When the other guy picks his cookies, he has by that picked the other two for you. Conversely, once you have grabbed your two, you have chosen the remaining cookie for him.

20 …
   (a) $\binom{8}{3} = \underline{\underline{56}}$.
   (b) $\binom{8}{5} = \underline{\underline{56}}$.
   (c) When you pick 3 women to invite to your party, you have by that chosen the remaining 5 for Bram. Conversely, when he has chosen his 5, he has thereby chosen the remaining 3 for you.

21 …
   (a) $\binom{52}{30} = \underline{\underline{270\,533\,919\,634\,160}}$.
   (b) $\binom{52}{22} = \underline{\underline{270\,533\,919\,634\,160}}$.
   (c) If you see two piles of cards of 20 and 32 cards, respectively, there is no way to tell which pile was actively chosen, and which one got the remainder.

22 …
   (a) $\binom{70}{47} = \underline{\underline{1\,791\,608\,261\,879\,217\,600}}$.
   (b) $\binom{70}{23} = \underline{\underline{1\,791\,608\,261\,879\,217\,600}}$.
   (c) If you see a rosebush of 47 red and 23 white roses, respectively, there is no way to tell which color was actively chosen, and which one got the remainder.

23 …
   (a) $\binom{n}{s}$.
   (b) $\binom{n}{n-s}$.
   (c) Because $\binom{n}{s} = \frac{n!}{s!(n-s)!} = \frac{n!}{(n-s)!s!} = \binom{n}{n-s}$.

24 $\binom{34}{8,5,9,12} = 351\,045\,037\,084\,341\,600$.

25 Repeated sampling: …

(a) $12^5 = \underline{248\,832}$.

(b) $\binom{12+5-1}{5} = \underline{4\,368}$.

(c) $\frac{12!}{(12-5)!} = \underline{95\,040}$.

(d) $\binom{12}{5} = \underline{792}$.

(e) …

B.5 Probability

1 Review: Read the chapter itself until you have found your answers.

Exploration

2 Coin flipping: …

3 Resistors: …

Definitions

4 The cooler: *your local corner supermarket*

(a) 0 or 100%, but you do not know which until you have checked the carton.

(b) $\frac{1}{5} = 20\%$.

5 The cooler II: …

(a) Estimate: $\frac{1}{5} = 20\%$, since a frequentist regards future occurrences to be ruled by randomness, and thus subject to probability.

(b) $\frac{1}{5} = 20\%$; i.e. the same answer as in the previous problem, since objective Bayesian probability is degree of knowledge.

Basic Probability

6 $1 - p = 0.63$.

Conditional Probability

7 (a) $P(A \cup B) = 0.625$

(b) $P(A|B) = 0.5$

(c) $P(B|A) = 0.25$

8 $P(B) = 0.25$

9 $P(3) = 0.25$

10 (a) $P(AB) = \frac{1}{15}$
   (b) $P(A \cup B) = \frac{31}{60}$
   (c) $P(A|B) = \frac{4}{15}$

11 (a) $P(AB) = 0.075$
   (b) $P(A \cup B) = 0.425$
   (c) $P(A|B) = 0.375$

12 $P(A|B) = \frac{1}{3}$

13 $P(B|A) = 0.2$

14 ...

Repeated Sampling

15 $P(HHTHTHHHTTTHHHHTTHTHTHHT) = 8.876 \times 10^{-9}$

16 $P(14 \text{ heads in } 37 \text{ flips}) = 0.133\,5$

17 $P(21 \text{ heads in } 47 \text{ flips}) = 0.060\,7$

18 $P(30 \text{ heads in } 60 \text{ flips}) = 0.023\,2$

19 *Assorted Candies I:* you are tidying up after a party, ...
   (a) 0.35.
   (b) The respective probabilities are $P(0 \text{ Almond Joy}) = 0.116\,029$, $P(1 \text{ Almond Joy}) = 0.312\,386$, $P(2 \text{ Almond Joy}) = 0.336\,416$, $P(3 \text{ Almond Joy}) = 0.181\,147$, $P(4 \text{ Almond Joy}) = 0.048\,770\,3$, $P(5 \text{ Almond Joy}) = 0.005\,252\,19$.
   (c) The respective probabilities are $P(0 \text{ Almond Joy}) = 0.083\,010\,8$, $P(1 \text{ Almond Joy}) = 0.322\,82$, $P(2 \text{ Almond Joy}) = 0.387\,384$, $P(3 \text{ Almond Joy}) = 0.176\,084$, $P(4 \text{ Almond Joy}) = 0.029\,347\,3$, $P(5 \text{ Almond Joy}) = 0.001\,354\,49$.

20 You have a box of 40 fuses ...
   (a) $P(YBBROBROV) = 3.980\,65 \times 10^{-7}$ (with replacement).
   (b) $P(YBBROBROV) = 6.257\,75 \times 10^{-7}$ (without replacement).
   (c) $P(1V, 3B, 2R, 2O, 1Y) = 0.006\,018\,75$ (with replacement).
   (d) $P(1V, 3B, 2R, 2O, 1Y) = 0.009\,461\,71$ (without replacement).

21 $P = 0.112\,668$

22 $P = 0.000\,853\,512$

23 You have flipped a coin 50 times.

(a) $\left(\frac{1}{2}\right)^{50} \approx 8.881\,78 \times 10^{-16}$

(b) $\left(\frac{1}{2}\right)^{50} \approx 8.881\,78 \times 10^{-16}$

(c) $\binom{50}{27} = 108\,043\,253\,365\,600$

(d) $1\,125\,899\,906\,842\,624$

(e) We can calculate this in two ways, since $P(H) = P(T) = 0.5$:

- $\dfrac{108\,043\,253\,365\,600}{1\,125\,899\,906\,842\,624} \approx 0.095\,961\,7$

- $\binom{50}{27} \times \left(\frac{1}{2}\right)^{50} \approx 108\,043\,253\,365\,600 \times 8.881\,78 \times 10^{-16} = 0.095\,961\,7$

24 You've been flipping coins again.

(a) $\left(\frac{1}{2}\right)^{8} = 0.003\,906\,25$

(b) $\left(\frac{1}{2}\right)^{8} = 0.003\,906\,25$

(c) $\binom{8}{5} = 56$

(d) $2^8 = 256$

(e) We can calculate this in two ways, since $P(H) = P(T) = 0.5$:

- $\frac{56}{256} = 0.218\,75$

- $\binom{8}{5} \times \left(\frac{1}{2}\right)^{8} = 56 \times 0.003\,906\,25 = 0.218\,75.$

25 (a) $P(TTHTHTHT) = P(H)^2 \times P(T)^6 = \left(\frac{1}{3}\right)^2 \times \left(\frac{2}{3}\right)^6 = 0.009\,754\,61$

(b) $P(H)^2 \times P(T)^6 = \left(\frac{1}{3}\right)^2 \times \left(\frac{2}{3}\right)^6 = 0.009\,754\,61$

(c) No

(d) The probability of the sequence equals the product of the probabilities of the elements of the sequence. A coin flip is "with replacement", so the probability of the elements depends only on whether it is heads or tails, not on its place in the sequence. The probability of the sequence is then the probability of heads to a power equal to the number of heads, times the probability of tails to a power equal to the number of tails.

(e) $\binom{8}{6}$

(f) $\binom{8}{6} \times P(T)^6 \times P(H)^2 = 28 \times \left(\frac{1}{3}\right)^2 \times \left(\frac{2}{3}\right)^6 = 0.273\,129.$

26 Gold Digger Airlines

(a) DC-3: Half of 2 engines is 1. The probability that half the engines fail is therefore $P = \binom{2}{1} 0.001^1 \times (1 - 0.001)^{2-1} = 0.001\,998.$

DC-6B: Half of 4 engines is 2. The probability that half the engines fail is therefore $P = \binom{4}{2} 0.001^2 \times (1 - 0.001)^{4-2} = 5.988\,01 \times 10^{-6}.$

(b) DC-3: $P = \binom{2}{1} p^1 \times (1 - p)^{2-1} = 2(p - p^2).$

DC-6B: $P = \binom{4}{2} p^2 \times (1 - p)^{4-2} = 6(p - p^2)^2.$

(c) The probability of engine failure in at least half the engines is

DC-3: $P(1 \text{ or } 2 \text{ fails}) = \binom{2}{1} p^1 \times (1 - p)^{2-1} + \binom{2}{2} p^2 \times (1 - p)^{2-2} = 2p - p^2.$

DC-6B: $P(2, 3, \text{ or } 4 \text{ fail}) = \binom{4}{2} p^2 \times (1 - p)^{4-2} + \binom{4}{3} p^3 \times (1 - p)^{4-3} + \binom{4}{4} p^4 \times (1 - p)^{4-4} = 6(p - p^2)^2 = 6(1 - p)^2 p^2 + 4(1 - p)p^3 + p^4 = 6p^2 - 8p^3 + 3p^4.$

We solve the inequality $2p - p^2 \le 6p^2 - 8p^3 + 3p^4$ and see that it holds for $p \in (\frac{2}{3}, 1)$. If you really *have to* fly, you should choose the two-engine DC-3 if $p > \frac{2}{3}$.

Combinatorics

27 Birthdays:

(a) Musa can be born on 365 days, but Ibrahim must be born on one of the 364 others. So there are 365×364 ways they can be born on different days. Since the total possible ways they can be born without this restriction is 365×365, and so the probability they are not born on the same day is $\frac{365}{365} \times \frac{364}{365} \approx 0.997\,26$

(b) $\frac{365}{365} \times \frac{364}{365} \times \frac{363}{365} \approx 0.991\,796$

(c) $1 - 0.991\,796 = 0.008\,204\,17$

(d) $0.983\,644$

(e) $1 - 0.983\,644 = 0.016\,355\,9$

(f) $P(22 \text{ students all have different birthdays}) = \frac{\frac{365!}{(365-22)!}}{365^{22}} \approx 0.524\,305,$

whereas $P(23 \text{ students all have different birthdays}) = \frac{\frac{365!}{(365-23)!}}{365^{23}} \approx 0.492\,703$. In other words: when more than 23 are present, the probability that at least two of them share a birthday exceeds $\frac{1}{2}$.

28 http://en.wikipedia.org/wiki/List_of_poker_hands

B.6 Bayes' Theorem

1 Read the chapter.

2 ... **the a-ha version:**
   (a) 50%
   (b) 0% – the probability changed even though you did not observe the marble itself. All you did to update the probability was to get more information.

3 **Just calculation:**

(a)

k	$P(A_k)$	$P(B\|A_k)$	$P \times L$	$P(A_k\|B)$
1	$\frac{1}{2}$	0.7	0.35	0.7
2	$\frac{1}{2}$	0.3	0.15	0.3
			0.05	

(b)

k	$P(A_k)$	$P(B\|A_k)$	$P \times L$	$P(A_k\|B)$
1	$\frac{1}{2}$	0.07	0.035	0.7
2	$\frac{1}{2}$	0.03	0.015	0.3
			0.05	

(c)

k	$P(A_k)$	$P(B\|A_k)$	$P \times L$	$P(A_k\|B)$
1	$\frac{1}{3}$	0.12	0.04	0.12
2	$\frac{1}{3}$	0.87	0.29	0.87
3	$\frac{1}{3}$	0.01	$\frac{1}{300}$	0.01
			$\frac{1}{3}$	

(d)

k	$P(A_k)$	$P(B\|A_k)$	$P \times L$	$P(A_k\|B)$
1	$\frac{1}{3}$	0.001 2	0.000 4	0.12
2	$\frac{1}{3}$	0.008 7	0.002 9	0.87
3	$\frac{1}{3}$	0.000 1	$\frac{1}{30\,000}$	0.01
			$\frac{1}{300}$	

(e)

k	$P(A_k)$	$P(B|A_k)$	$P \times L$	$P(A_k|B)$
1	0.3	0.001	0.000 3	0.037 5
2	0.7	0.011	0.007 7	0.962 5
			0.008	

(f)

k	$P(A_k)$	$P(B|A_k)$	$P \times L$	$P(A_k|B)$
1	0.1	$\frac{2}{3}$	$\frac{2}{30}$	$\frac{22}{103}$
2	0.9	$\frac{3}{11}$	$\frac{27}{110}$	$\frac{81}{103}$
			$\frac{103}{330}$	

(g)

k	$P(A_k)$	$P(B|A_k)$	$P \times L$	$P(A_k|B)$
1	$\frac{1}{11}$	0.012	$\frac{12}{11\,000}$	$\frac{12}{1420}$
2	$\frac{8}{11}$	0.001	$\frac{8}{11\,000}$	$\frac{8}{1420}$
3	$\frac{2}{11}$	0.7	$\frac{1400}{11\,000}$	$\frac{1400}{1420}$
			$\frac{1\,420}{11\,000}$	

(h)

k	$P(A_k)$	$P(B|A_k)$	$P \times L$	$P(A_k|B)$
1	0.237 1	0.8	0.189 68	0.670 11
2	0.455 4	0.07	0.031 878	0.112 62
3	0.307 5	0.2	0.061 5	0.217 27
			0.283 058	

(i)

k	$P(A_k)$	$P(B|A_k)$	$P \times L$	$P(A_k|B)$
1	0.07	0.8	0.056	0.904 057
2	0.07	0.07	0.004 9	0.079 105
3	0.07	0.007	0.000 49	0.007 910 5
4	0.79	0.000 7	0.000 553	0.008 927 56
			0.061 943	

(j)

k	$P(A_k)$	$P(B\|A_k)$	$P \times L$	$P(A_k\|B)$
1	$\frac{1}{55}$	0.1	$\frac{1}{550}$	$\frac{1}{385}$
2	$\frac{2}{55}$	0.2	$\frac{4}{550}$	$\frac{4}{385}$
3	$\frac{3}{55}$	0.3	$\frac{9}{550}$	$\frac{9}{385}$
4	$\frac{4}{55}$	0.4	$\frac{16}{550}$	$\frac{16}{385}$
5	$\frac{5}{55}$	0.5	$\frac{25}{550}$	$\frac{25}{385}$
6	$\frac{6}{55}$	0.6	$\frac{36}{550}$	$\frac{36}{385}$
7	$\frac{7}{55}$	0.7	$\frac{49}{550}$	$\frac{49}{385}$
8	$\frac{8}{55}$	0.8	$\frac{64}{550}$	$\frac{64}{385}$
9	$\frac{9}{55}$	0.9	$\frac{81}{550}$	$\frac{81}{385}$
10	$\frac{10}{55}$	1.0	$\frac{100}{550}$	$\frac{100}{385}$
			$\frac{385}{550}$	

(k)

k	$P(A_k)$	$P(B\|A_k)$	$P \times L$	$P(A_k\|B)$
1	$\frac{1}{55}$	0.1	$\frac{10}{5500}$	$\frac{10}{220}$
2	$\frac{2}{55}$	0.09	$\frac{18}{5500}$	$\frac{18}{220}$
3	$\frac{3}{55}$	0.08	$\frac{24}{5500}$	$\frac{24}{220}$
4	$\frac{4}{55}$	0.07	$\frac{28}{5500}$	$\frac{28}{220}$
5	$\frac{5}{55}$	0.06	$\frac{30}{5500}$	$\frac{30}{220}$
6	$\frac{6}{55}$	0.05	$\frac{30}{5500}$	$\frac{30}{220}$
7	$\frac{7}{55}$	0.04	$\frac{28}{5500}$	$\frac{28}{220}$
8	$\frac{8}{55}$	0.03	$\frac{24}{5500}$	$\frac{24}{220}$
9	$\frac{9}{55}$	0.02	$\frac{18}{5500}$	$\frac{18}{220}$
10	$\frac{10}{55}$	0.01	$\frac{10}{5500}$	$\frac{10}{220}$
			$\frac{220}{5500}$	

(l)

k	$P(A_k)$	$P(B\|A_k)$	$P \times L$	$P(A_k\|B)$
1	$\frac{1}{55}$	0.1	$\frac{100}{55\,000}$	$\frac{100}{1210}$
2	$\frac{2}{55}$	0.081	$\frac{162}{55\,000}$	$\frac{162}{1210}$
3	$\frac{3}{55}$	0.064	$\frac{192}{55\,000}$	$\frac{192}{1210}$
4	$\frac{4}{55}$	0.049	$\frac{196}{55\,000}$	$\frac{196}{1210}$
5	$\frac{5}{55}$	0.036	$\frac{180}{55\,000}$	$\frac{180}{1210}$
6	$\frac{6}{55}$	0.025	$\frac{150}{55\,000}$	$\frac{150}{1210}$
7	$\frac{7}{55}$	0.016	$\frac{112}{55\,000}$	$\frac{112}{1210}$
8	$\frac{8}{55}$	0.009	$\frac{72}{55\,000}$	$\frac{72}{1210}$
9	$\frac{9}{55}$	0.004	$\frac{36}{55\,000}$	$\frac{36}{1210}$
10	$\frac{10}{55}$	0.001	$\frac{10}{55\,000}$	$\frac{10}{1210}$
			$\frac{1210}{55\,000}$	

(m) …

4 **(a)** The alternatives A_k correspond to the urns, so A_k corresponds to the event that you picked urn k.

(b) The *prior* probability of each urn must be equal, since there are no grounds to prefer one urn to another, so $P(A_k) = \frac{1}{3}$.

(c) The probability of drawing a red ball from the urns is, respectively, $P(R\|A_1) = \frac{91}{91+34} = \frac{91}{125}$, $P(R\|A_2) = \frac{14}{14+25} = \frac{14}{39}$, and $P(R\|A_3) = \frac{40}{40+25} = \frac{40}{65}$. The probability of blue is 1 minus the probability of red.

(d) $P(S\|A_1) = \left(\frac{91}{125}\right)^{20} \cdot \left(1 - \frac{91}{125}\right)^{10} \approx 3.875\,40 \cdot 10^{-9}$, $P(S\|A_2) = \left(\frac{14}{39}\right)^{20} \cdot \left(1 - \frac{14}{39}\right)^{10} \approx 1.479\,19 \cdot 10^{-11}$, $P(S\|A_3) = \left(\frac{40}{65}\right)^{20} \cdot \left(1 - \frac{40}{65}\right)^{10} \approx 4.297\,34 \cdot 10^{-9}$.

(e) Table:

k	$P(A_k)$	$P(B\|A_k)$	$P \times L$	$P(A_k\|B)$
1	$\frac{1}{3}$	$3.875\,40 \cdot 10^{-9}$	$1.291\,8 \cdot 10^{-9}$	$0.473\,329$
2	$\frac{1}{3}$	$1.479\,19 \cdot 10^{-11}$	$4.930\,63 \cdot 10^{-12}$	$0.001\,806\,64$
3	$\frac{1}{3}$	$4.297\,34 \cdot 10^{-9}$	$1.432\,45 \cdot 10^{-9}$	$0.524\,864$
			$2.729\,18 \cdot 10^{-9}$	

(f) i. $P(A_1) = P(A_2) = \frac{1}{2}$

ii. $P(H|A_1) = P(T|A_1) = \frac{1}{2}$

iii. $P(T|A_2) = 1 - P(H|A_2) = 0.593$

iv. (see table below)

v. Let B be the result TT. The table is then

k	$P(A_k)$	$P(B|A_k)$	$P \times L$	$P(A_k|B)$
1	0.5	0.25	0.125	0.415 525
2	0.5	0.351 649	0.175 825	0.584 477
			0.300 825	

vi. Let C be that you got HTH. The table is then

k	$P(A_k|B)$	$P(C|A_kB)$	$P \times L$	$P(A_k|CB)$
1	0.415 525	0.125	0.051 940 6	0.474 977
2	0.584 477	0.098 229 9	0.057 413 1	0.525 021
			0.109 354	

5 Text exercises

(a) You have 6 dice, D_1, D_2, D_3, D_4, D_5, and D_6. ...

i. The alternatives are which die was picked: $A_k =$ "he picked die D_k".

ii. The pivotal event B is that the die landed on 3.

iii. The prior probability $P(A_k)$ is the probability that your friend picked the D_k die. Since no die is preferred, each die is equally probable, so $P(A_k) = \frac{1}{6}$. The probability of getting 3 on die D_k is $P(B|A_k)$, so $P(B|A_1) = 0$, $P(B|A_2) = 0$, $P(B|A_3) = \frac{1}{3}$, $P(B|A_4) = \frac{1}{4}$, $P(B|A_5) = \frac{1}{5}$, $P(B|A_6) = \frac{1}{6}$.

iv. We get the posterior probabilities $P(A_k|B)$ from the table:

k	$P(A_k)$	$P(B|A_k)$	$P \times L$	$P(A_k|B)$
1	$\frac{1}{6}$	0	0	0
2	$\frac{1}{6}$	0	0	0
3	$\frac{1}{6}$	$\frac{1}{3}$	$\frac{1}{18}$	$\frac{20}{57}$
4	$\frac{1}{6}$	$\frac{1}{4}$	$\frac{1}{24}$	$\frac{5}{19}$
5	$\frac{1}{6}$	$\frac{1}{5}$	$\frac{1}{30}$	$\frac{4}{19}$
6	$\frac{1}{6}$	$\frac{1}{6}$	$\frac{1}{36}$	$\frac{10}{57}$
			$\frac{19}{120}$	

(b) **i.** Your task is to find the probability that the outcome was one of the top six values on the die, given that the outcome was an even number. We will use the Bayes apparatus, with the first alternative A_1 being that the outcome is in the top six, and the second alternative, A_2, being that it wasn't.

    ii. The pivotal event B is that you know the outcome was an even number.

    iii. You already know that $P(A_1) = \frac{4}{5}, P(B) = \frac{9}{15}$, and $P(B|A_1) = \frac{2}{3}$. With all this information, Bayes' *formula* will do the job.

    iv. Bayes' formula says that $P(A_1|B) = \frac{P(A_1) \cdot P(B|A_1)}{P(B)} = \frac{\frac{4}{5} \cdot \frac{2}{3}}{\frac{9}{15}} = \frac{8}{9}$.

(c) • $0.5 \cdot 0.7 + 0.5 \cdot 0.1 = \underline{0.4}$.

    • We use Bayes' theorem as follows.

    i & iii. The alternatives are which team was sent to the Olympics. A_1 = The Clean Team was sent to the Olympics, whereas A_2 = The Steroid Team was sent to the Olympics. The *prior* probabilities are then $P(A_1) = \frac{1}{3}$ and $P(A_2) = \frac{2}{3}$.

    ii & iii. The pivotal event B is that a lifter from the team tested negatively for steroids. The likelihood is the probability of the negative test, given the team, and equals the proportion of *non-*steroid users on the team, so $P(B|A_1) = 1 - 10\% = 1 - 0.1 = 0.9$ and $P(B|A_2) = 1 - 70\% = 1 - 0.7 = 0.3$.

    iv. Since we are after only one posterior probability, $P(A_1|B)$, it is quicker to use the formula in *Bayes' theorem* than to use a table. We get:

$$P(A_1|B) = \frac{P(A_1) \cdot P(B|A_1)}{P(A_1) \cdot P(B|A_1) + P(A_2) \cdot P(B|A_2)}$$

$$= \frac{\frac{1}{3} \cdot 0.9}{\frac{1}{3} \cdot 0.9 + \frac{2}{3} \cdot 0.3} = \underline{0.6}.$$

(d) **i.** The question to be answered is "What is the probability that Sofia was assigned to Halfling?" So Halfling is one alternative, A_1. The others are then the alternatives to Halfling: A_2 = Freequen, and A_3 = Gnormal.

    ii. The pivotal event B is the crash when tested with a 100-variable problem.

    iii. One out of ten testers were assigned to Halfling, so the prior probability Sophie was assigned there is $P(A_1) = \frac{1}{10}$. Six out of ten were assigned to Freequen, meaning $P(A_2) = \frac{6}{10}$, and 3 out of 10 got Gnormal, so $P(A_3) = \frac{3}{10}$.

The likelihood is the probability of a crash, given the software. Since Halfling crashes 30% of the time, $P(B|A_1) = 0.3$, and similarly for the others, $P(B|A_2) = 0.1$ and $P(B|A_3) = 0.2$.

iv. We calculate the posterior probabilities $P(A_k|B)$ in a table:

k	$P(A_k)$	$P(B|A_k)$	$P \times L$	$P(A_k|B)$
1	0.1	0.3	0.03	0.2
2	0.6	0.1	0.06	0.4
3	0.3	0.2	0.06	0.4
			0.15	

6 **(a)** $A_1 = $ "Claus picked the Danish bag", $A_2 = $ "Claus picked the Norwegian bag", $A_3 = $ "Claus picked the Swedish bag".

(b) Since no bag is preferred, $P(A_1) = P(A_2) = P(A_3)$. Together with $P(A_1) + P(A_2) + P(A_3) = 1$, this means that $P(A_k) = \frac{1}{3}$ for $k = 1, 2, 3$.

(c) The pivotal event B is the squeezing of 8 gifts, where it turned out that 4 were soft.

(d) We do our calculations for sampling *with* replacement, so
- $P(B|A_1) = 0.7^4 \cdot 0.3^4 = 0.001\,944\,81$
- $P(B|A_2) = 0.4^4 \cdot 0.6^4 = 0.003\,317\,76$
- $P(B|A_3) = 0.2^4 \cdot 0.8^4 = 0.000\,655\,36$.

(e) We do this in a table:

k	$P(A_k)$	$P(B|A_k)$	$P \times L$	$P(A_k|B)$
1	$\frac{1}{3}$	0.001 944 81	0.000 648 27	0.32863
2	$\frac{1}{3}$	0.003 317 76	0.001 105 92	0.560 628
3	$\frac{1}{3}$	0.000 655 36	0.000 218 453	0.110 741
			0.001 972 64	

(f) We found, using our total knowledge up to know, new and updated probabilities of which bag Mr. Claus had taken on board his sleigh. The new probabilities are $P(\text{Danish bag}) \approx 33\%$, $P(\text{Norwegian bag}) \approx 56\%$, $P(\text{Swedish bag}) \approx 11\%$.

(g) $C = $ sampling 42 new gifts, 17 are soft.
- $P(C|A_1 B) = 0.7^{17} \cdot 0.3^{25} = 1.971\,05 \cdot 10^{-16}$
- $P(C|A_2 B) = 0.4^{17} \cdot 0.6^{25} = 4.884\,29 \cdot 10^{-13}$
- $P(C|A_3 B) = 0.2^{17} \cdot 0.8^{25} = 4.951\,76 \cdot 10^{-15}$.

The table is then

k	$P(A_k)$	$P(B\|A_k)$	$P \times L$	$P(A_k\|B)$
1	0.328 63	$1.971\,05 \cdot 10^{-16}$	$6.477\,47 \cdot 10^{-17}$	0.000 236 025
2	0.560 628	$4.884\,29 \cdot 10^{-13}$	$2.738\,27 \cdot 10^{-13}$	0.997 766
3	0.110 741	$4.951\,76 \cdot 10^{-15}$	$5.483\,65 \cdot 10^{-16}$	0.001 998 12
			$2.744\,4 \cdot 10^{-13}$	

meaning the new probabilities are P(Danish bag) $\approx 0.2‰$, P(Norwegian bag) $\approx 99.8\%$, P(Swedish bag) $\approx 2‰$.

7 **(a)** $P(A) = \frac{3}{3+1} = 0.75$ and $P(G) = \frac{1}{3+1} = 0.25$.
(b) We answer by using the Bayes table:

k	$P(A_k)$	$P(B\|A_k)$	$P \times L$	$P(A_k\|B)$
A	0.75	$\frac{7}{69}$	$\frac{21}{276}$	$\frac{21}{51}$
G	0.25	$\frac{10}{23}$	$\frac{30}{276}$	$\frac{30}{51}$
			$\frac{51}{276}$	

(c) Let $C =$ you get marzipan in your calendar on December 3rd. When we are now calculating the probability of getting marzipan, we must not forget that we have already eaten 2, so the remainder is either 6 or 14, depending on which calendar we have, so $P(C|AB) = \frac{6}{22}$ and $P(C|AB) = \frac{14}{22}$. We calculate the total probability without using a table this time, and

$$P(C|B) = P(C|AB) \cdot P(A|B) + P(C|GB) \cdot P(G|B)$$

$$= \frac{6}{22} \cdot \frac{21}{51} + \frac{14}{22} \cdot \frac{30}{51} = \frac{91}{187}.$$

8 Let $A_1 =$ classical, $A_2 =$ milk chocolate only, and $A_3 =$ has 15 of each.
(a) The principle of symmetry says that your prior probabilities should equal their proportions, so

$$P(A_1) = \frac{10\,000}{10\,000+4\,000+6\,000} = 0.5;$$

$$P(A_2) = \frac{4\,000}{10\,000+4\,000+6\,000} = 0.2;$$

$$P(A_3) = \frac{6\,000}{10\,000 + 4\,000 + 6\,000} = 0.3.$$

(b) We calculate the likelihoods of our sequence by the formula of ordered sampling *without* replacement.

- $P(B|A_1) = \frac{\binom{30-5}{22-3}}{\binom{30}{22}} = \frac{308}{10179}$

- $P(B|A_2) = 0$ (Why can we tell this without using the formula?)

- $P(B|A_3) = \frac{\binom{30-5}{15-3}}{\binom{30}{15}} = \frac{35}{1044}.$

(c) We find the *posterior* probabilities by means of the Bayes table:

k	$P(A_k)$	$P(B|A_k)$	$P \times L$	$P(A_k|B)$
1	0.5	$\frac{308}{10\,179}$	$\frac{154}{10\,179}$	$\frac{176}{293}$
2	0.2	0	0	0
3	0.3	$\frac{35}{1044}$	$\frac{7}{696}$	$\frac{117}{293}$
			$\frac{2\,051}{81\,432}$	

9 Here, we use Bayes' *formula*, and let S = member of secret society, and L = lied to protect client.

$$P(S|L) = \frac{P(S) \cdot P(L|S)}{P(L)} = \frac{\frac{1}{4} \cdot \frac{2}{3}}{\frac{1}{2}} = \frac{1}{3}.$$

10 (a) We name the bags $A_1, \ldots, A_4$, in the order from the question.

i. $P(A_k) = \frac{1}{4}.$

ii. Ordered sampling (B) with replacement gives us:

- $P(B|A_1) = \left(\frac{5}{8}\right)^3 \cdot \left(\frac{3}{8}\right)^4 = \frac{10\,125}{2\,097\,152}$

- $P(B|A_2) = \left(\frac{7}{15}\right)^3 \cdot \left(\frac{8}{15}\right)^4 = \frac{1\,404\,928}{170\,859\,375}$

- $P(B|A_3) = \left(\frac{2}{11}\right)^3 \cdot \left(\frac{9}{11}\right)^4 = \frac{52\,488}{19\,487\,171}$

- $P(B|A_4) = \left(\frac{11}{24}\right)^3 \cdot \left(\frac{13}{24}\right)^4 = \frac{38\,014\,691}{4\,586\,471\,424}.$

iii. The Bayes table gives us:

k	$P(A_k)$	$P(B\|A_k)$	$P \times L$	$P(A_k\|B)$
1	$\dfrac{1}{4}$	$\dfrac{10\,125}{2\,097\,152}$	0.001 206 994 057	0.200 893
2	$\dfrac{1}{4}$	$\dfrac{1\,404\,928}{170\,859\,375}$	0.002 055 678 829	0.342 148
3	$\dfrac{1}{4}$	$\dfrac{52\,488}{19\,487\,171}$	0.000 673 366 083	0.112 075
4	$\dfrac{1}{4}$	$\dfrac{38\,014\,691}{4\,586\,471\,424}$	0.002 072 109 880	0.344 883
			$P(B) = 0.006\,008\,15$	

(b) Observation C gives new likelihoods ….

- $P(C\|A_1B) = \left(\dfrac{5}{8}\right)^2 \cdot \left(\dfrac{3}{8}\right)^1 = 0.146\,484$
- $P(C\|A_2B) = \left(\dfrac{7}{15}\right)^2 \cdot \left(\dfrac{8}{15}\right)^1 = 0.116\,148$
- $P(C\|A_3B) = \left(\dfrac{2}{11}\right)^2 \cdot \left(\dfrac{9}{11}\right)^1 = 0.027\,0473$
- $P(C\|A_4B) = \left(\dfrac{11}{24}\right)^2 \cdot \left(\dfrac{13}{24}\right)^1 = 0.113\,788$

… and a new table:

k	$P(A_k)$	$P(B\|A_k)$	$P \times L$	$P(A_k\|B)$
1	0.200 893	0.146 484	0.029 427 6	0.264 061
2	0.342 148	0.116 148	0.039 739 8	0.356 595
3	0.112 075	0.027 047 3	0.003 031 33	0.027 200 9
4	0.344 883	0.113 788	0.039 243 5	0.352 142
			$P(B) = 0.111\,442$	

(c) Remember that you returned your bag, and that your brother has picked a new one. Your old *posterior* applied to your bag, not his, so your brother starts afresh with prior $P(A_k) = 0.25$. His sampling, D, is *without* replacement …

- $P(D\|A_1) = \dfrac{\binom{8-3}{5-1}}{\binom{8}{5}} = 0.089\,285\,7$
- $P(D\|A_2) = \dfrac{\binom{15-3}{7-1}}{\binom{15}{7}} = 0.143\,59$
- $P(D\|A_3) = \dfrac{\binom{11-3}{2-1}}{\binom{11}{2}} = 0.145\,455$

- $P(D|A_4) = \dfrac{\binom{24-3}{11-1}}{\binom{24}{11}} = 0.141\,304$

... and his table is:

k	$P(A_k)$	$P(D|A_k)$	$P \times L = P(DA_k)$	$P(A_k|D)$
1	0.25	0.089 285 7	0.022 321 4	0.171 824
2	0.25	0.143 59	0.035 897 5	0.276 329
3	0.25	0.145 455	0.036 363 8	0.279 918
4	0.25	0.141 304	0.035 326	0.271 929
			$P(D) = 0.129\,909$	

11 We see that the one-step update gives the same probabilities as the two-step update in the example.

12 **(a)** $P_0(A_1) = P_0(A_2)\frac{1}{2}$.

(b) $p = \frac{1}{2}$.

(c) $P = 0$, because either the die is all black, and cannot land white on the first toss, or it is all white, and cannot land black on the second toss.

(d) $p = P(W) = 0.5$ and $P(WB) = p \cdot (1 - p) = 0.25$.

(e) As we saw in the exercises above: no. The probability is equal for a single toss, but different for more than one toss.

13 **(a)** $P_1(D_4) = 0$, $P_1(D_6) = 0.311\,745$, $P_1(D_8) = 0.263\,035$, $P_1(D_{10}) = 0.202\,011$, $P_1(D_{12}) = 0.155\,872$, $P_1(D_{20}) = 0.067\,336\,9$.

(b) $p = 0.485\,577$.

(c) $P(RRRRR) = p^5 = 0.026\,995\,3$.

(d) $p = 0.052\,004\,3$.

(e) As we saw in assignment 12, the probabilities are equal for one trial, but different for repeated trials.

14 **(a)** 0.5.

(b) $P(B) = 0.804\,718$. (So then, $P(A) = 0.195\,282$.)

(c) $p = P(A) \cdot 0.025 + P(B) \cdot 0.005 = 0.008\,905\,64$.

(d) This is where you need the tool. The probability that exactly k of the 3000 cars has brake problems is

$$P(k) = \binom{3000}{k} \left(P(A) \cdot 0.025^k (1 - 0.025)^{3000-k} + P(B) \cdot 0.005^k (1 - 0.005)^{3000-k} \right).$$

To find the probability that 50 or more have problems, you must either sum over all the k from 50 and up to 3000, like this, $\sum_{k=50}^{3000} P(k)$, or you must make use of the complementary probability and calculate $1 - \sum_{k=0}^{49} P(k)$. This is not to be done by hand, whichever way you choose. The answer is $P(50$ or more of the cars have brake problems$)$ = 0.195 126.

(e) You could have calculated as if the origin of each car was independently sampled. Then, each car would have an independent probability $p = 0.008\,905\,64$ of brake error, and the probability of k cars with brake errors would be

$$P(k) = \binom{3000}{k} p^k (1-p)^{3000-k}.$$

This would give $P(50$ or more of the cars have brake problems$)$ = 0.000 033 643 6, which is wrong. It is wrong because the origin of each car is *not independent*, but rather the opposite, since they all come from the same factory.

B.7 Stochastic Variables on $\mathbb{R}$

1 Read the chapter.

Discrete Stochastic Variables

2 (a) • Table:

x	$P(X = x)$
0	0.5
1	0.5

• $E[X] = 0.5$, $Var(X) = 0.25$, $\sigma_X = 0.5$, $\tau_X = 4$

(b) • Table:

x	$P(X = x)$
0	0.25
1	0.5
2	0.25

• $E[X] = 1$, $Var(X) = 0.5$, $\sigma_X = \sqrt{0.5} \approx 0.707$, $\tau_X = 2$

(c) • Table:

x	$P(X = x)$
0	0.125
1	0.375
2	0.375
3	0.125

• $E[X] = 1.5$, $Var(X) = 0.75$, $\sigma_X = \sqrt{0.75} \approx 0.866$, $\tau_X = \frac{4}{3}$

(d) • Table:

x	$P(X = x)$
1	0.25
2	0.25
3	0.25
4	0.25

• $E[X] = 2.5$, $Var(X) = 1.25$, $\sigma_X = \sqrt{1.25} \approx 1.118$, $\tau_X = 0.8$

(e) • Table:

x	$P(X = x)$
2	0.062 5
3	0.125
4	0.187 5
5	0.25
6	0.187 5
7	0.125
8	0.062 5

• $E[X] = 5$, $Var(X) = 2.5$, $\sigma_X = \sqrt{2.5} \approx 1.58$, $\tau_X = 0.4$

(f) • Table:

x	$P(X = x)$
1	$\frac{1}{6}$
2	$\frac{1}{6}$
3	$\frac{1}{6}$
4	$\frac{1}{6}$
5	$\frac{1}{6}$
6	$\frac{1}{6}$

- $E[X] = 3.5$, $Var(X) = \frac{35}{12} \approx 2.92$, $\sigma_X = \sqrt{\frac{35}{12}} \approx 1.71$, $\tau_X = \frac{12}{35}$

(g) • Table:

x	$P(X = x)$
2	$\frac{1}{36}$
3	$\frac{2}{36}$
4	$\frac{3}{36}$
5	$\frac{4}{36}$
6	$\frac{5}{36}$
7	$\frac{6}{36}$
8	$\frac{5}{36}$
9	$\frac{4}{36}$
10	$\frac{3}{36}$
11	$\frac{2}{36}$
12	$\frac{1}{36}$

- $E[X] = 7$, $Var(X) = \frac{35}{6} \approx 5.83$, $\sigma_X = \sqrt{\frac{35}{6}} \approx 2.42$, $\tau_X = \frac{6}{35}$

3 (a) • Table:

x	$f(x)$
1	0.25
2	0.5
3	0.75
4	1

- The sum of the values of $f(x)$ is $2.5 \neq 1$, so f is not a discrete probability distribution.

(b) • Table:

x	$f(x)$
1	0.1
2	0.2
3	0.3
4	0.4

- The sum of the values of $f(x)$ is 1, and $f(x) \geq 0$ for all x, so it is a discrete probability distribution.
- $\mu_X = 3$, $Var(X) = 1$, $\sigma_X = 1$, $\tau_X = 1$, $P(X \in \{1, 4\}) = 0.5$

(c) • Table:

x	$f(x)$
1	$\frac{1}{k}$
2	$\frac{2}{k}$
3	$\frac{3}{k}$
4	$\frac{4}{k}$
5	$\frac{5}{k}$
6	$\frac{6}{k}$

- The sum of the values of $f(x)$ is $\frac{21}{k}$, and $f(x) \geq 0$ for all x, which means it is a discrete probability distribution iff $k = 21$. We set $k = 21$, and then
- $\mu_X = \frac{13}{3}$, $Var(X) = \frac{20}{9}$, $\sigma_X = \frac{2\sqrt{5}}{3}$, $\tau_X = \frac{9}{20}$, $P(X \in \{2, 4, 6\}) = \frac{4}{7}$

4 (a) $\sum_{n=1}^{4} \frac{k}{(n-1)!} = \frac{8}{3}k$, which is 1 when $k = \frac{3}{8}$.

(b)

x	$P(X = x)$
1	0.375
2	0.375
3	0.187 5
4	0.062 5

(c) $\mu_Y = 1 \cdot 0.375 + 2 \cdot 0.375 + 3 \cdot 0.187\,5 + 4 \cdot 0.062\,5 = 1.937\,5$.

(d) $E[Y^5] = 1^5 \cdot 0.375 + 2^5 \cdot 0.375 + 3^5 \cdot 0.187\,5 + 4^5 \cdot 0.062\,5 = 121.938$.

Continuous Stochastic Variables

5 • Yes. This is a continuous probability distribution.
- No, this is *not* a continuous probability distribution.
- Yes, this is a continuous probability distribution.
- No, this is *not* a continuous probability distribution.

6 •

$$\mu_X = \int_{-10}^{0} x \cdot \frac{1}{40}\, dx + \int_{0}^{10} x \cdot \frac{3}{40}\, dx = \frac{5}{2}$$

$$\sigma_X^2 = \int_{-10}^{0} x^2 \cdot \frac{1}{40}\, dx + \int_{0}^{10} x^2 \cdot \frac{3}{40}\, dx - \left(\frac{5}{2}\right)^2 = \frac{100}{3}$$

$$P(X \geq 5) = \int_{5}^{10} \frac{3}{40}\, dx = 0.375.$$

- This is not a continuous probability distribution.
-

$$\mu_X = \int_{\pi}^{2\pi} -\frac{x}{2}\sin(x)\, dx = \frac{3\pi}{2}$$

$$\sigma_X^2 = \int_{\pi}^{2\pi} -\frac{x^2}{2}\sin(x)\, dx - \left(\frac{3\pi}{2}\right)^2 = \frac{\pi^2}{4} - 2$$

$$P(X \geq 5) = \int_{5}^{2\pi} -\frac{1}{2}\sin(x)\, dx = \sin^2\left(\frac{5}{2}\right) \approx 0.358\,169.$$

- This is not a continuous probability distribution.

7 (a) • Graph ... draw an ordinary function graph.
- $\int_{-\infty}^{\infty} f(x)\, dx = \int_{0}^{1} 2x\, dx = 1$, and $f(x) \geq 0$ for all x, so f is a continuous probability distribution.
- $\mu_X = \frac{2}{3}$, $Var(X) = \frac{1}{18}$, $\sigma_X = \frac{1}{3\sqrt{2}}$, $\tau_X = 18$, $P(X \in [0.5,\, 1.5]) = 0.75$.

(b) • Graph ... draw an ordinary function graph.
- $\int_{-\infty}^{\infty} f(x)\, dx = \int_{-\frac{\pi}{2}}^{\frac{\pi}{2}} \cos(x)\, dx = 2$, so f is *not* a continuous probability distribution.

(c) • Graph ... draw an ordinary function graph.
- $\sin(x)$ is not positive in the interval $[-\frac{\pi}{2}, 0]$, so f is *not* a continuous probability distribution.

(d) • Graph ... draw an ordinary function graph.
- $\int_{-\infty}^{\infty} f(x)\, dx = \int_{0}^{1} 3x^2\, dx = 1$, and $f(x) \geq 0$ for all x, so f is a continuous probability distribution.
- $\mu_X = \frac{3}{4}$, $Var(X) = \frac{3}{80}$, $\sigma_X = \frac{\sqrt{3}}{4\sqrt{5}}$, $\tau_X = 22.5$, $P(X \in [0,\, 0.5]) = 0.125$.

(e)
   i. $\mu_X = 0$
   ii. $\sigma_X = \frac{1}{\sqrt{6}}$
   iii. $P(X \in (0.5, 1.5)) = 0.125$.

(f)
   i. $\bar{f}$ is not a probability distribution, because $\int_{-\infty}^{\infty} \bar{f}(x)\, dx = \frac{4}{3} \neq 1$
   ii. $k = \frac{3}{4}$
   iii. The stochastic variable X:
     A. $\mu_X = 0$
     B. $P(X \in [-2, -0.5] \cup [0.5, 2]) = 0.453\,125$.

B.8   Stochastic Variables II

Mixed Distributions

1 **(a)** $\mu_X = \underline{3.4}$, $Var(X) = \underline{24.84}$, $P(X \geq 0) = \underline{0.76}$.

   (b) $\mu_X = \underline{7.4}$, $Var(X) = \underline{51.44}$, $P(X \geq 0) = \underline{0.11}$.

   (c) $\mu_X = \underline{0}$, $Var(X) = \underline{50}$, $P(X \geq 0) = \underline{0.5}$.

   (d) $\mu_X = \underline{5}$, $Var(X) = \underline{14.2}$, $P(X \geq 0) = 0.\underline{930\,12}$.

   (e) $\mu_X = \underline{2.08}$, $Var(X) = \underline{97.933\,6}$, $P(X \geq 0) = \underline{0.855\,163}$.

2 **(a)** We find $\mu_c = 5$ and $\mu_d = 4$, so $\mu_X = \underline{4.7}$. Further, $Var(X_c) = E[X_c^2] -$

   $\mu_c^2 = \int_0^{10} x \times \frac{1}{10}\,dx - 5^2 = \frac{25}{3}$, so $Var(X) = 0.7\,Var(X_c) + 0.3\,Var(X_d) +$
   $0.7 \times 0.3(5-4)^2 = \underline{6.043\,33}$.

   (b) We find $\mu_c = \int_0^1 x \times 2x\,dx = \underline{2/3}$ and $\mu_d = 0$, so $\mu_X = \underline{0.1}$. Further,

   $Var(X_c) = \int_0^1 x^2 \times 2x\,dx - (\frac{2}{3})^2 = \frac{1}{18}$, so $Var(X) = \underline{0.065}$.

   (c) We find $\mu_c = \frac{\pi}{2}$ and $\mu_d = \pi$, so $\mu_X = \underline{\frac{7\pi}{10}}$. Further, $Var(X_c) = \frac{\pi^2-8}{4}$, so

   $Var(X) = \underline{\frac{21\pi^2-120}{100}}$.

3 We find $\mu_c = 5$ and $\mu_d = 6.75$, so $\mu_X = \underline{6.4}$. Further, $Var(X_c) = \frac{25}{3}$ and
   $Var(X_d) = 8.437\,5$, so $Var(X) = \underline{14.552\,5}$.

4 We first decompose X: the probability of no loss is $p_0 = 0.88$. That gives us
   $w_d = p_0 = 0.88$, and $w_c = 1 - w_d = 0.12$.
   Further,

$$f_d(x) = \begin{cases} 1 \text{ for } x = 0 \\ 0 \text{ otherwise.} \end{cases}$$

   The distribution of the payment, given that the tour runs at a loss, equals
   the conditional distribution of the loss,

$$f_c(x) = g(x) = 5 \times 10^{-7} e^{-5 \times 10^{-7} x} \text{ (for } x > 0\text{)}.$$

   This gives us $\mu_d = 0$, $Var(X_d) = 0$, $\mu_c = 2 \times 10^6$, and $Var(X_c) = 4 \times 10^{12}$.
   Then

$$\mu_X = 0.88 \times 0 + 0.12 \times 2 \times 10^6 = 240\,000.00$$
$$Var(X) = 0.88 \times 0 + 0.12 \times 4 \times 10^{12} + 0.88 \times 0.12(2 \times 10^6 - 0)^2$$
$$= 9.024 \times 10^{11}$$

$$\sigma_X = \sqrt{Var(X)} = 949\,947.00.$$

The charge for Sandra's policy is then

$$charge = 1.7 \times \mu_X + 0.1 \times \sigma_X = 1.7 \times 240\,000.00 + 0.1 \times 949\,947.00$$
$$= \underline{502\,995.00}.$$

5 We decompose X: The probability of no loss is $p_0 = 0.88$. Loss in excess of 5 million gives a payment of 5 million, so $p_{5\text{mill}} = P(\text{loss}) \times P(\text{loss} > 5\,000\,000 | \text{loss}) = 0.12 \times \int_{5\,000\,000}^{\infty} f(x)\,dx = 0.009\,850\,2$. Outside of these two, X has no positive point probabilities. So $w_d = p = p_0 + p_{5\text{mill}} = 0.889\,850\,2$, and $w_c = 1 - w_d = 0.110\,15$.

This means

$$f_d(x) = \begin{cases} \dfrac{0.88}{0.889\,850\,2} = 0.988\,93 & \text{for } x = 0 \\[2mm] \dfrac{0.009\,850\,2}{0.889\,850\,2} = 0.011\,069\,5 & \text{for } x = 5 \text{ million.} \end{cases}$$

The continuous distribution applies to a loss $x \in I = [0, 5\,000\,000]$. We use the distribution of the loss, $f(x) = 5 \times 10^{-7} e^{-5 \times 10^{-7} x}$, so $\int_0^{5\,000\,000} f(x)\,dx = 0.917\,915$. For $x \in I$, we have

$$f_c(x) = \frac{5 \times 10^{-7} e^{-5 \times 10^{-7} x}}{0.917\,915} = 5.447\,13 e^{-5 \times 10^{-7} x}.$$

We get $\mu_d = 55\,347.50$, $Var(X_d) = 2.736\,74 \times 10^{11}$, $\mu_c = 1\,552\,872$, and $Var(X_c) = 1.564\,44 \times 10^{12}$. Then

$$\mu_X = 0.889\,850\,2 \times 55\,347.50 + 0.110\,15 \times 1\,552\,872$$
$$= 220\,300.00$$
$$Var(X) = 0.889\,850\,2 \times 2.736\,74 \times 10^{11} + 0.110\,15 \times 1.564\,44 \times 10^{12}$$
$$+ 0.889\,850\,2 \times 0.110\,15 \times (1\,552\,872 - 55\,347.50)^2$$
$$= 6.356\,62 \times 10^{11}$$
$$\sigma_X = \sqrt{Var(X)} = 797\,284.00.$$

For this policy, Sandra will be charged

$$charge = 1.7 \times \mu_X + 0.1 \times \sigma_X = 1.7 \times 220\,300.00 + 0.1 \times 797\,284.00$$
$$= \underline{454\,238.00}.$$

Two- and Multi-variable Probability Distributions

6 Table, with the marginal probabilities in red:

	$X = 1$	$X = 2$	$X = 3$	
$Y = 1$	0.05	0.05	0.3	0.4
$Y = 2$	0.05	0.25	0.05	0.35
$Y = 3$	0.2	0.05	0	0.25
	0.3	0.35	0.35	1

$P(X + Y = 4) = 0.75$ and $\rho_{XY} = -0.694\,203$.

7 The philosophers Max Stirner and Karl Werder:

(a)

	$X = 1$	$X = 2$	$X = 3$
$Y = 2$	0	0.1	0.3
$Y = 3$	0.05	0.3	0.6
$Y = 4$	0.2	0.65	1

(b) $P(X + Y = 5) = P(X = 1 \ \& \ Y = 4) + P(X = 2 \ \& \ Y = 3) + P(X = 3 \ \& \ Y = 2) = 0.15 + 0.15 + 0.2 = \underline{0.5}$.

(c) • $P(X = 1 | X + Y = 5) = \frac{P(X=1 \& X+Y=5)}{P(X+Y=5)} = \frac{P(X=1 \& Y=4)}{P(X+Y=5)} = \frac{0.15}{0.5} = \underline{\underline{0.3}}$

• $P(X = 2 | X + Y = 5) = \frac{P(X=2 \& X+Y=5)}{P(X+Y=5)} = \frac{P(X=2 \& Y=3)}{P(X+Y=5)} = \frac{0.15}{0.5} = \underline{\underline{0.3}}$

• $P(X = 3 | X + Y = 5) = \frac{P(X=3 \& X+Y=5)}{P(X+Y=5)} = \frac{P(X=3 \& Y=2)}{P(X+Y=5)} = \frac{0.2}{0.5} = \underline{\underline{0.4}}$.

(d) The probability of getting a questions about Stirner, given that you get five questions in total about the two philosophers.

8 No, since $P(Y = y)$ is 0 iff $y > X$, and $\frac{1}{X}$ otherwise, so $P(Y = y)$ is *dependent* on the value of X.

9 $f_x(x) = \frac{1}{5}(7 - 4x)$ for $x \in [0, 1]$, whereas $f_y(y) = \frac{2}{5}(3 - 2y^3)$. Further, the covariance is $\sigma_{xy} = -\frac{1}{250}$, and since $\sigma_{xy} \neq 0$, then X and Y are *not* independent.

10 **(a)** $f_x(x) = \frac{2}{\pi}\sqrt{1 - x^2}$ when $x \in (-1, 1)$, and 0 otherwise. Same function for f_y.
(b) No, since $f_x \times f_y \neq f_Z(x, y)$.

(c) $f_{|Y=y}(x) = \frac{1}{2\sqrt{1-y^2}}$ if $|x| < \sqrt{1-y^2}$, and 0 otherwise. Correspondingly for $f_{|X=x}(y)$.

11 They are not independent. $\rho_{xy} = 0$.

The Sum of Independent Stochastic Variables

12 **(a)** $\mu_Z = 7, \sigma_Z = \sqrt{13}$
 (b) $\mu_Z = 6, \sigma_Z = 13$
 (c) $\mu_Z = 21, \sigma_Z = \sqrt{21}$
 (d) $\mu_Z = 7, \sigma_Z = 2$
 (e) $\mu_Z = k\mu, \sigma_Z = k\sigma$
 (f) $\mu_Z = 6, \sigma_Z = \dfrac{7}{2\sqrt{5}}$

13 The standard deviation of aX is $a\sigma_X$, and the standard deviation of $(1-a)Y$ is $(1-a)\sigma_Y$. The standard deviation of Z is then $\sigma_Z^2 = a^2\sigma_X^2 + (1-a)^2\sigma_Y^2$. This is a function of a, so let us name it $f(a)$. Then, $f'(a) = 2a\sigma_X^2 - 2(1-a)\sigma_Y^2$ and $f''(a) = 2\sigma_X^2 + 2\sigma_Y^2 > 0$, so we find a critical point (a minimum) when $f'(a) = 0$. That is, when $2a\sigma_X^2 - 2(1-a)\sigma_Y^2 = 0$. The minimum value of σ_Z is thus realized when $a = \dfrac{\sigma_Y^2}{\sigma_X^2 + \sigma_Y^2}$.

B.9 Discrete Distributions

B.9.1 Discrete Uniform Probability Distribution

The discrete uniform distribution has been omitted as a section so that the student may have at hand a tractable probability distribution to build up and study its properties. The first assignment is the key.

1 The general properties of a discrete uniform distribution with parameters a and b, and $n = b - a + 1$, are then:
 (a) $\mu_X = \dfrac{a+b}{2}$
 (b) $\sigma_X^2 = \dfrac{n^2-1}{12}$
 (c)

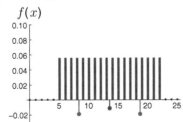

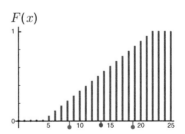

2 $P(X \in \{2, 3, 5, 7, 11, 13, 17, 19\}) = 0.08.$

3 This is identical to the previous question, and hence the answer is identical too.

4 $\mu_X = 49.5$ and $\sigma_X = \sqrt{\frac{9999}{12}} \approx 28.866\,1.$

5 $\mu_X = 29.5$ and $\sigma_X \approx 17.3$. There are 15 elements in $A = \{0, \dots, 59\}$ with one of its digits equal to 4, so $P(A) = 0.25.$

B.9.2   Bernoulli Distribution, $bern_p$

6 A Bernoulli distribution applies to the outcome of a single Bernoulli trial, and measures the probability of success in that single trial. The distribution is not used on its own for single trials, though, but is a component of composite distributions for repeated, independent Bernoulli trials – typically, the binomial or negative binomial distributions.

7 Since $\mu_X = p$, the maximal value is obtained at the highest value of p, which is $p = 1.$

8 Since $\sigma_X^2 = p(1-p)$, a function whose maximum is reached at $p = \frac{1}{2}$, the variance is maximized for $p = 0.5.$

9 The precision τ_X is the inverse of the variance, $\tau_X = \frac{1}{p(1-p)}$, which grows arbitrarily large (to infinity) as p approaches 0 and 1.

10 **(a)** $o_f = 0.812\,5/(1 - 0.812\,5) = 4.333.$
 (b) $o_f = 13 : 3.$
 (c) $o_a = 1/o_f = 3 : 13 = 0.230\,8.$
 (d) Continental European odds $= o_a + 1 = 1.230\,8.$
 (e) $p \cdot G = 0.812\,5 \cdot 100 = 81$ euro and 25 cents.
 (f) $100/0.812\,5 = 123$ euro and 8 cents.
 (g) With this bookmaker, your winnings will be $\frac{1+10}{10} = 1.1$ times your bet, which is less than the fair rate of 1.230 8, so this bet is worse for you than a fair bet, and you should not take it.

B.9.3   Binomial Distribution, $bin_{(n,p)}$

11 $X \sim f(x) = \binom{5}{x} 0.55^x \cdot 0.45^{5-x}$, and $\mu_X = 2.75$ and $\sigma_X = 1.112\,43$. Further, $P(X \in \{0, 1, 2\}) = 0.406\,873.$

12 This is identical to the last part of the previous problem, and hence the answer is also the same.

13 $X \sim f(x) = \binom{14}{x} 0.4^x \cdot 0.6^{14-x}$, and $\mu_X = 5.6$ and $\sigma_X = 1.833\,03$. Further, $P(X \in \{3, 4, 5\}) = 0.446\,063$.

14 This is identical to the last part of the previous problem, and hence the answer is also the same.

15 $Z \sim f(z) = \binom{20}{z} 0.8^z \cdot 0.2^{20-z}$, and $\mu_Z = 16$ and $\sigma_Z = 1.788\,85$. Further, $P(Z \in \{0, 1, 2, 3, 4, 5\}) = 1.8 \cdot 10^{-7}$.

16 This is identical to the last part of the previous problem, and hence the answer is also the same.

17 If you don't have a good calculator, use the Normal approximation for large numbers.

(a) $bin_{(10, 0.3)}(3) = \binom{10}{3} 0.3^3 0.7^7 = 0.266\,828 \approx 27\%$

(b) $\sum_{k=27}^{33} bin_{(100, 0.3)}(k) = 0.554\,859 \approx 55\%$

(c) $\sum_{k=270}^{330} bin_{(1\,000, 0.3)}(k) = 0.964\,751 \approx 96\%$

(d) $\sum_{k=2700}^{3\,300} bin_{(10\,000, 0.3)}(k) = 0.999\,999\,999\,942\,169\,532\,741 \approx 100\%$

18 Let the number of errors be $X \sim bin_{(70\,368\,744\,177\,664,\, 50 \times 10^{-15})}$.

(a) $E[X] = np = 3.518\,44$.

(b) $Var(X) = np(1 - p) = 3.518\,44$.

(c) Here, it pays to use an approximation – even if you have a decent calculating tool.

• Exact answer:

$$P(X \le 5) = BIN_{(70\,368\,744\,177\,664,\, 50 \times 10^{-15})}(5) = 0.855\,167\,136\,678 \ldots$$

Even with a good tool like Mathematica, you must help the tool a bit. Do not allow it to use its built-in binomial distribution, but ask it instead to calculate using the formula for a binomial distribution, but with the replacement $(1 - p)^a \approx e^{a(-p - \frac{1}{2}p^2 - \frac{1}{3}p^3 - \cdots - \frac{1}{n}p^n - \cdots)}$, stopping at some n. Note that you get high accuracy even when you stop at $n = 1$.

- Poisson approximation: $P(X \le 5) = POIS_{3.518\,437}(5) = 0.855\,167$.
- Normal approximation: $P(X \le 5) = \Phi_{(3.518\,437,\,3.518\,437)}(5 + \frac{1}{2}) = 0.713\,349$.

19 ...

(a) $-10, -8, -6, -4, -2, 0, 2, 4, 6, 8, 10$ units.

(b) We sum the probabilities $bin_{(10,\,0.51)}\left(\frac{x+10}{2}\right)$ for $x = 4, 6, 8, 10$, as in the following formula:

$$\sum_{k=2}^{5} bin_{(10,\,0.51)}(k+5) = \sum_{k=7}^{10} bin_{(10,\,0.51)}(k) \approx 19\%.$$

(c) $E[X] = (2p - 1)n = (2 \cdot 0.51 - 1) \cdot 10\,000 = 200$.

(d) $E[X] = 4np(1 - p) = 4 \cdot 10\,000 \cdot 0.51 \cdot 0.49 = 999\,6$.

(e) Calculate exactly (as in the problem with 10 steps), or employ the Normal approximation. Exact answer:

$$\sum_{k=50}^{5\,000} bin_{(10\,000,\,0.51)}(k+5\,000) = \sum_{k=5\,050}^{10\,000} bin_{(10\,000,\,0.51)}(k) \approx 84\%.$$

B.9.4 Hypergeometric Distribution, $hyp_{(n,S,N)}$

20 $f(x) = \dfrac{\binom{13}{x}\binom{7}{5-x}}{\binom{20}{5}}$ and $\mu_X = n \cdot \frac{S}{N} = 5 \cdot \frac{13}{20} = 3.25$, $\sigma_X = 0.947\,643$ and $P(X \in \{0, 1, 2\}) = 0.206\,785$.

21 This is identical to the last part of the previous problem, and hence the answer is also the same.

22 $f(x) = \dfrac{\binom{7}{x}\binom{23}{9-x}}{\binom{30}{9}}$ and $\mu_X = n \cdot \frac{S}{N} = 9 \cdot \frac{7}{30} = 2.1$, $\sigma_X = 1.079\,75$ and $P(X \in \{0, 1\}) = 0.297\,011$.

23 This is identical to the last part of the previous problem, and hence the answer is also the same.

24 Let X be the number of aces among your new cards. X follows a hypergeometric distribution with $N = 47$ (the remaining cards), $S = 3$ (the remaining aces), and $n = 3$ (the number of new cards), and thus

$$f(x) = \frac{\binom{3}{x}\binom{44}{3-x}}{\binom{47}{3}},$$

so

$$P(\text{at least 1 ace}) = 1 - P(\text{no ace}) = 1 - hyp_{(3,3,47)}(0)$$
$$= 1 - 0.817 = 0.183\,225.$$

B.9.5 Poisson Distribution, $pois_\lambda$

25 $\mu_X = \lambda = 2.8$, $\sigma_X = \sqrt{\lambda} = \sqrt{2.8} = 1.673\,32$, $P(X = 1) = \frac{2.8^1}{1!} \cdot e^{-2.8} = 0.170\,268$, and $P(X \neq 1) = 1 - P(X = 1) = 0.829\,732$.

26 Since $\mu_X = 3.5$, we have $\lambda = 3.5$. Then $\sigma_X = \sqrt{3.5} = 1.870\,83$, $P(X = 5) = \frac{3.5^5}{5!}e^{-3.5} = 0.132\,169$, and $P(X > 0) = 1 - P(X = 0) = 1 - \frac{3.5^0}{0!}e^{-3.5} = 0.969\,803$.

27 $\mu_X = \sigma_X^2 = 25$, which gives $\lambda = 25$. Then $P(X = 7) = \frac{25^7}{7!}e^{-25} = 0.000\,016\,8$ and $P(X \leq 2) = P(X = 0) + P(X = 1) + P(X = 2) = 4.7 \cdot 10^{-9}$.

28 $P(X = 0) = e^{-\lambda} = 0.1$, so $\lambda = -\ln(0.1) \approx 2.303$. Then $\mu_X = 2.303$ and $\sigma_X = \sqrt{2.303} = 1.517$. Further, $P(X = 1) = \frac{2.303^1}{1!}e^{-2.303} = 0.230\,3$.

29 $f(x) = \frac{2.37^x}{x!}e^{-2.37}$ and $\mu_X = \lambda = 2.37$, whereas $\sigma_X = \sqrt{2.37} = 1.539$. Then

$$P(X \in \{0, 1\}) = P(X = 0) + P(X = 1) = \frac{2.37^0}{0!}e^{-2.37} + \frac{2.37^1}{1!}e^{-2.37} = 0.315.$$

30 The same as the end of the above problem. The probability of 1 click or less is 0.315.

31 $f(x) = \frac{e^{-1}}{x!}$ and $\mu_X = \lambda = 1$, whereas $\sigma_X = \sqrt{\lambda} = 1$. Further, $P(X \in \{0\}) = P(X = 0) = \frac{e^{-1}}{0!} = e^{-1} \approx 0.367\,879$.

32 The same as the end of the above problem. The probability of one eviction or fewer is $0.367\,879 \approx 37\%$.

33 **(a)** $\lambda = \mu_X = \frac{30}{7} \approx 4.285\,7$.
(b) $\sigma_X = \sqrt{\lambda} = 2.070\,2$.
(c) $P(X = 5) = 0.165\,8$.

B.9.6 Overview and General

34 The difference between the binomial and uniform distributions.
(a) $P = \frac{3}{12} = 0.25$.

(b) $P = \left(\binom{12}{0} + \binom{12}{1} + \binom{12}{2} + \binom{12}{3} \right) \cdot \left(\frac{1}{2} \right)^{12} = \frac{299}{4096}.$

35 **The PirateBay problem:** No numerical answers, but the best choice of distribution is indicated. Discuss the solutions in groups; which approximations apply when the numbers are sufficiently large, and so on?
  (a) Uniform distribution.
  (b) Hypergeometric distribution.
  (c) Binomial distribution, as an approximation to hypergeometric distribution.
  (d) Poisson distribution works best for these calculations.

36 **The Autopol problem:** No numerical answers, but the best choice of distribution is indicated. Discuss the solutions in groups; which approximations apply when the numbers are sufficiently large, and so on?
  (a) Uniform distribution.
  (b) Binomial distribution.
  (c) Binomial distribution.
  (d) Binomial distribution.
  (e) Hypergeometric distribution.
  (f) Poisson distribution.

B.10 Continuous Distributions

B.10.1 Normal Distribution, $\phi_{(\mu,\sigma)}$

1 Remember that this includes both your calculators and online tools such as Wolfram Alpha and software like R that you might have on your computer

2 Cumulative Normal distribution $\Phi_{(\mu,\sigma)}$ and probability
  (a) $P(X \leq 1.43) = \Phi_{(0,1)}(1.43) = 0.923\,641.$
  (b) $P(X > 1.43) = 1 - \Phi_{(0,1)}(1.43) = 0.076\,359.$
  (c) $P(X \leq -2.38) = \Phi_{(0,1)}(-2.38) = 0.008\,656\,32.$
  (d) $P(X > -2.38) = 1 - \Phi_{(0,1)}(-2.38) = 0.991\,344.$
  (e) $P(X \leq 4.35) = \Phi_{(3,\,1.9)}(4.35) = 0.761\,311.$
  (f) $P(X > 4.35) = 1 - \Phi_{(3,\,1.9)}(4.35) = 0.238\,689.$
  (g) $P(X \leq 0.2) = \Phi_{(-7,\,4)}(0.2) = 0.964\,07.$
  (h) $P(X \in (-2.38,\ 1.43)) = \Phi_{(0,1)}(1.43) - \Phi_{(0,1)}(-2.38) = 0.914\,985.$
  (i) $P(X \in (8,\ 13)) = \Phi_{(10,5)}(13) - \Phi_{(10,5)}(8) = 0.381\,169.$

3 Inverse cumulative Normal distribution z
  (a) $z_{0.05} = -1.644\,85.$
  (b) $z_{0.95} = 1.644\,85.$
  (c) $a = 0.355\,146.$
  (d) $a = 3.644\,85.$

4 The Normal approximation

(a) $P(X \le 4) \approx \Phi_{(3,1.2)}\left(4 + \frac{1}{2}\right) = 0.894\,35.$

(b) $P(X \le 4) \approx \Phi_{(3,1.2)}(4) = 0.797\,672.$

(c) $P(X \in \{6,7\}) \approx \Phi_{(5.1,2.2)}\left(7 + \frac{1}{2}\right) - \Phi_{(5.1,2.2)}\left(6 - \frac{1}{2}\right) = 0.290\,206.$

(d) $P(X \in (5,7)) \approx \Phi_{(5.1,2.2)}(7) - \Phi_{(5.1,2.2)}(5) = 0.324\,234.$

(e) $P \approx \Phi_{(12.1,4.7)}\left(19 + \frac{1}{2}\right) - \Phi_{(12.1,4.7)}\left(12 - \frac{1}{2}\right) = 0.493\,101.$

5 Sums of Normally distributed stochastic variables

(a) $X \sim \phi_{(3,\sqrt{30})}(x).$

(b) $X \sim \phi_{(20,\sqrt{160})}(x).$

(c) $X \sim \phi_{(7,2)}(x).$

B.10.2 Binormal Distribution, $\phi_{(\mu,\Sigma)}$

X and Y independent

6 $P(X < 7, Y < 0) = P(X < 7) \cdot P(Y < 0) = \Phi_{(5,10)}(7) \cdot \Phi_{(-3,9)}(0)$
$$= 0.579\,26 \cdot 0.630\,559 = \underline{0.365\,257}.$$

X and Y dependent

7 **(a)** $f_x(x) = \phi_{(20,7)}(x)$ and $f_y(y) = \phi_{(12,8)}(y).$

(b) $f_{x|y}(x) = \phi_{(14.75+0.4375y,\,16)}(x)$
$f_{y|x}(y) = \phi_{\left(\frac{4}{7}(x+1),\,12.25\right)}(y).$

(c) $\theta = \frac{1}{2}\tan^{-1}\left(\frac{2\cdot 28}{49-64}\right) = -0.654/,541$

$$R_\theta = \begin{bmatrix} 0.793\,327 & 0.608\,795 \\ -0.608\,795 & 0.793\,327 \end{bmatrix}$$

$$\hat{\mu} = R_{-\theta}\mu = \begin{bmatrix} 8.561 \\ 21.695\,8 \end{bmatrix}$$

$$\hat{\Sigma} = R_{-\theta}\Sigma R_\theta = \begin{bmatrix} 27.5129 & 0 \\ 0 & 85.4871 \end{bmatrix}.$$

B.10.3 Gamma Distribution, $\gamma_{(k,\lambda)}$, With Family

8 $\mu_T = 0.2$, $\sigma_T = 0.2$, and $P(T \le 4) = 1.000\,0.$

9 $\mu_T = 0.434\,783$, $\sigma_T = 0.434\,783$, and $P(T \le 3.2) = 0.999\,364.$

10 $\mu_T = 0.322\,581$, $\sigma_T = 0.322\,581$, and $P(T > 1.2) = 1 - P(T \le 1.2) = 0.024\,234.$

11 $\mu_T = 0.227\,273$, $\sigma_T = 0.227\,273$, and $P(T \in \langle 0.15, 0.28]) = 0.225\,143$.

12 $\mu_T = E[T] = 2.9$. Then $\lambda = \frac{1}{2.9} \approx 0.344\,8$ and $\sigma_T = \frac{1}{\lambda} = \frac{1}{1/\lambda} = 2.9$.

13 $\sigma_T = \sqrt{Var(T)} = \sqrt{2.25} = 1.5$. Then $\lambda = \frac{1}{\sigma_T} = \frac{1}{1.5} = \frac{2}{3}$ and $\mu_T = \frac{1}{\lambda} = \frac{1}{1/\sigma_T} = 1.5$.

14 $\mu_T = 0.4$, $\sigma_T = 0.282\,843$, and $P(T \leq 1) = 0.959\,57$.

15 $\mu_T = 1.304\,35$, $\sigma_T = 0.753\,066$, and $P(T \leq 1.7) = 0.748\,411$.

16 $\mu_T = 1.612\,9$, $\sigma_T = 0.721\,312$, and $P(T > 2.3) = 0.161\,46$.

17 $\mu_T = 0.454\,5$, $\sigma_T = 0.321\,412$, and $P(T \in \langle 0.2, 2]) = 0.778\,315$.

18 $k = 2$ and $\mu_T = 2.9$, so $\lambda = \frac{k}{\mu_T} = \frac{2}{2.9} = 0.689\,7$, and then $\sigma_T = \frac{\sqrt{k}}{\lambda} = \frac{\mu_T}{\sqrt{k}} = \frac{2.9}{\sqrt{2}} = 2.050\,6$.

19 $k = 3$ and $\sigma_T^2 = 2.25$, so $\sigma_T = 1.5$. Then $\lambda = \frac{\sqrt{k}}{\sigma_T} = \frac{\sqrt{3}}{1.5} = 1.154\,7$, so $\mu_T = \frac{k}{\lambda} = \sigma_T \times \sqrt{k} = 2.598$.

20 $\lambda = \frac{1}{\mu_T} = \frac{1}{23}$, so

(a) $P(\text{one sympathizer within 30 minutes}) = EXP_{\frac{1}{23}}(30) = 1 - e^{-\frac{30}{23}} \approx 0.728\,7$;

(b) the waiting time T for *two* is then Erlang distributed, with $k = 2$ and $\lambda = \frac{1}{23}$,

$$P(T \leq 30) = 1 - \left(\frac{1}{0!} \times e^{-30\lambda} + \frac{30\lambda}{1!} \times e^{-30\lambda} \right) = 1 - \frac{53}{23} e^{-\frac{30}{23}}$$
$$= \underline{0.374\,716}.$$

21 **(a)** $P(\text{the first sale within 2 weeks}) = EXP_\lambda(2) = 1 - e^{-2\lambda}$.

(b) The waiting time for 2 sales is then Erlang distributed $erl_{(2,\lambda)}$. Let T be the waiting time in weeks for 2 sales.

$$P(T \leq 4) = 1 - \sum_{n=0}^{1} \left(\frac{(4\lambda)^n}{n!} e^{-4\lambda} \right) = 1 - e^{-4\lambda}(1 + 4\lambda).$$

This gives the equation

$$1 - e^{-4\lambda}(1 + 4\lambda) = 0.6,$$

which in turn gives $\lambda = 0.505\,6$. Then you need $\lambda > 0.505\,6$ in order to have more than a 60% probability of permanent employment.

22 Let T be the waiting time for 10 clicks.

(a) Then $T \sim erl_{(10,0.25)}$, so $E[T] = \mu_T = \frac{10}{0.25} = 40$ minutes.

(b) The probability of at least 3 clicks equals the probability that the third click has occurred, in other words

$$f(3) = ERL_{(3,0.25)}(2) \approx \underline{0.014\,39}.$$

(c) We find the *number* of clicks by means of Rule A.4.6, that is, by the Poisson distribution. Here, the parameter is $0.25 \times 2 = \underline{0.5}$, so

$$P(T = 3) = \frac{0.5^3}{3!}e^{-0.5} \approx \underline{0.012\,64}.$$

23 Call the waiting time T. Then

(a) $\mu_T = \frac{1}{\lambda} = \underline{15}$;

(b) $P(T \in (5, 20)) = EXP_{1/15}(20) - EXP_{1/15}(5)$
$$= \left(1 - e^{-\frac{20}{15}}\right) - \left(1 - e^{-\frac{5}{15}}\right) = \underline{0.452\,9}.$$

24 (a) $\lambda = \frac{1}{15}$ and $n = 3$, so $f(t) = \frac{\lambda^n t^{n-1}}{(n-1)!}e^{-\lambda t} = \frac{t^2}{6750}e^{-\frac{t}{15}}$.

(b) Probability of waiting longer than 30 minutes:

- direct: the corresponding *cumulative* distribution becomes

$$ERL_{(3,1/15)}(t) = 1 - \sum_{k=0}^{2}\frac{t^2}{15^k \times k!}e^{-\frac{t}{15}}$$

so

$$P(T > 30) = 1 - ERL_{(3,1/15)}(30) = \sum_{k=0}^{2}\frac{30^k \times e^{-2}}{15^k \times k!} = \underline{\frac{5}{e^2} \approx 0.68}.$$

- by the Normal approximation: $\mu = \frac{n}{\lambda} = 45$ and $\sigma^2 = \frac{n}{\lambda^2} = 675$, giving $\sigma = 15\sqrt{3}$. This means that

$$P(T > 30) \approx 1 - \Phi_{(45,\,15\sqrt{3})}(30) = \underline{0.72}.$$

(c) The *number* of customers have asked about the book within a limited time period is Poisson distributed, with parameter $30\lambda = 30 \times \frac{1}{15} = 2$, so

$$P(m) = \frac{2^m}{m!}e^{-2}.$$

(d) The expected value for a Poisson distribution equals the parameter λ, so the expected number of customers who have asked about the book during a period of 30 minutes is $\underline{2}$.

25 (a) $\Gamma_{(16,4)}(t) = \mathbb{X}^2_{32}(8t)$.
(b) $P(\tau < 6) = \Gamma_{(16,4)}(6) = 0.965\,6$.
(c) $\tau_0 = \Gamma^{-1}_{(16,4)}(0.9) = 5.323\,09$, so $P(\tau < 5.323\,09) = 90\%$.
(d) $P(\sigma > 0.408\,248) = 0.965\,6$ and $P(\sigma > 0.309\,205) = 90\%$.

26 (a) $\Gamma_{(19.5,\,44)}(t) = \mathbb{X}^2_{39}(88t)$.
(b) $P(\tau > 0.64) = 1 - \Gamma_{(19.5,\,44)}(0.64) = 0.035\,797\,1$.
(c) $\tau_0 = \Gamma^{-1}_{(19.5,\,44)}(1 - 0.95) = 0.291\,993$, so $P(\tau > 0.291\,993) = 95\%$.
(d) $P(\sigma < 1.25) = 0.035\,797\,1$ and $P(\sigma < 1.850\,6) = 95\%$.

27 (a) $\Gamma_{(101.5,\,15)}(t) = \mathbb{X}^2_{203}(30t)$.
(b) $P(\sigma > 2) = P(\tau < 0.25) = 4.7 \times 10^{-105} \approx 0$.
(c) This corresponds to finding a value τ_0 such that $P(\tau > \tau_0) = 98\%$, or equivalently, $P(\tau < \tau_0) = 2\%$. The solution is $\tau_0 = \Gamma^{-1}_{(101.5,\,15)}(0.02) = 5.459\,77$, so $\sigma_0 = 0.427\,969$. Thus, $P(\sigma < 0.427\,969) = 98\%$.

B.10.4 Student's t Distribution, $t_{(\mu,\sigma,\nu)}$

28 $t_{4,0.1} = -1.533\,2$.

29 $x = -3.355\,4$.

30 $x = 1.439\,8$.

31 $x = 12.39$.

32 $x = 21.95$.

33 $Z \sim t_{(4.5,\,3.8,10)}$.

B.10.5 Beta Distribution, $\beta_{(a,b)}$

34 $\mu_X = 0.5$, $\sigma_X = \frac{1}{2\sqrt{3}}$, $P(X \leq 0.4) = 0.4$.

35 $\mu_X = 0.5$, $\sigma_X = \frac{1}{2\sqrt{5}}$, $P(X > 0.6) = 0.352$.

36 $\mu_X = 0.4$, $\sigma_X = 0.2$, $P(X \in \langle 0.3, 0.6]) = 0.472\,5$.

37 $\mu_X = 0.6$, $\sigma_X = 0.2$, $P(X \in \langle 0.4, 0.65]) = 0.383\,781$.

38 $P(X \leq 0.7) = I_{(53,22)}(0.7) = 0.436\,117$ and $p = I^{-1}_{(53,22)}(0.8) = 0.751\,395$.

39 $P(X \leq 0.7) = I_{(108,72)}(0.7) = 0.003\,480\,6$ and $p = I^{-1}_{(108,72)}(0.8) = 0.659\,279$.

40 $P(X \leq 0.2) = I_{(17,42)}(0.2) = 0.058\,5361$ and $p = I^{-1}_{(17,42)}(0.1) = 0.214\,525$.

41 $P(X \geq 0.8) = 1 - I_{(43,19)}(0.8) = 0.026\,194\,2$ and $p = I^{-1}_{(43,19)}(1 - 0.9) = 0.617\,472$.

42 $P(X \geq 0.6) = 1 - I_{(128,81)}(0.6) = 0.647\,272$, and $p = I^{-1}_{(128,81)}(1 - 0.99) = 0.532\,88$.

43 $\mu_X = 0.553\,551$, $\sigma_X = 0.016\,682\,4$, Normal approximation: $P(X \in [0.45, 0.5\rangle) = 0.000\,664$. (Exact: $P(X \in [0.45, 0.5\rangle) = 0.000\,7007\,15$.)

44 **(a)** $P = \frac{12}{29}$.

    (b) $P = \frac{17}{29}$.

    (c) $P = \frac{34}{145} \approx 0.234\,483$.

    (d) $P = \frac{34}{145} \approx 0.234\,483$.

    (e) $P = 0.003\,122\,16$.

    (f) $P = 0.003\,122\,16$.

    (g) $P = 0.218\,551$.

    (h) $P = 0.097\,082\,8$.

    (i) $P = 0.388\,331$.

    (j) $P = 0.444\,963$.

    (k) $P = 0.140\,231$.

    (l) $P = 0.024\,694\,7$.

    (m) $P = 0.001\,780\,87$.

    (n) $P = 0.833\,294$.

    (o) $P = 0.166\,706$.

45 **(a)** $\mu_X = 0.323\,944$.

    (b) The probability is the same as μ_X, that is, $0.323\,944$.

    (c) $P(X \in \langle \mu_X - 0.05, \mu_X + 0.05]) = 0.632\,215$.

46 **(a)** $\mu_X = 0.326\,797$.
   (b) The probability is the same as μ_X, that is, $0.326\,797$.
   (c) $P(X \in \langle \mu_X - 0.05, \ \mu_X + 0.05]) = 0.813\,675$.

47 You are finding the probability of 0, 1, or 2 factual errors. Notice that there are $\binom{10}{k}$ possible sequences of k factual errors in 10 tries, so

$$P(0) = \binom{10}{0} \cdot \frac{\binom{17+64}{17} \cdot \frac{17\cdot64}{17+64}}{\binom{17+64+0+10}{17+0} \cdot \frac{(17+0)(64+10)}{17+64+0+10}} = 0.108\,611$$

$$P(1) = \binom{10}{1} \cdot \frac{\binom{17+64}{17} \cdot \frac{17\cdot64}{17+64}}{\binom{17+64+1+9}{17+1} \cdot \frac{(17+1)(64+9)}{17+64+1+9}} = 0.252\,93$$

$$P(2) = \binom{10}{2} \cdot \frac{\binom{17+64}{17} \cdot \frac{17\cdot64}{17+64}}{\binom{17+64+2+8}{17+2} \cdot \frac{(17+2)(64+8)}{17+64+2+8}} = 0.284\,546.$$

So the probability of 2 or fewer errors in her next 10 factual claims is $0.646\,087$.

B.10.6 Weibull Distribution, $weib_{(k,\lambda)}$

48 $\mu_T = 3.672\,7$ and $\sigma_T^2 = 0.707\,68$, whereas $P(T \leq 4) = 0.632\,1$.

49 $\mu_T = 240$ and $\sigma_T = 3\,802.32$, whereas $P(T \geq 1) = 0.418\,721$.

50 $\mu_T = 4.431\,1$ and $\sigma_T = 2.316\,3$, whereas $P(T \in (1, 3)) = 0.263\,1$.

51 $\mu_T = 6$ and $\sigma_T = 13.416$, whereas $P(T \in (2, 4)) = 0.126\,8$.

52 **(a)** $\mu_T = 1.772\,45$.
   (b) $P(T < \mu_T) = 0.544\,062$.

53 **(a)** $\mu_T = 2.5$.
   (b) $\sigma_T = 2.5$.
   (c) $P(T \in (0.5, 1.5)) = 0.269\,919$.

54 $\lambda = 0.5$ and $k = 1$.
   (a) We use the formulas for the Weibull distribution, and get

$$\mu_T = \frac{\lambda}{k}\Gamma\left(\frac{1}{k}\right) = \frac{0.5}{1}\Gamma\left(\frac{1}{1}\right) = \underline{0.5}.$$

$$\sigma_T^2 = 0.5^2 \left(\Gamma \left(1 + \frac{2}{1} \right) - \left(\Gamma \left(1 + \frac{1}{1} \right) \right)^2 \right) = \underline{0.25}.$$

(b) $P(T > 1) = 1 - P(T \le 1) = 1 - F(1)$

$$= 1 - \left(1 - e^{-\left(\frac{x}{\lambda} \right)^k} \right) = e^{-\left(\frac{x}{\lambda} \right)^k}$$

$$= e^{-\left(\frac{1}{0.5} \right)^1} = \underline{e^{-2} \approx 0.135\,335}.$$

(c) $$f(x) = \frac{1}{0.5} \cdot \left(\frac{x}{\lambda} \right)^{1-1} e^{-\left(\frac{x}{0.5} \right)^1} = \frac{1}{0.5} e^{-\frac{x}{0.5}}$$

$$= \frac{1}{\lambda} e^{-\frac{x}{\lambda}} = exp_\lambda.$$

Exponential distribution, with parameter $\lambda = 0.5$.

B.10.7   Continuous Uniform Distribution

The continuous uniform distribution has been omitted as a section so that the student should have at hand a tractable probability distribution to build up and study its properties. The first assignment is the key.

55 **(a)** $\mu_X = \frac{a+b}{2}$.

  (b) $\sigma_X^2 = \frac{1}{12} \cdot (b - a)^2$.

  (c)

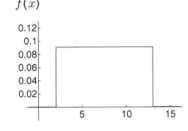
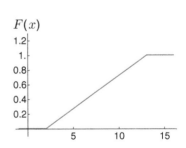

56 $P(X \in M) = \frac{w}{b-a}$.

B.11   Inference: Introduction

1 Bayesian/frequentist
  (a) The frequentists.
  (b) Both frequentists and Bayesians.
  (c) The Bayesians.
  (d) Both frequentists and Bayesians.
  (e) The Bayesians.
  (f) The frequentists.

2-6   These assignments are for reflection, to be discussed in groups. There is no fixed right answer.

B.12   Bayes' Theorem for Distributions

1   –   The total number of sacks is $\sum_{x=1}^{10} 17x = 935$. The probability that he picked a sack of type x is $P(A_x) = 17x/935 = \underline{x/55}$.

   –   We will use Bayes' theorem. The prior is the answer from the previous sub-problem, $P(A_x) = x/55$, whereas the likelihood is the probability of B being "picking soft + soft". The exact answer is hypergeometric, and gives

$$P(B|A_x) = \binom{1\,000\,000 - 2}{10\,000x^2 - 2} \Big/ \binom{1\,000\,000}{10\,000x^2}.$$

However, since the number sampled is miniscule compared to the total, we will employ the binomial approximation, yielding

$$P(B|A_x) = (10\,000x^2/1\,000\,000)^2 = x^4/10\,000.$$

The Bayes table then becomes

x	$f_{pre}(x)$	$g(x)$	$f_{pre} \times g$	$f_{post}(x)$
$1, \ldots, 10$	$\dfrac{x}{55}$	$\dfrac{x^4}{10\,000}$	$\dfrac{x^5}{550\,000}$	$\dfrac{x^5}{220\,825}$
			$\dfrac{220\,825}{550\,000}$	

   –   Set $x = 8$ in the posterior expression, $f_{post}(x) = x^5/220\,825$, and you get $f(8) = 8^5/220\,825 \approx \underline{0.148\,4}$.

2

k	$f_{pre}(k)$	$g(k)$	$f_{pre} \times g$	$f_{post}(k)$
$1, \ldots, 100$	$\dfrac{k^2}{338\,350}$	$\dfrac{1}{k}$	$\dfrac{k}{338\,350}$	$\dfrac{k}{5050}$
			$\dfrac{1}{67}$	

3   **(a)** This is a geometric probability distribution with parameter $p = \frac{1}{2}$, so $f(k) = \frac{1}{2^k}$ for all positive integers k.

   (b) A_k means that there are k chocolates in the bag. There is always precisely one dark chocolate, whereas the rest are milk chocolates, so when $B =$ "you got a dark chocolate", then $P(B|A_k) = \frac{1}{k}$.

(c)

k	$f_{pre}(k)$	$g(k)$	$f_{pre} \cdot g$	$f_{post}(k)$
$k \in \mathbb{N}$	$\frac{1}{2^k}$	$\frac{1}{k}$	$\frac{1}{k \cdot 2^k}$	$\frac{1/\ln 2}{k \cdot 2^k}$
			$\ln 2$	

4 **(a)** This is a geometric probability distribution with $p = 0.5$, so $f(n) = 0.5^n$.

(b) After n trials, there are 2^n rocks in the bag, whereof $2^n - 1$ are gold nuggets. The probability of sampling a gold nugget is then $g(n) = 1 - \frac{1}{2^n}$.

(c)

k	$f_{pre}(k)$	$g(k)$	$f_{pre} \cdot g$	$f_{post}(k)$
$k \in \mathbb{N}$	$\frac{1}{2^k}$	$1 - \frac{1}{2^k}$	$\frac{1}{2^k} - \frac{1}{2^{2k}}$	$\frac{1.5}{2^k} - \frac{1.5}{2^{2k}}$
			$\frac{2}{3}$	

(d) There is now one more rock and one gold nugget less than at the start. The probability of sampling a gold nugget, given A_n (that the genie got n heads), is then $g_2(n) = P(C|A_nB) = 1 - 2/2^n = 1 - 1/2^{n-1}$. We find the probability of a gold nugget in the next try thus:

k	$f_{pre_2} = f_{post_1}$	$g_2(k)$	$f_{pre_2} \cdot g_2$
$k \in \mathbb{N}$	$\frac{1.5}{2^k} - \frac{1.5}{2^{2k}}$	$1 - \frac{2}{2^k}$	$\frac{1.5}{2^k} - \frac{4.5}{2^{2k}} + \frac{3}{2^{3k}}$
			$\frac{3}{7}$

5 **(a)** $f(x) = \frac{1}{15}$.

(b) $g(x) = \frac{x}{15} \cdot \left(1 - \frac{x}{15}\right)$.

(c)

k	$f_{pre}(x)$	$g(x)$	$f_{pre} \cdot g$	$f_{post}(x)$
$1, \ldots, 15$	$\frac{1}{15}$	$\frac{x}{15}\left(1 - \frac{x}{15}\right)$	$\frac{x}{15^2}\left(1 - \frac{x}{15}\right)$	$\frac{3x}{112}\left(1 - \frac{x}{15}\right)$
			$\frac{112}{675}$	

(d)

k	$f_{pre}(x)$	$g(x)$	$f_{pre} \cdot g$
$1, \ldots, 15$	$\frac{3x}{112} \cdot \left(1 - \frac{x}{15}\right)$	$\frac{x}{15}$	$\frac{x^2}{560} \cdot \left(1 - \frac{x}{15}\right)$
			$\frac{1}{2}$

6 **(a)** The total number of handles is $1 + 2 + \cdots + 7 = 28$. Then, $f(x) = x/28$.
(b) $g(x) = (50 - x^2)/50$.

(c)

k	$f_{pre}(x)$	$g(x)$	$f_{pre} \cdot g$	$f_{post}(x)$
$1, \ldots, 7$	$\dfrac{x}{28}$	$\dfrac{50-x^2}{50}$	$\dfrac{x(50-x^2)}{1400}$	$\dfrac{x(50-x^2)}{616}$
			$\dfrac{11}{25}$	

(d) Given that you picked sack x, it is $g_2(x) = \dfrac{x^4}{2500}$, which gives

k	$f_{pre}(x)$	$g(x)$	$f_{pre} \cdot g$
$1, \ldots, 7$	$\dfrac{x(50-x^2)}{616}$	$\dfrac{x^4}{2500}$	$\dfrac{x^5(50-x^2)}{1\,540\,000}$
			$\dfrac{203}{1250}$

7 **(a)** The total number of nut packs is $\sum_{x=1}^{5}(30 - 5x) = 150 - 5\sum_{x=1}^{5}x = 75$, so $f(x) = (30 - 5x)/75$.
(b) $g(x) = (10x/100)^2 = x^2/100$.

(c)

k	$f_{pre}(x)$	$g(x)$	$f_{pre} \times g$	$f_{post}(x)$
$1, \ldots, 5$	$\dfrac{30-5x}{75}$	$\dfrac{x^2}{100}$	$\dfrac{x^2(30-5x)}{7500}$	$\dfrac{x^2(30-5x)}{525}$
			$\dfrac{7}{100}$	

(d) Given that the picked pack was of type x, you have $g_2(x) = \dfrac{10x}{100}$, which means

k	$f_{pre}(x)$	$g(x)$	$f_{pre} \times g$
$1, \ldots, 5$	$\dfrac{x^2(30-5x)}{525}$	$\dfrac{x}{10}$	$\dfrac{x^3(30-5x)}{5250}$
			$\dfrac{53}{150}$

8 We see that the total set of dividing points for the piecewise defined functions is $\{-2, -1, 0, 2\}$, so we divide the table according to indicated intervals so as to see the functional expressions inside each interval:

x values	$f_{pre}(x)$	$g(x)$	$f_{pre} \times g$	$f_{post}(x)$
$(-\infty, -2]$	0	14	0	0
$(-2, -1]$	$0.5 + 0.25x$	14	$7 + 3.5x$	$1.75 + 0.875x$
$(-1, 0]$	$0.5 + 0.25x$	2	$1 + 0.5x$	$0.25 + 0.125x$
$(0, 2]$	$0.5 - 0.25x$	3	$1.5 + 0.75x$	$0.375 + 0.1875x$
$(2, \infty)$	0	3	0	0
			4	

Detailed calculation of S:

$$S = \int_{-\infty}^{-2} 0\, dx + \int_{-2}^{-1} 7 + 3.5x\, dx + \int_{-1}^{0} 1 + 0.5x\, dx$$
$$+ \int_{0}^{2} 1.5 - 0.75x\, dx + \int_{2}^{\infty} 0\, dx$$
$$= 0 + 1.75 + 0.75 + 1.5 + 0 = 4.$$

(In this example, the rows for the values $x \leq 2$ and $x > 2$ are included for the sake of completeness. In general, rows where $f_{pre} \times g = 0$ for all x values in the row's interval may be omitted.)

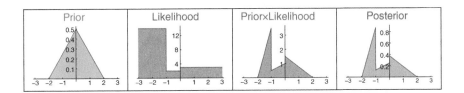

| Prior | Likelihood | PriorxLikelihood | Posterior |

9 **(a)** The beta distribution is a *conjugate prior* for Bernoulli processes, which makes calculations easier. Save the results of your manual calculations here, and return to this problem after having read Section 13.2, and compare your calculations here to how you would solve this same assignment by employing the methods there.

(b) Congratulations! You have just performed a *hypothesis test*. Save your results, and revisit this assignment to compare to the calculation you make by following the method in Section 14.2.2.

(c) Congratulations! You have just found an interval with an 80% probability of containing the true value of π. That is: $P(\pi \in [A_1, B_1]) = 0.8$. This is your first interval estimate. Save your results, and revisit this assignment to compare to the calculation you would do following the methods of Section 15.7.

(d) The likelihood is $g(x) = x^k(1-x)^l$.

(e) Write your prior as $f_{pre}(x) = K_1 x^{a-1} \cdot (1-x)^{b-1}$. Then the posterior is

$$
\begin{aligned}
f_{post}(x) &= \frac{f_{pre}(x)g(x)}{\int_0^1 f_{pre}(x)g(x)\,dx} \\
&= \frac{(K_1 x^{a-1}(1-x)^{b-1}) \cdot x^k(1-x)^l}{\int_0^1 f_{pre}(x)g(x)\,dx} \\
&= K_2 x^{a+k-1}(1-x)^{b+l-1} \\
&= \beta_{(a+k,b+l)}(x).
\end{aligned}
$$

You might wonder how we got K_2, but think: could it be any other value than the constant for $\beta_{(a+k,b+l)}(x)$, when we know that $\int_0^1 \beta_{(a+k,b+l)}(x)\,dx = 1$?

(f) You have now sharpened your hypothesis test by including more data.

(g) You have now sharpened your interval estimate by including more data. Notice that the new interval is narrower than the first.

(h) Compare your guess to Section 15.7.

(i) If you made it this far, you have done a good piece of work and deserve a decent coffee break. The rest of the book will not just be easy for you, but plain sailing!

B.13 Bayes' Theorem with Hyperparameters

Bayes' Theorem for Gaussian Processes

1 Unknown μ, known σ.

(a) $\mu \sim \phi_{(4.686,\,1.11803)}$, and $X_+ \sim \phi_{(4.686,\,2.738\,61)}$.
Probabilities: $P(\mu < 6) = 88.01\%$ and $P(X_+ < 6) = 68.43\%$.

(b) $\mu \sim \phi_{(45.414\,3,\,2.267\,79)}$, and $X_+ \sim \phi_{(45.414\,3,\,6.414\,27)}$.
Probabilities: $P(\mu < 40) = 0.85\%$ and $P(X_+ < 40) = 19.93\%$

(c) $\mu \sim \phi_{(338.667,\,9.179\,87)}$, and $X_+ \sim \phi_{(338.667,\,18.359\,7)}$.
Probabilities: $P(\mu < 350) = 89.15\%$ and $P(X_+ < 350) = 73.15\%$.

(d) $\mu \sim \phi_{(2976.97,\,57.735)}$, and $X_+ \sim \phi_{(2976.97,\,208.167)}$.
Probabilities: $P(\mu < 3125) = 99.48\%$ and $P(X_+ < 3125) = 76.15\%$.

(e) $\kappa_0 = 4$ and $\Sigma_0 = 20$, so $\mu \sim \phi_{(5.3,\,0.557\,735)}$ and $X_+ \sim \phi_{(5.3,\,2.081\,67)}$.
Probabilities: $P(\mu < 6) = 88.732\,7\%$ and $P(X_+ < 6) = 63.166\,6\%$.

(f) $\kappa_0 = 4$ and $\Sigma_0 = 2040$, so $\mu \sim \phi_{(509.8,9.0)}$ and $X_+ \sim \phi_{(509.8,31.3)}$.
Probabilities: $P(\mu < 500) = 13.81\%$ and $P(X_+ < 500) = 37.71\%$.

(g) $\kappa_0 = 1$ and $\Sigma_0 = 0.19$, so $\mu \sim \phi_{(0.200,0.013)}$ and $X_+ \sim \phi_{(0.200,0.033)}$.
Probabilities: $P(\mu < 0.15) = 0.01\%$ and $P(X_+ < 0.15) = 6.49\%$.

Known μ, unknown σ.

(h) $\tau \sim \gamma_{(2,\,3.341\,3)}(t)$, while $X_+ \sim t_{(23,\,1.292\,54,\,4)}(x)$.
Probabilities: $P(\sigma < 1.25) = P(\tau > 0.64) = 1 - P(\tau < 0.64) = 36.98\%$
and $P(X_+ < 25) = 90.17\%$.

(i) $\tau \sim \gamma_{(3.5,\,41.106\,2)}(t)$, while $X_+ \sim t_{(50,\,3.427\,04,\,7)}(x)$.
Probabilities: $P(\sigma < 10) = 99.72\%$ and $P(X_+ < 45) = 9.4\%$.

(j) $\tau \sim \gamma_{(6.,\,0.088\,95)}(t)$, while $X_+ \sim t_{(1.2,\,0.125\,505,\,12)}(x)$.
Probabilities: $P(\sigma < 0.1) = 12.22\%$ and $P(X_+ < 1) = 6.85\%$.

Unknown μ, unknown σ.

(k) $\tau \sim \gamma_{(3.5,\,41\,409)}(t)$ and $\mu \sim t_{(3\,979,\,38.456\,4,\,7)}(x)$, while
$X_+ \sim t_{(3979,\,115.369,\,7)}(x)$. Probabilities: $P(\sigma < 100) = 30.84\%$, whereas
$P(\mu < 4\,000) = 69.90\%$ and $P(X_+ < 4\,000) = 56.96\%$.

(l) $\tau \sim \gamma_{(3.,\,1\,761.89)}(t)$ and $\mu \sim t_{(11.528\,6,\,9.159\,66,\,6)}(x)$, while
$X_+ \sim t_{(11.528\,6,\,25.907\,4,\,6)}(x)$. Probabilities: $P(\sigma < 15) = 1.57\%$, whereas
$P(\mu < 0) = 12.75\%$ and $P(X_+ < 0) = 33.60\%$.

(m) $\tau \sim \gamma_{(2.,\,1.612\,37)}(t)$ and $\mu \sim t_{(-0.583\,5,\,0.448\,94,\,4)}(x)$, whereas
$X_+ \sim t_{(-0.583\,5,\,1.003\,86,\,4)}(x)$. Probabilities: $P(\sigma < 0.5) = 1.18\%$, while
$P(\mu < 0) = 86.82\%$ and $P(X_+ < 0) = 70.39\%$.

(n) $\tau \sim \gamma_{(6,\,87.482)}(t)$ and $\mu \sim t_{(22.756\,8,\,1.059\,04,\,12)}(x)$, whereas
$X_+ \sim t_{(22.756\,8,\,3.962\,56,\,12)}(x)$. Probabilities: $P(\sigma < 5) = 85.77\%$, while
$P(\mu < 20) = 1.15\%$ and $P(X_+ < 20) = 24.99\%$.

(o) $\tau \sim \gamma_{(5,\,5.401\,91)}(t)$ and $\mu \sim t_{(19.737\,5,\,0.328\,692,\,10)}(x)$,   whereas
$X_+ \sim t_{(19.737\,5,\,1.090\,15,\,10)}(x)$.   Probabilities: $P(\sigma < 1.179) = 65.11\%$,
while $P(\mu < 17.9) = 0.01\%$ and $P(X_+ < 17.9) = 6.14\%$.

(p) $\tau \sim \gamma_{(3.5,\,3302.45)}(t)$ and $\mu \sim t_{(92.137\,5,\,10.860\,2,\,7)}(x)$,   whereas
$X_+ \sim t_{(92.137\,5,\,32.580\,7,\,7)}(x)$. Probabilities: $P(\sigma < 25) = 15.86\%$, while
$P(\mu < 50) = 0.30\%$ and $P(X_+ < 50) = 11.85\%$.

(q) $\tau \sim \gamma_{(17.5,\,7.5\cdot10^{-8})}(t)$ and $\mu \sim t_{(3.305\,3\cdot10^{-6},\,0.000\,010\,910\,9,\,35)}(x)$, whereas
$X_+ \sim t_{(3.305\,3\cdot10^{-6},\,0.000\,066\,368\,4,\,35)}(x)$.
Probabilities: $P(\sigma < 0.000\,04) = 12.22\%$, while $P(\mu < -0.000\,04) =$
0.02% and $P(X_+ < -0.000\,04) = 25.92\%$.

2 $X_+ = \mu + (X_+ - \mu)$, and since these two stochastic variables are indepen-
dent, we employ Rule 10.1.10, and get that $\mu_{X_+} = E[\mu] = \frac{\Sigma_1}{\kappa_1}$ while $\sigma_{X_+}^2 =$

$$\sigma_\mu^2 + \sigma_{(X_+-\mu)}^2 = \sigma^2 + \left(\sigma\sqrt{\tfrac{1}{\kappa_1}}\right)^2 = \sigma^2\left(1 + \tfrac{1}{\kappa_1}\right)^2.$$

3 The neutral prior hyperparameters are $\kappa_0 = \Sigma_0 = SS_0 = n_0 = 0$, so $v_0 =$
$n_0 - 1 = -1$. Updating these by means of the data then gives $\kappa_1 = 25$,
$v_1 = 24$, $\Sigma_1 = 1\,229.75$, $SS_1 = 110.94$, so then

$$\tau \sim \gamma_{(12,\,55.47)}$$

$$\mu \sim t_{(49.19,\,0.43,\,24)}$$

$$X_+ \sim t_{(49.19,\, 2.192\,6,\, 24)}$$

$$P(X_+ > 50) = 1 - T_{(49.19,\, 2.192\,6,\, 24)}(50) = 0.357\,5.$$

4 Non-informative = neutral prior means $\kappa_0 = \Sigma_0 = SS_0 = n_0 = 0$, so $v_0 = n_0 - 1 = -1$. We see from the posterior of μ that $v_1 = 6$, which means that the number of measurements is $n = v_1 - v_0 = 6 - (-1) = 7$. That again means that $\kappa_1 = 0 + 7 = 7$. We now look at μ's second parameter, and first notice that $\sqrt{\frac{SS_1}{v_1}} = 11 \cdot \sqrt{\kappa_1} = 11 \cdot \sqrt{7}$. Further, the next parameter for

X_+ equals $\sqrt{\frac{SS_1}{v_1}} \cdot \sqrt{1 + \frac{1}{\kappa_1}} = 11 \cdot \sqrt{7} \cdot \sqrt{1 + \frac{1}{7}} = 11 \cdot \sqrt{7+1} \approx 31.112\,7$.

This gives us a predictive probability distribution

$$X_+ \sim \mu \sim t_{(943,\, 31.1127,\, 6)},$$

which allows us to calculate the specific probability,

$$P(X_+ < 900) = T_{(943,\, 31.1127,\, 6)}(900) \approx 0.1081 = 10.8\%.$$

Bayes' Theorem for Bernoulli Processes

5 *Posterior:*

(a) Posterior: $\pi \sim \beta_{(19,31)}$. $P(\pi < 0.4) = 0.622\,259$. Normal approximation: $\pi \sim \phi_{(0.38,\, 0.067\,967\,7)}$. $P(\pi < 0.4) = 0.615\,719$.

(b) Posterior: $\pi \sim \beta_{(5,96)}$. $P(\pi < 0.07) = 0.836\,836$. Normal approximation: $\pi \sim \phi_{(0.049\,505,\, 0.021\,478\,3)}$. $P(\pi < 0.07) = 0.830\,015$.

(c) Posterior hyperparameters: $a_1 = 42$, $b_1 = 13$. $P(\pi < 0.7) = 0.134\,791$. Parameters for Normal approximation: $\mu = 0.763\,636$, $\sigma = 0.056\,772\,7$. $P(\pi < 0.7) = 0.131\,166$.

(d) Posterior hyperparameters: $a_1 = 434.5$, $b_1 = 177.5$. $P(\pi < 0.7) = 0.290\,458$. Parameters for Normal approximation: $\mu = 0.709\,967$, $\sigma = 0.018\,327\,9$. $P(\pi < 0.7) = 0.293\,278$.

6 *Predictive:*

(a) $K_{+5} \sim \beta b_{(19,31,5)}$ and $L_{+5} \sim \beta nb_{(19,31,5)}$.
Probabilities: $P(K_{+5} \le 3) = \text{\ss B}_{(19,31,5)}(3) = 0.917\,656$ and $P(L_{+5} \le 7) = \text{\ss NB}_{(19,31,5)}(7) = 0.499\,604$.

(b) $K_{+20} \sim \beta b_{(5,96,20)}$ and $L_{+4} \sim \beta nb_{(5,96,4)}$.
Probabilities: $P(K_{+20} \le 3) = \text{\ss B}_{(5,96,20)}(3) = 0.974\,169$ and $P(L_{+4} \le 12) = \text{\ss NB}_{(5,96,4)}(12) = 0.012\,409\,5$.

(c) $K_{+7} \sim \beta b_{(42,13,7)}$ and $L_{+14} \sim \beta nb_{(42,13,14)}$.
Probabilities: $P(K_{+7} \le 4) = \text{\ss B}_{(42,13,7)}(4) = 0.225\,352$ and $P(L_{+14} \le 5) = \text{\ss NB}_{(42,13,14)}(5) = 0.699\,216$.

(d) $K_{+32} \sim \beta b_{(434.5,177.5,32)}$ and $L_{+20} \sim \beta nb_{(434.5,177.5,20)}$.

Probabilities: $P(K_{+32} \leq 23) = ßB_{(434.5, 177.5, 32)}(23) = 0.605\,77$
and $P(L_{+20} \leq 8) = ßNB_{(434.5, 177.5, 20)}(8) = 0.573\,26$.

7 **(a)** Prior: $\beta_{(0.5, 0.5)}$. Posterior: $\pi \sim \beta_{(9.5, 1.5)}$.
   (b) $\pi \sim \beta_{(21.5, 3.5)}$.
   (c) $\pi \sim \beta_{(24.5, 3.5)}$.
   (d) $P(\pi > 0.75) = 0.961\,184$. (For this assignment, use a good tool that has the Beta distribution available in some form. In Mathematica/Wolfram Alpha, you write 1-CDF[BetaDistribution[24.5, 3.5],0.75].)
   (e) $K_{+3} \sim \beta b_{(24.5, 3.5, 3)}(x)$.
   (f) $P(K_{+3} = 2) = \beta b_{(24.5, 3.5, 3)}(2) = 0.269\,289$.

8 **(a)** $a_0 = np = 21 \cdot \frac{3}{7} = 9$ and $b_0 = n(1-p) = 21 \cdot \frac{4}{7} = 12$.
   (b) $a_1 = 32$ and $b_1 = 30$.
   (c) New prior = old posterior: $a_1 = 32$ and $b_1 = 30$.
   (d) New posterior: $a_2 = 490$ and $b_2 = 396$.

9 **(a)** Posterior: $a_1 = 41.5$ and $b_1 = 9.5$, so $\pi \sim \beta_{(41.5, 9.5)}$.
   (b) We make use of the cumulative β distribution, and get

$$P(\pi > 75\%) = 1 - I_{(41.5, 9.5)}(0.75) = 0.876.$$

   If you calculate this by means of the Normal approximation, $\beta_{(41.5, 9.5)} \approx \phi_{(0.814, 0.0540)}$, which gives $P(\pi > 0.75) = 1 - P(\pi \leq 0.75) = 1 - \Phi_{(0.814, 0.0540)}(0.75) = 0.882$.

10 You are estimating the proportion of Macintoshes among the laptops of a rather large company. Posterior: $a_1 = 15$ and $b_1 = 5$.

11 **(a)** Flat prior (Bayes, $u = 1$) gives $\beta_{(20, 154)}$. Jeffreys' prior ($u = 0.5$) gives $\beta_{(19.5, 153.5)}$. Novick and Hall ($u = 0$) give $\beta_{(19, 153)}$.
   (b) We first calculate the exact answer. Then flat prior gives $P(\pi > 0.1) = 71.9\%$. Jeffreys' gives $P(\pi > 0.1) = 68.6\%$. Novick and Hall give $P(\pi > 0.1) = 65.0\%$. If we resort to the Normal approximation, our calculations give that for the flat prior, $P(\pi > 0.1) = 73.2\%$. Jeffreys' gives $P(\pi > 0.1) = 70.2\%$. Novick and Hall give $P(\pi > 0.1) = 67.0\%$. We are therefore best off to report that the probability that these diamond mines yield more than $\frac{1}{10}$ high grade fancy diamonds is $\frac{2}{3}$.

12 Let π be the proportion of games won by white.
   (a) Prior hyperparameters: $a_0 = 7$, $b_0 = 7$.
   Posterior hyperparameters: $a_1 = 20$, $b_1 = 14$. Then $\pi \sim \beta_{(20, 14)}$.

(b) $\pi \sim \beta_{(31,23)}$.
(c) $P(\pi \geq 0.5) = 0.864\,16$.
  Using the Normal approximation will give $P(\pi \geq 0.5) = 0.867$.

Bayes' Theorem for Poisson Processes

13 **(a)** Posterior: $\lambda \sim \gamma_{(7,5)}$.
  (b) Posterior hyperparameters: $\kappa_1 = 8$, $\tau_1 = 17$.
  (c) Posterior: $\lambda \sim \gamma_{(29,8)}$.
  (d) Posterior hyperparameters: $\kappa_1 = 9$, $\tau_1 = 192$.

14 **(a)** Predictive: $N_{+1} \sim nb_{(7,\frac{5}{6})}(\eta)$ and $T_{+2} \sim g\gamma_{(2,7,5)}$.
  Probabilities: $P(T_{+2} \leq 3) = G\Gamma_{(2,7,5)}(3) = 0.864\,958$ and
  $P(N_{+1} \leq 4) = NB_{(7,0.833\,333)}(4) = 0.975\,494$.
  (b) Predictive: $N_{+1} \sim nb_{(8,\frac{17}{18})}(\eta)$ and $T_+ = T_{+1} \sim g\gamma_{(1,8,17)}$.
  Probabilities: $P(T_+ \leq 2) = G\Gamma_{(1,8,17)}(2) = 0.589\,264$ and
  $P(N_{+1} \leq 1) = NB_{(8,0.944\,444)}(1) = 0.914\,349$.
  (c) Predictive: $N_{+1} \sim nb_{(29,\frac{8}{9})}(\eta)$ and $T_{+3} \sim g\gamma_{(3,29,8)}$.
  Probabilities: $P(T_{+3} \leq 1) = G\Gamma_{(3,29,8)}(1) = 0.684\,853$ and
  $P(N_{+1} \leq 5) = NB_{(29,0.888\,889)}(5) = 0.829\,829$.
  (d) Predictive: $N_{+1} \sim nb_{(9,\frac{192}{193})}(\eta)$ and $T_{+2} \sim g\gamma_{(2,9,192)}$.
  Probabilities: $P(T_{+2} \leq 30) = G\Gamma_{(2,9,192)}(30) = 0.400\,007$ and
  $P(N_{+1} \leq 0) = P(N_{+1} = 0) = nb_{(9,0.994\,819)}(0) = 0.954\,323$.

15 Prior: $\kappa_0 = 0$, $\tau_0 = 0$. Data: $n = 111$, $t = 23$. Posterior: $\kappa_1 = 111$, $\tau_1 = 23$.
  Then $\lambda \sim \gamma_{(111,23)}$.

16 Prior: $\kappa_0 = 0$, $\tau_0 = 0$. Data: $n = 53$, $t = 26$. Posterior: $\kappa_1 = 53$, $\tau_1 = 26$.
  Then $\lambda \sim \gamma_{(53,26)}$.

17 Prior: $\kappa_0 = 0$, $\tau_0 = 0$. Data: $n = 13$, $t = 10$. Posterior: $\kappa_1 = 13$, $\tau_1 = 10$.
  Then $\lambda \sim \gamma_{(13,10)}$.

18 This corresponds to three updates in one go, with $\tau_0 = 0$, $t_1 = 1.9$, $t_2 = 0.7$,
  and $t_3 = 1.2$, so $\tau_3 = 3.8$, while $\kappa_0 = 0$, $n_1 = 0$, $n_2 = 3$, and $n_3 = 1$, so $\kappa_3 = 4$. Then $\lambda \sim \gamma_{(4,3.8)}$.

19 Prior: $\kappa_0 = 0$, $\tau_0 = 0$. Data: $n = 157$, $t = 100$ (1 deciliter $= 100\,\text{cm}^3$). Posterior: $\kappa_1 = 157$, $\tau_1 = 100$. Then $\lambda \sim \gamma_{(157,100)}$.

B.14 Bayesian Hypothesis Testing

Utility Functions

1 In the problems below, you are given the probability distribution of a stochastic variable X and a utility function $u(x)$. Find the expected utility U.

(a) $U = 3E[X] + 2 = 3 \cdot \frac{17}{17+9} + 2 = \underline{3.961\,54}$.

(b) $U = 9 \cdot P(X < 0.3) - 4 \cdot P(X > 0.3) = 9 \cdot \Gamma_{(7,21)}(0.3) - 4 \cdot (1 - \Gamma_{(7,21)}(0.3)) = \underline{1.742\,97}$.

(c) $U = -1 \cdot P(X < 3) + 1.5 \cdot P(3 < X < 6) + 4 \cdot P(X > 6) = -1 \cdot \Phi_{(5.3,1.9)}(3) + 1.5 \cdot (\Phi_{(5.3,1.9)}(6) - \Phi_{(5.3,1.9)}(3)) + 4 \cdot (1 - \Phi_{(5.3,1.9)}(6)) = \underline{2.108\,1}$.

(d) $U = -1 \cdot P(X < 15) + 7 \cdot P(X > 15) = -1 \cdot T_{(10,5,2)}(15) + 7 \cdot (1 - T_{(10,5,2)}(15)) = \underline{0.690\,599}$.

(e) $U = -2E[X] + 90 = -2 \cdot 41.3 + 90 = \underline{7.4}$.

(f) $U = -5E[X] + 8 = -5 \cdot (-2.73) + 8 = \underline{21.65}$.

2 You are going to decide whether $A: \Theta < \theta_0$ or $B: \Theta > \theta_0$. The gain in utility of choosing A instead of B is

$$u(x) = \begin{cases} w_A & x < \theta_0 \\ -w_B & x > \theta_0. \end{cases}$$

In the first three subproblems below, you are given θ_0, w_A, and w_B. Formulate the decision problem as a hypothesis test by indicating significance level α and stating the alternative hypothesis H_1.

(a) Since $w_B < w_A$, then H_1 is the same as B, which is that $\Theta > 7$, and the significance level is $\alpha = \frac{w_B}{w_B + w_A} = \frac{1}{1+9} = 0.1$.

(b) $w_A < w_B$, so then $H_1 = A: \Theta < -3$, and $\alpha = \frac{25}{25+175} = 0.125$.

(c) $w_A < w_B$, so then $H_1: \Theta < 100$, and $\alpha = \frac{1}{1+100} \approx 0.01$.

(d) $\theta_0 = 17$, $A =$ "not produce" $(\Theta < 17)$, $B =$ "produce" $(\Theta > 17)$. $w_A = 19$, $w_B = 1$. Since $w_B < w_A$, then H_1 is the same as B, which is that $\Theta > 17$, and the significance level is $\alpha = \frac{w_B}{w_B + w_A} = \frac{1}{1+19} = 0.05$.

Hypothesis Test for Gaussian Processes

3 You are given the *posterior* distribution $\theta \sim f(x)$, the significance α, and alternative hypothesis H_1. Test, and decide between the competing hypotheses.

(a) $H_0: \theta \leq 3$, and so $P(H_0) = \Phi_{(7,2)}(3) = 0.022 < \alpha$, so we *reject* H_0.

(b) $H_0: \theta \le 7$, and so $P(H_0) = \Phi_{(9,2)}(7) = 0.158 \ge \alpha$, so we *don't reject H_0.*

(c) $H_0: \theta \ge 16$, and so $P(H_0) = 1 - \Phi_{(8,3)}(16) = 0.004 < \alpha$, so we *reject H_0.*

(d) $H_0: \theta \le 3$, and so $P(H_0) = T_{(7,2,3)}(3) = 0.069\,663 > \alpha$ so we *don't reject H_0* in favor of H_1.

(e) Here, $H_0: X \le 4$, and so $P(H_0) = T_{(9,2,5)}(4) = 0.027 < \alpha$, so we *reject H_0* in favor of H_1.

(f) Here, $H_0: X \ge 16$, and so $P(H_0) = 1 - T_{(24,3,1)}(16) = 0.885\,8 > \alpha$ so we *don't reject H_0* in favor of H_1.

4 $\mu \sim t_{(0.984, 0.007\,18, 9)}$, so then $P(\mu \ge 1.0) = 0.026\,4 < \alpha$, which means you *reject H_0*, and conclude that the mean serving contains less than 1.0 pint.

5 In the corresponding problem, problem (3) in Chapter 13, we found that the *posterior* distribution was $\mu \sim t_{(49.19, 0.43, 24)}$, so that $P(H_0) = 1 - T_{(49.19, 0.43, 24)}(50) = 0.035\,9 > \alpha$, which means that we *don't reject H_0*.

Hypothesis Test for Bernoulli Processes

6 You are given a (posterior) distribution for $\pi \sim \beta_{(a,b)}$, a significance α, and H_1. Test the following competing hypotheses, to decide between them, both by direct calculation and by Normal approximation.

(a) $H_0: \quad \pi \le 0.5$, which gives us $P(H_0) = I_{(35,24)}(0.5) = 0.074 < \alpha$, which means that we *reject H_0*. Normal approximation gives $\pi \sim \phi_{(0.593, 0.063\,4)}$, and so $P(H_0) = \Phi_{(0.593, 0.063\,4)}(0.5) = 0.071 < \alpha$, so therefore we *reject H_0*.

(b) $H_0: \pi \ge 0.85$, and so $P(H_0) = 0.055 > \alpha$, which means that we *don't reject H_0*. Normal approximation gives $\pi \sim \phi_{(0.788, 0.040\,9)}$, and so $P(H_0) = 0.064 > \alpha$, so therefore we *don't reject H_0*.

(c) Then the posterior is $\pi \sim \beta_{(23,52)}$. Further, $H_0: \pi \le 0.2$, and so $P(H_0) = 0.015\,9 < \alpha$, which means that we *reject H_0*. Normal approximation gives $\pi \sim \phi_{(0.307, 0.052\,9)}$, and so $P(H_0) = 0.021\,9 > \alpha$, so therefore we *don't reject H_0*. We therefore take note of this: that in cases where we are very close to the limit set by the significance, the Normal approximation may yield the opposite conclusion of what exact calculation does.

7 $P(H_0) = P(\pi \le 0.75) = I_{(41.5\ 9.5)}(0.75) = 0.124 > \alpha$, so you *don't reject H_0*. MegaCola does not launch their campaign. (Using Normal approximation, you get $P(\pi \le 0.75) = 0.118$.)

8 H_0 is then $\theta < 0.1$. The *posterior* distribution is $\pi \sim \beta_{(19.5, 153.5)}$, and so $P(H_0) = 0.314 > 0.1$, and you *don't reject H_0*. The mining company will *not* buy the new equipment.

Hypothesis Test for Poisson Processes

9 You are given a probability distribution for $\tau \sim \gamma_{(k,\lambda)}$, a significance α, and H_1. Determine the hypothesis test, both by direct calculation and by using the Normal approximation.

(a) H_0: $\tau \geq 4$, and so $P(H_0) = 1 - \Gamma_{(6,3)}(4) = 0.020\,341 < \alpha$, which means that we *reject* H_0. Normal approximation gives $\pi \sim \phi_{(2,0.816\,497)}$, and so $P(H_0) = \Phi_{(2,0.816\,497)}(4) = 0.007 < \alpha$, so therefore we *reject* H_0.

(b) H_0: $\tau \leq 0.2$, and so $P(H_0) = \Gamma_{(10,29.7)}(0.2) = 0.079\,9 > \alpha$, which means that we *don't reject* H_0. Normal approximation gives $\pi \sim \phi_{(0.336\,7,0.106\,474)}$, and so $P(H_0) = \Phi_{(0.336\,7,0.106\,474)}(0.2) = 0.099\,6 > \alpha$, so therefore we *don't reject* H_0.

(c) H_0: $\tau \leq 0.5$, and so $P(H_0) = \Gamma_{(17.5,53.4)}(0.5) = 0.976\,023 > \alpha$, which means that we *don't reject* H_0. Normal approximation gives $\pi \sim \phi_{(0.327\,7,0.078\,34)}$, and so $P(H_0) = \Phi_{(0.327,7,0.078\,34)}(0.5) = 0.986\,1 > \alpha$, so therefore we *don't reject* H_0.

(d) H_0: $\tau \geq 3.2$, and so $P(H_0) = \Gamma_{(20/3,\sqrt{32})}(3.2) = 0.000\,818 < \alpha$, which means that we *reject* H_0. Normal approximation gives $\pi \sim \phi_{(1.197\,4,0.463\,7)}$, and so $P(H_0) = 1 - \Phi_{(1.1974,0.4637)}(3.2) = 0.000\,007\,86 < \alpha$, so therefore we *reject* H_0.

10 Prior: $\kappa_0 = 0$, $\tau_0 = 0$. Data: $n = 53$, $t = 26$. Posterior: $\kappa_1 = 53$, $\tau_1 = 26$. Then $\lambda \sim \gamma_{(53,26)}$. Then $P(H_0) = P(\lambda \leq 1.5) = \Gamma_{(53,26)}(1.5) = 0.019 < \alpha$, which means that we *reject* H_0.

11 Prior: $\kappa_0 = 0$, $\tau_0 = 0$. Data: $n = 157$, $t = 100$ (1 deciliter = 100 cm³). Posterior: $\kappa_1 = 157$, $\tau_1 = 100$. Then $\lambda \sim \gamma_{(157,100)}$. Then $P(H_0) = P(\lambda \geq 1.75) = 1 - \Gamma_{(157,100)}(1.75) = 0.079 > \alpha$, which means that we *don't reject* H_0.

Pairwise Comparison for Gaussian Processes

12 The *posterior* probability distributions are $\mu_K \sim t_{(15.571\,4,2.784\,8,6)}$ and $\mu_O \sim t_{(20.625,0.323\,899,7)}$, so with $\Theta = \mu_O - \mu_K$, we get $\Theta \sim t_{(5.053\,57,2.803\,57,6)}$. The hypothesis that Odd coughs longest in the mean is then H_1: $\Theta > 0$, whereas H_0: $\Theta \leq 0$. Then $P(H_0) = T_{(5.053\,57,2.803\,57,6)}(0) = 0.061 < 0.2$, which means that we *reject* the null hypothesis H_0 in favor of the alternative hypothesis H_1. Odd may therefore conclude that his coughing bouts are indeed the longest lasting in the mean.

13 The *posterior* distributions become $\mu_B \sim t_{(72.5,7.093,5)}$ and $\mu_P \sim t_{(48.8,1.827,4)}$, so $\mu_B - \mu_P = \Theta \sim t_{(23.7,7.325,5)}$. The claim that Baggins purrs the longest is then H_1: $\Theta > 0$. You *decide the hypothesis test* by calculating: $P(H_0) = P(\Theta \leq 0) = T_{(23.7,7.325,5)}(0) = 0.0115 < \alpha$, so you *reject* H_0. Baggins purrs the longest.

Pairwise Comparison for Poisson Processes

14 **(a)** $P(H_0) = F_{(2k, 2m)}\left(\frac{ml}{kn}\right) = F_{(2\cdot7, 2\cdot4)}\left(\frac{4\cdot70}{7\cdot80}\right) = 0.123 > \alpha$,

which means we *don't reject* H_0 and *don't* say that $\Theta > \Psi$.

(b) $P(H_0) = 1 - F_{(2k,2m)}\left(\frac{ml}{kn}\right) = 1 - F_{(2\cdot9,2\cdot11)}\left(\frac{11\cdot20}{9\cdot20}\right) = 0.324 > \alpha$,

which means that we *don't reject* H_0 and *don't* say that $\Theta < \Psi$.

(c) The assignment says that $\Theta \sim \gamma_{(90, 200)}$, and $\Psi \sim \gamma_{(110, 200)}$. Significance $\alpha = 0.1$, and $H_1: \Theta < \Psi$. Then

$$P(H_0) = 1 - F_{(2k, 2m)}\left(\frac{ml}{kn}\right) = 1 - F_{(2\cdot90, 2\cdot110)}\left(\frac{110 \cdot 200}{90 \cdot 200}\right) = 0.078 < \alpha,$$

which means that we *reject* H_0 and say that $\Theta < \Psi$.

15 $\lambda_{\text{Highlands}} \sim \gamma_{(65, 8)}$. You already know that $\lambda_{\text{Lowlands}} \sim \gamma_{(111, 23)}$, so

$$P(H_0) = P(\lambda_{\text{Highlands}} \leq \lambda_{\text{Lowlands}}) = P\left(\frac{\lambda_{\text{Highlands}}}{\lambda_{\text{Lowlands}}} \leq 1\right)$$

$$= F_{(2k,2m)}\left(\frac{ml}{kn}\right) = F_{(2\cdot65,2\cdot111)}\left(\frac{111 \cdot 8}{65 \cdot 23}\right) = 0.06\% < \alpha.$$

You therefore reject H_0, and conclude that the catch rate is higher in the Highlands than in the Lowlands.

Pairwise Comparison for Bernoulli Processes

16 **(a)** $P(H_0) = P(\psi \geq \pi) = P(\pi \leq \psi)$. Exact (Rule A.3.3):

$$P(H_0) = \sum_{k=0}^{2-1} \frac{B(4+k, 3+5)}{(5+k)\cdot B(k+1, 5)\cdot B(4,3)} = 0.121\,212 > \alpha,$$

which means that we *don't reject* H_0.
Normal approximation:

$$P(H_0) = \Phi_{\left(\frac{4}{4+3} - \frac{2}{2+5} \sqrt{\frac{4\cdot3}{(4+3)^2(4+3+1)} + \frac{2\cdot5}{(2+5)^2(2+5+1)}}\right)}(0) = 0.113\,9 > \alpha,$$

which again means that we *don't reject* H_0.

(b) $P(H_0) = P(\psi \leq \pi)$. Exact (Rule A.3.3):

$$P(H_0) = \sum_{k=0}^{17-1} \frac{B(23+k, 17+23)}{(23+k)\cdot B(k+1, 23)\cdot B(23,17)} = 0.086\,941 < \alpha.$$

Exact *reject* H_0.
Normal approximation: $P(H_0) = 0.084\,743\,9 < \alpha$, which means that we *reject* H_0.

(c) $P(H_0) = P(\psi \leq \pi)$. Exact (Rule A.3.3):

$$\sum_{k=0}^{17-1} \frac{B(20+k, 20+23)}{(23+k)\cdot B(k+1, 23)\cdot B(20,20)} = 0.247\,963 > \alpha,$$

which means that we *don't reject* H_0.

Normal approximation: $P(H_0) = 0.247\,301 > \alpha$, which again means that we *don't reject* H_0.

(d) In other words: $\psi \sim \beta_{(200,200)}$ and $\pi \sim \beta_{(170,230)}$, $\alpha = 0.05$, $H_1: \psi > \pi$. Then $P(H_0) = P(\psi \le \pi)$. Exact calculation (Rule A.3.3):

$$P(H_0) = \sum_{k=0}^{170-1} \frac{B(200+k,\,200+230)}{(230+k)\cdot B(k+1,\,230)\cdot B(200,200)} = 0.016\,552\,4 < \alpha,$$

which means that we *reject* H_0.
Normal approximation: $P(H_0) = 0.016\,338\,9 < \alpha$, which again means that we *reject* H_0.

17 (Recall that "success" in this context is an *erroneous* delivery, not a delivery well made.) Imperial Deliveries: 196 correct, 4 errors, making the *posterior* probability distribution $\pi_{ID} \sim \beta_{(5,197)}$. Centurium Falcon Freight: 199 correct, 1 error, making the *posterior* probability distribution $\pi_{CFF} \sim \beta_{(2,200)}$. For our problem, we have that $H_0: \pi_{ID} \le \pi_{CFF}$, which means

$$P(\pi_{ID} \le \pi_{CFF}) = \sum_{k=0}^{2-1} \frac{B(5+k,\,197+200)}{(200+k)\cdot B(k+1,\,200)\cdot B(5,197)} = 0.107\,617 < \alpha.$$

Calculating by means of Normal approximation gives $P(H_0) = 0.125\,374 < \alpha$. In both cases, you *reject* the null hypothesis that Imperial Deliveries are at least as good as Centurium Falcon Freight when it comes to error rate, and you therefore choose Centurium Falcon Freight as your freight company from now on.

B.15 Estimates

Gaussian Processes With Known σ

1 From distribution to interval.
  (a) $I^{\mu}_{0.025,l} = (-\infty, 19.879\,9)$ and $I^{\mu}_{0.025,r} = (8.12011, \infty)$, and
  $I^{\mu}_{0.05} = (8.1201\,1, 19.879\,9)$.
  (b) $I^{\mu}_{0.005,r} = (-22.846, \infty)$ and $I^{\mu}_{0.005,l} = (-\infty, 14.246)$, and
  $I^{\mu}_{0.01} = (-22.846, 14.246)$.
  (c) $I^{\mu}_{0.1} = (40.830\,2, 55.169\,8)$.
  (d) $I^{+}_{0.005} = (0.017\,623\,9, 0.018\,376\,1)$.
  (e) Does not exist, since $\sigma \le 0$.

2 From data + prior to interval. Find $I^{\mu}_{2\alpha}$ and $I^{+}_{2\alpha}$.
  (a) $I^{\mu}_{0.05} = (3.64, 6.36)$ and $I^{+}_{0.05} = (0.29, 9.71)$.

(b) $I_{0.1}^{\mu} = (179.7, 188.7)$ and $I_{0.1}^{+} = (159.1, 209.3)$.

(c) $I_{0.02}^{\mu} = (5.04, 5.18)$ and $I_{0.02}^{+} = (4.87, 5.35)$.

3 Sample size:

(a) $n = 2\,655$.

(b) $n = 146$.

4 $I_{\alpha} = (17.2, 17.8)$.

5 (Did you remember to convert 3 cm to 0.03 m?)

$I_{0.1}^{\mu} = (1.554\,4, 1.585\,6)$ and $I_{0.1}^{+} = (1.518\,25, 1.621\,75)$.

Gaussian Processes With Unknown σ

6 From distribution to interval.

(a) $I_{0.001}^{\mu} = (20.915, 115.285)$.

(b) $I_{0.001}^{\mu} = (-34.3626, 170.563)$ and $I_{0.1}^{\mu} = (42.731, 93.469)$.

(c) $I_{0.02}^{+} = (1.40246, 8.59754)$.

7 From data + prior to interval. Find $I_{2\alpha}^{\mu}$ and $I_{2\alpha}^{+}$. In addition, find $I_{2\alpha}^{\tau}$ and $I_{2\alpha}^{\sigma}$.

(a) $I_{0.05}^{\tau} = (0.919\,908, 14.201\,8)$, so $I_{0.05}^{\sigma} = (0.265\,355, 1.042\,62)$.

$I_{0.05}^{\mu} = (0.201\,377, 1.093\,62)$, whereas $I_{0.05}^{+} = (-0.532\,83, 1.827\,83)$.

(b) $I_{0.1}^{\tau} = (0.002\,895\,68, 0.007\,780\,3)$, so $I_{0.1}^{\sigma} = (11.337\,1, 18.583\,4)$.

$I_{0.1}^{\mu} = (142.799, 152.609)$, whereas $I_{0.1}^{+} = (123.181, 172.228)$.

(c) $I_{0.05}^{\tau} = (0.051\,064\,2, 0.122\,647)$, so $I_{0.05}^{\sigma} = (2.855\,44, 4.425\,29)$.

$I_{0.05}^{\mu} = (18.759, 20.921\,9)$, whereas $I_{0.05}^{+} = (12.749, 26.932)$.

8 Sample size:

(a) $n \geq \dfrac{4t_{12,0.05}^2}{1.3^2} \cdot \dfrac{17}{6} - 7 = 14.3$, so the smallest number is $n = 15$.

(b) $n \geq \dfrac{4t_{8,0.07}^2}{2^2} \cdot \dfrac{50}{4} - 5 = 28.5$, so the smallest number is $n = 29$.

9 $I_{0.04}^{\mu} = (365.5, 390.5)$.

10 $I_{0.1}^{\mu} = (29.0, 32.4)$.

11 $I_{0.1}^{\mu} = (4.44, 6.44)$, whereas $I_{0.1}^{+} = (2.12, 8.76)$. Further, $I_{0.1}^{\tau} = (0.124, 0.629)$, so $I_{0.1}^{\sigma} = (1.26, 2.84)$.

12 $I_{0.05}^{\mu} = (7.4, 8.1)$.

Poisson Process

13 From distribution to interval.
    (a) $I^\lambda_{0.05} = (0.064,\ 0.516)$.
    (b) $I^\lambda_{0.001} = (0.010\,53,\ 0.148\,9)$.
    (c) $I^\lambda_{0.1} = (0.044\,420\,2,\ 0.592\,983)$.

14 From data + prior to interval. Find $I^\mu_{2\alpha}$ and $I^+_{2\alpha}$.
    (a) $I^\lambda_{0.02} = (0.327\,244,\ 1.069\,72)$.
    (b) $I^\lambda_{0.1} = (0.377\,166,\ 0.596\,314)$.
    (c) $I^\lambda_{0.001} = (0.010\,53,\ 0.1489)$.

15 $n \geq \dfrac{4z^2_{0.1}}{0.2^2} - 5 = 159.237$, so the smallest number is $n = 160$.

16 $I^\lambda_{0.15} = (4.18, 5.50)$.

17 $I^\lambda_{0.2} = (1.69,\ 2.40)$.

18 $I^\lambda_{0.08} = (0.74,\ 1.99)$.

19 $I^\lambda_{0.1} = (0.36,\ 2.04)$.

20 $I^\lambda_{0.05} = (1.33,\ 1.82)$.

Bernoulli Process

21 From distribution to interval.
    (a) Exact: $I^\pi_{0.1} = (0.246\,666, 0.375\,169)$.
        Normal approximation: $I^\pi_{0.1} = (0.245\,096, 0.373\,609)$.
    (b) Exact: $I^\pi_{0.001} = (0.010\,241\,5, 0.135\,421)$.
        Normal approximation: $I^\pi_{0.001} = (-0.010\,711, 0.114\,415)$.
    (c) $I^\pi_{0.05} = (0.094\,299\,3, 0.905\,701)$.
    (d) $I^\pi_{0.12} = (0.06, 0.94)$.

22 From data + prior to interval. Find I^π_α.
    (a) Exact: $I^\pi_{0.07} = (0.258\,303, 0.530\,382)$.
        Normal approximation: $I^\pi_{0.07} = (0.253\,862, 0.526\,626)$.
    (b) Exact: $I^\pi_{0.1} = (0.122\,851, 0.527\,327)$.
        Normal approximation: $I^\pi_{0.1} = (0.104\,797, 0.510\,587)$.
    (c) Exact: $I^\pi_{0.05} = (0.578\,125, 0.800\,939)$.
        Normal approximation: $I^\pi_{0.05} = (0.583\,418, 0.807\,207)$.

(d) Exact: $I^\pi_{0.02} = (0.226\,254, 0.351\,71)$.
Normal approximation: $I^\pi_{0.02} = (0.223\,866, 0.349\,611)$.

23 $n = \frac{z^2_{0.1}}{0.05^2} - 12 - 31 = 613.95$, so at least 614 new observations.

24 **(a)** $I^\pi_{0.1} = (0.412\,2, 0.619\,5)$.
(b) $I^\pi_{0.1} = (0.525\,5, 0.580\,4)$.
(c) $I^\pi_{0.1} = (0.548\,9, 0.564\,6)$.
(d) Since we are looking at the *total* number of flips, we do not subtract our
previous flips, a and b, so $n = \dfrac{z^2_{0.05}}{0.01^2} = 27\,055.4$, that is, at least 27 056.

25 **(a)** $I^\pi_{0.2} = (0.479\,5, 0.694\,7)$.
(b) $I^\pi_{0.2} = (0.487\,5, 0.659\,4)$.

(c) $a_2 = 31$ and $b_2 = 23$, and so $n = \dfrac{z^2_{0.1}}{0.05^2} - 31 - 23 = 602.95$,
in other words, he needs 603 more observations.
(d) $I^\pi_{0.2} = (0.552\,1, 0.601\,5)$.

B.16 Frequentist Inference

Interval Estimates

1 Find the $P\% = (1 - \alpha)100\%$ interval estimates $\widehat{I^\mu_\alpha}$ and $\widehat{I^+_\alpha}$; σ is known.

(a) $\widehat{I^\mu_{0.1}} = 4.32 \pm z_{0.05} \times 2 \times \sqrt{\frac{1}{10}} = (3.279\,7, 5.360\,3)$ and $\widehat{I^+_{0.1}} = 4.32 \pm$
$z_{0.05} \times 2 \times \sqrt{1 + \frac{1}{10}} = (0.869\,726, 7.770\,27)$.

(b) $\widehat{I^\mu_{0.05}} = 66.008\,5 \pm z_{0.025} \times 12.1 \times \sqrt{\frac{1}{47}} = (62.549\,2, 69.467\,8)$ and
$\widehat{I^+_{0.05}} = 66.008\,5 \pm z_{0.025} \times 12.1 \times \sqrt{1 + \frac{1}{47}} = (42.042, 89.975)$.

(c) $\widehat{I^\mu_{0.001}} = 27.404\,6 \pm z_{0.000\,5} \times 3.73 \times \sqrt{\frac{1}{10\,000}} = (27.281\,9, 27.527\,4)$ and
$I^+_{0.001} = 27.404\,6 \pm z_{0.000\,5} \times 3.73 \times \sqrt{1 + \frac{1}{100\,00}} = (15.130\,4, 39.678\,9)$.

2 Find the $P\% = (1 - \alpha)100\%$ interval estimates $\widehat{I^\mu_\alpha}$, $\widehat{I^\sigma_\alpha}$ (use that $\sigma = 1/\sqrt{\tau}$)
and $\widehat{I^+_\alpha}$; σ is unknown.
(a) $\widehat{I^\tau_{0.1}} = (0.212\,132, 0.518\,018)$,   so $I^\sigma_{0.1} = (1.389\,4, 2.171\,18)$.
$I^\mu_{0.1} = 8.206\,9 \pm t_{28,0.05} \times 1.688\,18 \times \sqrt{\frac{1}{29}} = (7.673\,61, 8.740\,18)$, while
$\widehat{I^+_{0.1}} = 8.206\,9 \pm t_{28,0.05} \times 1.688\,18 \times \sqrt{1 + \frac{1}{29}} = (5.285\,99, 11.127\,8)$.

(b) $\widehat{I^{\tau}_{0.02}} = (0.037\,7407,\,0.052\,533\,8)$, so $I^{\sigma}_{0.02} = (4.362\,95,\,5.147\,48)$.

$I^{\mu}_{0.02} = 29.688\,6 \pm t_{397,0.01} \times 4.724\,27 \times \sqrt{\frac{1}{398}} = (29.135\,5,\,30.241\,7)$,

while $\qquad \widehat{I^{+}_{0.02}} = 29.688\,6 \pm t_{397,\,0.01} \times 4.724\,27 \times \sqrt{1 + \frac{1}{398}} =$

$(18.639\,9,\,40.737\,3)$.

(c) We calculate and find the statistics $n = 8$, $S_x = 466.4$, $SS_x = 196.28$, and get: $\widehat{I^{\tau}_{0.02}} = (0.006\,312\,63,\,0.094\,127\,3)$, so $I^{\sigma}_{0.02} = (3.26,\,12.59)$.

$I^{\mu}_{0.02} = 58.3 \pm t_{7,0.01} \times 5.295\,28 \times \sqrt{\frac{1}{8}} = (52.687\,3,\,63.912\,7)$, while

$\widehat{I^{+}_{0.02}} = 58.3 \pm t_{7,\,0.01} \times 5.295\,28 \times \sqrt{1 + \frac{1}{8}} = (41.462,\,75.138)$.

3 Find the $(1 - \alpha)100\%$ confidence interval $\widehat{I^{p}_{\alpha}}$.

(a) $\widehat{I^{\pi}_{0.05}} = 0.404\,762 \pm z_{0.025} \times \sqrt{\frac{0.404\,762(1-0.404\,762)}{42}} =$
$(0.256\,316,\,0.553\,208)$.

(b) The numbers are $k = 46$ positives and $l = 54$ negatives. $\alpha = 0.1$. Then
$\widehat{I^{\pi}_{0.1}} = 0.46 \pm z_{0.05} \times \sqrt{\frac{0.46(1-0.46)}{100}} = (0.378\,021,\,0.541\,979)$.

Hypothesis Testing

4 Determine the hypothesis test outcome about the parameter π for a Bernoulli process; significance α.

(a) Since $k + l < 30$, we do an exact calculation. $p = P(X \geq 8) =$
$\sum_{m=8}^{14} \binom{14}{m} \times 0.5^m (1 - 0.5)^{14-m} = 0.395\,264 > \alpha$, so we *don't reject* the null hypothesis.

(b) $p = P(X \leq 2) = \sum_{m=0}^{2} \binom{20}{m} \times 0.25^m (1 - 0.25)^{20-m} = 0.091\,3 < \alpha$, so we *reject* the null hypothesis.

(c) $k + l > 30$, so we may use a normal approximation: $w = \frac{0.04-0.1}{0.019\,595\,9} = -3.061\,86$. $\Phi(w) = \Phi(-3.06\,186) = 0.0011 < \alpha$, so we *reject* the null hypothesis. By comparison, exact calculation gives $p = P(X \leq 4) = \sum_{m=0}^{4} \binom{100}{m} \times 0.1^m (1 - 0.1)^{100-m} = 0.023\,711\,1 < \alpha$, which also implies that we reject H_0.

(d) $k + l > 30$, so we may use a normal approximation: $w = \frac{0.4-0.5}{0.048\,989\,8} = -2.041\,24$. $\Phi(-|w|) = \Phi(-2.041\,24) = 0.021 < \frac{0.05}{2}$, so we *reject* the null hypothesis, and may say $\pi \neq 0.5$.

We may check this by direct calculation as well, by checking if the probability one way is less than $\frac{\alpha}{2}$. We see that $p = P(X \leq 40) = \sum_{m=0}^{40} \binom{100}{m} \times 0.5^m (1 - 0.5)^{100-m} = 0.028 < \frac{\alpha}{2}$, so we reject the null hypothesis when we go via direct calculation as well.

(e) The numbers of this problem are: $H_1 : \pi \neq 0.5$, $\alpha = 0.1$. You have $n = 100$ trials with $k = 46$ successes. This means that $w = \frac{0.46-0.5}{0.049\,839\,7} = -0.802\,572$, so $\Phi(-|w|) = \Phi(-0.802\,572) = 0.211\,111 > \alpha$, which means we *don't reject* the null hypothesis.

5 Determine the hypothesis test outcome about the mean μ for a Gaussian process; significance α.

(a) $w = \frac{26.493\,8-25.}{\frac{3.73}{\sqrt{27}}} = 2.080\,99$, so $\Phi(-|w|) = \Phi(-2.080\,99) = 0.019 < \frac{\alpha}{2}$, so we *reject* the null hypothesis.

(b) $w = \frac{80.674\,1-80.}{\frac{5.1}{\sqrt{200}}} = 1.869\,22$, so $\Phi(-w) = \Phi(-1.869\,22) = 0.030\,796\,4 > \alpha$, so we *don't reject* the null hypothesis.

(c) $w = \frac{31.9-25}{\frac{7.997\,86}{\sqrt{8}}} = 2.440\,17$, so $\Phi(-|w|) = \Phi(-2.440\,17) = 0.022\,4 > \frac{\alpha}{2}$, so we *don't reject* the null hypothesis.

(d) $w = \frac{99.3-100.}{\frac{10.383\,8}{\sqrt{500}}} = -1.507\,39$, so $T(w) = T(-1.507\,39) = 0.066\,2 > \alpha$, so we *don't reject* the null hypothesis.

6 $w = \frac{60\,376.3-60\,000}{\frac{423.645}{\sqrt{15}}} = 3.439\,85$. The null hypothesis is that the mean is 60 000, so we calculate $T_\nu(-w) = T_{14}(-3.439\,85) = 0.001\,99 < \alpha$, and see that we must *reject* the null hypothesis.

7 Determine the hypothesis test outcome concerning the variance σ^2 of a Gaussian process; significance α.

(a) $n = 15$, $SS_x = 768.773$, so $\frac{SS_x}{\sigma_0^2} = \frac{768.773}{10^2} = 7.687\,73$. Since $\chi^2_{14,0.05} = 6.570\,63$, it follows that $\frac{SS_x}{\sigma_0^2} > \chi^2_{\nu,\alpha}$, and hence we *don't reject* the null hypothesis.

(b) This gives us $SS_x = 4070.72$, so $\frac{SS_x}{\sigma_0^2} = \frac{4070.72}{5^2} = 162.829$. Since $\chi^2_{99,0.98} = 129.996$, it follows that $\frac{SS_x}{\sigma_0^2} > \chi^2_{\nu,\alpha}$, and hence we *reject* the null hypothesis.

8 From the data, we get that $n = 15$, $SS_x = 2.512\,65 \times 10^6$. Then $\frac{SS_x}{\sigma_0^2} = \frac{2.512\,65 \times 10^6}{666^2} = 5.66\,478$. The comparison values are $\chi^2_{\nu,\alpha/2} = \chi^2_{14,0.05} = 6.570\,63$ and $\chi^2_{\nu,1-\alpha/2} = \chi^2_{14,0.95} = 23.684\,8$. We see that $\frac{SS_x}{\sigma_0^2} < \chi^2_{\nu,\alpha/2}$, and hence we *reject* the null hypothesis that $\sigma = 666$.

B.17 Linear Regression

1 Volume training 1:

(a) $- \tau \sim \gamma_{\left(\frac{8}{2}, \frac{132.296}{2}\right)}$

$- y(x) \sim t_{\left(244.08+7.19091(x-31.968), 4.066\,57 \times \sqrt{\frac{1}{5}+\frac{1}{520.561}(x-31.968)^2}, 8\right)}$

$- Y_+(x) \sim t_{\left(244.08+7.19091(x-31.968), 4.06657 \times \sqrt{\frac{6}{5}+\frac{1}{520.561}(x-31.968)^2}, 8\right)}$

$- I_{0.05} = 244.08 + 7.190\,9(x-31.968) \pm 9.3775$

$\qquad \times \sqrt{\frac{1}{5}+\frac{1}{520.56}(x-31.968)^2}$

$- I_{0.1}^+ = 244.08 + 7.190\,9(x-31.968) \pm 7.562$

$\qquad \times \sqrt{\frac{6}{5}+\frac{1}{520.56}(x-31.968)^2}.$

(b) $- \tau \sim \gamma_{(0.5, 0.75)}$

$- y(x) \sim t_{\left(3+0.5(x-2), 1.224\,74 \times \sqrt{\frac{1}{3}+\frac{1}{2}(x-2)^2}, 1\right)}$

$- Y_+(x) \sim t_{\left(3+0.5(x-2), 1.224\,74 \times \sqrt{\frac{4}{3}+\frac{1}{2}(x-2)^2}, 1\right)}$

$- I_{0.1} = 3 + 0.5(x-2) \pm 7.732\,7 \times \sqrt{\frac{1}{3}+\frac{1}{2}(x-2)^2}$

$- I_{0.1}^+ = 3 + 0.5(x-2) \pm 7.732\,7 \times \sqrt{\frac{4}{3}+\frac{1}{2}(x-2)^2}.$

(c) $- \tau \sim \gamma_{(3, 0.735\,714)}$

$- y(x) \sim t_{\left(8.25-0.657\,143(x-3.75), 0.495\,215 \times \sqrt{\frac{1}{4}+\frac{1}{8.75}(x-3.75)^2}, 6\right)}$

$- Y_+(x) \sim t_{\left(8.25-0.657\,143(x-3.75), 0.495\,215 \times \sqrt{\frac{5}{4}+\frac{1}{8.75}(x-3.75)^2}, 6\right)}$

$- I_{0.05} = 8.25 - 0.657\,143(x - 15/4) \pm 1.211\,75$

$\qquad \times \sqrt{\frac{1}{4}+\frac{1}{8.75}(x-3.75)^2}$

$- I_{0.05}^+ = 8.25 - 0.657\,143(x-3.75) \pm 1.211\,75$

$\qquad \times \sqrt{\frac{5}{4}+\frac{1}{8.75}(x-3.75)^2}.$

(d) $- \tau \sim \gamma_{(1, 4.5)}$

$- y(x) \sim t_{\left(3.5+2.(x-1.5), 2.121\,32 \times \sqrt{\frac{1}{4}+\frac{1}{5}(x-1.5)^2}, 2\right)}$

$- Y_+(x) \sim t_{\left(3.5+2.(x-1.5), 2.121\,32 \times \sqrt{\frac{5}{4}+\frac{1}{5}(x-1.5)^2}, 2\right)}$

$- I_{0.1} = 3.5 + 2(x - 1.5) \pm 6.194\,23 \times \sqrt{\frac{1}{4}+\frac{1}{5}(x-1.5)^2}$

$- I_{0.05}^+ = 3.5 + 2(x - 1.5) \pm 9.127\,3 \times \sqrt{\frac{5}{4}+\frac{1}{5}(x-1.5)^2}.$

(e) $- \tau \sim \gamma_{(0.5, 0.914\,557)}$

$- y(x) \sim t_{\left(6-0.056\,962(x-2.33\,33), 1.352\,45 \times \sqrt{\frac{1}{3}+\frac{1}{52.667}(x-2.333\,3)^2}, 1\right)}$

$- Y_+(x) \sim t_{\left(6-0.056\,962(x-2.333\,3), 1.352\,45 \times \sqrt{\frac{4}{3}+\frac{1}{52.667}(x-2.333\,3)^2}, 1\right)}$

– $I_{0.02} = 6 - 0.056\,962(x - 2.333\,3) \pm 43.036$

$\times \sqrt{\dfrac{1}{3} + \dfrac{1}{52.667}(x - 2.333\,3)^2}$

– $I_{0.02}^+ = 6 - 0.056\,962(x - 2.333\,3) \pm 43.036$

$\times \sqrt{\dfrac{4}{3} + \dfrac{1}{52.667}(x - 2.333\,3)^2}.$

(f) – $\tau \sim \gamma_{(8,\,305.331)}$

– $y(x) \sim t_{\left(39.625+0.847\,613(x-37.625),\,6.177\,9 \times \sqrt{\frac{1}{8} + \frac{1}{4363.88}(x-37.625)^2},\,16\right)}$

– $Y_+(x) \sim t_{\left(39.625+0.847\,613(x-37.625),\,6.177\,9 \times \sqrt{\frac{9}{8} + \frac{1}{4363.88}(x-37.625)^2},\,16\right)}$

– $I_{0.2} = 39.625 + 0.847\,61(x - 301/8) \pm 8.258\,4$

$\times \sqrt{\dfrac{1}{8} + \dfrac{1}{4363.9}(x - 37.625)^2}$

– $I_{0.2}^+ = 39.625 + 0.847\,61(x - 37.625) \pm 8.258\,4$

$\times \sqrt{\dfrac{9}{8} + \dfrac{1}{4363.9}(x - 37.625)^2}.$

(g) – $\tau \sim \gamma_{(6,\,147.677)}$

– $y(x) \sim t_{\left(26.562\,5+0.903\,14(x-28.225),\,4.961\,14 \times \sqrt{\frac{1}{8} + \frac{1}{1910.46}(x-28.225)^2},\,12\right)}$

– $Y_+(x) \sim t_{\left(26.562\,5+0.903\,14(x-28.225),\,4.961\,14 \times \sqrt{\frac{9}{8} + \frac{1}{1910.46}(x-28.225)^2},\,12\right)}$

– $I_{0.02} = 26.563 + 0.903\,14(x - 28.225) \pm 13.301$

$\times \sqrt{\dfrac{1}{8} + \dfrac{1}{1910.46}(x - 28.225)^2}$

– $I_{0.02}^+ = 26.563 + 0.903\,14(x - 28.225) \pm 13.301$

$\times \sqrt{\dfrac{9}{8} + \dfrac{1}{1910.46}(x - 28.225)^2}$

(h) – $\tau \sim \gamma_{(2.5,\,57.074\,9)}$

– $y(x) \sim t_{\left(10.957\,1+0.092\,891\,5(x-205.171),\,4.778\,07 \times \sqrt{\frac{1}{7} + \frac{1}{124\,54.4}(x-205.171)^2},\,5\right)}$

– $Y_+(x) \sim t_{\left(10.957\,1+0.092\,891\,5(x-205.171),\,4.778\,07 \times \sqrt{\frac{8}{7} + \frac{1}{12\,454.4}(x-205.171)^2},\,5\right)}$

– $I_{0.05} = 10.957 + 0.092\,892(x - 205.171) \pm 12.282$

$\times \sqrt{\dfrac{1}{7} + \dfrac{1}{12\,454}(x - 205.17)^2}$

– $I_{0.1}^+ = 10.957 + 0.092\,892(x - 205.171) \pm 9.628\,1$

$\times \sqrt{\dfrac{8}{7} + \dfrac{1}{12\,454}(x - 205.17)^2}.$

2 Volume training 2:

(a) – $\tau \sim \gamma_{(9,\,302.561)}$

– $y(x) \sim t_{\left(11.2+0.503\,382(x-84.05),\,5.798\,1 \times \sqrt{\frac{1}{20} + \frac{1}{1278.95}(x-84.05)^2},\,18\right)}$

– $Y_+(x) \sim t_{\left(11.2+0.503\,382(x-84.05),\,5.798\,1 \times \sqrt{\frac{21}{20} + \frac{1}{1\,278.95}(x-84.05)^2},\,18\right)}.$

(b) – $\tau \sim \Upsilon_{(50\,002,\,450\,019)}$

– $y(x) \sim t_{\left(-36.6667+-3.913\,65(x-11.5),\,3\times\sqrt{\frac{1}{6}+\frac{1}{787.5}(x-11.5)^2},\,100\,004\right)}$

$\approx \phi_{\left(-36.6667-3.913\,65(x-11.5),\,3\times\sqrt{\frac{1}{6}+\frac{1}{787.5}(x-11.5)^2}\right)}$

– $Y_+(x) \sim t_{\left(-36.6667-3.913\,65(x-11.5),\,3\times\sqrt{\frac{7}{6}+\frac{1}{787.5}(x-11.5)^2},\,100\,004\right)}$

$\approx \phi_{\left(-36.6667-3.913\,65(x-11.5),\,3\times\sqrt{\frac{7}{6}+\frac{1}{787.5}(x-11.5)^2}\right)}.$

(c) – $\tau \sim \Upsilon_{(502,\,2004.75)}$

– $y(x) \sim t_{\left(62.6667+0.998\,67(x-57.1667),\,1.998\,38\times\sqrt{\frac{1}{6}+\frac{1}{5\,640.83}(x-57.1667)^2},\,1004\right)}$

$\approx \phi_{\left(62.6667+0.998\,67(x-57.1667),\,2\times\sqrt{\frac{1}{6}+\frac{1}{5\,640.83}(x-57.1667)^2}\right)}$

– $Y_+(x) \sim t_{\left(62.6667+0.998\,67(x-57.166\,7),\,1.998\,38\times\sqrt{\frac{7}{6}+\frac{1}{5\,640.83}(x-57.166\,7)^2},\,1004\right)}$

$\approx \phi_{\left(62.6667+0.998\,67(x-57.166\,7),\,2\times\sqrt{\frac{7}{6}+\frac{1}{5\,640.83}(x-57.166\,7)^2}\right)}.$

3 Volume training 3:

(a) – $I_{0.08} = 11 - 1.047\,25(x - 9.9) \pm 6.717\,26 \times \sqrt{\dfrac{1}{20} + \dfrac{1}{469.8}(x - 9.9)^2}$

– $I_{0.01}^+ = 11 - 1.047\,25(x - 9.9) \pm 10.2607 \times \sqrt{\dfrac{21}{20} + \dfrac{1}{469.8}(x - 9.9)^2}$

(b) – $I_{0.07} = 20.6 + 0.529\,68(x - 21.4) \pm 2.861\,36 \times \sqrt{\dfrac{1}{5} + \dfrac{1}{175.2}(x - 21.4)^2}$

– $I_{0.07}^+ = 20.6 + 0.529\,68(x - 21.4) \pm 2.861\,36 \times \sqrt{\dfrac{6}{5} + \dfrac{1}{175.2}(x - 21.4)^2}.$

(c) – $I_{0.07} = 20.6 + 0.529\,68(x - 21.4) \pm 13.349 \times \sqrt{\dfrac{1}{5} + \dfrac{1}{175.2}(x - 21.4)^2}$

– $I_{0.07}^+ = 20.6 + 0.529\,68(x - 21.4) \pm 13.349 \times \sqrt{\dfrac{6}{5} + \dfrac{1}{175.2}(x - 21.4)^2}.$

4 – Regression line: $y = 1\,263.2 + 21.89(x - 1\,995)$

– $I_{0.1} = 1\,263.2 + 21.89(x - 1995) \pm 293.629 \times \sqrt{\dfrac{1}{5} + \dfrac{1}{1\,000}(x - 1\,995)^2}$

– $I_{0.1}^+ = 1\,263.2 + 21.89(x - 1\,995) \pm 293.629 \times \sqrt{\dfrac{6}{5} + \dfrac{1}{1\,000}(x - 1\,995)^2}.$

5 (a) $y = 114.4 + 7.9(x - 2)$

(b) $b \sim \phi_{(7.9,\,1.264\,91)}(x).$

(c) $a_* = y(\bar{x}) \sim \phi_{(114.4,\,1.788\,85)}(x).$

(d) $Y_+ \sim \phi_{(256.6,\,23.186)}.$

(e) $I = (218.5,\,294.7).$

(f) This question is open to many correct answers, and the key is that you
have reflected upon the question and have reasons for your answer.
Concerning why it is a *bad* model, you may for instance demonstrate
that this model becomes absurd if the number of pumpings, x, is set

very high, as the ball will then be predicted to rebound higher than the height it was dropped from. As for its being a good model, you may show how it actually fits very well within a narrower range of x values.

6 **(a)** $y = -18.2 - 23.2829(z - 6.50881)$.

(b) $b \sim t_{(-23.2829, 1.73784, 3)}$.

(c) H_1 is that $b > -25$. Then, $P(H_0) = T_{(-23.2829, 1.73784, 3)}(-25) = 0.197983$, which is greater than the significance α, so we *don't reject* H_0, and will therefore not claim that this singer's *spectral slope* is higher than -25.

7 The linear regression line is $y = 78\,000 + 6\,761.62(x - 14.75) = -21\,733.9 + 6\,761.62x$, which has a positive slope, $6\,761.62$. The *posterior* probability distribution of the slope is $b \sim t_{(6\,761.62, 4\,380.15, 2)}$. The alternative hypothesis H_1 is that $b > 0$. We test the hypothesis:

$$P(H_0) = P(b \leq 0) = T_{(6\,761.62, 4\,380.15, 2)} = 0.131\,323 < \alpha.$$

With significance $\alpha = 0.2$, you may conclude that the salary increases with the number of goals scored.

8 The regression line is $y(x) = 2184.67 + 23.4437(x - 103.333) = +23.4437x - 237.838$. We indicate the uncertainty through the *posterior* distribution: $y(x) \sim t_{\left(2\,184.67 + 23.4437(x-103.333),\, 123.2 \times \sqrt{\frac{1}{9} + \frac{1}{24\,528}(x-103.333)^2},\, 7\right)}$.

For a temperature of $100°C$, we have $y(100) \sim t_{(2\,106.5, 41.2, 7)}$, so $I_{0.1} = (2\,028.5, 2\,184.5)$. The predictive distribution is $Y_+(100) \sim t_{(2\,106.5, 129.9, 7)}$, so $I_{0.1}^+(100) = 2\,106.5 \pm 246.1 = (1\,860.4, 2\,352.6)$. The explanation of the difference is that, for the measurement, we must add the uncertainty of the measurement itself to the uncertainty in our knowledge of the solubility.

C

Tables

CONTENTS
C.1 z_p (Left Tail), 490
C.2 Percentiles for t Distribution with v Degrees of Freedom, 491
C.3 Percentiles for χ^2 Distribution with v Degrees of Freedom, 492
C.4 The Γ Function, 493
C.5 $\Phi(x) = \int_0^x \phi_{(0,1)}(t)\,dt,\ \ x \leq 0,\ 494$
C.6 $\Phi(x) = \int_0^x \phi_{(0,1)}(t)\,dt,\ \ x \geq 0,\ 495$

The following tables were generated by Mathematica. Use Mathematica or some similar tool directly if you need more accurate values. The commands used are

- z_p is InverseCDF[NormalDistribution[0,1], p]
- $t_{v,p}$ is InverseCDF[StudentTDistribution[0,1,v], p]
- $\chi_v^2(x)$ is InverseCDF[ChiSquareDistribution[v, x]]
- $\Gamma(x)$ is Gamma[x]
- $\Phi(x)$ is CDF[NormalDistribution[0,1], x].

These commands also work in Wolfram Alpha (http://alpha.wolfram.com). See also the sections for the different probability distributions to find the commands for TI, Casio, and HP calculators. If you are interested in more powerful statistics tools, it might be worth your time to learn the free and freely available statistics tool R. See http://bayesians.net for more information. Even though modern tools have made tables obsolete, such tools are sometimes unavailable. For this reason, extended or other tables will be added at http://bayesians.net upon request.

The Bayesian Way: Introductory Statistics for Economists and Engineers, First Edition.
Svein Olav Nyberg.
© 2019 John Wiley & Sons, Inc. Published 2019 by John Wiley & Sons, Inc.

C.1 z_p (Left Tail)

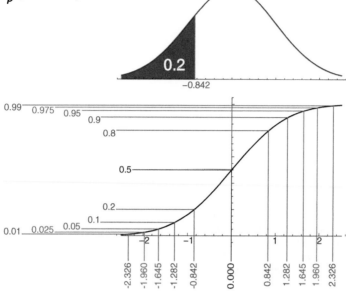

p	z_p	p	z_p	p	z_p
0.00000	$-\infty$	0.07	−1.476	0.94	1.555
0.0001	−3.719	0.08	−1.405	0.95	1.645
0.00025	−3.481	0.09	−1.341	0.955	1.695
0.0005	−3.290	0.10	−1.282	0.96	1.751
0.001	−3.090	0.15	−1.036	0.965	1.812
0.0025	−2.807	0.20	−0.842	0.97	1.881
0.005	−2.576	0.30	−0.524	0.975	1.960
0.01	−2.326	0.40	−0.253	0.98	2.054
0.015	−2.170	**0.50**	**0**	0.985	2.170
0.02	−2.054	0.60	0.253	0.99	2.326
0.025	−1.960	0.70	0.524	0.995	2.576
0.03	−1.881	0.80	0.842	0.9975	2.807
0.035	−1.812	0.85	1.036	0.999	3.090
0.04	−1.751	0.90	1.282	0.9995	3.290
0.045	−1.695	0.91	1.341	0.99975	3.481
0.05	−1.645	0.92	1.405	0.9999	3.719
0.06	−1.555	0.93	1.476	1.00000	∞

Values for the right tail have the opposite sign, since z is antisymmetric around $p = 0.5$:

$$z_{1-p} = -z_p.$$

C.2 Percentiles for t Distribution with v Degrees of Freedom

The table shows $t_{v,p}$ for the left tail. Values for the right tail have the opposite sign, since t is anti-symmetric around $p = 0.5$:

$$t_{v,1-p} = -t_{v,p}.$$

v \ p	0.01	0.025	0.05	0.1	0.9	0.95	0.975	0.99
1	−31.82	−12.71	−6.314	−3.078	3.078	6.314	12.71	31.82
2	−6.965	−4.303	−2.920	−1.886	1.886	2.920	4.303	6.965
3	−4.541	−3.182	−2.353	−1.638	1.638	2.353	3.182	4.541
4	−3.747	−2.776	−2.132	−1.533	1.533	2.132	2.776	3.747
5	−3.365	−2.571	−2.015	−1.476	1.476	2.015	2.571	3.365
6	−3.143	−2.447	−1.943	−1.440	1.440	1.943	2.447	3.143
7	−2.998	−2.365	−1.895	−1.415	1.415	1.895	2.365	2.998
8	−2.896	−2.306	−1.860	−1.397	1.397	1.860	2.306	2.896
9	−2.821	−2.262	−1.833	−1.383	1.383	1.833	2.262	2.821
10	−2.764	−2.228	−1.812	−1.372	1.372	1.812	2.228	2.764
11	−2.718	−2.201	−1.796	−1.363	1.363	1.796	2.201	2.718
12	−2.681	−2.179	−1.782	−1.356	1.356	1.782	2.179	2.681
13	−2.650	−2.160	−1.771	−1.350	1.350	1.771	2.160	2.650
14	−2.624	−2.145	−1.761	−1.345	1.345	1.761	2.145	2.624
15	−2.602	−2.131	−1.753	−1.341	1.341	1.753	2.131	2.602
16	−2.583	−2.120	−1.746	−1.337	1.337	1.746	2.120	2.583
17	−2.567	−2.110	−1.740	−1.333	1.333	1.740	2.110	2.567
18	−2.552	−2.101	−1.734	−1.330	1.330	1.734	2.101	2.552
19	−2.539	−2.093	−1.729	−1.328	1.328	1.729	2.093	2.539
20	−2.528	−2.086	−1.725	−1.325	1.325	1.725	2.086	2.528
21	−2.518	−2.080	−1.721	−1.323	1.323	1.721	2.080	2.518
22	−2.508	−2.074	−1.717	−1.321	1.321	1.717	2.074	2.508
23	−2.500	−2.069	−1.714	−1.319	1.319	1.714	2.069	2.500
24	−2.492	−2.064	−1.711	−1.318	1.318	1.711	2.064	2.492
25	−2.485	−2.060	−1.708	−1.316	1.316	1.708	2.060	2.485
26	−2.479	−2.056	−1.706	−1.315	1.315	1.706	2.056	2.479
27	−2.473	−2.052	−1.703	−1.314	1.314	1.703	2.052	2.473
28	−2.467	−2.048	−1.701	−1.313	1.313	1.701	2.048	2.467
29	−2.462	−2.045	−1.699	−1.311	1.311	1.699	2.045	2.462
30	−2.457	−2.042	−1.697	−1.310	1.310	1.697	2.042	2.457

In Mathematica®, $t_{v,p}$ is InverseCDF[StudentTDistribution[0, 1, v], p].

C.3 Percentiles for χ^2 Distribution with v Degrees of Freedom

The table shows χ_p^2 for the left tail. You find the right tail for p as χ_{1-p}^2. Note that there are no symmetries to make use of in the case of χ^2.

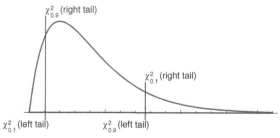

v \ p	0.01	0.025	0.05	0.1	0.9	0.95	0.975	0.99
1	0.000	0.001	0.004	0.016	2.706	3.841	5.024	6.635
2	0.020	0.051	0.103	0.211	4.605	5.991	7.378	9.210
3	0.115	0.216	0.352	0.584	6.251	7.815	9.348	11.345
4	0.297	0.484	0.711	1.064	7.779	9.488	11.143	13.277
5	0.554	0.831	1.145	1.610	9.236	11.07	12.833	15.086
6	0.872	1.237	1.635	2.204	10.645	12.592	14.449	16.812
7	1.239	1.690	2.167	2.833	12.017	14.067	16.013	18.475
8	1.646	2.180	2.733	3.490	13.362	15.507	17.535	20.090
9	2.088	2.700	3.325	4.168	14.684	16.919	19.023	21.666
10	2.558	3.247	3.940	4.865	15.987	18.307	20.483	23.209
11	3.053	3.816	4.575	5.578	17.275	19.675	21.920	24.725
12	3.571	4.404	5.226	6.304	18.549	21.026	23.337	26.217
13	4.107	5.009	5.892	7.042	19.812	22.362	24.736	27.688
14	4.660	5.629	6.571	7.790	21.064	23.685	26.119	29.141
15	5.229	6.262	7.261	8.547	22.307	24.996	27.488	30.578
16	5.812	6.908	7.962	9.312	23.542	26.296	28.845	32.000
17	6.408	7.564	8.672	10.085	24.769	27.587	30.191	33.409
18	7.015	8.231	9.390	10.865	25.989	28.869	31.526	34.805
19	7.633	8.907	10.117	11.651	27.204	30.144	32.852	36.191
20	8.260	9.591	10.851	12.443	28.412	31.410	34.170	37.566
21	8.897	10.283	11.591	13.240	29.615	32.671	35.479	38.932
22	9.542	10.982	12.338	14.041	30.813	33.924	36.781	40.289
23	10.196	11.689	13.091	14.848	32.007	35.172	38.076	41.638
24	10.856	12.401	13.848	15.659	33.196	36.415	39.364	42.980
25	11.524	13.120	14.611	16.473	34.382	37.652	40.646	44.314
26	12.198	13.844	15.379	17.292	35.563	38.885	41.923	45.642
27	12.879	14.573	16.151	18.114	36.741	40.113	43.195	46.963
28	13.565	15.308	16.928	18.939	37.916	41.337	44.461	48.278
29	14.256	16.047	17.708	19.768	39.087	42.557	45.722	49.588
30	14.953	16.791	18.493	20.599	40.256	43.773	46.979	50.892

C.4 The Γ Function

	0.0	0.1	0.2	0.3	0.4	0.5	0.6	0.7	0.8	0.9
0	1	9.513 51	4.590 84	2.991 57	2.218 16	1.772 45	1.489 19	1.298 06	1.164 23	1.068 63
1	1	0.951 351	0.918 169	0.897 471	0.887 264	0.886 227	0.893 515	0.908 639	0.931 384	0.961 766
2	1	1.046 49	1.101 8	1.166 71	1.242 17	1.329 34	1.429 62	1.544 69	1.676 49	1.827 36
3	2	2.197 62	2.423 97	2.683 44	2.981 21	3.323 35	3.717 02	4.170 65	4.694 17	5.299 33
4	6	6.812 62	7.756 69	8.855 34	10.136 1	11.6317	13.3813	15.4314	17.837 9	20.667 4
5	24	27.931 8	32.578 1	38.078	44.598 8	52.342 8	61.553 9	72.527 6	85.6217	101.27
6	120	142.452	169.406	201.813	240.834	287.885	344.702	413.408	496.606	597.494
7	720	868.957	1 050.32	1 271.42	1 541.34	1 871.25	2 275.03	2 769.83	3 376.92	4 122.71
8	5 040	6 169.59	7 562.29	9 281.39	11 405.9	14 034.4	17 290.2	21 327.7	26 340.	32 569.4
9	40 320	49 973.7	62 010.8	77 035.6	95 809.5	119 292	148 696	185 551	231 792	289 868

	0	1	2	3	4	5	6	7	8	9
10	3.63×10^{5}	3.63×10^{6}	3.99×10^{7}	4.79×10^{8}	6.23×10^{9}	8.72×10^{10}	1.31×10^{12}	2.09×10^{13}	3.56×10^{14}	6.40×10^{15}
20	1.22×10^{17}	2.43×10^{18}	5.11×10^{19}	1.12×10^{21}	2.59×10^{22}	6.2×10^{23}	1.55×10^{25}	4.03×10^{26}	1.09×10^{28}	3.05×10^{29}
30	8.84×10^{30}	2.65×10^{32}	8.22×10^{33}	2.63×10^{35}	8.68×10^{36}	2.95×10^{38}	1.03×10^{40}	3.72×10^{41}	1.38×10^{43}	5.23×10^{44}
40	2.04×10^{46}	8.16×10^{47}	3.35×10^{49}	1.41×10^{51}	6.04×10^{52}	2.66×10^{54}	1.2×10^{56}	5.5×10^{57}	2.59×10^{59}	1.24×10^{61}
50	6.08×10^{62}	3.04×10^{64}	1.55×10^{66}	8.07×10^{67}	4.27×10^{69}	2.31×10^{71}	1.27×10^{73}	7.11×10^{74}	4.05×10^{76}	2.35×10^{78}
60	1.39×10^{80}	8.32×10^{81}	5.08×10^{83}	3.15×10^{85}	1.98×10^{87}	1.27×10^{89}	8.25×10^{90}	5.44×10^{92}	3.65×10^{94}	2.48×10^{96}
70	1.71×10^{98}	1.2×10^{100}	8.5×10^{101}	6.1×10^{103}	4.5×10^{105}	3.3×10^{107}	2.5×10^{109}	1.9×10^{111}	1.5×10^{113}	1.1×10^{115}
80	8.9×10^{116}	7.2×10^{118}	5.8×10^{120}	4.8×10^{122}	3.9×10^{124}	3.3×10^{126}	2.8×10^{128}	2.4×10^{130}	2.1×10^{132}	1.9×10^{134}
90	1.7×10^{136}	1.5×10^{138}	1.4×10^{140}	1.2×10^{142}	1.2×10^{144}	1.1×10^{146}	1×10^{148}	9.9×10^{149}	9.6×10^{151}	9.4×10^{153}
100	9.3×10^{155}	9.3×10^{157}	9.4×10^{159}	9.6×10^{161}	9.9×10^{163}	1×10^{166}	1.1×10^{168}	1.1×10^{170}	1.2×10^{172}	1.3×10^{174}
110	1.4×10^{176}	1.6×10^{178}	1.8×10^{180}	2×10^{182}	2.2×10^{184}	2.5×10^{186}	2.9×10^{188}	3.4×10^{190}	$4. \times 10^{192}$	4.7×10^{194}
120	5.6×10^{196}	6.7×10^{198}	8.1×10^{200}	9.9×10^{202}	1.2×10^{205}	1.5×10^{207}	1.9×10^{209}	2.4×10^{211}	$3. \times 10^{213}$	3.9×10^{215}
130	$5. \times 10^{217}$	6.5×10^{219}	8.5×10^{221}	1.1×10^{224}	1.5×10^{226}	$2. \times 10^{228}$	2.7×10^{230}	3.7×10^{232}	$5. \times 10^{234}$	6.9×10^{236}
140	9.6×10^{238}	1.3×10^{241}	1.9×10^{243}	2.7×10^{245}	3.9×10^{247}	5.6×10^{249}	$8. \times 10^{251}$	1.2×10^{254}	1.7×10^{256}	2.6×10^{258}

For $n \geq 10$, *Stirling's approximation* to Γ yields a value less than 1% off from the true value: $\Gamma(x) \approx \sqrt{\dfrac{2\pi}{x}} \left(\dfrac{x}{e}\right)^{x}$.

For positive integers n, you may use that Γ is a generalization of faculty to get an exact answer: $\Gamma(n) = (n - 1)!$

C.5 $\Phi(x) = \int_0^x \phi_{(0,1)}(t)\, dt, \;\; x \leq 0$

	0.00	0.01	0.02	0.03	0.04	0.05	0.06	0.07	0.08	0.09
−0.0	0.500	0.496	0.492	0.488	0.484	0.480	0.476	0.472	0.468	0.464
−0.1	0.460	0.456	0.452	0.448	0.444	0.440	0.436	0.433	0.429	0.425
−0.2	0.421	0.417	0.413	0.409	0.405	0.401	0.397	0.394	0.390	0.386
−0.3	0.382	0.378	0.374	0.371	0.367	0.363	0.359	0.356	0.352	0.348
−0.4	0.345	0.341	0.337	0.334	0.330	0.326	0.323	0.319	0.316	0.312
−0.5	0.309	0.305	0.302	0.298	0.295	0.291	0.288	0.284	0.281	0.278
−0.6	0.274	0.271	0.268	0.264	0.261	0.258	0.255	0.251	0.248	0.245
−0.7	0.242	0.239	0.236	0.233	0.230	0.227	0.224	0.221	0.218	0.215
−0.8	0.212	0.209	0.206	0.203	0.200	0.198	0.195	0.192	0.189	0.187
−0.9	0.184	0.181	0.179	0.176	0.174	0.171	0.169	0.166	0.164	0.161
−1.0	0.159	0.156	0.154	0.152	0.149	0.147	0.145	0.142	0.140	0.138
−1.1	0.136	0.133	0.131	0.129	0.127	0.125	0.123	0.121	0.119	0.117
−1.2	0.115	0.113	0.111	0.109	0.107	0.106	0.104	0.102	0.100	0.099
−1.3	0.097	0.095	0.093	0.092	0.090	0.089	0.087	0.085	0.084	0.082
−1.4	0.081	0.079	0.078	0.076	0.075	0.074	0.072	0.071	0.069	0.068
−1.5	0.067	0.066	0.064	0.063	0.062	0.061	0.059	0.058	0.057	0.056
−1.6	0.055	0.054	0.053	0.052	0.051	0.049	0.048	0.047	0.046	0.046
−1.7	0.045	0.044	0.043	0.042	0.041	0.040	0.039	0.038	0.038	0.037
−1.8	0.036	0.035	0.034	0.034	0.033	0.032	0.031	0.031	0.030	0.029
−1.9	0.029	0.028	0.027	0.027	0.026	0.026	0.025	0.024	0.024	0.023
−2.0	0.023	0.022	0.022	0.021	0.021	0.020	0.020	0.019	0.019	0.018
−2.1	0.018	0.017	0.017	0.017	0.016	0.016	0.015	0.015	0.015	0.014
−2.2	0.014	0.014	0.013	0.013	0.013	0.012	0.012	0.012	0.011	0.011
−2.3	0.011	0.010	0.010	0.010	0.010	0.009	0.009	0.009	0.009	0.008
−2.4	0.008	0.008	0.008	0.008	0.007	0.007	0.007	0.007	0.007	0.006
−2.5	0.006	0.006	0.006	0.006	0.006	0.005	0.005	0.005	0.005	0.005
−2.6	0.005	0.005	0.004	0.004	0.004	0.004	0.004	0.004	0.004	0.004
−2.7	0.003	0.003	0.003	0.003	0.003	0.003	0.003	0.003	0.003	0.003
−2.8	0.003	0.002	0.002	0.002	0.002	0.002	0.002	0.002	0.002	0.002
−2.9	0.002	0.002	0.002	0.002	0.002	0.002	0.002	0.001	0.001	0.001
−3.0	0.001	0.001	0.001	0.001	0.001	0.001	0.001	0.001	0.001	0.001
−3.1	0.001	0.001	0.001	0.001	0.001	0.001	0.001	0.001	0.001	0.001
−3.2	0.001	0.001	0.001	0.001	0.001	0.001	0.001	0.001	0.001	0.001

If $x \leq -3.3$, then $\Phi(x) = 0.000$.

C.6 $\Phi(x) = \int_0^x \phi_{(0,1)}(t)\,dt,\ x \geq 0$

	0.00	0.01	0.02	0.03	0.04	0.05	0.06	0.07	0.08	0.09
0.0	0.500	0.504	0.508	0.512	0.516	0.520	0.524	0.528	0.532	0.536
0.1	0.540	0.544	0.548	0.552	0.556	0.560	0.564	0.567	0.571	0.575
0.2	0.579	0.583	0.587	0.591	0.595	0.599	0.603	0.606	0.610	0.614
0.3	0.618	0.622	0.626	0.629	0.633	0.637	0.641	0.644	0.648	0.652
0.4	0.655	0.659	0.663	0.666	0.670	0.674	0.677	0.681	0.684	0.688
0.5	0.691	0.695	0.698	0.702	0.705	0.709	0.712	0.716	0.719	0.722
0.6	0.726	0.729	0.732	0.736	0.739	0.742	0.745	0.749	0.752	0.755
0.7	0.758	0.761	0.764	0.767	0.770	0.773	0.776	0.779	0.782	0.785
0.8	0.788	0.791	0.794	0.797	0.800	0.802	0.805	0.808	0.811	0.813
0.9	0.816	0.819	0.821	0.824	0.826	0.829	0.831	0.834	0.836	0.839
1.0	0.841	0.844	0.846	0.848	0.851	0.853	0.855	0.858	0.860	0.862
1.1	0.864	0.867	0.869	0.871	0.873	0.875	0.877	0.879	0.881	0.883
1.2	0.885	0.887	0.889	0.891	0.893	0.894	0.896	0.898	0.900	0.901
1.3	0.903	0.905	0.907	0.908	0.910	0.911	0.913	0.915	0.916	0.918
1.4	0.919	0.921	0.922	0.924	0.925	0.926	0.928	0.929	0.931	0.932
1.5	0.933	0.934	0.936	0.937	0.938	0.939	0.941	0.942	0.943	0.944
1.6	0.945	0.946	0.947	0.948	0.949	0.951	0.952	0.953	0.954	0.954
1.7	0.955	0.956	0.957	0.958	0.959	0.960	0.961	0.962	0.962	0.963
1.8	0.964	0.965	0.966	0.966	0.967	0.968	0.969	0.969	0.970	0.971
1.9	0.971	0.972	0.973	0.973	0.974	0.974	0.975	0.976	0.976	0.977
2.0	0.977	0.978	0.978	0.979	0.979	0.980	0.980	0.981	0.981	0.982
2.1	0.982	0.983	0.983	0.983	0.984	0.984	0.985	0.985	0.985	0.986
2.2	0.986	0.986	0.987	0.987	0.987	0.988	0.988	0.988	0.989	0.989
2.3	0.989	0.990	0.990	0.990	0.990	0.991	0.991	0.991	0.991	0.992
2.4	0.992	0.992	0.992	0.992	0.993	0.993	0.993	0.993	0.993	0.994
2.5	0.994	0.994	0.994	0.994	0.994	0.995	0.995	0.995	0.995	0.995
2.6	0.995	0.995	0.996	0.996	0.996	0.996	0.996	0.996	0.996	0.996
2.7	0.997	0.997	0.997	0.997	0.997	0.997	0.997	0.997	0.997	0.997
2.8	0.997	0.998	0.998	0.998	0.998	0.998	0.998	0.998	0.998	0.998
2.9	0.998	0.998	0.998	0.998	0.998	0.998	0.998	0.999	0.999	0.999
3.0	0.999	0.999	0.999	0.999	0.999	0.999	0.999	0.999	0.999	0.999
3.1	0.999	0.999	0.999	0.999	0.999	0.999	0.999	0.999	0.999	0.999
3.2	0.999	0.999	0.999	0.999	0.999	0.999	0.999	0.999	0.999	0.999

If $x \geq 3.3$, then $\Phi(x) = 1.000$.

$$C.6 \quad \Phi(x) = \int_x^\infty \varphi_{0,1}(t)\,dt, \quad x \ge 0$$

Index

The Bayesian Way: Introductory Statistics for Economists and Engineers, First Edition.
Svein Olav Nyberg.
© 2019 John Wiley & Sons, Inc. Published 2019 by John Wiley & Sons, Inc.

Printed and bound by CPI Group (UK) Ltd, Croydon, CR0 4YY

16/04/2025

14658518-0001